Queen of Scots

THE TRUE LIFE OF
MARY STUART

John Guy

A MARINER BOOK
HOUGHTON MIFFLIN COMPANY
BOSTON · NEW YORK

First Mariner Books edition 2005

Copyright © 2004 by John Guy
ALL RIGHTS RESERVED

Library of Congress Cataloging-in-Publication Data

Guy, J. A. (John Alexander)
Queen of Scots : the true life of Mary Stuart / John Guy.
p.   cm.
Includes bibliographical references and index.
ISBN-13: 978-0-618-25411-8   ISBN-10: 0-618-25411-0
ISBN-13: 978-0-618-61917-7 (pbk.)   ISBN-10: 0-618-61917-8 (pbk.)
1. Mary, Queen of Scots, 1542–1587. 2. Scotland—History—
Mary Stuart, 1542–1567. 3. Great Britain—History—Elizabeth,
1558–1603. 4. Queens—Scotland—Biography. I. Title.
DA787.A1G89 2004
941.105'092—dc22 [B] 2003067592

Title-page calligraphy by Bernard Maisner
Maps by Jacques Chazaud and Richard Guy
Book design by Robert Overholtzer

DOC 10
4500738918

In memory of my mother

Princes at all times have not their wills, but my heart being my own is immutable.

— Mary to Thomas Randolph, English ambassador to Scotland, March 8, 1564

Contents

Acknowledgments

WRITING THIS BOOK has been an exciting, invigorating experience, one of the most thrilling of my life, an adventure even for someone who had already worked on the historical records for a quarter of a century. I had no idea when I began that so much fresh material could be found in the archives about a woman who has been the daughter of debate for four centuries. Then, when I steadily began to uncover this material, I felt a sense of elation. I simply could not stop working on the book until I got to the bottom and the end of the story.

I'm deeply grateful for all the help and support I've received from the archivists and curators whose repositories and libraries I've ransacked for so many weeks and months. Monique Cohen and her staff at the Département des Manuscrits, Bibliothèque Nationale de France, Paris, showed me how to find what I needed in a library I'd never used before. In more familiar haunts, Dr. Sarah Tyacke and her team at the National Archives (Public Record Office), London, and the staff of the University Library at Cambridge were as helpful and courteous as ever. Dr. Andrea Clarke and her colleagues in the Department of Manuscripts at the British Library were always willing to assist me, supplying microfilms of key volumes of the Cottonian and Additional Manuscripts so that I could read them at home. I also thank the staff of the Rare Books Department for producing every copy in the collection of certain titles, including multiple copies of the same edition. Dr. Richard Palmer and his staff at Lambeth Palace Library offered me the opportunity to read newly acquired documents concerning Mary's trial and execution, some of which had been out of the public domain for decades. I'm most grateful to the Trustees for access to this material.

In Edinburgh, my path was greatly eased by the reading room staff of the National Archives of Scotland, HM General Register House, and of the Department of Special Collections, National Library of Scotland. At St. Andrews University Library, Christine Gascoigne and her colleagues in the Rare Books and Manuscripts Department repeatedly came to my aid. For access to and permission to quote from the manuscripts of the old Advocates Library and other documents held at the George IV Bridge repository of the National Library of Scotland, I wish to thank the Trustees.

For access to the Cecil Papers at Hatfield House and for permission to cite them, I am most grateful to The Marquess of Salisbury, and to Robin Harcourt Williams, librarian and archivist. For access to and permission to quote from the manuscripts and rare books at the Henry E. Huntington Library, San Marino, California, I gladly thank Dr. Mary Robertson, chief curator of manuscripts, whom by a happy coincidence I first met in Sir Geoffrey Elton's Tudor seminar in Cambridge some thirty years ago. For permission to read the manuscripts and rare books at the Folger Shakespeare Library, Washington, D.C., I acknowledge the generosity of Dr. Gail Kern Paster, director, and the Trustees.

Preliminary drafts of the maps and genealogical tables were drawn and digitized by Richard Guy of Orang-Utan Productions. For undertaking the picture research and obtaining loans of transparencies, I thank Sheila Geraghty, whose expertise was invaluable. My colleague Stephen Alford at Cambridge University read the entire manuscript in draft and I relished all of our lengthy conversations. Professor Michael Lynch, Department of Scottish History, University of Edinburgh, read and most generously commented on the uncorrected proofs. I'm grateful for his suggestions and list of corrections on the Scottish side, and for corrections supplied by Rachel Guy, who also read the page proofs. I accept full responsibility for such errors as may still remain.

Some academic historians may regret my spelling of "Stuart" in preference to "Stewart" for the dynasty. But Mary called herself "Stuart"; her motto, "Sa virtu m'atire," works as a near-perfect anagram only if the family name is spelled "Stuart"; and it seemed likely to irritate readers if both "Stuart" and "Stewart" were used. I also prefer "Ker of Fawdonside" to the alternative "Kerr," adopting the orthography of the manuscripts. And I've followed the example of Elizabeth I and William Cecil in styling James Hamilton, Third Earl of Arran, as "Arran," after his father, the second earl, was made Duke of Châtelherault, even though he was not strictly Earl of Arran until his father died.

I've nothing but thanks and admiration for Peter Robinson and Emma Parry, my agents in London and New York, for their constant encouragement and for persuading me that I could write this book and make it work. Both read the manuscript and gave helpful advice. In preparing a book in which the interpretation counts for just as much as the archival research, I've also realized how much I've learned from the BBC producers with whom I've been privileged to work during the past four years, in particular Catrine Clay, Dick Taylor and Jane McWilliams.

I owe an immense debt to Eamon Dolan, my editor at Houghton Mifflin. His comments on my drafts were pitched exactly right, always helpful and to the point. I feel privileged to be published by Houghton Mifflin, whose magnanimity in allowing me to get on with my work uninterrupted for almost three years created the closest thing to ideal conditions. For assistance in the editorial and publicity stages, I also wish to thank Larry Cooper, Bridget Marmion, Lori Glazer, Whitney Peeling and Carla Gray.

I express heartfelt gratitude to my former students at the University of St. Andrews, and those I currently teach at Cambridge, for their contributions to seminars and supervisions where Mary made her appearance more often than she should have. Other debts are to Fiona Alexander, who saw instantly that the mysterious "object" Mary holds in her left hand in the placard of the mermaid and the hare, previously defying explanation, is a rolled-up net. Frances and David Waters offered constant encouragement, uncannily predicting the date on which I'd deliver the final manuscript, and making sure we had tickets for Mozart's *Le Nozze di Figaro* for the very next night.

Most importantly, Julia accepted Mary's presence in what must increasingly have seemed like a ménage à trois, showing infinite patience. She pored over innumerable drafts, reading some chapters as many as a dozen times and discussing Mary at all hours. I can never adequately thank her or repay her love. Emma was just as tolerant, never complaining that she hardly saw her father, and merely teasing him about when he'd finish "the book." Lucy, Susie and Gemma sometimes got their paws into Mary's affairs more than I might have liked, but in doing so kept me in touch with normality.

London
October 24, 2003

THE TUDORS AND THE STUARTS

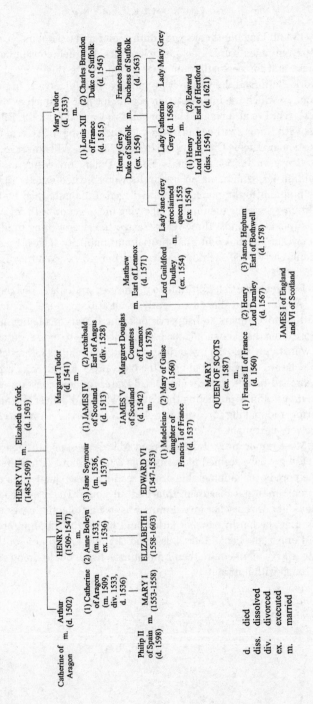

Catherine of Aragon — m. — Arthur (d. 1502)

HENRY VII (1485–1509) — m. — Elizabeth of York (d. 1503)

HENRY VIII (1509–1547)
m.
(1) Catherine of Aragon (m. 1509, div. 1533, d. 1536)
(2) Anne Boleyn (m. 1533, ex. 1536)
(3) Jane Seymour (m. 1536, d. 1537)

Philip II of Spain (d. 1598) — m. — MARY I (1553–1558)

ELIZABETH I (1558–1603)

EDWARD VI (1547–1553)

Margaret Tudor (d. 1541)
m.
(1) JAMES IV of Scotland (d. 1513)
(2) Archibald Earl of Angus (div. 1528)

JAMES V of Scotland (d. 1542)
m.
(1) Madeleine daughter of Francis I of France (d. 1537)
(2) Mary of Guise (d. 1560)

Margaret Douglas Countess of Lennox (d. 1578)
m.
Matthew Earl of Lennox (d. 1571)

MARY QUEEN OF SCOTS (ex. 1587)
m.
(1) Francis II of France (d. 1560)
(2) Henry Lord Darnley (d. 1567)
(3) James Hepburn Earl of Bothwell (d. 1578)

JAMES I of England and VI of Scotland

Mary Tudor (d. 1533)
m.
(1) Louis XII of France (d. 1515)
(2) Charles Brandon Duke of Suffolk (d. 1545)

Frances Brandon Duchess of Suffolk (d. 1563)
m.
Henry Grey Duke of Suffolk (ex. 1554)

Lady Jane Grey proclaimed queen 1553 (ex. 1554)
m.
Lord Guildford Dudley (ex. 1554)

Lady Catherine Grey (d. 1568)
m.
(1) Henry Lord Herbert (diss. 1554)
(2) Edward Earl of Hertford (d. 1621)

Lady Mary Grey

d.   died
diss. dissolved
div.   divorced
ex.   executed
m.   married

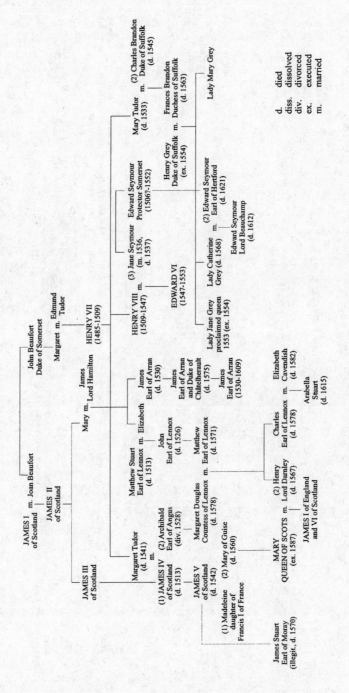

The Stuarts and the English Succession

The Guise Family

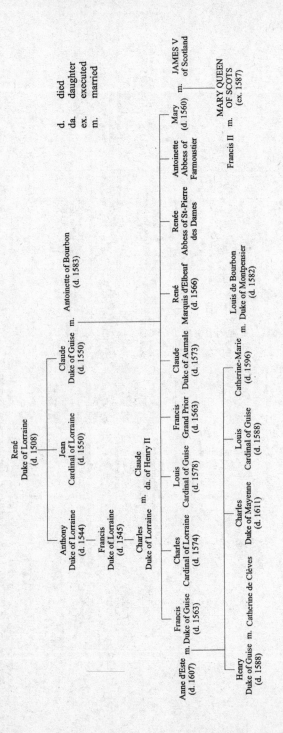

d.   died
da.   daughter
ex.   executed
m.   married

Scotland
in the Sixteenth Century

Orkney
Islands

North
Sea

Western Isles
The Minch

Moray Firth

Elgin   Banff
Inverness

Spey

Aberdeen

SCOTLAND

Tay   Dundee
Perth

Firth of Forth

Glasgow   Edinburgh   Berwick-
                                               upon-
Clyde   Melrose   Tweed   Tweed

Jedburgh   Kelso
                      Roxburgh

Hermitage □

Firth of Clyde   Tyne

Dumfries •

                  Carlisle

Solway Firth   ENGLAND

100 km   Workington   Eden

75 miles

Chazaud

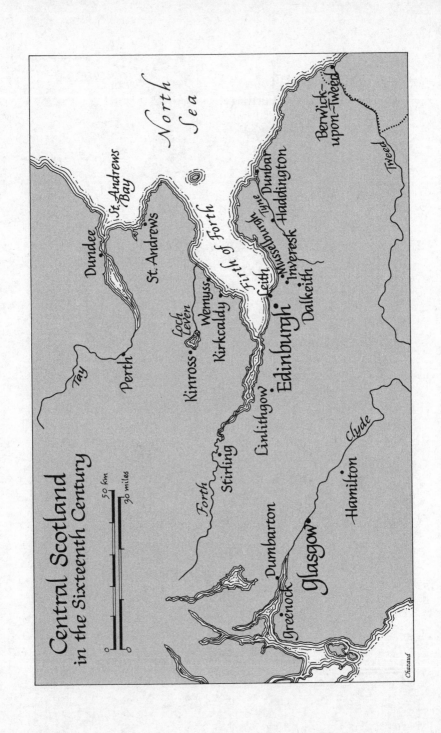

Central Scotland in the Sixteenth Century

50 km
30 miles

North Sea

Tay
Dundee
St. Andrews Bay
St. Andrews
Perth
Loch Leven
Kinross
Wemyss
Kirkcaldy
Firth of Forth
Leith
Musselburgh
Tyne
Dunbar
Haddington
Inveresk
Edinburgh
Dalkeith
Berwick-upon-Tweed
Tweed
Linlithgow
Stirling
Forth
Dumbarton
Greenock
Clyde
Glasgow
Hamilton

Chazaud

France under Henry II

showing the principal places connected to
Mary Queen of Scots

English Channel

Calais

NETHERLANDS

HOLY ROMAN EMPIRE

Cateau-Cambrésis

Dieppe

St.Quentin

PICARDY

Le Havre

Rouen

Rheims

LORRAINE

Roscoff

NORMANDY

Seine

CHAMPAGNE

Paris

Fontainebleau

BRITTANY

Orléans

Blois

Angers

BURGUNDY

Ancenis

Amboise

Dijon

Nantes

Tours

Bourges

Loire

POITOU

Poitiers

SWITZ.

F R A N C E

ATLANTIC OCEAN

Lyons

Bordeaux

Rhône

GUYENNE

PROVENCE

Garonne

Toulouse

Mediterranean Sea

SPAIN

300 km

0

200 miles

Chazaud

QUEEN OF SCOTS

AROUND EIGHT O'CLOCK in the morning on Wednesday, February 8, 1587, when it was light enough to see without candles, Sir Thomas Andrews, sheriff of the county of Northamptonshire, knocked on a door. The place was Fotheringhay Castle, about seventy-five miles from London. All that remains there now beneath the weeds is the raised earthen rampart of the inner bailey and a truncated mound, or "motte," on the site of the keep, a few hundred yards from the village beside a sluggish stretch of the River Nene.

But in the sixteenth century the place was bustling with life. Fotheringhay was a royal manor. Richard III had been born at the castle in 1452. Henry VII, the first of the Tudor kings, who had slain Richard at the battle of Bosworth, gave the estate as a dowry to his wife, Elizabeth of York, and Henry VIII granted it to his first bride, Catherine of Aragon, who extensively refurbished the castle. In 1558, Elizabeth I inherited the property when she succeeded to the throne on the death of her elder sister, Mary Tudor.

Despite its royal associations, nothing had prepared Fotheringhay, or indeed the British Isles, for what was about to happen there. Andrews was in attendance on two of England's highest-ranking noblemen, George Talbot, Earl of Shrewsbury, and Henry Grey, Earl of Kent. The door on which he knocked was the entrance to the privy chamber of

Mary Queen of Scots, dowager queen of France and for almost nineteen years Elizabeth's prisoner in England.

The door opened to reveal Mary on her knees, praying with her bed-chamber servants. Andrews informed her that the time was at hand, and she looked up and said she was ready. She rose, and her gentlewomen stood aside.

She was only forty-four. Born and brought up to be a queen, she walked confidently through the doorway as if she were once more pro-cessing to a court festival. Almost six feet tall, she had always looked the part. She had been fêted since her childhood in France for her beauty and allure. "Charmante" and "la plus parfaite" were the adjectives most commonly applied to her singular blend of celebrity. Not just physically mesmerizing with her well-proportioned face, neck, arms and waist, she had an unusual warmth of character with the ability to strike up an in-stant rapport. Always high-spirited and vivacious, she could be unre-servedly generous and amiable. She had a razor-sharp wit and was a natural conversationalist. Gregarious as well as glamorous, she could be genial to the point of informality as long as her "grandeur" was re-spected. Many contemporaries remarked on her almost magical ability to create the impression that the person she was talking to was the only one whose opinion really mattered to her.

As a result of premature aging caused by the inertia and lack of exer-cise of which she had so bitterly complained during her long captivity, her beauty was on the wane. Her features had thickened and she had rounded shoulders and a slight stoop. Her face, once legendary for its soft white skin and immaculate, marble-like complexion, had filled out and become double-chinned. But captivity did not alter all things. Her small, deep-set hazel eyes darted as restlessly as ever, and her ringlets of auburn hair seemed as lustrous.

Mary had been awake for most of the night and had carefully pre-pared herself. This was to be her grandest performance, her greatest tri-umph; she had considered every detail.

Her clothes set the tone. She appeared to be dressed entirely in black apart from a white linen veil. Lace-edged and as delicate as gauze, it flowed down from her hair over her shoulders to her feet in the French style. Fastened to the top of the veil was a small white cambric cap. It just touched the tip of her forehead and was also edged with lace, leav-ing room for her curls to peek out at the sides. Her gown of thick black satin reached almost to the ground, where it was attached to her train.

Trimmed with gold embroidery and sable, it was peppered with acorn buttons of jet, set with pearl.

A closer look revealed an outer bodice of crimson velvet and an underskirt of embroidered black satin, both visible where the gown was fashionably cut away. To bedeck it, Mary wore long, richly embroidered slashed sleeves in the Italian style, under which could be seen uncut inner sleeves of purple velvet. Her shoes were of the finest Spanish suede. Later someone observed that she wore sky-blue stockings embroidered with silver thread and held up by green silk garters, these on top of soft white stockings that she used to protect her skin from chafing.

She carried a crucifix of ivory in one hand and a Latin prayer book in the other. A string of rosary beads with a golden cross hung from a girdle at her waist. Around her neck lay a silver or gold chain on which hung a pendant, a medallion bearing the image of Christ as the Lamb of God.

Led by Andrews and followed by the two earls, Mary walked along the corridor and into a larger room where her household was waiting to greet her and bid her farewell. An eyewitness (perhaps the Earl of Kent himself) wrote that she exhorted her servants to fear God and live in obedience. She kissed her women servants and gave her hand to her menservants to kiss. She asked them not to grieve for her, but "to rejoice and pray for her." One of them afterward reported that she showed no fear and even smiled.

Mary then descended the stairs toward the great hall on the ground floor. Her legs were so swollen and inflamed by rheumatism, she leaned for support on the arms of two soldiers. When the procession reached the anteroom of the hall, they encountered Andrew Melville, her steward, who knelt and fighting back tears cried out, "Madam, it will be the sorrowfullest message that I ever carried, when I shall report that my queen and dear mistress is dead."

Mary answered, also weeping, "You ought to rejoice rather than weep for that the end of Mary Stuart's troubles is now come." "Carry this message," she continued, "and tell my friends that I die a true woman to my religion, and like a true Scottish woman and a true French woman."

As Mary recovered her composure, her mood abruptly changed. She glanced back up the stairs and exclaimed that she was "evil attended." She demanded "for womanhood's sake" that her own servants should escort her. She harangued the earls, who became fearful that she would cause an even bigger scene and have to be dragged violently into the great hall.

Shrewsbury feebly claimed that he and Kent were simply following orders. Hearing this, Mary bridled: "Far meaner persons than myself have not been denied so small a favor." "Madam," replied Kent, "it cannot well be granted, for that it is feared lest some of them would with speeches both trouble and grieve Your Grace and disquiet the company . . . or seek to wipe their napkins in some of your blood, which were not convenient."

"My lord," said Mary, "I will give my word and promise for them that they shall not do any such thing." She could not stop herself adding, "You know that I am cousin to your queen, and descended from the blood of Henry VII, a married queen of France and the anointed queen of Scotland."

The earls huddled together, whispering inaudibly, then gave in to Mary, who was used to getting her own way. Her two favorite gentlewomen, Jane Kennedy and Elizabeth Curle, and four of her gentlemen, including Melville, were allowed to join the procession. "Allons donc," said Mary, smiling again — "Now let us go." She spoke in French because this and Lowland Scots were her native tongues; English she had learned only with difficulty in her captivity.

Her retinue now made ready, she strode purposefully into the great hall with Melville carrying her train. It was self-consciously a royal entry; Mary walked before the hundred or so spectators straight toward the focal point, a wooden stage that had been hastily constructed over the previous two days beside an open fireplace in which a great pile of logs blazed. She mounted the two steps that led up to the platform and sat down on a low stool that was offered to her, after which the earls seated themselves on her right while the sheriff stood on her left.

There was of course no throne. The stage was a scaffold two feet high and twelve feet square, shrouded with black cotton sheets that hung low over the sides to camouflage the rough joinery, with a rail eighteen inches high around three sides and the unenclosed fourth side in full view of the spectators in the lower end of the hall. There was a cushion for Mary to kneel on, this beside an execution block also swathed in black.

Two masked men stood in readiness on the platform, one "Bull," the headsman of the Tower of London, and his assistant. They were dressed in long black gowns with white aprons, their ax laid casually against the rail. In the lower end of the space, the knights and gentlemen of Northamptonshire and its neighboring counties looked toward the stage flanked by a troop of soldiers, their view unrestricted because the platform had been set at the right height. Outside in the courtyard, beyond the passageway at the main entrance to the great hall, a large crowd of another thousand or so waited for news.

The sheriff called for silence, after which Robert Beale, the clerk of Elizabeth's Privy Council and the man responsible for delivering the execution warrant to Fotheringhay, read it out. As he spoke — the warrant would have taken about ten minutes to read — Mary sat completely still. She showed no emotion, listening, as Robert Wingfield of Upton, Northamptonshire, who was within ten yards of her, reported, "with as small regard as if it had not concerned her at all; and with as cheerful a countenance as if it had been a pardon." Her nerve was to be tested, however, when Dr. Richard Fletcher, Dean of Peterborough, and at this time one of Elizabeth's favorite preachers, stepped forward at the Earl of Shrewsbury's signal.

Fletcher, the father of the dramatist John Fletcher, who was Shakespeare's collaborator on *Henry VIII*, had been brought in to deliver a set-piece "admonition" to Mary that strictured her for her traitorous Catholicism, and to lead the assembly in prayers. He was one of Elizabeth's chaplains in ordinary, renowned for his "comely person" and "courtly speech."

But his admonition backfired spectacularly; the attempted sermon — for that is all it was — was the greatest faux pas of his career. When the moment came, he started to stammer nervously. "Madam," he began, "the queen's most excellent majesty"; "Madam, the queen's most excellent majesty . . ." Three times he stumbled, but when he started for the fourth, Mary cut him off. In a clear and unwavering voice, she said, "Mr. Dean, I will not hear you. You have nothing to do with me, nor I with you."

Fletcher, somewhat abashed, countered, "I say nothing but that I will justify before the majesty of the mighty God." He was not at first willing to give way to her, believing that God would never abandon the just, but would minister to them through his angels. If Mary had been condemned to die, it was God's work and the preacher would be called to account for his sermon only before God.

Hearing this, Mary got into her stride, as she always did in an argument. "I am settled," she said, "in the ancient Roman Catholic religion, and mind to spend my blood in defense of it."

Fletcher unwisely responded, "Madam, change your opinion and repent you of your former wickedness, and settle your faith only in Jesus Christ, by him to be saved." This was not the way to speak to a queen. Mary, visibly coloring, ordered him to be silent. There was an awkward pause. Then the earls gave way. Fletcher was told to omit the sermon, which in a fit of pique he insisted be transcribed from his notes into a report of the day's proceedings.

A bizarre, even farcical scene ensued. The Earl of Kent urged Fletcher to begin the prayers, but as the dean started speaking again, Mary prayed loudly and in Latin with her crucifix before her eyes.

There followed a battle of wills, because as the knights and gentlemen in the hall joined Fletcher in his versicles and responses, Mary and her six servants shouted louder and louder until the queen, in tears, slipped off her stool, at which point she knelt and continued as before.

Even after Fletcher had ceased praying, Mary carried on, in English now to cause maximum embarrassment. She prayed for the Church, for an end to religious discord, for her son, the twenty-year-old James VI of Scotland — whom her enemies had brought up as a Protestant — that he might be converted to the true Catholic faith. She prayed that Elizabeth might prosper and long continue to reign, serving God aright. She confessed that she hoped to be saved "by and in the blood of Christ at the foot of whose crucifix she would willingly shed her blood." She petitioned the saints to pray for her soul, and that God would in his great mercy and goodness avert his plagues from "this silly island."

To the Earl of Kent, himself a staunch Protestant, this was highly offensive. "Madam," he said, "settle Christ Jesus in your heart and leave those trumperies." But Mary ignored him. Eventually she finished, kissing the crucifix and making the sign of the cross in the Catholic way.

This was largely contrived. Mary had never truly been the ideological Catholic that she now wished to appear to the world. She was far too political for that. As a ruler in Scotland, she had sensibly accepted a compromise based on the religious status quo and the inroads made by the Protestant Reformation. Only after her imprisonment in England had she reinvented herself as a poor Catholic woman persecuted for her religion alone. What happened in the great hall at Fotheringhay was for show, and it worked. By humiliating Fletcher, Mary won a propaganda victory that resounded around Catholic Europe.

Satisfied, she calmly turned to Bull, who meekly knelt and sought her forgiveness. She answered, "I forgive you with all my heart, for now, I hope, you shall make an end of all my troubles."

The executioners helped Mary's gentlewomen to undress her down to her petticoat. As they unbuttoned her, she smiled broadly and joked that she "never had such grooms before to make her unready" nor did she "ever put off her clothes before such a company."

She laid her crucifix and prayer book on her stool, and one of the executioners took the medallion from around her neck, since custom allowed that such personal items were a perquisite. But Mary interposed, saying that she would give these things to her servants and that he would receive money in lieu of them.

As Mary's veil and black outer garments were removed, stifled cries of shock and astonishment reverberated around the hall. Her petticoat was of tawny velvet, her inner bodice of tawny satin. One of her gentlewomen handed her a pair of tawny sleeves with which she immediately covered her arms. A metamorphosis had occurred.

For several minutes Mary stood stock still on the stage, clad in the color of dried blood: the liturgical color of martyrdom in the Roman Catholic Church. It was a sight so melodramatic, so abhorrent to the earls, that they omitted all reference to it from their official report to the Privy Council. The incident is known only from a contemporary French account based on the reports of Mary's attendants, which is confirmed by two independent English accounts, one by Shrewsbury's servant, who was writing to a friend and had no reason to lie.

Mary kissed her gentlewomen, who burst into uncontrolled fits of sobbing. "Ne criez vous," she said, "j'ai promis pour vous." Or as one of the English eyewitness accounts renders it, "Peace, peace, cry not, I have promised the contrary, cry not for me but rejoice."

She raised her hands and blessed them, and turning to her other servants, Melville especially, who were weeping aloud and continually crossing themselves, she prayed in Latin and blessed them too, bade them farewell, and asked them to remember her in their prayers.

She knelt down "most resolutely" on the cushion while Jane Kennedy covered her eyes with a white Corpus Christi cloth embroidered in gold that Mary had chosen the previous night. Jane kissed the cloth, tied it around Mary's face in the shape of a triangle and pinned it securely to her cap. The two gentlewomen then left the platform.

As Mary knelt, she recited in Latin the psalm *In te Domino confido*, "In thee, O Lord, have I put my trust." Reaching out for the block, she laid down her head, positioning her chin carefully with her hands and holding them there, so that if one of the executioners had not moved them, they would have been cut off. She stretched out her arms and legs and cried, "*In manus tuas, Domine, commendo spiritum meum*" — "Into your hands, O Lord, I commend my spirit." She repeated these words three or four times until, with one executioner holding down her body, the other severed her head.

Except it was the headsman's turn to blunder. It should have taken only a single blow, but the strain was too great even for England's most experienced executioner. His first strike was misaligned, and the blow fell on the knot of the blindfold, missing the neck and hacking into the back of the head. One account says Mary made a "very small noise," but another says she cried out in agony, "Lord Jesus receive my soul." A second strike severed the neck, but not completely, and the executioner sliced through the remaining sinews, using the ax as a cleaver. At length he raised the head, shouting "God save the queen." An audible gasp went up from the hall, because Mary's lips were still moving as if in prayer, and continued to do so for a quarter of an hour.

And then the final twist. As the executioner lifted up the head, Mary's auburn curls and white cap became detached from her skull. The illusion of monarchy dissolved as the executioner found himself clutching a handful of hair while the head fell back to the floor, rolling like a misshapen football toward the spectators, who saw that it was "very grey and near bald."

Suddenly everything was clear. The Queen of Scots had worn a wig. The assembly was struck dumb, until the Earl of Shrewsbury could stand it no longer and burst into tears.

As the executioner retrieved the skull, Dr. Fletcher recovered his wits. He bellowed, "So perish all the queen's enemies," to which the Earl of Kent, standing over the corpse, echoed, "Such be the end of all the queen's and the gospel's enemies." But it was a gruesome finale, a harrowing catharsis. Even in the London theaters, where revenge plays and tragedies were newly in vogue, no one had seen anything quite like this.

Mary's distraught servants were led from the scene and locked in their rooms. The executioners were disrobing the corpse when one of them saw that her favorite pet dog, a Skye terrier, had hidden itself in the folds of her petticoat and sneaked onto the stage. When detected, it ran about wailing miserably and lay down in the widening pool of blood between her severed head and shoulders. Since it could not be coaxed away, it was forcibly removed and washed, whereupon it refused to eat. One of Mary's servants claimed it soon died, but this is not corroborated.

In the afternoon, by order of the earls, the black cotton sheets, the execution block and cushion, Mary's clothes and ornaments, and anything else with blood on it were burned in the open fireplace so that no relics of the "martyrdom" she had so conspicuously sought to evoke could be obtained by her Catholic supporters. Still present in the great hall to observe these cleansing operations were the knights and gentlemen of the county, and when the earls wrote their official account of the execution, these men signed their names to the report as solemn witnesses.

The Earl of Shrewsbury's fourth son, Henry Talbot, was sent posthaste to London to deliver the report to the Privy Council that same night. When he had departed, the mortal remains of the dead queen were put on a stretcher and carried back upstairs to be embalmed. The scaffold was demolished and everyone except the sheriff, who had the job of burying the heart and inner organs in a secret place within the foundations of the castle, was sent home. Some of Mary's ornaments must also have been buried in the deep recesses of the castle, because the ring she was given at her betrothal to her second husband, Henry Lord Darnley, was later unearthed in the ruins and exhibited at Peterborough in 1887.

No one who had witnessed Mary's last day could ever have forgotten it. Whatever view is taken of her character, whatever credence is given to the stories told about her as a way of justifying her forced abdication and execution, the business on that day was regicide. Mary was an anointed queen. Elizabeth, her fellow sovereign as much as her rival for the past thirty years, was herself all too anxious to defend the ideal of monarchy: the principle that rulers were accountable to God alone. She had done everything possible to prevent Mary's execution until she felt it could no longer be avoided, and then to shift the blame for it onto the shoulders of others.

Elizabeth had a firm grasp of the issues. She knew that Mary's death would alter the way that monarchy was regarded in the British Isles. A regicide would give a massive boost to Parliament, diminishing forever the "divinity that hedges a king." It would help to propagate the theory of popular sovereignty — the belief that political power lies in the people and not in the ruler — and the idea that the representatives of the people were those they elected to Parliament. This was the ideology invoked by Mary's rebel lords in Scotland to depose her. And the same theory would be instilled there, and more subversively in parts of France, for 250 years after her death, finally to cross the Atlantic when Dr. William

Small, a Scot, taught ethics and political science to the young Thomas Jefferson at the College of William and Mary in Virginia.

How did so versatile a queen as Mary, one so beautiful and intelligent, so convivial and down-to-earth, so full of life and irresistible, end up disgraced and deposed? One of the reasons is that Elizabeth's chief minister and leading adviser for forty years, William Cecil, was her antagonist. More than anyone else, he was her great nemesis. Unlike Elizabeth, Mary was a Catholic, and Cecil's overriding ambition was to remold the whole of the British Isles into a single Protestant community. He had little room for an independent Scotland, hence his intermittent clashes with his Scottish allies over the extent of English domination. Whereas Elizabeth did all she could to protect the ideal of divine-right monarchy irrespective of the religion of its incumbent, Cecil believed that Parliament had the right to settle the succession to the throne on religious grounds, meaning that Mary's dynastic claim had at all costs to be discounted.

In death as in life, Mary always aroused the strongest feelings. To her apologists she was an innocent victim. She was mishandled and traduced: a political pawn in the hands of those perfidious Scottish lords and ambitious French and English politicians who found her inconvenient and in their way. To her critics she was fatally flawed. She was far too affected by her emotions. She ruled from the heart and not the head. She was a femme fatale, a manipulative siren, who flaunted her sexuality in dancing and banqueting and did not care who knew it.

Her enemies largely won the argument. Mary has come down to us not as a shrewd and charismatic young ruler who relished power and, for a time, managed to hold together a fatally unstable country, but rather as someone who cared more about her luxuries and pets. She knew how to play to the gallery. One of the accounts of her execution dismissed her as "transcending the skills of the most accomplished actress." But a sense of theater was essential to the exercise of power in the sixteenth century, and there was far more to Mary than so cynical a judgment implies.

This book tries to get to the truth about her, or as close to the truth as possible: to see her not merely as a bundle of stereotypes or as a convenient and tenuously linked series of myths, but as a whole woman whose choices added up and whose decisions made sense. The rationale relates closely to the method: to write Mary's life and tell her story using the original documents rather than relying on the familiar printed collections or edited abstracts, themselves often compiled to perpetuate rather than to engage with the legends. It may come as a surprise to learn that such documents survive in voluminous quantities, preserved in archives and research libraries as far apart as Edinburgh, Paris, London, the stately homes of England, and Washington, D.C., and Los Angeles. Some of them have not been read by a historian since 1840. Many have not been freshly examined since the 1890s, and among these are unrecognized handwritten transcripts of two of the famous Casket Letters.

The aim is to tell Mary's story, where possible letting her speak for herself in her own words, but also to consider why the stories of others about the very same events are often so strikingly different. Only when this is done can the myriad of facts be properly sifted, the sequence of events be explained and understood, and a searchlight cast on a turbulent life.

1

The First Year

MARY STUART was born in the coldest of winters. Snow blanketed the ground, and the narrow pathways and rough winding tracks between England and Scotland were completely blocked. The cattle that roamed the Lowlands and the valleys of the border region during the summer months were crouching in their low stone byres. The River Tweed, often a raging torrent as it flowed to the sea at Berwick-upon-Tweed on the eastern side of the border, was frozen over. Whereas it normally took a rider five or six days to carry important dispatches from Edinburgh to London, the news of Mary's birth took four days to reach Alnwick in Northumberland, only a few miles south of Berwick.

The new baby was the only daughter and sole surviving heir of James V of Scotland and his second queen, Mary of Guise. She was born at Linlithgow Palace, some seventeen miles west of Edinburgh, on Friday, December 8, 1542.

The deep frost scarcely troubled the occupants of the queen's suite on the third floor of the northwest tower of the palace. Recent construction had transformed Linlithgow into a luxurious residence. James V had lavish tastes and sought to introduce the latest Renaissance styles. The windows of the palace were glazed, the ceilings painted, the stonework and woodwork intricately carved with crowns and thistles. In the great hall and throughout the dozen or so rooms of the royal apartments, logs blazed in the fireplaces. The finest Flemish tapestries and hangings of rich arras and cloth of gold covered the stone walls to keep out drafts.

Linlithgow, along with Falkland in Fife, was a favorite lodging of Mary of Guise. She had helped to redesign both palaces like French châteaux. This was hardly surprising, because she was herself French. She was the widowed Duchess of Longueville, the eldest daughter of Claude, Duke of Guise, and his wife, Antoinette of Bourbon. The Guises were one of the most powerful noble families in France. Their patrimonial seat was at Joinville in the Champagne region, their estates scattered across strategically important areas of northern and eastern France.

The family of her first husband, Louis d'Orléans, Duke of Longueville, owned significant estates in the Loire region, so Mary of Guise knew all about Renaissance palaces. She compared Linlithgow for its elegance and picturesque setting to the châteaux of the Loire, where the French royal family lived when not near Paris. Like Chenonceaux, the jewel of the Loire, Linlithgow was a pleasure palace partly surrounded by water. The outer walls stood on a semicircular knoll extending into the loch on the north side, overlooking St. Michael's parish church and the town of Linlithgow to the south.

Mary Stuart was born at a turning point in history. Only two weeks before, on November 24, her father's forces had been routed by the English at the battle of Solway Moss. To the Scots, England was the "auld enemy." Relations between the two neighbors had smoldered since Edward I had claimed the feudal overlordship of Scotland and tried to annex the country in the 1290s. The Scots sought French and papal support, and fostered a hardy patriotism in defense of their kingdom's independence. A score of English invasions after 1296 ushered in a period of hostility that lasted for five or more generations.

Border skirmishes were the norm. Outright war was the exception, not least because the two countries were so unequally matched. England was so much richer and more powerful than its northern neighbor. Its population was around 3.5 million, Scotland's barely 850,000. The only Scottish town of any size was Edinburgh, where 13,000 people lived. This was at most a fifth of London's population. It was far easier to raise taxes and levy troops in England than in Scotland, since the machinery of government was more centralized and the chain of command more efficient. A set-piece battle would almost inevitably end in a crushing defeat for the Scots.

There were regional inequalities within Scotland. Between a third and a half of the population lived in the border region and the Highlands, while the rest occupied the more prosperous and cosmopolitan Lowlands. The king was advised by the lords in Parliament, but although the Scottish Parliament was supposed to represent the whole country, it tended to stereotype highlanders and borderers as chancers and criminals. The Highland clans stood aloof from the rest of the country, and as a rule the highlanders and lowlanders had a tacit agreement to ignore one another. Many highlanders spoke Gaelic rather than Lowland Scots, exacerbating cultural differences. The language of the lowlanders was in fact much closer to northern English than to anything spoken by highlanders.

The politics of Scotland were tribal: blood ties and kin culture were predominant. Behind the feudal lord lay the more ancient status of chief of a clan or kindred. Loyalty to kin placed the Scottish lords at the head of networks sometimes covering entire regions and shaping the structures of power at every level. The monarchy itself relied on these structures and on what it could redistribute from the patronage of the Church.

The wars within the British Isles resumed under Henry VIII, who acceded to the English throne in 1509. Henry was a strong leader. He saw himself as an English patriot and also as a military strategist. His ambition was to resume the Hundred Years War against France and to win conquests there. Of his royal predecessors, those he admired most were the Black Prince and Henry V, whose glorious victories in France brought them lands and reputation. Repeatedly the efforts of his councilors were bedeviled by his chivalric dreams. But war was the "sport of kings." And if Henry sought to conquer French territory, he had to deal first with Scotland, France's "auld ally" and England's back door. A popular rhyme quipped: "Who that intendeth France to win, with Scotland let him begin." Henry was fond of quoting it, and he put its lessons into practice.

Typically, the defeat of the Scots at Solway Moss was less the result of a full-scale English invasion than of a border skirmish that went tragically wrong. The disaster stemmed less from Henry VIII's aggression than from James V's decision to launch a counterattack on an epic scale without choosing the ground or the moment carefully enough.

In reaction to the incursions of English forces led by the Duke of Norfolk, James sent an army to pillage the disputed territory to the north and east of Carlisle known as the Debatable Land. His troops forded the River Esk at low tide. When they returned, it was high tide and they were caught between the river and a bog. Forced to retreat by a smaller but better-disciplined English battalion, the Scots were snared. Around 1200 were taken prisoner, including 23 important nobles and lairds, who were dispatched as hostages to London, where they were put in the Tower.

James V felt a deep psychological blow. He had been militarily and personally humiliated, his loss of face the greater in that he had been ensconced safely at a distance and was not leading his troops. Cowardice was not the issue. James was a brave warrior, but he misjudged the risks. The result was a disintegration of his forces. It was a more damaging loss to his reputation than that suffered thirty years before by his father, James IV, whose own army had been cut to pieces at Flodden Field by the father of the very same English commander. The political effects of both defeats — a long royal minority — were identical. But in 1513, at least the Scots had been scythed down in hand-to-hand combat during a set-piece battle. They died honorably rather than like rats in a trap.

James V rode to Linlithgow to see his wife begin her confinement, but almost instantly left for Edinburgh and then Falkland, where he took to his bed. It is unlikely that he loved his wife, since he had so many mistresses. But he cared greatly about his baby, and his sudden departure tells us more about his mental state than about his family ties. He went to pieces when told that his heir was not a boy. His two infant sons had died the previous year, and now his thoughts turned to Marjorie Bruce, King Robert I's daughter and the founder of the Stuart dynasty. He exclaimed: "The devil go with it! It will end as it began. It came from a woman, and it will end in a woman." Or as a more colloquial source says, "It came with a lass, and it will pass with a lass."

James died at midnight on December 14. He was only thirty, but had suffered recurrent illnesses. A life of sexual dissipation, leading to "pox" and endemic "fevers," and a serious hunting accident had weakened his immune system. His last symptoms, a "marvelous vomit" and "a great lax," suggest dysentery as the cause of death, perhaps the result of drinking contaminated water. Other possible causes were "pestilence," or cholera, caught from the Earl of Atholl, with whom James had been carousing and who had just died.

James V died of natural causes, unlike his father, who had perished at Flodden in the murkiest of circumstances. Although seemingly killed by the English in the battle, it is just as likely that he was murdered in the closing stages of the fight by one of his rebellious lords.

His son had succeeded him at the age of seventeen months. Now history had repeated itself. His granddaughter, Mary Stuart, was queen at the age of six days.

She was baptized as soon as it was safe to take her into the cold outside air. She traveled the short distance from the south-side gateway of Linlithgow Palace into St. Michael's Church in the arms of her nurse, Janet Sinclair. She was named Mary after her mother, but also because her birthday was the day celebrated by the Roman Catholic Church as the day the Virgin Mary had been conceived.

After baptism at the font, Mary was anointed with chrism and wrapped in a robe of white taffeta of Genoa that had been specially made for the occasion. Almost certainly (for such was the practice with royal children) she was then brought to the high altar and confirmed, although she did not take the sacrament at Mass until she was nine years old. A report reached Henry VIII that she was "a very weak child and not like to live." This was wide of the mark, and what shortly became a more insidious threat to her security and peaceful succession would be dispelled by her mother's courage.

James V's death was to set in motion a complex chain of events in which political, religious and factional maneuvers relentlessly combined. England and France were competing to assert a hegemony over Scotland, which became a pawn in the struggle between the two larger countries and their ruling dynasties. As a child, Mary played no role herself in these intrigues, but all of them were about her. The aim of each and every plot was either to secure physically the person of the infant queen or else to marry her into the English or French royal family as a guarantee of future influence. Such machinations helped to shape the dynastic legacy Mary would inherit as she grew older, and all combined to set the agenda she would bravely confront when she reached the age of majority.

Throughout Mary's formative years, her mother was her example. In Scotland for less than five years when her daughter was born, Mary of Guise was politically astute if on a steep learning curve. She quickly turned her mind to politics, keeping the obsequies for her late husband to a minimum.

She was unusually tall, with auburn hair and delicate features. Her manner was regal, her cheekbones high, her eyebrows raised and arched, her forehead elevated. Her lips were slightly compressed, her nose tending to appear aquiline when viewed from the side. Her deportment was confident and dignified; she was intelligent and attentive, generous to friends and supporters, with easy yet polished manners, affable to equals and inferiors alike.

These were all consummate Guise qualities: James V's widow quickly won admiring hearts in Scotland. She was popular with the ordinary people and was able to inspire fervent loyalty. All these same qualities would later be visible in her daughter, who came to resemble her mother closely in looks and personality.

Mary of Guise learned her political skills from her family. Comparative *arrivistes*, the Guises had risen at the French court through a combination of shrewd marriages and military prowess. Closely linked in Francis I's reign to the triumvirate comprising Anne de Montmorency, Constable of France, the Dauphin Henry, heir to the throne, and his beautiful and sophisticated mistress Diane de Poitiers, they were equally influential in the hierarchy of the Catholic Church. Duke Claude's brother Jean was a pluralist who managed to accumulate nine bishoprics and six abbeys. His pickings included the cardinal-archbishopric of Rheims, the most important diocese in France; it was at the great Gothic cathedral of Rheims that the kings of France were crowned. Moreover, the Guises kept it in the family by having it bestowed before Jean's death on Claude's second son, Charles, Cardinal of Lorraine, who took possession of it at the age of fourteen.

In all, Claude had ten surviving children, each of whom held a significant position in state or church. When Mary of Guise married James V, she began fifty years of her family's involvement at the hub of Scottish, French and English affairs. This was because diplomatic alliances were sealed by marriage pacts in the sixteenth century. International politics centered around families, children and the succession to hereditary rights, and in dynastic circles such concerns took precedence over religious affiliations.

Mary of Guise understood her own role perfectly. She set out to protect her daughter's birthright and safeguard the traditional "auld alliance" between France and Scotland. She was used to the call of duty. The most eligible widow in France at the age of twenty-one, she was selected by Francis I to succeed his own daughter Madeleine, James V's first wife, who fell victim to a viral infection within weeks of landing at the port of

Leith. She had been reluctant to leave France, but honor required her to do so. She was not unaccustomed to misfortune. Three of her five children died in infancy: one son by her first husband (another boy, Francis, survived until 1551) and both her sons by James.

After her wedding in Scotland, Mary of Guise regularly wrote to her own mother, Antoinette of Bourbon, a matriarchal figure of whom it was said even the French king was in awe. Her letters show that she quickly adjusted to her new life. She tolerated her husband's infidelities — he sired seven, and probably nine, illegitimate children — and spent her time supervising building projects at the royal palaces and laying out the gardens, where she grew an exotic range of ornamental fruit trees from cuttings sent from France. At Falkland Palace, where her stylish improvements to the façade may still be seen, she personally inspected the work, climbing a ladder to take a closer look before authorizing payment to the stonemasons. She had her own French domestic staff, as it was unthinkable that her intimate servants could be Scots: the French frequently made ribald jokes about the vulgarity of the Scots behind their backs. Her servants adored her, and after James V's death, several of his own domestic staff tried to negotiate a transfer to her employment because she paid higher wages than anyone else.

Technically, the infant Mary was Queen of Scots from the moment her father died. In practice, this was a fiction. A governor or regent would have to be appointed to rule until she was declared to be "of age" and able to govern herself. What was uppermost in everyone's mind was who exactly would be chosen, since a long royal minority was an invitation to noble infighting. This especially applied to Scotland, where the monarchy was so much weaker than in England or France, and the crown relied on the kinship networks of the lords to help maintain law and order.

The most binding way to appoint a regent was in the will of the dying ruler. Although James V had not done this — probably because he did not know his death was close at hand — his "last will and testament" was manufactured on his behalf. It was conveniently framed by David Beaton, Cardinal-Archbishop of St. Andrews, the magnate who had exercised the greatest influence on the living king and was determined to keep himself in power.

Beaton led the pro-French faction and was a staunch opponent of the Protestant Reformation. He sought to claim the posts of "tutors testamentary" to Mary and "governors of the kingdom" for himself and three of his allies. His aim was to preempt the claim of the strongest alternative candidate, James Hamilton, Earl of Arran, who was pro-English and heir apparent to the crown should Mary die. Arran, for his part, loudly proclaimed that Beaton had forged the will, saying that as James V had lapsed into semiconsciousness, Beaton had "caused him to subscribe a blank paper" on which it was to be inscribed.

The nobles voted for Arran as governor, but he was hardly the ideal candidate. The best thing ever said about him is that he was a survivor. Weak, vacillating and cowardly, he was also exceptionally greedy. His legitimacy was questioned by Beaton, but as the grandson of James II's eldest daughter, he had a right by blood to the regency.

Although Mary of Guise did not challenge his appointment, she was wary of Arran. She knew that his policy was less the protection of the infant queen than advancement for himself and his family, hitting the mark when she described him as "a simple and the most inconstant man in the world, for whatsoever he determineth today, he changeth tomorrow."

Arran spent six weeks moving into his new palaces and surrounding himself with his friends and relations, to whom he awarded pensions. In an attempt at reconciliation, he nominated Beaton to the chancellorship, one of the greatest offices of state; but, proving the accuracy of the dowager queen's assessment, he changed his mind two weeks later and threw Beaton into prison.

The new governor was steeped in Scottish tradition. He was well aware of the role played by violence in society and politics. He quickly seized all of James V's castles except Stirling, which was part of Mary of Guise's dowry and still her property. Then, as soon as he felt sufficiently confident, he began to pay attention to Henry VIII's efforts to influence Scottish affairs.

Henry was determined to outwit and outmaneuver France. He wanted Arran to nurture a pro-English faction in Scotland that would displace French influence there. He meant to use the hostages taken at Solway Moss to this end. Henry was about to declare war on France. He had already agreed with his main European ally, Charles V, king of Spain as well as Holy Roman Emperor, that they would join forces in a coordinated attack, seeking to partition France between them.

Henry had an ambitious dynastic plan. He wanted to rule the whole of the British Isles. Naturally he seized his opportunity as soon as James V died, leaving a baby as queen. From the beginning of his reign, he had reiterated Edward I's claim to the feudal overlordship of Scotland. It was a formula he had learned by heart. Later, when he quarreled with the pope and broke finally with Rome in order to divorce his first wife, Catherine of Aragon, he evolved a theory of "imperial" kingship, designed to justify his title of Supreme Head of the English Church, but in which he also envisaged himself as king and emperor of the whole of the British Isles.

According to Henry's version of law and history, Scotland, despite being an independent state and a sovereign realm, would become a satellite of England, a jewel within the orb of Henry's imperial crown. It was the beginning of the Tudor claim to an "Anglo-British" empire that, while Mary was still a child, was to provoke the French counterclaim to a "Franco-British" one.

Henry first released the twenty-three Scots hostages from the Tower, then invited them to admire his recently completed palace at Hampton Court. They were to join him there as his guests for the sumptuous Christmas revels.

Before allowing them to leave, Henry bound them to support his dynastic plan. All were expected to sign articles obliging them to send their infant queen to England. Mary was to be "kept" there, to be brought up by Henry until she could be married to Prince Edward, then a boy of five, his son and heir by his marriage to Jane Seymour. In addition, ten of the most important hostages agreed to uphold Henry's immediate claim to the throne of Scotland should Mary unexpectedly die.

In March 1543, Sir Ralph Sadler was sent to Edinburgh as Henry's ambassador to negotiate a treaty. The centerpiece was to be Mary's marriage to Edward. She was less than four months old, but Henry was in earnest. Such a union could enable him to achieve his goal at a single stroke. Sadler, who did not quite know what to expect, began by meeting Arran in the garden of the palace of Holyroodhouse at Edinburgh. He found him evasive and ill briefed. Arran took the narrowest view of the negotiations: he simply wanted to know what was in it for him. He demanded bribes and rewards, and a promise that he would continue to rule as regent if Henry had his way.

After these inconclusive talks, Sadler rode to meet Mary and her mother at Linlithgow. He wanted to see Mary for himself, because Henry kept asking him about her. Her mother played along, asking a nurse to "unwrap her out of her clothes" so she could be inspected. Sadler dandled Mary on his knee and reported, "I assure Your Majesty, it is as goodly a child as I have seen of her age, and as like to live, with the grace of God."

Sadler then broached the topic of Mary's betrothal to Henry's son. To his amazement, Mary of Guise was positive, offering to help and even endorsing the English plan to take Mary to London for "safe-keeping." Her reaction was so different from Sadler's expectation, he was at first nonplused. He suspected some "juggling," and he was right. Mary of Guise dissembled. But she had a clear purpose, which now un-folded. Whereas Henry VIII wanted to subordinate Scotland to England through a dynastic marriage that detached the country permanently from French influence, Mary of Guise was equally determined to protect French and Guise interests there. And if Arran tried to marginalize her by negotiating with England behind her back, she would pretend to ally with England too, lulling Sadler into a false sense of security and so out-flanking Arran, who would remain at a disadvantage as long as he lacked physical custody of Mary.

Mary of Guise had good reason to suspect Arran, who had severed her channels of communication with France. He had planted spies in her household and sought to intercept her letters. She did get one message through. Antoinette of Bourbon had received news of her daughter's troubles by June 10, sending word to her sons at Francis I's court to see if any pressure could be applied to assist her.

Arran's actions, swiftly following Beaton's arrest, were meant to un-dermine the pro-French party. The governor had decided to ally with Henry VIII. As if to cement his links to the English king, he issued a sur-prise declaration in support of Henry VIII's break with Rome and the dissolution of the monasteries. Arran's action was the more bizarre in that he was still a Catholic who acknowledged the authority of the pope. But he was increasingly on the defensive, caught between the pro-French and pro-English lords, who were evenly divided: the moment was right for a coup.

In charming and beguiling Henry VIII's ambassador Sadler, Mary of Guise showed her political skill. A plan had formed in her mind. She would move Mary from Linlithgow, a pleasure palace that could not withstand a siege, to the security of her own castle of Stirling, an almost impregnable fortress at the top of a steep rock that was also near enough to the coast to restore her links to France by sea.

Already she had sent a trusted servant to Stirling with coffers packed with clothes and household goods. Larger consignments of beds and furniture followed, with further deliveries of silver plate, tableware, linen, dry foodstuffs and kitchen utensils such as pots, pans and roasting spits.

Arran insisted that no one was to leave Linlithgow. Mary of Guise ignored him, playing her cards brilliantly. She spun Sadler the yarn that Arran had no intention to have Mary marry in England. Arran would bargain with England to send Mary south to win rewards for himself and his allies, but then break his word and keep her a prisoner in Scotland, biding his time until Henry VIII was dead, when he planned to usurp the Scottish throne. Sadler was to report this to Henry, but not to disclose his sources, or else she and her daughter would be in danger.

In England and France alike, plotting what might or might not happen in the future if and when the king died was a serious crime. It seemed to make his death more likely as a contingency, and was called "imagining or encompassing" his death, a branch of the law of treason and punishable by death. Sadler had to distance himself from Arran if he was scheming in this way. Otherwise he might himself be indictable as an accomplice.

And worse was to come. Arran, the dowager continued, planned to marry Mary to his own son. Henry VIII should take care to prevent this by ordering Arran to release Beaton, who had been maligned over the business of the forged will and who should replace Arran as governor. Unlike his enemies, Cardinal Beaton "could better consider the benefit of the realm."

It was a classic bluff. Mary of Guise was maligning Arran based on her low opinion of his character. She knew she was in a deadly struggle for the custody of her daughter and was determined to get Sadler's support for her move to bring Mary safely to Stirling. Furthermore, Sadler, who by now deeply mistrusted the scheming, vacillating Arran, was taken in.

There were other reasons why Sadler was receptive to Mary of Guise. Her own mother's lobbying at the French court had paid off. Francis I had decided to intervene. He wanted to thwart Henry VIII and distract him from his plans to invade France. It was already the talk of Paris that the Duke of Guise was preparing to embark for Scotland. His commission was said to have been issued, and Sadler badly needed to know exactly what was about to happen and what the French really intended.

In the end, the duke never arrived. Francis I revoked his commission, sending the young exiled Scottish lord Matthew Stuart, Earl of Lennox, in his place. This was almost as worrying to Sadler. Lennox, head of a minor branch of the Stuart family and a naturalized French subject, had the best claim to the regency if Arran was toppled. He landed in April at Dumbarton, his family's ancestral stronghold on the north bank of the Clyde, and went straight to see Mary of Guise.

She had become the linchpin. Whether or not Sadler was fully persuaded by her blandishments, simply to have unhindered access to her was to his advantage.

Arran strove to recover the initiative. He felt more and more beleaguered when a group of (mainly) Catholic lords congregated around Lennox and Beaton escaped from prison. With Lennox back in Scotland, Henry VIII suspected a plot to kidnap the infant Mary and carry her off to Dumbarton. He ordered Arran to muster his forces and bring Mary into the "safety" of Edinburgh Castle without delay.

But Arran was in a quandary. He sensed his own weakness, and by the end of April was temporarily reconciled to Lennox. Many of his allies had resisted Mary's betrothal to Henry VIII's son and heir. It was a step too far, and Lennox capitalized on this. Arran finally told Sadler that England's terms were so unreasonable, "every man, woman and child in Scotland would liever die in one day than accept them."

Henry VIII issued an ultimatum. If Arran refused the English dynastic plan, he would face outright war. Henry's bullying was deeply resented. On July 1, 1543, however, he was minimally rewarded by the terms of the treaty of Greenwich. Ostensibly, he got what he wanted. Mary was to stay in Scotland until she was ten, at which age she would marry Prince Edward in England. Her dowry, to be paid by Henry, was to amount to the considerable sum of £2000 per annum, which would double automatically if she duly became queen consort of England.

The rest of the treaty was a compromise. Until she was married, Mary's education was to be left to the Scots, except that for her "better care," Henry might, at his own expense, send an English nobleman with his wife or other governesses and attendants, not exceeding twenty in total, to live with her. This was a clause designed to ensure that Mary would speak English as her native language rather than Scots or French.

But Henry had to make concessions. He wanted a speedy settlement to free him to concentrate fully on his planned invasion of France. To achieve a quick result, he found himself accepting terms that guaranteed Scotland's independence. A key clause confirmed the country "shall continue to be called the kingdom of Scotland and retain its ancient laws and liberties." The Scots also insisted that if the marriage was childless,

Mary might return home as an independent queen. This was to be a dynastic "union" of England and Scotland with the core stripped out.

Mary of Guise was jubilant; the loopholes in the treaty were obvious. Arran started to panic. He warned Sadler that the infant Mary had to be closely watched, being "a little troubled with the breeding of teeth." Not even Sadler was this credulous. Writing to Henry VIII, he attempted to fathom why Arran should suddenly want to protect Mary "as if she were his own child." Of course, Arran's greatest fear was that she would be secretly conveyed to France and brought up there by her mother's Guise family.

The treaty of Greenwich was a dead letter from the start. Mary of Guise had no intention of honoring it; she had used the period of negotiation simply to face down Arran and Henry VIII and to win time to build a new, more comprehensive coalition. Now she revealed her true hand. She allied with the pro-French Beaton and Lennox, whose joint forces mustered at Linlithgow on July 24. There, a bond* was signed to prevent Mary's removal to England and for mutual defense against the pro-English Arran.

Two days later, Arran himself arrived, but with a much smaller retinue. From that moment, his capitulation was assured. Sadler had already sanctioned Mary's removal to Stirling, where she was to be guarded by a group of Scottish lords officially nominated in Parliament. Mary of Guise now contrived that her daughter was taken there under very different circumstances.

The child and her mother made their journey on the 27th. Lennox provided their bodyguards, mustering an army of 2500 cavalry and 1000 infantry to protect a baggage train extending for almost a mile. Although barely out of swaddling clothes, the young Mary traveled with all the pomp and circumstance she would one day take as her due.

Stirling, always Mary of Guise's chosen destination, would become their home for the next four years. It was an appropriate setting, as its magnificent great hall and royal apartments had been lavishly reconstructed as part of a massive royal rebuilding program. The great hall alone could seat three hundred people for dinner.

A fortnight after their arrival, Sadler was summoned for an audience.

* A bond (or "band") was a feudal contract or indenture promising loyalty, support, protection and service, usually for life.

Mary of Guise still dissembled, yet the ambassador was unable to figure out why. She said she had all along been willing to allow Mary to be taken to England for "safekeeping." Only Arran's duplicity had prevented her. In fact, she hoped the better to perform her true intentions now that she had escaped from his clutches. She was "in good plight" to deliver Mary to Henry's nominees if he still so wished.

She used this interview to win more time. Sadler, who had a soft spot for children, was taken to see Mary, who was growing fast. She "soon would be a woman," he said, "if she took after her mother." She had suffered a mild bout of chickenpox, but was fully recovered. As Sadler noted, she was "a right fair and goodly child."

He had been duped again. Mary's removal to Stirling was for her mother the beginning and the end of the matter. Arran, ever the survivor, cut his losses and made peace with his rivals. On September 3, the governor left Edinburgh, apparently to visit his sick wife, but met Beaton secretly at Falkirk. The two men embraced and rode to Stirling, where Arran disclosed the full extent of his dealings with Henry VIII and recanted his support for the Reformation.

On the 8th, Arran agreed to revise the terms of his regency, promising to share power with Beaton and to follow the advice of a council comprising representatives of the pro-French and pro-English factions and headed by none other than Mary of Guise. The effect was to reconcile the nobles, who closed ranks against Henry VIII's aggression.

The climax ensued. Next day — by a delicious irony the thirtieth anniversary of the battle of Flodden — Mary was carried in procession from her nursery at Stirling and crowned Queen of Scots in the adjacent Chapel Royal. It was an event of the utmost significance. The coronation was the most solemn ritual known to church and state: its symbolism was sacramental and conferred religious as well as civil legitimacy on her. In the course of the ceremony, a nine-month-old child was transformed into an anointed queen, possessed of those sacred powers of majesty that God alone could bestow or call to account.

In the procession, Arran bore the crown, Lennox the scepter, and the Earl of Argyll, the most powerful of all the Scottish lords and Arran's brother-in-law, the sword of state. These regalia, known collectively as the "honors of Scotland" and still on display at Edinburgh Castle, had been obtained by James IV and his son in their tireless efforts to trumpet their prestige. They were first used together on this day. The crown, originally worn by James V at the coronation of Mary of Guise, was far too big and heavy for a child. It was held over Mary's head by Beaton, who was dressed in the full panoply of a cardinal. He blessed her and anointed her with holy oil. She howled volubly and kept it up while every bishop and peer present knelt in turn to recite his oath of allegiance.

By tradition, heralds read aloud the royal genealogy, a roster of titles and dignities that could take up to half an hour to recite. In view of Mary's lusty interventions, this part of the proceedings was omitted. The pro-English lords were conspicuous by their absence, but otherwise the day passed "with great solemnity" and was rounded out with banquets, masques, dramatic interludes and other entertainments in the great hall, followed by "great dancing before the queen with great lords and French ladies."

Mary's coronation concluded a remarkable interlude that had begun when her mother first turned her attention to Henry VIII's ambassador. It signified a reversal of the balance of power. The pro-English lords had been marginalized and Arran reconciled to Beaton, who was restored to office as chancellor. The treaty of Greenwich was all but renounced: the pro-French faction was ascendant in Scotland.

Henry VIII had played his opening hand and lost. He had also learned an important lesson. He would never again, as he told almost anyone who would listen, trust the Scots. Under the watchful eye of her mother, Mary had ascended to her throne. And if she was now queen, her mother was indisputably queenmaker.

2

❦

The Rough Woolings

W HEN MARY OF GUISE pulled off the coup that made possible her daughter's coronation, she knew the breathing space would be short. Arran and Lennox were rival claimants to the succession. Their families were old enemies; it was impossible to believe they would stay on the same side for long. Even before Arran agreed to share power with Cardinal Beaton and restored him to office, Lennox was growing disaffected. In any case, Arran's reconciliation to Beaton was only the prelude to his efforts to stage a comeback. Mary of Guise knew she must retain his support for her pro-French policy until the treaty of Greenwich was officially renounced by Parliament.

To this end, business could be combined with pleasure. With Mary crowned queen, her mother could afford some fun and dalliance. Scarcely had the great hall of Stirling Castle been cleaned up after the coronation festivities than it was Mary of Guise's birthday. She was only twenty-eight. It was, wrote a chronicler, although late autumn, "like Venus and Cupid in the time of fresh May, for there was such dancing, singing, playing and merriness . . . that no man would have tired therein."

Such celebration had a political point, because Mary of Guise, herself barely a year older than Lennox and still one of the most beautiful women in Scotland, planned to assure his loyalty by thoughts of marriage. She had accurately judged his ambition. It would soon become a fixation, to the point where Lennox scarcely distinguished between a marriage in Scotland or England as long as it brought him closer to a crown.

Lennox was lissome and urbane, intelligent if duplicitous, "a strong man, of personage well proportioned in all his members, with lusty and manly visage." Tall and svelte, he oozed sophistication and was "very pleasant in the sight of gentlewomen." His savoir-faire had been acquired in France, where he served as a lieutenant in the Garde Écossaise, the king's personal bodyguard.

His rival in love if not in lineage was Patrick Hepburn, Earl of Bothwell. Exiled by James V to Venice and Denmark for his unruliness, he had recently returned to Scotland. His ancestors were lords of Hailes in East Lothian, an affluent Lowland region to the east and south of Edinburgh. His grandfather had risen on the battlefield, was promoted to an earldom and given Bothwell Castle in Lanarkshire. Later he exchanged it for the Hermitage, a vast and isolated border citadel in the valley of the Hermitage Water in Liddesdale, midway between Hawick and Carlisle, close to the Debatable Land. The exchange turned the Hepburns into powerful border lords. The Hermitage was a key location, the fortress from which the western and central sectors of the border with England were controlled. James IV also gave the family Crichton Castle, some eleven miles southeast of Edinburgh, enabling them to improve their position at court and in Parliament.

Patrick was a Scottish patriot: pro-French and anti-English, but also an opportunist who flirted with England when it suited him. Gossip said he was a royal bastard. This is unlikely, although his mother was briefly one of James IV's mistresses. The family's fortune derived from his grandfather, who had accumulated a cluster of offices retained by his heirs. As a result Patrick was hereditary Lord Admiral, a lucrative post, since it entitled the Hepburns to the profits of all ships wrecked off the coast of Scotland, making them one of the few noble families to enjoy financial independence from the king. He was also sheriff of Edinburgh, which gave the family influence with the legal profession in the Court of Session and in Parliament.

The Hepburns stood for the values of chivalry and warfare. They saw themselves as "men of honor," which in their eyes justified dueling and even treachery as acts of self-defense. Their code of ethics flourished among military men on the Continent and in Ireland, but was considered repellent by civilian administrators and diplomats. Sadler, Henry

VIII's ambassador, described Patrick as "the most vain and insolent man in the world, full of pride and folly, and here nothing at all esteemed." He might have said the same about Lennox, except the values of the Lennoxes were civilian, based on courtly manners and polite society: perfidy was just as rampant but cloaked by the veil of gentility.

Patrick was sandy-haired, of medium build, with a fair complexion and a slight stoop. He had a broad smile and a winning manner. Like Lennox, he could captivate women. He had numerous affairs, and had no scruple about abandoning his wife to advance his suit for Mary's mother, using his influence in the Catholic Church to get an annulment of his marriage without prejudice to his children's legitimacy.

Lennox and his rival calculated that whoever became Mary's stepfather would be able to displace Arran as regent and rule in Mary's name. So they dogged her mother's steps from Stirling to Edinburgh and back, posturing like peacocks to catch her eye. They danced and sang and recited poems. They engaged in shooting and jousting matches, wearing the most fashionable clothes and running up massive bills with the jewelers and haberdashers of Edinburgh. They followed her to St. Andrews, where she was Beaton's guest in his castle. She handled both men as befitted an accomplished practitioner of courtly love, making encouraging noises but offering "nothing but fair words."

Patrick tried an end run by letting it be known that she had promised to marry him. Lennox took this rumor as truth and retired to his stronghold at Dumbarton. Sulking, he decided to change sides, petitioning Henry VIII for the hand of Lady Margaret Douglas, the King's niece and daughter of his sister Margaret, widow of James IV, by her second marriage, to the Earl of Angus. He also sought Henry's aid in recovering "his right and title" to the regency in Scotland, which he claimed Arran had usurped. His overture delighted Henry, who saw Lennox as a surrogate for Arran now that the latter had defected to Beaton.

Lennox's *volte-face* upset the balance between the noble factions in Scotland. Now he was pro-English rather than pro-French. In an attempt to counter this, Francis I redeemed his promise to the Guise family. In October 1543, six of his ships sailed up the River Clyde and landed at Greenock. On board were the new French ambassadors, Jacques de la Brosse and Jules de Mesnage, who brought money and artillery to help Mary of Guise and her supporters.

The flotilla sailed on to Dumbarton, where Lennox overreached himself. He seized the ambassadors' money and most of their guns. A stalemate was averted only when he was warned that as a naturalized French subject, he could be tried for treason in France. He grudgingly submitted and returned to Stirling, where he was briefly reconciled to the French cause.

La Brosse then used the money at his disposal to provide pensions for the leading lords. Amounting to 59,000 crowns of the sun (more than James V's usual revenue in the last year of his reign), it was a substantial windfall. Nothing greased the wheels of Scottish politics better than pensions, and when Parliament reassembled in December, it took less than a week to exonerate Beaton of all the charges against him and enact, in the infant Mary's name, a renewal of the "auld alliance" between Scotland and France.

Parliament's next step was to repudiate the treaty of Greenwich, tearing up the marriage contract between Scotland and England. The revocation of the treaty left Henry VIII incensed. According to La Brosse, he was threatening revenge as if he had lost a great battle.

Mary of Guise was exultant. The reinstatement of the "auld alliance" was a personal triumph. She spent a joyous Christmas with her little daughter at an ever-bustling Stirling. The entertainments for the French ambassadors were lavish and unstinting. Once more there was music, dancing and feasting. And she won £100 from Arran at the card tables.

Lennox, however, posed a threat. On March 21, 1544, he met Mary of Guise for the last time. She had no intention of making him her daughter's stepfather. When she finally told him so, he left in a fury. A week later he set sail for England, where he signed an indenture to marry Margaret Douglas. By this deed, equivalent to a legal conveyance, he promised to strive for a dynastic union between England and Scotland, and to govern Scotland, if he was ever to obtain the regency, at the direction of the English king, to whom he even assigned his own claim to the Scottish throne.

On June 29, the nuptial Mass for Lennox and Margaret Douglas was celebrated in the presence of Henry VIII and Catherine Parr, his sixth and last queen. The bride's dowry was provided by Henry, who granted the couple substantial estates in Yorkshire.

Even before the wedding, Henry had decided to take revenge on the Scots. He was so incensed by their disavowal of the treaty, he had warned the citizens of Edinburgh that he would "exterminate" them "to the third and fourth generations" if they got in his way. Outraged by what he saw as stark treachery, he was also anxious lest Mary be shipped to France, out of his reach.

While Lennox had been working out how best to play his cards, Henry was mustering troops for an invasion of Scotland. His paranoia was plain. Not only did he aim to unleash the biggest invasion since Edward I's reign, he went so far as to compile hit lists of individual Scots. His plans did not balk at assassination. Beaton was a leading target, as Henry blamed him more than Mary of Guise for detaching Arran from the pro-English party. Soon Sadler, himself a supporter of the Reformation, was seeking out Beaton's Protestant enemies, infiltrating their secret networks to see if a plot could be devised.

The main aim of Henry VIII's invasion was to force the Scots to reinstate the dynastic marriage clauses of the treaty of Greenwich. For this reason, it was afterward said to have begun the "Rough Wooings" of Mary Queen of Scots. The campaign started in the first week of May, under the command of Edward Seymour, Earl of Hertford, Prince Edward's uncle.

Since Henry had already committed himself to a summer invasion of France in alliance with Spain's Charles V, he was potentially fighting on two fronts. To avoid this, he gave Hertford only a month, later reduced to three weeks, to fulfill his Scottish mission. Hertford was to besiege the town and castle of Edinburgh, destroy the port of Leith, the deepest harbor on the Firth of Forth and the gateway to Edinburgh, and then turn to the central Lowland belt between Edinburgh and Stirling. Once these had been laid waste, he was to cross the Forth into Fife, the breadbasket of Scotland, where he was to "extend like extremities and destructions in all towns and villages," and especially at Beaton's stronghold of St. Andrews.

Hertford traveled by sea, sailing up the Firth of Forth and disembarking 15,000 men at Granton, two miles or so beyond Leith. Advancing toward the town, he found 6000 Scots lined up on the inside bank of the Water of Leith, the strategic position from which Edinburgh was defended from the north. Arran, Beaton and Patrick, Earl of Bothwell, commanded these forces, but after half an hour's fighting, they were overwhelmed and fled.

Hertford, however, failed to seize Edinburgh Castle. Its position was well-nigh impregnable; it was heavily fortified with artillery, and anyone who approached it was vulnerable to attack. So the order was given to ignore the castle and burn and pillage the rest of the town and its suburbs. The fires raged for three days: almost every house and church within the walls was looted or destroyed. The palace of Holyroodhouse and the adjacent Abbey-Kirk were ransacked. A detachment of troops was then sent over to Fife, burning Kinghorn and the villages around Kirkcaldy, but soon returning. Time was short, and Hertford's troops were unable to come within twenty miles of St. Andrews.

Throughout these terrifying events, Mary was protected by the high walls of Stirling. Sadler learned that her attendants, charged with her security on pain of their lives, would if necessary whisk her off to the Highlands, where "it is not possible to come by her." Hertford did advance into the central Lowland zone. On May 15, he reported that the region had been ravaged to "within six miles of Stirling," and Leith would be flattened the next day. The result was that Mary was taken to Dunkeld, one of the main approaches to the Highlands some thirty miles north of Stirling. She was safe there, and Hertford's deadline was near.

Ordered to return south so that his crack troops could be shipped to Calais to begin the French campaign, Hertford marched from Edinburgh to Berwick-upon-Tweed, down the east coast route, burning the market towns on the way and flattening as many other fortified towers, villages, churches and houses as he could manage.

It was a catastrophe for Scotland, and Arran got the blame. He, even more than Beaton, was held accountable. The nobles argued that he should, in future, share the regency with Mary of Guise. Her popularity had soared, because her pro-French policy was held to be synonymous with Scottish freedom from its "auld enemy." She managed to escape all the blame for the fire and brimstone brought down on the population by Henry VIII. It seemed that she alone had the interests of Scotland at heart. Certainly her family held the key to the French alliance: without her, Francis I would be less inclined to defend Scotland's cause against England.

Mary's mother saw her opportunity. She wanted to be sole regent, not co-regent. Arran objected and each side summoned rival Parliaments: the deadlock lasted for months. Finally, Beaton hammered out a compromise whereby Arran promised to take her "counsel and advice." Thereafter, she sat regularly with the lords of the Privy Council and in Parliament, where she strove to maintain the appearance of unity while shifting Scotland as far as possible into a French orbit.

Her morale, daunted somewhat by this distasteful compromise, was raised by a victory. In February 1545, an English raiding party crossed the Tweed to pillage the ancient abbey town of Melrose and its magnificent church. On their return, they were ambushed by a smaller Scots force, which took many prisoners. It was a blow to English prestige at a time when Henry VIII was briefly vulnerable. Paradoxically, this was the result of his success in France, where he had captured Boulogne and defended it against a countersiege. But to secure his much-vaunted conquest, Henry was forced to dig in, which tied him down. Powerful as England was relative to Scotland, the country was weak compared to France. Henry had been ditched by his ally, Charles V, who had made a separate peace with Francis I. He was fighting alone, running up vast debts and stretching his forces to the limit.

In Paris, the Dauphin Henry was beginning to take the lead in Scottish affairs. He and the Guise family warmly congratulated Mary's mother on the ambush, offering to assist her further. By May 1545, fresh reinforcements were ready to embark and more pensions granted to the Scottish nobles. Nothing seemed to be too much trouble. When Jacques de Lorges, Sieur de Montgommery, who was commissioned to lead the expedition, heard that Mary of Guise was "ill provided with wines," he ordered a consignment of "good ones" to be sent to her.

The troops arrived within a month. Whether Francis I wholeheartedly supported them is another matter. Perhaps no more than five hundred men disembarked. And there was a sting in the tail. In order to pay his soldiers, Francis had melted down 10,000 crowns of the sun and mixed in copper and lead to manufacture 150,000 crowns. The debased coinage had been given to Lorges's men, but the deception was apparent from the moment they arrived and the canny shopkeepers of Edinburgh refused to accept the false coins.

As the campaigning season drew to its close at Boulogne, Henry felt safe enough to turn again to Scotland. He sent Hertford over the border in September to continue the Rough Wooings. Leaving his base at Newcastle-upon-Tyne, Hertford led twelve thousand men toward Berwick, rested for three nights, then advanced twenty or so miles inland over the hilly terrain toward Kelso. From there, he turned south again toward Jedburgh, burning and looting everything in his path and petrifying the local inhabitants.

Next he attacked the frontier villages in a slash-and-burn raid covering almost two hundred square miles. He began by claiming he would inflict as much damage as on his previous campaign, but ended up boasting it would be twice as bad. Still, this was a far less crushing invasion than its predecessor. Hertford did not venture deeper into Scotland, because he was wary of the French reinforcements.

Henry VIII, meanwhile, had turned to diplomacy with France. Scotland was at the top of his agenda: his latest gambit was to offer to return Boulogne to Francis I in return for a marriage between Mary and Prince Edward. The following spring, Hertford traveled to Paris to resume the negotiations. Both kingdoms were financially exhausted, and a truce was agreed that included Scotland. It might have given both sides breathing space had not the unthinkable happened.

On May 29, 1546, Beaton was assassinated. Three months earlier, the zealous cardinal had ordered the burning of George Wishart, a leading Protestant reformer, lashing him to the stake with ropes and strapping bags of gunpowder to his body to ensure a spectacular show. Since burnings for heresy were almost as unpopular with Catholics as with Protestants, the effect of Beaton's display was to turn opinion sharply against him.

The assassins, a group of lairds from Beaton's home base of Fife, chiefly resented his social and political power. Nothing moved without his say-so, and the chief conspirator, Norman Leslie, sheriff of Fife, had challenged his jurisdictional claims. In the resulting feud, Leslie was backed by his friends. Sadler, Henry VIII's ambassador, was directly involved. Not content to offer them unqualified English support, he had also bankrolled them, calculating that with Beaton dead, the pro-French party would collapse.

Between five and six o'clock in the morning, a band of assassins arrived at the main entrance to St. Andrews Castle. They slipped inside with the stonemasons who were then reporting for work, passing Marion Ogilvy, Beaton's mistress and the mother of his eight children, as she left the castle by a postern or side gate as she usually did on her way to do her shopping. Once inside, they snatched the keys from the porter. Leslie was then able to enter the courtyard and order the servants and workmen to leave, then run back to secure the postern gate in case Beaton fled that way.

Beaton heard a noise and tried to escape, but seeing his path blocked, returned to his bedroom and bolted the door. He reopened it only when

Leslie's men stacked burning coals outside. He fell into a chair and prayed. This cut no ice with the conspirators, who preached a long-winded sermon, calling on the "vile papist" to repent, before stabbing him. "Fye, all is gone" were his last words. Leslie then hung his naked body from the castle walls by knotting a pair of sheets to make a rope.

When the people crowded to view this spectacle, a man called Guthrie undid his breeches and "pissed" into Beaton's open mouth. The Castilians, as they were nicknamed, since they were forced to barricade themselves in the castle to evade the authorities, then packed the corpse into a salt chest, which they cast into a deep, bottle-shaped dungeon. This was a very specific act of revenge — the body could easily have been thrown into the sea from the rear wall of the castle — because friends of the assassins had themselves been imprisoned there by Beaton. This unusual dungeon ranked for its terrors with anything to be found in Europe, as there could be no escape except through the neck of the "bottle," requiring the use of a rope or ladder lowered from above. And because the dungeon was carved out of the cliff beneath sea level, the roar of the waves could be heard inside. This was a conspiracy fully in keeping with the tribal politics of blood feud.

Henry VIII was overjoyed. He saw the murder as a breakthrough in his campaign to defeat the "auld alliance." He could not have been more wrong. If anything, Beaton had been a stabilizing influence in Scotland. Now opinion veered even more sharply toward France.

The lords hurriedly met in council and chose the Earl of Huntly, head of the Gordon clan, the leading Catholic and pro-French family in the eastern Highlands, to replace Beaton as chancellor. Several pro-English lords took this opportunity to defect to the pro-French faction. Within Fife itself there was an abrupt shift of grassroots opinion: those lairds and their dependents who had helped the Castilians were turned almost overnight from local heroes into targets of spontaneous assaults.

What Arran could not manage was to retake St. Andrews Castle. He began a siege, but the fortress could be supplied by sea: Henry VIII dispatched food and munitions all the way from England. When Arran's men struggled valiantly to mine their way in by hewing a passage through the solid rock, they were thwarted by a counter-mine cut by the Castilians.

Such failure angered Mary of Guise. She was already irked at the backhanded way Francis I had "aided" her. The use of debased coin to pay the wages of Lorges's troops especially rankled. She wanted St. Andrews Castle retaken and the Castilians punished.

Then, momentous events occurred at breathtaking speed that conjured up exhilaration and fear in almost equal measure. Henry VIII died in January 1547, followed two months later by Francis I. These titans had dominated the affairs of the British Isles and northern Europe for thirty years. Suddenly there was a vacuum. And it was the Earl of Hertford and the Guise family who moved instantly to fill it.

Henry VIII's son and heir, Edward VI, was only nine. A regent or protector (as the office was called in England) would be needed to govern until the king was eighteen, but Henry had shied away from giving so much power to any one person. Instead, he had used his will to appoint a Council of Regency to rule during his son's minority. Despite this, Hertford made himself Duke of Somerset within a week. He took viceregal powers to govern as protector: his overriding aim was to realize Henry VIII's dynastic plan by imposing on Scotland the defunct treaty of Greenwich and uniting the two crowns through Mary's marriage to Edward. To this end, he aided the Castilians. Among his first official acts was to make a pact with them, distributing pensions and wages and shipping food and munitions to St. Andrews.

In France, the dauphin succeeded his father as King Henry II. The Guise family were among his chief advisers; the result was that Henry at once declared himself to be the "protector" of Scotland. He decided to spare no expense to safeguard the "auld alliance" and ensure that Mary would marry no one except his own son, the Dauphin Francis. He flatly countered Somerset's idea of an Anglo-British union with his own master plan for a Franco-British empire. Moreover, his level of commitment far exceeded anything shown by his father, Francis I, whose chief concern had been to frustrate and rival Henry VIII.

Henry II called on Leone Strozzi, Prior of Capua, a brilliant naval officer trained in Italy, to lead an expedition against the Castilians. Strozzi sailed into St. Andrews Bay on July 16, laying siege to the castle on the 24th. Serious firepower was used. On the 30th, he bombarded the castle from the roof of the ancient abbey to the east and from the tower of the university chapel to the west. The assault began at daybreak and was over by three in the afternoon. Before Arran could even cross the Forth and ride the fifty or so miles from North Queensferry to St. Andrews, the castle had been retaken and its occupants imprisoned or forced to board the French galleys.

When the Castilians were finally led out in chains, they included a young Protestant reformer named John Knox, who had abandoned his previous career for the life of a preacher after meeting Wishart, the man burned by Beaton, in his native East Lothian. Knox had no hand in the cardinal's murder, but approved of it. He came to St. Andrews, where he preached in the castle chapel and in the town. His teaching was that "the pope is antichrist" and "the Mass is abominable idolatry": his oratory was so compelling that he won many converts.

Knox had to row in the French galleys for eighteen months. Then Somerset arranged for his release and safe-conduct to London. It was to be a fateful move, because Somerset's secretary and master of requests was a young Cambridge graduate and rising star named William Cecil. He was the man destined to be Mary's nemesis, and now he met Knox for the first time. It was partly through Cecil's influence that Knox received job offers in England and was appointed one of Edward VI's chaplains.

Somerset's response to Strozzi's recapture of St. Andrews Castle was to order a final round of the Rough Wooings. In late August, an army of fifteen thousand men arrived at Newcastle. Somerset marched at the head of his troops to Berwick and continued into East Lothian. In a change of tactics, he intended to settle permanent English garrisons in Scotland. He meant to occupy and subdue the country, and so force it into submission. His army was shadowed by an accompanying English fleet as it made its way north, ready to open fire if his troops were ambushed by the Scots.

For once, Arran had his own forces ready. When on September 10 the English infantry surmounted a hill close to Inveresk, near Musselburgh, on the eastern approaches to Edinburgh, they were confronted by the largest Scottish army in history. Some twelve thousand troops were skillfully positioned behind defensive trenches on the west bank of the Esk. They were put there to block the road into Edinburgh, and since on one side of them lay the sea and on the other an impenetrable marsh, Somerset would have to either launch a frontal attack or wheel his army around.

As Somerset turned in search of a defensive position, Arran attacked.

The armies clashed on the hills above the hamlet of Pinkie. For an awesome moment, it looked as if the Scots pikemen could win. Then the English ordered a shock cavalry charge and fired their heavy guns to deadly effect. As the Scots buckled, Arran fled. His troops panicked. In the bombardment and ensuing carnage, ten thousand Scots were scythed down and killed. It was a second Flodden: the way lay open to Edinburgh and Stirling. Mary, aged four and three quarters, was hastily carried in a litter by night to Inchmahome Priory, a remote spot on an island in the Lake of Menteith, some eighty-five miles from the battle.

Somerset did not reoccupy Edinburgh. Instead, he started to build a grid of quick, cheap forts from coast to coast in an attempt to hold the country at his will. If he expected this to work, he miscalculated. The backlash was so massive, it led to the one thing he had inexplicably overlooked: the removal of the Queen of Scots to France.

Far from tolerating union with England on the back of a military conquest, Arran made his terms with Henry II, accepting the offer of the duchy of Châtelherault in Poitou and the promise of a bride for his son. The bargain was sealed in January 1548, and within a month, negotiations were under way for Mary's betrothal to the dauphin.

Henry II could not have spoken more clearly. He promised to liberate Scotland from Somerset's garrisons. In keeping with his style, this was to be a Guise family affair. Francis, the eldest son of Claude Duke of Guise, already seen as the outstanding military strategist of his generation, was put in charge. He was to plan the campaign jointly with his brother Charles, Cardinal of Lorraine, the most gifted administrator at Henry II's court. The Guise brothers were, of course, the uncles of the young Queen of Scots.

In readiness for her voyage to France, Mary was moved to Dumbarton Castle, already confiscated from Lennox and secured to the pro-French party. With Somerset increasingly bogged down in provisioning his garrisons, the greatest danger to the young queen was from an attack of measles. Rumors that she was dead were quickly contradicted. The prospects for her safety were good, as an advance party of French troops had arrived to defend her, equipped with enough money and ammunition to last a year.

The main French expeditionary force disembarked at Leith on June 17. An armada of 130 ships transported 5500 infantry and 1000 cavalry, striking fear into the occupants of Somerset's forts. Many of the French soldiers were veterans who had served in Italy. Their captains were professional commanders, most of them retainers of the Guises. They were loyal to their commanders and dedicated to their country, unlike many of Somerset's soldiers, who were foreign mercenaries.

Early in July, the French forces laid siege to Haddington, Somerset's principal fort in the eastern Lowlands. A week later, Parliament convened in the nearby abbey and, after only a short discussion, approved the treaty between Scotland and France. The French lieutenant-general, André de Montalembert, Sieur d'Essé, explained that Henry II had "set his whole heart and mind for [the] defense of this realm" and sought to betroth Mary to his son.

Parliament quickly acceded to this as "very reasonable." The potential sticking point was national autonomy, but as Henry II had promised to defend Scotland's laws and liberties "as these be kept in all kings' times past," the terms were easily ratified.

The day after Parliament ended, Mary of Guise wrote to her brothers, "I leave tomorrow to send her [Mary] to him [Henry II], as soon as the galleys have completed the circuit." To confuse the English naval patrols, the commander of the French ships sent to fetch Mary sailed around the north coast of Scotland, skirting the Orkney Islands and back down the west coast to the Clyde. It was a dangerous, roundabout route.

On July 29, Mary kissed her mother goodbye and boarded her ship, which was Henry II's own royal galley. Although only five and a half years old, she carried herself like a queen. Her embarkation was watched by Jean de Beaugué, an army veteran and friend of d'Essé, who wrote: she is "one of the most perfect creatures that ever was seen, such a one as from this very young age with its wondrous and estimable beginnings has raised such expectations that it isn't possible to hope for more from a princess on this earth."

As Mary walked down the narrow steps from the castle to the pier, she registered her personal trademark. Whatever she did and wherever she went in the future, whether her actions and behavior were to be applauded or demonized, she would prove to have this gift of conjuring a sense of occasion.

There followed a week's delay caused by storms and a smashed rudder. The ships were tossed about violently at anchor, and Mary discovered that she was among the few on board immune to seasickness. It was perhaps the first occasion in her life when she found she could be strong while so many others around her were weak.

On August 7, the flotilla finally reached the open sea. It was Mary's first big undertaking, her debut on the international stage. She visibly relished the part. Her high spirits were observed by Artus de Maillé, Sieur de Brézé, a Guise retainer whom Henry II sent as his ambassador for the voyage. In a series of letters written on board ship, he informed her anxious mother, "The queen, your daughter, fares as well and is, thanks to God, as cheerful as you have seen her for a long time."

After a rough crossing lasting eighteen days, almost twice as long as had been estimated, the galleys reached St.-Pol-de-Léon, a small haven not far from the busy port of Roscoff in Brittany. From there the party was to travel on to St.-Germain-en-Laye, Henry II's favorite château, built on a cliff overlooking the River Seine on the outskirts of Paris.

Exhausted by the storms, everyone was glad to relax for a few days before starting the next stage of their journey. Mary needed to rest. Her voyage was the beginning of an adventure, and yet she must have felt some apprehension. She was unsure if she would ever see her mother again. And although from here on she was fêted wherever she went as "la petite Royne" ("the little queen"), it was not just because she was royal, but because she was pledged to marry the dauphin of France.

3

Arrival in France

I N SPITE OF Mary's understandable apprehension as she first set
foot in France, she was far from alone. At least a dozen familiar
faces surrounded her. Her mother had chosen a personal retinue
to escort her, balancing Henry II's representatives, the Sieur de
Brézé and his companions, with a roughly equal number of Scots. Lords
Erskine and Livingston came first. Parliament had made them Mary's
official guardians while she was away. Lady Fleming, an illegitimate
daughter of James IV and one of her mother's closest confidantes, was
next. She was appointed Mary's governess, taking charge of her female
staff. Janet Sinclair, Mary's old nurse, was still constantly at her side. In
addition, Lord James Stuart, one of James V's illegitimate sons and
Mary's half-brother, took his place in her party. Aged seventeen, he was
on his way to Paris to attend university. This was perhaps the first time
that Mary had met the sibling who would later come to play such a momentous role in her life. For now, however, he was crisscrossing between
Paris and Edinburgh, ostensibly training for a career in the Church, but
in reality just waiting for an opportunity to enter the limelight.

Mary's maids of honor and official playmates were her four best
friends: Mary Fleming, Mary Beaton, Mary Seton and Mary Livingston,
the so-called four Maries. All almost exactly her own age and the daughters of leading Scottish families, they had first appeared as Mary's companions when she was taken by her mother to the island priory of
Inchmahome. Mary Fleming, Lady Fleming's daughter, enjoyed preemi-

nence by virtue of their blood ties, and Mary treated her as her cousin. She was famous for her quick wit and love of life. Mary Beaton's beauty was second only to Mary's, with whom she later shared a love of literature and poetry. Mary Seton, who stayed by Mary's side for almost her entire life, was famous as a hairdresser, able to braid and crimp the always fashion-conscious Mary's auburn hair into a new style every day. Lastly, Mary Livingston, Lord Livingston's daughter, loved the outdoor life and dancing.

Their nickname was a joke. The "Three Maries" was the name given in France to a well-known Catholic devotional manual for noblewomen. The three saints were the Virgin Mary, Mary Magdalene and Mary the wife of Clopas, who stood by the cross at the crucifixion of Christ. Mary undoubtedly knew this book, probably found it tedious and her playmates a great deal more fun, and so invented the moniker. It reflected the wicked, mildly blasphemous sense of humor for which she was later renowned. It did not pass unnoticed that during the long and grueling crossing she had taunted those of her companions who, unlike herself, were seasick.

As soon as the galleys landed at St.-Pol-de-Léon, advance word was sent to Paris that Mary was on her way. Henry II had already given orders that all the towns and villages near his palace at St.-Germain be carefully checked to make sure that none of the stonemasons had been in contact with plague during the extensive rebuilding projects. Mary was to be welcomed by her grandparents Claude Duke of Guise and Antoinette of Bourbon, to whom an outrider was sent. Another messenger set out on the long journey across the Alps to Turin, where Henry II was visiting his northern Italian garrisons, to inform him of Mary's arrival. Such was the attention due to the princess pledged to the dauphin of France.

After everyone felt sufficiently rested, their baggage was loaded onto carts and they set out across country for Nantes, at the mouth of the Loire, where they boarded a river barge for Orléans. Illness, probably gastroenteritis, struck down some of the men. Lords Erskine and Livingston were violently sick, taking months to convalesce. Far worse, Mary Seton's brother, "le petit Ceton," died of a "flux of the stomach" at Ancenis, some twenty miles from Nantes on the way to Angers. Mary and her female attendants did not succumb, possibly because they took more care than the men about what they were drinking. It was considered a normal precaution for royal and aristocratic women to carry bottled water in their luggage, whereas the men would have consumed local wine and beer.

Little Seton's fate gave Mary her first close experience of mortality. After the funeral, the party returned to their barge to resume their journey through the lush, densely forested Loire Valley. Somewhere along this part of the route, the Sieur de Brézé, the Guise retainer who was Mary's official escort, received orders recalling him to Guyenne in the south of France, where a peasant revolt had broken out against the salt tax. He left Mary in the care of her grandmother, Antoinette of Bourbon, who had met the party and guided them home.

Antoinette recorded her first impressions of Mary in her letters. "I assure you," she began, "she is the prettiest and best for her age that you ever saw." "She has auburn hair, with a fine complexion, and I think that when she comes of age she will be a beautiful girl, because her skin is delicate and white." Antoinette noted that her face was well formed, especially the chin, although maybe it was a little long. In deportment "she is graceful and self-assured. When all is said and done, we may be well pleased with her." The duchess added with barely concealed condescension that the rest of Mary's retinue, Lady Fleming excepted, were less good-looking and "not even as clean as they might be."

Mary was expected to join Henry II's children at Carrières-sur-Seine, just a few miles outside St.-Germain, where they were staying during the renovation of the château. She reached Carrières on Sunday, October 14, and was received in style. By now there were already four children in the royal nursery: Francis the dauphin, his two sisters Elizabeth and Claude, and a younger brother Louis, who would shortly die of measles before his second birthday. Later, four more children were born: Charles, Henry, Marguerite and one more son, also named Francis.

As Mary was to lodge with them, the issue of protocol arose. Who should enjoy precedence among them? It was settled that the dauphin would take preeminence, as he was male and the heir to the French throne. But would Mary, a queen in her own right, enjoy precedence over the others, especially Princess Elizabeth, the king's eldest daughter, who was three and a half years old?

Henry II gave careful thought to this. In reaching his decision, his mistress, the redoubtable Diane de Poitiers, played a greater role than did his wife, Catherine de Medici. The domestic arrangements of Henry II were unorthodox but curiously serene. Catherine was the daughter of

Lorenzo de Medici, Duke of Urbino, and the niece of Pope Clement VII. Henry had been forced to marry her in 1533 by his father, Francis I, who needed Medici support for his Italian diplomacy. Both parties were just fourteen, and it was a *mésalliance*. Perhaps not least because Catherine was expected to consummate her marriage in front of her father-in-law — he joked that she had "shown valor in the joust" — she disliked sex. She attained puberty late, and was infertile for almost seven years. There was even talk of repudiating her; in an age of dynastic monarchy, a barren queen was disposable. Catherine was highly vulnerable to a mistress and an annulment until her first child, Francis, was born.

In the years of her "sterility," Catherine was supported by Diane, who saw every advantage in maintaining the status quo and every disadvantage in being displaced by a new and less accommodating wife. Diane forced Henry to spend more time in Catherine's bed. In return, Catherine allowed Diane the space to exert influence. By the time Mary arrived at St.-Germain, Diane had a role in bringing up the royal children. She had the full attention of Jean de Humières, the official who in 1546 was put in charge of the royal nursery. His wife, Françoise de Contay, Lady Humières, assisted him, and thanks to Diane was able to retain her position after her husband's death in 1550, when he was succeeded by Claude d'Urfé, formerly ambassador to Rome.

A few weeks before Mary reached Carrières, Diane sent a memo to Humières specifying the king's decision on protocol. It granted precedence over all except the dauphin to Mary, who was to share the best room in the house with Princess Elizabeth. Moreover, she was to "walk ahead of my daughters because her marriage to my son is agreed, and on top of that she is a crowned queen."

Catherine de Medici saw Mary for the first time at St.-Germain. She found her beautiful and vivacious: "our little Scottish queen has but to smile to turn all the French heads." Later in their relationship, Catherine was Mary's keen antagonist, her motive not so much jealousy of her allure as a desire to protect the status of her own children and a growing fear and dislike of the powerful Guises. But such competition was absent while Henry II was alive. This ménage à trois worked extremely well.

Mary was celebrated at court. Her half-brother the young Duke of Longueville, who rushed excitedly to meet her at St.-Germain, thought she was stunning. Likewise Jacques de Lorges, who had led the reinforcements to Scotland in 1545 that were paid in debased coin, wrote to Mary of Guise to say her daughter was so "charming and intelligent as to give everyone who sees her incomparable joy and satisfaction." Hungry for news of her daughter, Mary of Guise was thrilled to receive this information.

On December 9, Henry II returned to St.-Germain. He had by then commissioned an artist to draw all the children's portraits, which he received at Nevers in late October, and so was able to recognize Mary and greet her by name. When they finally met, he found her to be as flawless in manners as she was in looks. "She is the most perfect child that I have ever seen," he wrote joyously to Montmorency, the constable of France. From that moment, she was, as he constantly said, "his very own daughter."

Mary's "coming out" took place a few days later, when her uncle Francis married Anne d'Este, daughter of Hercules d'Este, Duke of Ferrara. It was a glittering event to which the diplomats of Europe were invited. At the reception, Henry II rose to make a speech. He drew everyone's attention to "my daughter the queen of Scotland," whom he had wanted to dance with the dauphin.

Already Mary loved dancing and knew many of the simpler courtly steps. When she and the dauphin took to the floor, the entire audience stood still to watch them. It was Mary's first serious test as a princess. She had to step forward in her stiff starched dress with its ornate strapwork and jeweled embroidery and in her tight flat shoes to perform a complex routine that she had learned by rote over the past few weeks, counting out the steps in time to the music and then practicing with the dauphin under the critical eye of Lady Fleming. Such dancing was a vital precursor to a betrothal, as it displayed to the court that the "lovers" were in good health and sound in all their limbs. At the end of the dance, the performers of whatever age were expected gently to kiss.

Since Mary had come to France for her betrothal, Henry was delighted by the dauphin's easy attachment to her. Francis was nearly five at Christmas 1548, while Mary was six. Henry persuaded himself that "from the very first day they met, my son and she got on as well together as if they had known each other for a long time."

This was largely wishful thinking. Although the dauphin was plucky and intelligent, he was physically weak. Even at this early age he must have looked incongruous beside Mary, because whereas she was unusually tall for her age, he was abnormally short. And while Mary was high-spirited and fluent in both speech and gesture, Francis was a clumsy stutterer. If he and Mary appeared to make a successful couple, it was because Diane de Poitiers had been at work. For several months she had been prompting Lady Humières to school the dauphin in the principles of elementary courtship. Lady Humières took the hint, and in reporting the news of her charge's initial success back to her, Diane urged, "If you want to please the king, go on teaching him these pretty little ways."

Mary picked up these signals and joined in on the act. She learned intuitively that it was important to handle Francis in a way that pleased the king, her new "father," and that the best way to relate to the dauphin was to seek to be his friend, while at the same time exploiting the conventions of courtly love to pretend to be his "beloved."

This worked like a charm. Both the constable and the Venetian ambassador remarked on her elegant and demure behavior. Encouraged by Lady Humières, the dauphin, in turn, took up hunting and martial sports. He was given a specially made suit of children's armor by Mary's uncle Francis, and in his carefully copied out letter of thanks depicted himself as the "gentle knight" of medieval chivalry who sought "to win the heart of the beautiful and honest lady who is your niece."

Shortly after Christmas, Antoinette of Bourbon briefed Mary's mother on the king's announcement that all the royal children would be educated together. This was a clear break with tradition, but Henry wished the children to "become used to each other's company."

In part, he really meant this. He wanted Mary and Elizabeth, his eldest daughter, to be brought up as sisters and the dauphin to meet Mary in a relaxed but supervised environment. The hidden agenda was that Mary's Scottish retinue, or at least those who were male, were becoming a nuisance. The Scots were not popular despite their high profile at the court, where they formed the royal bodyguard, the Garde Écossaise. By the standards of French court etiquette, they were seen as rude and uncouth. Now Henry wanted those officers who had entered Mary's service in Scotland to be sent home.

Language was a key factor. As yet, Mary spoke Scots and almost no French. This had to change quickly. The question of cost also arose, since Mary's mother had not provided salaries for her daughter's staff, and Henry was unwilling to fund individual households for all the royal children where one would better achieve his aims. Under the new arrangement, everyone would speak French and follow the correct protocol. Mary's household would merge with the dauphin's. His sisters and their attendants would be placed in the same establishment, sharing a luxurious suite of apartments with Mary and her gentlewomen, and everyone would be subject to Humières and his wife.

Mary's male attendants, with very few exceptions, were shunted aside, and even on the female side there were changes. At first the four Maries accompanied Mary everywhere, but when she had settled down and seemed more relaxed in her new surroundings, they were sent to a Dominican convent school at Poissy, about four miles from St.-Germain, to learn French, obliging Mary to speak only French in their absence. Even Sinclair, her old nurse, found herself under threat. She was reinstated after Mary's mother intervened in her favor, but her appeal against eating with Frenchwomen failed. Sinclair was tired of being patronized for her Scottish ways by the servants of the other royal children. Despite her pleas, she was forced to live and work alongside the staff whom Henry introduced to the roster of the household to attend to the needs of his own daughters.

Lady Fleming, in almost complete contrast, was secure. A fluent French speaker who had married a Scot, she was said to be "everything that could be desired." Once more there was a hidden agenda. On the surface, it looked as if Mary could not dispense with her governess, whom it was not considered proper for a young queen to share with the other children. Accordingly, Fleming was left in control of Mary's female staff, despite the scope this created for disputes over expenditures with Humières and, later, d'Urfé. But the truth was that the beautiful and voluptuous Fleming had become Henry II's latest lover. The king even wrote to Mary's mother to solicit favors for her. "I believe that you appreciate the care, trouble and great attention that my kinswoman the Lady Fleming shows from day to day about the person of our little daughter the Queen of Scots," he noted disingenuously. "I must continually remember her children and her family."

One night the jealous Diane de Poitiers surprised the king as he left her rival's embraces. She made a scene and accused Henry of dishonoring the Queen of Scots by carrying on an affair within her own household. By creeping in and out of Mary's apartments on his way to a rendezvous, Henry had brought the young queen's reputation into question.

Henry ignored Diane's protests, but as ever she had the last word. In 1551, Lady Fleming became pregnant. She gave birth to a son, Henry of Angoulême, which was considered to be a fatal mistake. It brought the king's affair out in the open, exciting ribald gossip. Diane acted

promptly to protect the royal family's good name, and the disgraced
Fleming was immediately sent back to Scotland, leaving her daughter,
the chief of the four Maries, behind.

So much was new and intoxicating, but Mary still missed her mother
badly. In April 1550, after a separation of almost two years, she was overjoyed to hear that her mother was planning to visit her. Mary's joy was
tinged with sorrow, because her grandfather, Claude Duke of Guise, had
recently died at Joinville, aged fifty-four. Mary's first official letter was a
formal diplomatic recommendation for the Sieur de Brézé, whom Henry
II now wished to send to Scotland with this sad news.

Mary was judged to be too young to attend her grandfather's funeral,
and so was represented by a proxy. Her mother was also unable to attend, which caused her to weep. "I have lost the best father that any child
lost," she told her brother Francis, who succeeded his father as duke.

Mary of Guise decided to expedite her visit, and in selecting her wardrobe, she consulted Diane de Poitiers — always the arbiter of fashion
rather than Catherine de Medici — about the protocol of mourning at
St.-Germain. Henry II took charge of her travel arrangements, obtaining a diplomatic passport from the English Privy Council, since unlike
her daughter, Mary's mother was prone to seasickness. She dreaded the
prospect of a long sea crossing from Scotland and much preferred to
take the slower route by land through England and from there sail the
short distance across the channel from the port of Dover.

Mary wrote animatedly to Antoinette of Bourbon about her mother's
plans. She simply had to write, she said, so that her grandmother could
hear the "joyous news that I've just received from the queen my mother."
In her haste to finish the letter, Mary lost her main verb in a cluster of
subordinate clauses, but the sense is clear. Her mother's arrival "will be
to me the greatest happiness that I can desire in this world . . . I pray you,
madam, that to increase my joy, you may find it convenient to visit me
soon, and in the meantime to arm yourself with all the patience you are
able to muster in such a case." Mary had become fluent in French in under two years. Her style is breathless, but grammatically unblemished.

Her mother's visit also had a political purpose. Henry II was planning
a gala celebration of the final expulsion of the English armies from their
forts and garrisons in Scotland and their fortifications at Boulogne.
Henry's whole court and extended family were to be present, and no ex-

pense was spared. Mary of Guise and her daughter, the Scottish queen, would be at the heart of the festivities.

Mary's mother did not travel alone. She brought in her train almost the entire Scottish court, notably the pro-English lords who had dealt with Henry VIII and Somerset over the years. Her aims were clearly defined. She sought to promote her own claim to the sole regency of Scotland now that the other candidates, both pro-French and pro-English, had accepted the treaty of Haddington, and to bind the nobility to the "auld alliance." Both goals were greatly assisted by Henry II's lavish hospitality and his generous distribution of fresh pensions.

The main event took place at Rouen, the capital of Normandy, in October. The highlight, in imitation of a Roman imperial victory, was a royal entry into the city in front of cheering and enthusiastic crowds through specially constructed triumphal arches. This ritual was preceded by pageants in which soldiers and actors dressed as classical heroes or victorious generals marched past the king and court, with Mary happily reunited with her mother, watching it all from a blue and gold viewing pavilion on the west bank of the Seine.

The procession was led by a chariot laden with trophies, followed by tableaux staged on floats drawn by "unicorns" — in fact they were white horses wearing headdresses — illustrating the martial victories of the Valois dynasty. Banners and models of forts captured in Scotland or near Boulogne were vaunted aloft on poles. In one of these pageants, Henry II became Hannibal, the Carthaginian general who led his army across the Alps to defeat the Romans with the aid of elephants, which duly made their appearance before the crowds. Many people thought they were real animals from the royal menagerie in Paris, but they were actually made of wood and papier-mâché, mounted on wheels and pushed along by men concealed inside.

Next came "prisoners" captured at Boulogne, who were led through the streets in chains. Altogether the most richly ornate float was reserved for a tableau depicting Henry II and his children. The king was shown as a Roman emperor surrounded by his heirs, while Calliope, a daughter of Zeus and a muse of the heroic age, held the "imperial" crown of France above his head.

There followed fifty or so actors of both sexes made up as Brazilian natives, who paraded naked through the streets and then staged a "war" between two indigenous tribes. The scene, set up as a tableau vivant before the viewing platform, was peppered with native huts, tents, palm trees and wild animals. The tribes hunted, cooked on open fires and traded with a French garrison. Then a great battle erupted, complete with bows and arrows, ending with the burning to the ground of the huts of the defeated tribe. This was pure mass entertainment, upstaging the earlier tableaux, and the crowds roared with appreciation.

In a final spectacle, a mock sea battle was staged on the Seine between rival "French" and "Portuguese" fleets in which real ships were packed with barrels of gunpowder. The sailors fired genuine cannons, most likely without shot. However, the inevitable happened and one of the barrels exploded, causing one ship to sink and its crew to lose their lives. Next day, the event was repeated with a substitute ship, but the same thing happened and more sailors were killed.

The fête was a dazzling visual manifesto. A central role was accorded to Scotland, which is why the Queen of Scots and her mother sat in glory. Henry II saw himself as the "protector" of Scotland; his victories had liberated Scotland and Boulogne. The fête's unifying theme was the dauphin. He and not Henry II was the figure around whom the mise en scène was choreographed, because he was portrayed as the husband of the already crowned Queen of Scots. He was the future king of Scotland and France, and by virtue of Mary's claim to the English throne as the great-granddaughter of Henry VII, the founder of the Tudor dynasty, could be the rightful king of England too. He was therefore heir to a triple monarchy. Through Mary's marriage to the dauphin, the Valois monarchy could realize its full potential, creating a Franco-British empire that would subsume the British Isles and then cross the Atlantic to Brazil, where French merchants were already making inroads and starting to challenge Portugal's commercial power.

Even Henry II's choice of Normandy as the venue of his fête was significant. It was the closest province to England with the greatest historical connections to the British Isles. It also happened to be the region in which the Guise family was fast accumulating land and building retinues. Considered in this light, the Franco-British project was simply the most audacious of their schemes. Claude Duke of Guise had begun to create a role for his family at the heart of the Valois state, and now that he was dead, his sons were the mainstay of the plan.

Mary was the cornerstone of the project. Henry II's logic was dynastic. Henry VIII's will had determined the order of succession to the English throne. If Edward VI, Henry VIII's only surviving son, died without heirs, then the king's will specified that the crown was to descend on the female side. Edward was to be succeeded by his elder sister, Mary Tudor, and if she also died without children, by his younger sister, Elizabeth Tudor.

But to the Catholics, Mary Queen of Scots was Mary Tudor's rightful successor. To them, Elizabeth was illegitimate. She was the daughter of Anne Boleyn, whom Henry VIII had married while his first wife, Catherine of Aragon, was still alive. The pope and the Catholic Church did not recognize Anne Boleyn as Henry's lawful wife, nicknaming her "the concubine." Henry had himself repudiated her, divorcing and then executing her in 1536, when Elizabeth was declared to be illegitimate by act of Parliament in a clause that had never been repealed. This left the way wide open for the claim of Mary Queen of Scots, even though Henry VIII had tried expressly to block it.

Henry always held that through his will he could establish the order of succession and eliminate the Stuart claim, and this he believed he had done. He had specified if all his children died without heirs, then the children of his younger sister, Mary Duchess of Suffolk, should inherit the throne. Henry was so angry with Scotland for the repudiation of the treaty of Greenwich, he vetoed the Stuart line. It was a calculated snub.

But Henry's settlement, although approved by Parliament in the Third Act of Succession of 1544, was still based on a significant assumption. If a claim to the throne could pass by the female line, then the lineage of Margaret Tudor, Henry's elder sister, also came into play. No one who followed the rules of dynastic succession could ignore Margaret's first marriage to James IV of Scotland. Their son was James V, the father of Mary Queen of Scots, who was Margaret's granddaughter and Henry VII's great-granddaughter.

Mary's claim to the English throne was no more than a speck on the horizon while Edward VI or Mary Tudor was still alive, because their legitimacy was never in question. But compared to Elizabeth Tudor's claim, Mary's was at least as strong if not stronger. Only the Third Act of Succession was unqualified in its defense of Elizabeth's claim, but Henry VIII had broken with the pope a decade before it was passed, and the legislation of a schismatic Parliament was not recognized by Catholics, who still comprised the vast majority of the English population.

Following the triumph of the fête and the consolidation of the Guise family's ambitions, Mary's mother lingered in France for just over a year, returning with her daughter to Paris and St.-Germain, Joinville and Blois, and then taking her Scottish entourage on a tour of the country. Her tour was tantamount to a royal progress; her Scottish nobles must have been exhausted, but more importantly they were impressed. They were certainly a lot richer, because Henry II had never been more generous, disbursing 5 million *livres tournois* on his Franco-British project between 1548 and 1551, five times more than Francis I spent on Scottish affairs during his entire reign.

The tour was interrupted by melodrama and ended tragically. In April 1551, a plot was uncovered at Amboise to murder the young Queen of Scots. One of the ex-Castilians, captured by the French after the bombardment of St. Andrews Castle, was determined to seek revenge. He changed his name and joined the Garde Écossaise. He obtained access to the royal apartments, where he planned to assassinate Mary by suborning her cook to poison her favorite dessert: frittered pears. But somebody talked and the plot was discovered. The chief conspirator escaped, fleeing to Ireland and from there to Scotland, where he was captured and sent back to France for trial and execution. Mary was either unaware of or untroubled by what looks like a narrow escape, but her mother was so worried she fell sick and took to her bed.

Her grief intensified in September, when the Duke of Longueville, her surviving son by her first marriage, fell victim to a mystery illness and died in her arms at Amiens. He was not quite sixteen. Once again, Mary was represented by a proxy at the funeral. Soon afterward, Mary of Guise returned to England on her way back to Scotland. All four of her sons were dead and she was heartbroken. She considered staying permanently in France, retiring to a convent or living with her family at Joinville. But she had committed herself to rule Scotland on behalf of her remaining child, and to secure the country militarily and dynastically to Henry II and the French alliance. It was her duty to return there and therefore to leave her sole living child behind. She kissed Mary for what would be the last time, and said goodbye. They would never meet again.

By the end of 1551, Mary was growing up. She was only nine years old, but her aunt Anne d'Este, Duchess of Guise, one of the most cosmopolitan and trendsetting women at Henry II's court, pointed out that "she can no longer be treated as a child." She did not look like a child; her conversation was not that of a child. Every day "she grows in charm and good manners" and "becomes more suited" to the "place" to which she was called and which she was never allowed to forget. Already she had a vast wardrobe stuffed with fine clothes and a collection of jewelry so large it needed three brass coffers and several additional boxes to contain it.

Since her arrival in France, Mary had learned a new language, a formal and sophisticated court etiquette and new tastes. So far, her Guise family had succeeded in protecting her birthright. But in the process, its significance had been redefined. In particular, Mary's future had acquired a compelling Franco-British dimension. More than just the heir to the Stuart succession, she had become the key to the entire Valois dynastic enterprise. As the processions and tableaux passed before her eyes at Rouen, they must, in their glamour and pageantry, have reinforced one driving idea in her mind. As well as being queen of Scotland and the next queen of France, she must also aspire to be a future queen of England.

Adolescence

AFTER MARY'S MOTHER had returned to Scotland, her uncle Charles, Cardinal of Lorraine, the boldest and most experienced politician at Henry II's court, took over as her mentor. Within a few years, he was exercising more authority over her household and upbringing than anyone else. The reorientation was gradual, but it had begun even as Charles disarmed his sister by urging her to allow "no one except yourself, or those to whom you delegate your authority, to have control over your daughter." The member of the Guise family most constantly by Henry II's side as the royal court traveled on its regular circuits from palace to hunting lodge to palace through the Loire Valley and then to and from Paris and its hinterland, Charles was ideally placed to exert influence. It was an arrangement designed to work to everyone's advantage, and he kept his sister informed in regular reports about his niece's health, education and finances.

In the 1550s, the Guise family had an unparalleled ascendancy. At a strategic level they were the champions of French dynastic claims in the British Isles and of war against Spain and the Holy Roman Empire. They worked indefatigably together and built up their networks at court and in the provinces. After the Rouen fête, they turned their attention to monopolizing power at court, seeking to ingratiate themselves with Catherine de Medici, whose dislike of Constable Montmorency, the premier nobleman and chief officeholder of France, was unmistakable.

The constable was appointed for life. His duties were almost exclusively military, and in wartime he commanded the army in the king's absence and the vanguard in his presence. On ceremonial occasions, he walked in front of the king carrying his sword of state. In the early stages of their rise to power, the Guises had been Montmorency's close allies, but this started to change after the fête. As their confidence and ambition grew, the Guises sought to oust and replace him. In attaining their goal, their trump card was the betrothal of Mary, their niece, to the dauphin.

Francis Duke of Guise, Charles's brother, was well positioned to lead the family to victory. He was a brilliant military commander who had distinguished himself in battle and was idolized by his troops. He saw himself more as a man of action than as a courtier, and when not on the battlefield was usually at Joinville or at the château of Meudon, one of the numerous properties the family were busily acquiring near Paris. He invited Mary there, and treated her as if she were his own daughter. He became a surrogate father to her, and she would always love him.

But it was her uncle Charles who advised her about protocol, about the letters she should write, to whom she should send them, and what she should say. If, however, Charles was unavailable, away on royal business or in retreat during Lent at a monastery, she consulted the duke. She deferred without question to their opinions, and after the scandal of Lady Fleming's affair with the king it became Charles's habit to make unannounced monthly inspections of Mary's household, "so as to find out in detail all that is going on." From then on, his grip was secure: Mary could be allowed to grow up, to pursue her studies and her pleasures almost as she chose, but within the limits of his advice.

Mary was fast approaching adolescence. She was outgrowing her household, but the reorganization of the royal children's living arrangements sprang in the event from Fleming's disgrace and Henry II's decision to create an independent household for the dauphin. Mary's new governess was Françoise d'Estamville, Lady Parois, an older woman incapable of threatening Catherine de Medici or Diane de Poitiers. She was sexually safe and correspondingly dull. To the young Mary she must have seemed stultifyingly boring. Her appearance was unremarkable, her morals irreproachable. She was known to be devout, an attribute mentioned by Charles in one of his reports. "I mustn't forget to tell you," he told Mary's mother in a postscript, "that Lady Parois is doing so well, she could hardly be expected to do better, and you may be sure that God is well served according to the old fashion."

With this remark, the cardinal gibed at Fleming, whose sympathy for the Huguenots, as the French Calvinists were called, was notorious. It was an isolated comment, often since misunderstood to mean that Mary was to be educated according to the most rigorous principles of the Catholic faith, to prepare her to rule as a Catholic icon. More generally, it is said that the Guise family were "ultra-Catholics" who aspired to win control of the politics of France and Scotland in order to save the Valois and Stuart monarchies from the threat of Protestant subversion.

This is a misapprehension. In the 1550s and for most of the 1560s, the Guises put their own interests above the cause of religion. It was only after 1567, when Charles V's son and heir, Philip II, ordered Spanish troops into the Netherlands to crush a militant Protestant revolt and when a crusading Catholic League against the Huguenots began to take shape in France, that the family became synonymous with the absolute defense of Catholicism. By then, it was clear to them that their interests could be preserved only with Philip II's support, and since he was a loyal Catholic with an unshakable vision of his divine mission to reunite Christendom under the papal banner, it was essential for them to do all they could to crush the Protestants.

This metamorphosis influenced Mary after her flight into exile in England. She would reinvent herself in the 1580s as a good Catholic woman persecuted for her religion alone. But in the 1550s, the Guises were *politiques*, or moderates, equally opposed to Protestant and Catholic extremism. Where religion mattered most to them was in relation to their dynastic project, because only the pope could make a definitive pronouncement of Elizabeth Tudor's illegitimacy, and so of Mary's claim to the English throne.

As the dauphin approached his ninth birthday, Henry II decided that his son should leave the consolidated household created for the royal children and live in his own establishment. This made little difference to his movements, as the new household, like that of the other children, continued to follow the court. It did, however, have a sudden and dramatic impact on the personnel and finances of the other children's household, because not only were the dauphin's officers and ancillary staff hived off, but their budgets were also transferred. In particular, it raised the question of whether Mary, herself a crowned queen and already almost ten, should have her own separate quarters.

The dauphin's new household came into effect in March 1553. It had been much discussed over the previous winter, and on February 25 was the main item of a report from the Cardinal of Lorraine to Mary's mother. He began with good news of Mary, whom he had just seen at Amboise. She "has grown so much and continues every day to grow in height, goodness, beauty, wisdom and virtues that she is as perfect and accomplished . . . as is possible." No one could be found in France to match her qualities. The king now liked her so much, "he spends much of his time in chatting with her, sometimes for the space of an hour." She "is as well able to entertain him with pleasant and sensible talk as if she were a woman of twenty-five." Anne d'Este, the cardinal's sister-in-law, had made a similar observation a few months earlier. As Mary was herself beginning to discover, her charm and conversational skills would be among her greatest assets.

Charles's letter then turned to the business at hand. Catherine de Medici had chosen not to give her daughters, Elizabeth and Claude, separate households. The lack of staff meant they would sleep temporarily in her own dressing room with Lady Humières in attendance. It was a mean decision. And it was embarrassing for Mary, who was about to leave the Loire and travel to St.-Germain, where the court had preceded her. She was bringing her usual attendants, but the matter of what "state" she ought to maintain, with the dauphin gone from the children's household, and which rooms she should occupy within the court had urgently to be resolved.

Charles did not simply have the interests of his niece in mind. His own reputation and that of his family would be adversely affected if Mary's "state" was to be diminished. As always, protocol was of the highest importance. He also had a political plan, one fully visible in August, when Henri Cleutin, Seigneur d'Oysel, the first resident French ambassador in Scotland and Henry II's lieutenant-general there, filed a report on English affairs. Edward VI, the only son of Henry VIII, had died of pulmonary tuberculosis on July 6 at the age of fifteen. Despite an attempt by his leading Protestant adviser, John Dudley, Duke of Northumberland, to exclude the Catholic Mary Tudor from the throne and replace her with the Protestant Lady Jane Grey, Mary Tudor had triumphed. Jane was a queen for just a few days before Northumberland's plot collapsed. On August 3, Mary Tudor processed into London to the cheers of a welcoming crowd.

This news was disturbing to France. Mary Tudor was the cousin of Spain's Charles V, with whom Henry II had been at war for two years on the frontier of the Netherlands and in Italy. Moreover, she intended to marry Philip II, in whose favor his father intended to abdicate. At the age of fifty-three, Charles was white-haired and exhausted. The arthritis in his hands was so bad he could hardly open a letter; the pain was so great that he prayed for death. His son, Philip, was young and energetic and keen to continue the war with France, where it was feared that he would use England as a military and naval base. Such alarm was justified, because when he married Mary Tudor in 1554 and assumed his role as king, he was heard to boast that for every fort built by the French in Scotland, he would construct three on the English side of the border.

The cardinal's plan was simple but audacious. He proposed that his niece should be declared fully of age, able to rule her realm as Queen of Scots in person or through an appointed deputy. Whereas the usual age of majority for a royal minor was between fifteen and eighteen years, Mary should "take up her rights" as queen "at the age of eleven years and one day." To this end, the existing regent, the Earl of Arran, now better known by his empty title of Duke of Châtelherault, should be dismissed. His self-serving duplicity and constant vacillation had worked against the French interest in Scotland, and with Mary Tudor on the English throne, no further risks could be taken.

It was quintessentially a Guise plan, because once Châtelherault had been removed, Mary of Guise was to be appointed sole regent to rule in her daughter's name. At a stroke, the Guise family would extend their hegemony to Scotland, improve their status at the French court as the relatives of a reigning queen, and win the gratitude of Henry II, who was deeply skeptical of Châtelherault and his allies, suspecting perfidy and the imminence of an Anglo-Scottish rapprochement at their hands.

D'Oysel, who made the journey to France for the debate, opposed the plan at a three-hour meeting of the Conseil des Affaires attended by the king and Mary's uncles. He suspected the motives of the Guise brothers. But the king overruled him and the plan was put into effect. Mary's mother was declared the sole regent, her appointment confirmed in April 1554 by a Parliament summoned to Edinburgh. One of Mary's earliest official acts was to write to congratulate her mother. As a reigning queen, Mary also had the right to choose her own guardians, and she squared the circle by naming Henry II and her uncles to the posts, thereby binding the Guise family even closer to the Valois monarchy.

Mary now wrote her first letter to a foreign ruler. Addressed to Mary Tudor, it recommended d'Oysel to her on his return journey and offered reassurances of friendship and "amity." "May it please God," penned the young Queen of Scots in her very best handwriting, "there shall be a per-petual memory that there were two queens in this isle at the same time, as united in inviolate amity as they are in blood and near lineage." Mary Tudor was addressed as "Madam, my good sister," and the letter ended with the standard valediction used in correspondence between sover-eigns, "Your good sister and cousin, Marie." Her letter, written in French, is neatly copied out between lines scribed into the paper with a stylus, which can still be detected in the original document.

None of this resolved the question of Mary's "state" at St.-Germain, Amboise, Fontainebleau and elsewhere, where the court had gone in its circuit and where confusion increasingly prevailed. Part of the problem was caused by illness. Claude d'Urfé, the officer in charge of the chil-dren's households, was genuinely sick. Lady Parois was unsympathetic, suggesting to Mary of Guise that he was merely overwhelmed by "feeble-ness." She could get nothing done as a result. The cardinal, meanwhile, chivvied his sister, reminding her that as soon as the dauphin's budgets were transferred, someone else would have to pay the salaries of Mary's staff. "I have had a list drawn up," he observed in a faux gesture of aid. It detailed everyone in Mary's service together with the estimated yearly expenses. "I think that in this list as it stands, there is nothing either superfluous or catchpenny."

What Charles never volunteered was money. Mary's mother was ex-pected to foot the bills, which Henry II refused to share because his Scottish expenses were already very high. Mary was herself a complica-tion. She was beginning to assert herself through lavish spending, partly on clothes and ponies, but mainly through her generosity, never more evident than in dealings with her servants, for whom she sought promo-tions, improved working conditions and pay increases. Janet Sinclair, her old nurse, was an early beneficiary, obtaining a recommendation for her son and a promotion for her husband. Money and gifts were show-ered on the small army of actors, dancers, singers, instrumentalists, balladeers and clowns who kept Mary and her companions entertained, and she maintained a steady stream of rewards for the keepers of her dogs and ponies and the keepers of the royal bears.

Despite the escalating estimates, Mary's uncles continued to press their case. Their niece, said the cardinal, was a queen "already possessed of a high and noble spirit that lets her annoyance be very plainly seen if she is unworthily treated." Her "grandeur" had to be respected. She wished to be grown up and "to exercise her independent authority." On the basis of her current spending, she would need around 24,000 *livres tournois*, or between 50,000 and 60,000 francs, a year to maintain her "state." True, this excluded the stables, themselves by far the biggest single expense; but economies might be found.

It was a substantial sum, less so by French than Scottish standards; in Scotland it was equivalent to half the regular annual income of the crown. It was possible for Mary's mother to contemplate such a huge commitment only because she had independent assets left over from her first and second marriages.

The negotiations dragged on for nine months, a delay that severely tested the morale of Mary's existing staff. Many were not paid in the interim, and most were owed money from the previous year. Tempers were rising, exacerbated by the insensitivity of Lady Parois, a jealous and greedy woman whose tactless demands and bad attitude led to friction. Her feud with Madame de Curel, one of Mary's most senior ladies-in-waiting and the candidate whom Antoinette of Bourbon had originally backed in preference to Lady Fleming as Mary's governess, ended with Curel's resignation. She walked out after a fight, shockingly conducted in Mary's presence. Others showed their frustration through straightforward absenteeism. Writing to Mary's mother, Parois grumbled that there were very few attendants left above stairs and no one to do Mary's hair except herself.

At last Mary secured her household. It was inaugurated on January 1, 1554, when her officers, attendants and domestic staff were put onto the roster. They included most of her previous servants, reinforced by some new faces to reflect her more exalted status. That evening, her uncle Charles was her first guest. As she wrote rapturously to her mother: "Madam, I'm thrilled . . . to tell you that today I entered into the household that you've been pleased to create for me; and this evening my uncle, the cardinal, comes to sup with me. I hope through your careful planning, everything will be well conducted."

But fresh quarrels and money problems quickly arose, especially in the stables, where corners had been cut. Since the court migrated so often from place to place, horses, mules, coaches, carts and litters were essential. Three baggage mules were needed to carry Mary's bed alone, and others were required for wall hangings, plate, kitchen utensils, bottles of wine and so on. Mary's establishment could not afford its transport costs, for which the cardinal sought to compensate by curtailing the number of "removes" that it made. Instead of shadowing the royal court at every stage, he wished Mary to make fewer journeys and have longer stops at each one, omitting certain of the intermediate châteaux visited by the king and the dauphin to limit costs. On this the young Mary asserted herself, adamantly refusing to be detached from the main body of the court for very long.

Within a year, the cardinal's estimates were shown to be woefully inadequate, and Parois was up in arms. Her litany of complaints was endless. For example, more baggage animals were needed because Mary's bed often arrived late at its destination. She had been forced to borrow beds, which was humiliating. On top of this, valuable objects had been damaged because, in the absence of sufficient mules to which her coffers could be securely strapped, they had been thrown onto carts, where they were jolted.

Mary, continued Parois, had nothing to wear. New fashions were in vogue, and she was in danger of being left behind. She urgently needed outfits for a wedding, and because she was growing so quickly, her *touret* (the jeweled frame on which the high collar at the back of the neck was stretched) needed lengthening. A pair of diamond settings mounted on thin metal would do the trick, but she lacked the money to buy them. Mary was also pleading to have monograms (most likely *M* or *MR*, for "Marie la Royne") sewn onto her dresses in the style that had lately become all the rage. But money was too tight.

One result of Parois's incessant complaints was a growing rumor that Mary's credit was bad. This was scandalous and also highly inconvenient, since it meant suppliers would not deliver anything or accept further orders unless paid first in cash. Regular supplies on credit were essential to the smooth operation of a royal court, but the merchants argued that if Mary defaulted, they would be forced to sue for compensation in Scotland, an unacceptable commercial risk.

Soon the staff were in revolt again. In the financial year 1555–56, salaries fell into arrears and absenteeism was rife. There were accusations of corruption and misappropriation, seemingly confirmed when a gift of cash that Henry II had given Mary to spend at the fair at St.-Germain went missing, allegedly stolen.

Parois was unable to cope and took sick leave, forcing Antoinette of Bourbon to step in and maintain basic services in Mary's suite. Antoinette took stock of the position. She did not believe Parois was up to the job and wanted her to resign. Unfortunately, it was one of the few occasions when both of Mary's uncles were absent. Antoinette considered replacing Parois unilaterally, but thought the matter too important and delicate to be handled on her own. She even hinted that Mary's mother might have to make a special visit to France.

The crunch came at Blois right after Christmas, while the cardinal was on a mission to Italy and unable to intervene. Mary, now thirteen, gave away some of her old dresses to two of her aunts. In performing this seemingly innocent gesture, she followed, as she afterward claimed, her mother's advice, since these aunts, her mother's younger sisters, were abbesses and the gifts were deeds of charity. The clothes, made of luxurious and costly fabrics, were to be cut up and turned into altar coverings. Such gifts, following hard on the heels of others to Mary's servants, had sparked Parois's jealousy. She spitefully snapped, "I see that you're afraid in case you enrich me! Obviously you mean to keep me poor."

This was no way to address a queen, but Mary answered regally. She was trying hard to control her emotions, something she always found difficult but could usually manage at least in the opening stages of a confrontation. "How sad," she replied, and then excused herself.

Yet this was no isolated skirmish: it marked the culmination of a clash of personalities. Mary was precocious and vivacious, able to conduct herself in ways far older than her age, still barely a teenager but one who expected to get her own way. Parois was a much older woman burdened by self-importance and an inflated sense of duty. She was at best fussy, at worst a killjoy and a prig. She had already reported Mary to her mother for insubordination when Mary had insisted on wearing too much jewelry.

Mary wrote to her mother, giving her side of the story and pouring out her heart. "Madam, I most humbly beseech you to believe that there's nothing in any of this, since . . . I've never stopped her from exercising authority over my wardrobe, because I knew very well that I shouldn't." However, she admitted: "I did tell John, my *valet de chambre*, that when she wished to remove anything he should inform me, because, if I wanted to give it away, I might find it gone."

The clash of wills had been expressed through a battle to control Mary's wardrobe. Now that it was in the open, Mary saw no reason to yield. "As to what she told you of my intransigent willfulness, madam, I've never had of her the credit of giving away a single pin, and thus I've acquired a reputation for being mean to the point where quite a few people have said to me how little I resemble you."

Mary meant this last remark to hit home, since her mother always prided herself on her generosity to her own servants. She then vented her indignation against Parois. "I'm actually quite amazed," she continued, "that she could tell you anything so far removed from the truth."

It was an impasse, and when the Cardinal of Lorraine returned from Italy, there was no more prevarication. As he warned his sister, Parois was "ill" in Paris. She had left Mary's household after the scene and it was unlikely she would return. Her "indisposition" was serious, possibly fatal. In any case, she wished to resign. The young Queen of Scots had gotten her way.

Mary herself, added the cardinal inscrutably, had been "sick," mainly because of "things she had discovered toward the end that were scarcely possible to bear." She had now fully recovered and was just as she was before, her looks and conversation holding everyone enthralled. But clearly some damage had been done.

A year later, Mary revisited the affair when she told her mother that Parois "had almost been the cause of my death for the fear I had of falling under your displeasure."

This was the first of several occasions in her life when Mary said she had been close to death or wished she were dead. This was not simply teenage melodrama. Mary really had been seriously ill. Not the least of the enigmas about her is her medical history. In many respects, her health was robust. She rode her ponies every day and was able to meet extreme physical demands. But her good health was punctuated by episodic illnesses, often triggered by anxiety or stress, sometimes lasting a few days, sometimes several weeks, in which acute physical pain and sickness were followed by rapid recovery. In her adult life, Mary complained of rheumatism and "spleen," the former common in an age without central heating and the latter caused by a gastric ulcer that eventually burst. But that does not fully explain her mood swings, which we know she felt, because in later life she wore a ring, an amethyst, that she believed had magical properties "contre la melancholie."

The symptoms were always the same. She would vomit, suffer abdominal pain, feel overwhelming sadness or depression, and burst into tears. She supposedly had a surfeit of "black bile," for which she was given medicines that, not surprisingly, failed to work. During one of her severest attacks in Scotland, witchcraft was blamed and a search instigated for the culprits and their charmed "bracelets."

A modern but disputed explanation is that Mary had inherited an illness known as porphyria. An overproduction of purple-red pigments in the blood intoxicates the nervous system, and in cases of acute intermittent porphyria, the commonest type, the victim suffers sudden bouts of vomiting, abdominal pain, weakness or paralysis of the limbs and psychogenic changes resembling hysteria. Despite the severity of these symptoms, the patient recovers speedily afterward.

Porphyria was first suggested when medical historians, claiming that George III was among those suffering from the disease, tried to trace its descent back to his Stuart ancestors, in particular to James VI and I. Their diagnosis of George III is still controversial. Many experts believe that his symptoms could easily have been caused by bipolar disorder, and even if James, Mary's son, could be shown to have had porphyria, it does not automatically follow that he acquired it on his mother's side.

The riddle cannot be resolved. From the viewpoint of her biographers, it does not greatly matter, since what she herself experienced were her symptoms, which are fully documented. As to the illnesses of her youth, there is more than enough evidence that they were quite unrelated to porphyria.

When Mary was nearly eight, she was laid low by a bout of gastroenteritis. Between the ages of ten and twelve, she was sick a number of times, once with a palpitation of the heart, but in each case the doctors blamed excessive eating. Once they prescribed a diet of rhubarb, a general remedy for patients thought to have an imbalance of the bodily "humors" ("black bile" was one of them) as stipulated in the medical textbooks, and especially for overexcitement.

At the age of eleven, Mary had a toothache. Much more seriously, at an unknown date she was struck down by smallpox. Henry II sent her his personal physician, Jean Fernel, a doctor with a legendary reputation who had first come to his attention as Catherine de Medici's gynecologist. Mary recalled her treatment in a letter to Elizabeth I, who herself had the disease in October 1562. "He would never tell me," she said, "the recipe of the lotion that he applied to my face, having punctured the pustules with a lancet." Although Elizabeth was her rival for the English throne, Mary's heart went out to her, as every noblewoman feared the ravages of smallpox even more than plague or childbirth. In the event, both women's skin was unmarked.

When Mary was thirteen and a half, she succumbed to an illness lasting several months. The symptoms included recurring attacks of fevers and chills, vomiting, headaches and abdominal pain. But this was not porphyria; it was a well-known viral disease prevalent in the summer called "the sweat" or "quartan ague." Mary caught it in August 1556 during a particularly hot spell. Her attacks were "wonderfully severe and sharp," and they were followed by remissions lasting between eight and ten days. "It is one of the maladies of this year which are affecting many," the cardinal reassured her worried mother. The dauphin had also fallen victim, as had the Duke of Guise's eldest son, Henry.

This was the illness to which Mary referred when she told her mother that Parois "had almost been the cause of my death for the fear I had of falling under your displeasure." Her sickness and Parois were firmly connected in her mind. Parois, the classic aggrieved ex-employee, had been smearing her reputation in Paris, insinuating that she had spoken ill of Catherine de Medici to Diane de Poitiers. The result was that Catherine froze her out. It was a setback, because the Guise family had been making steady progress in their efforts to win over Catherine, despite her privately expressed doubts about the feasibility of their Franco-British project.

No permanent damage was done. When Mary's uncles declared their niece to have reached the age of majority and arranged for her to have her own household, they had run a calculated risk. Overall it had worked out well, but after the debacle over Parois they were taking no chances. From this point on, Mary's establishment was overseen from Joinville and Meudon, with Anne d'Este playing a leading role. No new governess was appointed, and soon Mary could describe her aunt and uncles as equally responsible for her.

When war with Spain was resumed, the Guise family got their biggest opportunity so far. In February 1556, the truce of Vaucelles with Spain broke down. Although meant to last five years, it did not hold, and by the summer of 1557, Philip II had brought England as well as Spain, Savoy and the Netherlands into the war against France. In August, an army led by Duke Emmanuel-Philibert of Savoy invaded northern France and laid siege to the market town of St.-Quentin in Picardy, close to the frontier. A smaller French army under Constable Montmorency marched to the relief of the town, but was crushed, and six hundred nobles were captured, including the constable and four of his sons, who were imprisoned.

It was the most humiliating French military defeat of the sixteenth century. In a panic Henry II sent for the Duke of Guise, then in Italy, and appointed him lieutenant-general of the kingdom. The duke became the royal viceroy at the stroke of a pen. Not only did he have full command of all French military and naval forces, he also received the signet seal, enabling him to issue directives to the secretaries of state concerning such matters as foreign affairs, finance and the royal household.

Henry II wanted to launch a counterattack, especially against England. Having recaptured Boulogne in 1550, he now wished to liberate Calais and its surrounding forts, the last enclave of the English occupation of France left over from the Hundred Years War.

The duke instantly saw the possibilities. He led a masterful attack on Calais on January 1, 1558, when a severe frost made it possible for his troops almost literally to walk on water. The weakness of the town was its old-fashioned castle. The English garrison was complacent and there had been defectors, but this scarcely mattered because the Guise assault was so rapid and precise. The surrounding forts were taken after a brief bombardment, and on the 24th Henry II made a ceremonial entry into the town to the sounds of the anthem "When Israel Came Out of Egypt." The king was elated by the reconquest, which in his view more than compensated for the catastrophe of the constable's defeat at St.-Quentin.

As Henry walked in procession with his nobles and councilors through the streets of Calais in his azure and gold robes trimmed with ermine and wearing his "imperial" crown, the Guise brothers were poised to capitalize on their momentous victory. They had monopolized politics in 1557–58. Their success lay in their effective collaboration, meticulous planning and intelligence gathering; their command of technical detail was unsurpassed. To recover Calais from its English occupation had been the aim of the French monarchy for more than a century. Previous attempts had come to nothing. Suddenly the goal had been achieved. It seemed to be a miracle, one entirely due to the genius of the Guises.

The constable was still a prisoner in Brussels, and to underline the internal realignment of power, the Cardinal of Lorraine clapped his nephew into irons on a charge of heresy. The Guise family were at the pinnacle of their power. They seemed to be invincible. If ever there was a moment for them to reap their due reward, this was it.

5

Education

MARY'S EDUCATION was meticulously planned. The old medieval assumption that book learning was irrelevant to kings and nobles had been shattered in France when Louise of Savoy prescribed a course of study for her son, Francis I, modeled on the best practice of the Italian Renaissance. Francis studied biblical history, rhetoric, and Greek and Latin literature. He learned to speak Italian and Spanish with reasonable fluency. He grew up to be a keen artistic connoisseur, who built a unique collection of paintings and antique sculptures, and a patron of writers and musicians. At his invitation Leonardo da Vinci and Benvenuto Cellini took up residence in France, each for several years. Among his many treasures, Francis owned the *Mona Lisa,* which he kept with other prized paintings from his collection in his bathroom.

The status of women at the French court had been transformed by the traffic between France and Italy. Women in Italy were essential contributors to courtly society, valued especially as conversationalists. Women's education had been championed in Europe by the Spaniard Juan Luis Vives and the Englishman Thomas More. In Mary's case, her family believed it essential to her courtly training to prescribe a course of study that followed their ideals. A key element was a knowledge of languages, followed by rhetoric, history and poetry. Although these subjects seemed relatively esoteric, they were considered almost entirely practical.

The art of politics and of governing well, as the theorists believed, could be taught from the examples of history. Next, the art of speaking well and of political persuasion — rhetoric or eloquence — was studied from the texts of classical and modern languages and from poetry. The leading advice books, among them Baldassare Castiglione's *Book of the Courtier*, held that "the arts of speaking and of ruling well" were closely related and best acquired from studies of antiquity. A ruler or councilor unable to mold language "like wax after his own mind" would be sure to fail. The "greatness and gorgeousness of an oration" were that "at the first show" of the words, their dignity and brightness would appear like "tables of painting placed in their good and natural light." Governors had to be able to speak confidently in public: their audiences, whether select groups of advisers or parliaments, could be "moved" by the skillful use of oratory, hence they should be taught the best of the models and techniques that the Greek and Roman rhetoricians had perfected.

The most striking thing about Mary's education is that she followed a curriculum almost identical to that of her male counterpart, the dauphin. This was unusual for a girl. It cannot have been solely because she was a queen, because Henry II's daughters and those of a number of his leading councilors also took their places in the schoolroom. But without her royal connections, Mary could not have hoped for the caliber of tutors or the unrestricted access to books and rare manuscripts that she enjoyed.

Her set texts included Cicero's *On Duties*, Plato's *Laws*, Aristotle's *Politics* and *Rhetoric*, Quintilian's *Training of an Orator*, and Plutarch's *Parallel Lives* of famous Greeks and Romans. In studying them, Mary would have progressed from acquiring a basic knowledge of Latin grammar to a more intensive study and imitation of the leading authors, reading Greek writers in Latin translations at first. When she had mastered Latin, she would have moved on to Greek and Italian and possibly Spanish. Her vernacular was, of course, already French.

Like all students who followed this curriculum, Mary was expected to make translations from her own language into Latin and to write prose compositions imitating the techniques for which Cicero and others were famous. Along the way, she would have been introduced to texts, notably by Plato and Aristotle, that had an ethical and philosophical content, to Ptolemy's standard work on geography, and to histories such as those of Plutarch and Livy.

Taken together, these elements were regarded as a vocational course of study, the equivalent (for a prospective ruler) of a degree in business administration. The final stage, more or less akin to graduation, was for Mary to deliver an oration in Latin in front of her family and the entire royal court in the great hall at the Louvre. Most of those present would have understood it, depending on how grammatically fluent her oration was and how quickly she talked.

Mary's education was overseen by Catherine de Medici, Diane de Poitiers and the Cardinal of Lorraine, who were responsible for choosing her tutors. Diane's role was crucial; she herself was exceptionally well educated. She was a renowned patron of artists, writers and poets, the most influential purchaser of classical and Italian works of art after Francis I's death and the connoisseur who did more than anyone else to establish the taste of Henry II's court.

Mary studied elementary Latin and general subjects under Claude Millot and Antoine Fouquelin. Nothing is known about Millot beyond his glowing reports on her progress and the fact that she awarded his brother a pension, but Fouquelin was the author of a celebrated treatise on rhetoric, *La rhétorique françoise*, published in Paris in 1557, with a dedication to his young pupil in which he enthused over her abilities and potential.

When she had mastered the rudiments, Mary joined the dauphin for more advanced Latin lessons under Jacques Amyot, a classical expert whom Henry II had appointed as a tutor to his sons. While Amyot was teaching the royal children, he was preparing his new edition of Plutarch's *Lives*, finally published in 1559. He may have assigned Mary written exercises based on his own translations, but it is more likely she was first introduced to Plutarch in a concise edition, or crib, made by Georges de Selve, a former student of Pierre Danès, her Greek tutor.

Danès was a leading scholar specially chosen by Henry II to teach Greek to the dauphin. He served on a commission to reform the University of Paris and had been a student of Guillaume Budé, the foremost French intellectual under Francis I and a linguist so brilliant he had learned Greek in weeks without a teacher. Danès tried to imitate his mentor's inspirational methods, and Mary carefully studied *L'Institution du prince*, Budé's classic manual for rulers, which was based on his distillation of the works of ancient authors. She owned a handsome copy in manuscript, which she brought back to Scotland with her possessions after the death of Francis II.

Her Latin compositions, written when she was eleven and twelve, are still extant in a leather-bound exercise book in the Bibliothèque Nationale in Paris. She was required to write a series of sixty-four short essays in Latin on themes prescribed by her tutors. They reveal the range of her reading, which was far from comprehensive but included Aesop's fables, the works of Cicero and Plato, the plays of the Roman comic author Plautus and above all Plutarch's *Lives*.

Mary often cited the *Colloquies* of Erasmus, the best-selling anthology of essays by the preeminent Dutch humanist. Written in smooth and eloquent Latin with brilliant flashes of wit and irony, the *Colloquies* was designed to entertain and instruct students in ways that led them back to primary texts of Scripture and the classical tradition. She also owed a copy of Erasmus's masterpiece, *The Praise of Folly*, which included biting satires on monarchy and flattery, and may have left Mary with a keen sense of her duty to balance luxuries and pleasure with her role as a guardian of her people.

Mary's own essays are disappointing in view of her later ebullience: stilted and moralistic, they reflect less her own opinions than the views her tutors wished her to express. Many appear to be little more than her own translations of existing model answers, and most are in the form of letters, since an epistolary style was considered to be the easiest to acquire and the best way for a student to begin studying oratory.

A majority were to her best friend, Princess Elizabeth, Henry II's eldest daughter. Other "correspondents" included the dauphin, Princess Claude and her uncles. Doubtless the other royal children read the essays addressed to them over Mary's shoulder, but the purpose of the themes was to practice a rhetorical style. None of the letters was ever sent.

One to the Cardinal of Lorraine began, "Many people in these days, my uncle, fall into errors in the Holy Scriptures, because they do not read them with a pure and clean heart." This was typical of the banality of Mary's schoolroom exercises, but a few to Elizabeth and Claude lapse into informality: "I am going to the park to rest my mind a little." "The king has given me leave to take a deer in the park with Madame de Castres." "The queen has forbidden me to go to see you, my sister, because she thinks you have measles, for which I am very sorry."

About the time that Mary was writing her themes, she was studying Ptolemy's pioneering textbook on geography, a work written in the second century A.D. and rediscovered in the 1400s. Ptolemy had mistakenly claimed that the earth is the center of the universe and that the sun, planets and stars revolve around it. But he was the first cartographer to project the spherical surface of the earth onto a flat plane and to superimpose a grid of lines of latitude and longitude over it. He placed north at the top of his maps. And he or perhaps his editors provided pithy comments on various countries and cities, offering a thumbnail sketch of them and their inhabitants.

Mary possessed a fine copy of the *Geography*, printed in Rome in 1490, which had once belonged to the leading Florentine banking family the Frescobaldis. From her studies, she would have learned that Scotland was much farther north than France, though farther south (and therefore warmer) than it really is relative to the equator. She would have read that in summer Edinburgh had a maximum of nineteen hours of daylight, when Paris had only sixteen. According to Ptolemy, much of Scotland was flat, the exception being the dense forests and high mountains of Caledonia (his name for the Highlands), a region that he compressed into too small a space between Glasgow and Inverness. While accurate about the hours of daylight, Ptolemy's depiction of Scotland was spoiled by an elongated and inaccurate projection, and by his ignorance of the mountainous border region between England and Scotland. Nevertheless, Mary was fascinated by geography and kept this volume into her adulthood.

When she was nearly thirteen, she delivered an extemporaneous Latin declamation in the great hall of the Louvre to an audience including Henry II, Catherine de Medici and her uncles. Her speech defended the education of women and refuted a courtier's opinion that girls should forgo learning. It was a topic she had chosen herself, reflecting her conviction that the Italian view of learning as fundamental for both sexes was fully justified.

The near-contemporary description of the oration, by Pierre de Bourdeille, Abbé de Brantôme, is rapturous. "Only think," he wrote, "what a rare and wonderful thing it was to see that learned and beautiful queen declaiming thus in Latin, which she understood and spoke admirably." Unfortunately, Bourdeille's fulsome praise cannot be taken at face value. At the time he was only two years older than Mary, was almost certainly not present at her declamation, and is notorious for his hyperbole.

Her oration was competent, but no more. And she was heavily coached. She had already written fifteen essays on the same topic, and although she crammed in a formidable number of references to the achievements of learned women, it turns out nearly every one was taken from a single source: a letter by the admired Italian writer Angelo Ambrogini of Montepulciano. Mary either owned a copy or had borrowed one from the royal library at Fontainebleau, because the slavish repetition of these citations cannot be a coincidence.

Her forte lay elsewhere. According to her tutors and the reports of her governess Parois, she was attentive and industrious, more so than the dauphin, whose inability to concentrate and lack of motivation earned him a homily from Mary in one of her essays. But she was never a born classicist: she could understand Latin better than she could speak it, and was much more attracted to French vernacular poetry, which she studied under Pierre de Ronsard, the chief writer of the Pléiade.

This was a circle of seven poets at Henry II's court who wrote verses and panegyrics in return for patronage. They were in the vanguard of a literary movement that aimed to show that nationhood could be shaped by a common language and that the French language could stand on equal terms with Latin, Greek and Italian. They also wanted to prove that poetry on secular topics such as love, friendship, prowess in arms or the individual self could equal that on religious subjects. To this end, Ronsard had to show that the French language could satisfy the demands of the most elevated literary genres, such as Greek and Roman epics and lyrics, or the latest fashionable Italian sonnets and *canzoni*.

An intense rivalry developed between adherents of the Pléiade and those rhetoricians, classicists and historians whom Ronsard held to be stuffy and old-fashioned and to write in barbarous French. It was a battle that pitched the young Turks of the Pléiade against the old guard, a clash of styles that quickly captivated the young Mary, who fell in step behind the Pléiade. Whereas the old guard used complex metaphors and convoluted pedagogic constructions that their rivals likened to the tower of Babel, the young Turks followed Dante's advice to write using clear, elegant, simple words based on colloquial speech: language so lucid and graceful that it sparkled, achieving an effect so sublime it could melt people's hearts.

Ronsard came to Mary's attention because he knew Scotland. He had first served there as a page in the household of Madeleine, James V's first wife, staying on for more than two years. Mary was impressed by his campaign for vernacular poetry; if given the choice, she would sooner read French literature than the classics, and as she was already a queen, it was only natural that she would be offered the choice.

"Above all," remarked Bourdeille, "she loved poetry and poets, but especially M. de Ronsard, M. du Bellay and M. de Maisonfleur, who wrote charming poems and elegies for her, including those on her departure from France, which I have often seen her reading to herself in France and in Scotland, with tears in her eyes and sighs in her heart." Her selection of authors is easily explained, because du Bellay was the leading member of the Pléiade after Ronsard, and Maisonfleur was a soldier-poet attached to the Duke of Guise. But whether Mary wrote verses on the theme of her return from France to Scotland is a moot point.

She herself was never much of a poet. Her extant output is small, although she did attempt a few sonnets in French and Italian during her long years of captivity in England. After her trial and execution, poems were attributed to her that were written in her honor but which she did not write herself. Most blatantly, the verses she is said to have composed as she began her return voyage to Scotland are an outright forgery:

> Adieu, plaisant pays de France!
> O ma patrie,
> La plus chérie!
> Qui a nourri ma jeune enfance!

These lines, first appearing in the *Anthologie françoise* of 1765, are not Mary's but the work of an eighteenth-century French journalist eager to publish a scoop. If she made a significant contribution to French poetry, it was as a patron and not as an author.

Ronsard's bid for patronage came in 1556, the year in which his circle began to call itself the Pléiade. He had first vied to catch the attention of Mary's uncles, and when that failed, he turned instead to her. It was almost a year after her oration at the Louvre; she was nearly fourteen. He first apostrophized her as the Roman goddess of dawn, the "beautiful and more than beautiful and charming Aurora." His poem "À la Royne d'Écosse" was a personal and a political eulogy. He offered his services "to you, to your nation, and to your crown."

Mary agreed to be Ronsard's patron. She afterward helped him to publish the first edition of his works. And in return, he and his followers became a compassionate presence in her life, sending her verses and providing her with some of the emotional reassurance and support she needed at her bleakest and most anxious moments. The writers of the Pléiade lined up behind the monarchy in the political and religious crises that afflicted France during the Wars of Religion. They remained loyal to Mary, whose reputation they considered it their duty to defend. Ronsard admittedly hedged his bets, dedicating poems to Elizabeth I, whom he flattered by saying she was Mary's equal in beauty. But such lapses apart, they behaved honorably and retained Mary's lifelong affection.

And even if her own proficiency as a writer was comparatively modest, Mary's love of poetry tells us something about her. She reveled in the imaginative, the romantic, the thoughtful. She had also become idiomatically and culturally French: as immersed in the language, its mental patterns and associations, as any native speaker.

It was no longer simply that Mary spoke fluent French; her identity was altered if not completely transformed. Although her tastes and leisure preferences centered on music, dancing and embroidery, the pastimes of aristocratic women throughout Europe, the types she preferred were quite different from those a native Scotswoman would enjoy. They were qualitatively different from her mother's, even though Mary of Guise was born a Frenchwoman, since they were as much Italian as French, reflecting the latest styles at Henry II's court, where cultural imports from Florence, Milan and Mantua became the rage. In singing and dancing, Mary tended to prefer styles developed in Mantua and Milan, whereas in embroidery she valued most the opulent designs of Florence and Venice.

Her singing voice was soft but clear, and she was trained like a Mantuan singer to modulate her tone to suit the declamation of the text, the acoustics of the building or the timbre of the instrumental accompaniment. Sometimes she sang while accompanying herself on the lute. At other times she played the clavichord or the harp.

As a musician she was nothing out of the ordinary, but as a dancer she had real flair. She was agile enough to master the complex routines of the latest fashionable dances, and rhythmic enough always to appear graceful. By the use of simple gestures, it was said she could conjure up emotions to match the music.

Henry II knew talent when he saw it, and the potential for theater. He went to great lengths to find Mary a suitable Italian dancing master, and she and the dauphin practiced regularly together in anticipation of the balls that would follow their betrothal.

Mary adored dancing. She sought out every opportunity to perform, appearing at the family festivals organized by her Guise relations at

Joinville and Meudon as well as attending court events. At Holyrood after her return to Scotland, she danced almost every night with her four Maries. Often she danced until after midnight, for which she was castigated by the Calvinists, who saw it as "depravity . . . attending the practice of France," an invitation to "idolatry" and sexual transgression.

Mary also loved embroidery. It was thought to be an ideal form of relaxation when it was too cold or wet to go outdoors, and when she was nine, two pounds of twisted woolen yarn were ordered so she could begin her training under the expert eye of Pierre Danjou, Henry II's personal embroiderer. She gained satisfaction from what she learned, and in Scotland often sat sewing during meetings of the Privy Council or while receiving ambassadors. Later, when in exile in England, her needlework would give her a comforting, calming way of spending the long years. "All the day," she was reported then as saying, "she wrought with her needle, and that the diversity of the colors made the work seem less tedious, and [she] continued so long at it till very pain did make her to give it over."

Catherine de Medici encouraged Mary's interest. She herself was a skilled amateur embroiderer, trained as a child at the Murate Convent in Florence, where the nuns were famous for their needlework. Mary began with knitting and plain sewing, after which she progressed to the more decorative work of needlepoint: the minute stitches of petit point as well as gros point. The aim was to make gifts or to ornament the heavy, dark-paneled rooms with tapestry-like wall hangings, valances, or table and cushion covers. The outlines of the intended design would be sketched out first on canvas from a pattern taken from an emblem book, then filled in as delicately as possible with a variety of colored wools or silks.

Even as a child, Mary had a taste for luxury. She had a passion for *brodures* — the most sumptuous and expensive embroideries — and especially the jeweled or enameled sort used for headdresses or as strapwork on bodices or sleeves. Such accessories were greatly coveted, but were too difficult for amateurs to make, as they involved stitching directly onto silk or satin, using thread of fine metal or else silk yarn onto which the tiny pearls or other small jewels were attached. When Mary found as a teenager that she could not live without these items, she purchased them indiscriminately, ignoring the protests of Parois.

On a more mundane level, Mary adored making *cotignac,* a type of French marmalade, putting on an apron and boiling quinces and sugar with powder of violets in a saucepan for hours before laying out the slices of crystallized fruits. The four Maries were all required to help her, and a mockup kitchen was created in their apartments so they could play at cooking and housekeeping, pretending to be servants or bourgeois women organizing their domestic routine and doing their own shopping. It was a game that Mary always remembered and sometimes played in Scotland, usually in St. Andrews, where she had a house near the abbey.

She loved pets and wanted as many as possible around her. Dogs, especially terriers and spaniels, were her favorites, and she let them romp around freely as was the custom in royal and aristocratic households. At one time she kept sixteen, and her kitchen staff were given standing instructions to save table scraps for them.

Her second favorite animals were ponies. When she was about fifteen, she asked her mother to send her "some good *haqueneys*, which I have promised to Monsieur and the others who have asked me for them." By this she meant that she wanted Mary of Guise to send ponies from the Shetland Isles, at the northern tip of Scotland, as presents for the dauphin's younger brothers, Princes Charles, Henry and Francis, then aged eight, seven and four. These ponies were ideal for children to learn to ride, renowned for their small size, gentle temperament and shaggy coat.

When she first arrived in France, Mary was too young to hunt, but she could watch and showed visible excitement each time the hounds were let loose from their kennels. Falconry was, however, within her grasp, and within a few weeks of arriving at St.-Germain, she astonished the ladies of the court by dressing her own pet falcon, casting the bird off and fearlessly reclaiming it on her arm without help from the falconers.

Later she watched while the dauphin learned to hunt, and joined in herself when she was twelve or thirteen. Her two favorite horses, given to her by Henry II, were called Bravane and Madame la Réale. (Bravane was perhaps Mary's nickname for a fearless filly and la Réale for a Spanish mare — literally the "royal" one.) She loved riding and soon relished hunting, for which she adopted the daring habit of wearing breeches of Florentine serge underneath her skirts. The fashion was introduced by Catherine de Medici from Italy, and was risqué because it allowed the wearer to ride astride her horse and not sidesaddle as female protocol required, a habit for which Mary would be greeted with suspicion in Scotland.

On wet days or candlelit evenings in the royal apartments, Mary liked to play cards with the dauphin, whom she beat more often than not, although on the occasions when she lost, her stakes were characteristically higher than his. Already Mary was prepared to take risks. Her other indoor amusements included chess, backgammon and playing with a set of Italian puppets.

By the time she was fourteen, Mary was much taller than average. In an age when a woman was considered tall if she reached five feet four inches, Mary finally grew to almost six feet (perhaps five feet ten inches or so), which, along with her delicately formed breasts, slim waist, soft white skin, marble-like complexion, high forehead and auburn hair crimped into ringlets, made her a striking figure. One potential flaw was in her posture, because as a child she refused to hold herself up straight. It was probably laziness rather than embarrassment at her height. Either way, it was a fault corrected by Parois, if only after a struggle. In consequence, Mary's deportment in her prime lacked any trace of the rounded shoulders and slight stoop that were her hallmark in middle age.

When the dauphin was eight and Mary nine, Catherine de Medici commissioned portraits of the royal children, which she asked to be sent to her as soon as possible. When they arrived, she complained that the artist had not caught their features adequately. So she called for improved likenesses, even if these were only done in chalk. The new portraits were ready by December 1552. Mary's was sent to Henry II, who liked it so much he refused to part with it. This finished version has not survived, but a preparatory sketch in red and black chalk shows her at age nine and a half.

Despite her tense expression, perhaps the result of being required to sit still for the artist, her almond-shaped eyes are unmistakable. Her ears were disproportionately large. Her nose was childish and snubbed, and not yet aquiline. Although perhaps less attractive to us, the high forehead, imperceptible eyebrows and tight lips were considered elegant at the time. To smile for a portrait was the height of rudeness then.

She was dressed in the latest couture: a close-fitting outer bodice with slashed sleeves, puffed at the shoulders and clinging to the arms. Her crimped hair, centrally parted, was fitted into a richly embroidered caul banded with jewels. She sported earrings, a gold necklace with rubies and diamonds, a string of pearls that looped up and down across the shoulders, and a large jeweled pendant. As a queen keen to maintain her status at the royal court, she would have regularly dressed this way despite her youth.

When Mary was twelve or thirteen, a more detailed drawing was commissioned from François Clouet, one of the leading court artists. More than any other portrait, it is a mesmerizing image of the young woman known in France as "la plus parfaite." Her face and lips were fuller, her gaze less anxious, her nose snubbed no longer, her eyebrows more in evidence and delicately penciled, her charm and vivacity signaled by her escaping curls and the gleam in her eyes.

A third and final drawing shows Mary shortly before her first marriage. She was little more than fifteen and yet looked twenty. She was slim, confident and poised, her expression purposeful, even assertive, radiating charisma and savoir-faire. This is a drawing of someone who knows what she wants and is used to getting it.

When Mary reached the age of fifteen, her uncles, the Cardinal of Lorraine and the Duke of Guise, were satisfied that she was ready to take on the duties her momentous role entailed. Statecraft was the benchmark the cardinal had in mind. "Discretion sur tout" was a maxim he had always tried to teach her. She had been trained to keep her letters safe and not to leave them lying around or in unlocked cabinets, where they could be purloined or read by the servants. "I can assure you," she had knowingly informed her mother, "nothing that comes from you will ever be disclosed by me."

She had learned to mark the confidential passages of her letters for encoding in cipher. This was done by her private secretary after she had written or dictated her draft, although when she was excitedly trying this out for the first time, her secretary advised her that there was no need, since he was already sending her mother all sensitive information in code.

Despite these concerns, Mary could sometimes be too trusting. When she needed to reply as queen to certain letters and petitions delivered in Scotland, she sent her mother thirty-five blank sheets with her signature at the foot. Fourteen, simply signed "Marie," were for general use. Fifteen signed "La bien votre Marie" were for more favored or important recipients, and six signed "Votre bonne soeur Marie" were exclusively for sovereign rulers, to be used when covering letters were required for diplomatic credentials or other special purposes.

Another distinctive trait arises from this period. Mary was taught to

write in italic script in the newly fashionable Italian (or roman) manner
rather than in the Gothic handwriting of the later Middle Ages, using a
formulary to ensure her letters matched the norms of this elegant style.
When she put her mind to it, Mary could write impeccably: her best
writing is indistinguishable from that of Henry II's own daughters or of
her young aunt Anne d'Este, who also learned the italic hand in her
teenage years.

What often happened in practice was that Mary began neatly enough,
but on the second or third page started to rush. Her tutors were on the
whole tactful: "she formed her letters elegantly and, what is rare in a
woman, quickly" was a typical comment. In reality, she found herself
constantly apologizing for her untidiness. In a postscript to her mother,
written at the age of fourteen, Mary urged her to "excuse please my terrible handwriting, because I was in a great hurry." And about a year later,
"You must forgive me if my writing is so bad, but I've had no time to do it
properly."

Mary was, therefore, no academic genius. She was vivacious and
quick-witted, an increasingly sophisticated and confident pupil who accepted, even relished, her royal status. Precocious as a personality, if less
so in her studies, she followed her instincts and concentrated on what
she liked best. But she disliked pedagogy and did not agree that ancient
literature was the best training for queenship. When assigned written
exercises, she would try to finish them as quickly as possible.

According to Rabelais, the most celebrated French author of the early
sixteenth century, one of the main functions of an education was to enable young aristocrats to grow like plants in the sun. Mary liked this idea
so much, she took it as her emblem. She chose the marigold, a flower
that always turns to face the sun, and the motto "Sa virtu m'atire" ("Its
virtue draws me"), a near-perfect anagram of the name Marie Stuart as
spelled in roman letters, with the *u* represented by *v*.

In Henry II's France, the game of anagrams was greatly in vogue. It
was played with letters like Scrabble and often linked to pictorial puzzles
from Italy in which badges, or *imprese,* were drawn. Mary based her personal monogram on the Greek letter M (or *mu*), which she wrote twice
in an interlaced form: once right side up and once upside down, so it
could be read either way. Above her monogram she placed her anagram,
which was then illustrated with a drawing of the marigold.

Mary could hardly have chosen the marigold without knowing the reference to Rabelais, because it was one of the examples in the emblem

books she used for her embroidery. The motto given there was different: "Non inferiora secutus" ("Not following lower things") was the original Latin version, which she decided to rewrite.

It seems to be an instance of her intellectual ingenuity, until one realizes that the very same emblem had already been chosen by Princess Marguerite, Henry II's youngest daughter, who was also in the schoolroom. She had retained the Latin version of the motto, but altered the flower from the marigold to the daisy — or "marguerite," to suit her name — whereas Mary kept the flower but changed the motto, transposing her friend's idea.

No sooner had Mary lighted on the marigold as her *impresa* than her uncle the Duke of Guise entered Calais and the spotlight turned again to her family's political and dynastic ambitions. She was only too aware that her uncles had set their sights on acquiring the throne of England for her and the dauphin, so making themselves indispensable at the heart of the Valois state.

It would not be long before a quite unintended effect of Mary's education began to surface. She started to think independently of her uncles and to question what they told her. Under the curriculum they had chosen for her, she had acquired the same skills as a male student and was taught to think for herself. However unimpressed she may have been by classical rhetoric, it had trained her in how to argue a case and how to spot the strengths and flaws in the reasoning of others.

For the time being, this tension was latent. It was still hidden when, as a reward for their recapture of Calais from the English, the Guise family at last attained their goal. At a public ceremony held in the great hall of the Louvre on April 19, 1558, Henry II announced the date of Mary's marriage to the dauphin. She was still fifteen, and Francis fourteen. The Cardinal of Lorraine joined the hands of the couple, who plighted their troths and exchanged rings, promising they would give themselves in marriage, each to the other, on their wedding day.

Within three days, the city magistrates of Paris had been invited to the wedding. It was to take place almost immediately. Once Henry II had made up his mind, nothing was allowed to stand in his way. His son's marriage, he avowed, was to be the most regal and triumphant ever celebrated in the kingdom of France. The secret preparations had been under way for a month: the ceremonial officials, stewards, wardrobe staff, purveyors, carpenters, dressmakers, embroiderers and pastry cooks had been working night and day to be ready in time.

After she was married to the dauphin, Mary altered her monogram so that the Greek letter *mu* was inscribed within the letter *phi*, to be transliterated as *M* and *F*, for Mary and Francis. When her copy of Ptolemy's *Geography* was rebound by the royal bookbinder, she chose front and back covers of olive morocco on which palm branches were stamped in gold leaf with the newly intertwined monograms at the center. She kept her anagram, "Sa virtu m'atire," the same — but with one important difference. No longer adorned by the marigold, it was emblazoned with the crown of France.

6

❦

A Dynastic Marriage

MARY'S WEDDING was spectacular. The service took place on Sunday, April 24, 1558, at the cathedral of Notre-Dame, the spiritual heart of Paris. The cathedral stood at one end of the Cité, the slender island in the middle of the Seine. The excited citizens packed into the Place du Parvis, the square in front of the building, or leaned out of the windows of neighboring houses, determined to catch a glimpse of the ceremony, largely held outside. A temporary gallery or covered walkway, twelve feet high, connected the starting point of the grand procession, the nearby palace of the Archbishop of Paris, to Notre-Dame itself. The gallery, in the shape of an arch, was decorated in the antique classical style and led to an open pavilion on a stage across the west front of the cathedral, surmounted by a canopy of azure silk embossed with gold fleurs-de-lis. It continued into the church along the nave to the chancel, ending in the royal closet where the bride and bridegroom were to hear Mass. The sides of the walkway were open so everyone could watch the procession as it passed by. This was doubtless Henry II's idea, with his grasp of the way glorifying spectacle could consolidate power.

The nobles and foreign ambassadors were seated outside on the stage, close to where the bride and groom were to be married. At ten o'clock, the Swiss halberdiers led the procession, marching in their smart uniforms while showing off their weapons to the sound of tambourins and fifes. They entertained the crowd for half an hour, until the bride's uncle and the head of the Guise family, Francis Duke of Guise, appeared on the stage.

He struck an imposing figure: tall and handsome like the rest of his family, his skin was tanned by sun and war. He strode forward purposefully with his head carried high, his fair golden hair cut short beneath a black velvet cap festooned with a plume of white feathers, and his beard and mustache neatly trimmed. Officially the master of ceremonies, he had been temporarily promoted, if only for this special day, to the post of Grand Master of the King's Household, the highest court office and a position usually held by his absent rival Montmorency, still a captive in Brussels.

At a signal from the duke, troupes of musicians emerged, playing trumpets, bugles, oboes, flutes, viols, violins and more. Clad in lavish red and yellow costumes, they astonished the crowd by their virtuosity. Then came a hundred gentlemen of the king's household in their finery, then the princes of the blood, then the mitered abbots and bishops bearing their croziers. Next were the senior Church dignitaries: the archbishops; the cardinals, including Mary's uncle Charles, Cardinal of Lorraine; and the papal legate, Cardinal Trivulzio.

In pride of place were the royal party. The bridegroom was flanked by his younger brothers Charles and Henry, and by Anthony of Bourbon, Duke of Vendôme and titular king of Navarre, who had married one of Henry II's cousins. Mary walked between Henry II and her own cousin, Charles Duke of Lorraine, head of the branch of the Guise family whose lands lay in Lorraine and Bar on the eastern border of France. Catherine de Medici brought up the rear, escorted by Anthony's brother, Louis Prince of Condé, and attended by a dozen or so princesses, duchesses, ladies and maids of honor.

The crowds had eyes only for Mary. They virtually ignored Francis, whose short, weedy build must have presented a strange contrast to her height and womanly beauty. They craned their necks to catch sight of her, cheering and waving their hats in the air. She looked radiant in her shimmering white dress, itself a daring and unconventional choice because white was the traditional color of mourning for royalty in France. Mary, however, was not going to be bound by convention on her wedding day. She meant to make a dramatic gesture. She knew that white suited her delicate skin and auburn hair, and insisted on it. Her dress was "sumptuously and richly made," lustrous with diamonds and jeweled embroidery, its long, sweeping train carried by two maids of honor.

From her neck hung a magnificent jeweled pendant, the one she called "Great Harry," a gift from her father-in-law and engraved with his initials, which she valued so much that she later placed it with the Scottish crown jewels. Her hair hung down loose — another bold choice — and on her head she wore a gold crown studded with diamonds, pearls, rubies, sapphires and emeralds. A huge gemstone at its center flashed in the sunlight and caught everyone's gaze: the rumor went around that it cost half a million crowns.

We have a good idea what Mary was thinking as she walked in the procession toward the stage. Early that morning, she had written to her mother to say she was so excited, "all I can tell you is that I account myself one of the happiest women in the world."

The Archbishop of Paris greeted the royal family as they reached the cathedral's great doors. Henry II drew a ring from his finger and gave it to the Cardinal-Archbishop of Rouen, who performed the marriage there and then on the stage. After a short sermon from the Archbishop of Paris, it was time for the bride and bridegroom to withdraw to the royal closet for the nuptial Mass.

Before going inside, the Duke of Guise, eager to win popularity and seeing the crowd's view was obstructed, with a gesture of his hand ordered the heralds to shunt the guests off the stage and into the overflow seating in the church. The spectators roared their approval, the duke acknowledging their cheers.

No ordinary Parisian was likely to forget the day. As the royal family turned to enter the church, heralds cried out three times "Largesse" and began throwing gold and silver coins of all types into the crowd. Money showered down like confetti, causing something of a riot. People jumped or dived for the coins, pushing and elbowing their neighbors to grab a share of the spoils. Several people were knocked over, receiving cuts and bruises. Others fainted in the crush, while those nearest the stage were jostled or had their clothes torn. The melee was so intense, there was a risk of a serious accident: those within earshot of the stage begged the heralds to stop before someone was killed.

The royal family withdrew to the closet, where the bride and bridegroom knelt on cushions of cloth of gold to receive the sacrament. During the offertory, heralds once again threw money, this time inside the church from one side of the nave to the other. After the Mass, the royal party reappeared, but before retracing their steps to the archbishop's palace, Henry II ordered the bride and bridegroom to make another circuit of the stage to please the crowd. Another roar of approval went up, louder than before and audible a mile away.

The procession returned to the archbishop's palace, where a private banquet was prepared. During the meal, Mary found her solid-gold crown had become too heavy to wear. She signaled to Henry II, who ordered one of his *gentilshommes* to hold it over her head. A magnificent ball followed at which she danced without her crown, letting her hair flow freely. She took to the floor with her father-in-law, reveling in everyone's delight and admiration.

At five o'clock, there was a new procession to the Palais, the official residence of the Parlement of Paris, where the state banquet was to be held. Henry II and the dauphin rode on horseback, and Mary and Catherine de Medici traveled in a golden litter. As the Palais was only at the other end of the Cité, a few hundred yards away, the procession took an indirect route. It crossed the main Pont Notre-Dame into the business and residential districts of Paris and returned by the Pont au Change, a small wooden bridge lined with goldsmiths' shops and moneychangers' stalls. As the master of ceremonies, the Duke of Guise had planned this route to maximize the numbers able to catch a glimpse of his ravishing niece.

The state banquet was designed to impress foreign ambassadors and the magistrates of the Parlement and the Hôtel de Ville. It was followed by masques, revels, dancing and entertainments on a breathtaking scale. The banquet sprawled across several different halls, and as no fête was complete without exotic floats or clockwork devices, six mechanical ships had been constructed. They were decked with cloth of gold and crimson velvet, with silver masts and sails of silver gauze that billowed in an artificial wind created by hidden bellows. The ships rocked from side to side and moved backward and forward. Painted canvas had been laid on the floor of the great hall to imitate waves, which undulated gently to complete the effect. On each deck were two seats of state, one empty, the other occupied by the ship's captain, who was in each case played by a male member of the royal family or a prince of the blood.

The ships made several circuits of the hall until each in turn stopped by the lady of the captain's choice; she was then lifted into the vacant seat of state. Henry II chose Mary, the dauphin Catherine de Medici, and the Prince of Condé took Anne d'Este, Duchess of Guise. The festivities lasted all night and continued the next day at the Louvre, culminating in a spectacular three-day tournament at the Tournelles, the most important royal palace in Paris and the one Henry II used most often as a family home.

Mary was exultant. She had made what seemed to everyone to be the ideal dynastic marriage, and the mise en scène was choreographed to reflect this. As the clockwork ships circumnavigated the hall, a narrator explained how the scene depicted Jason, the Greek hero who led the Argonauts in the quest for the Golden Fleece. Henry II, he announced, was Jason. By capturing the Golden Fleece, he would conquer an empire and create a universal monarchy. His son Francis was already dauphin of France; the wedding made him king consort of Scotland and therefore a king-dauphin. But that was only the beginning. The king-dauphin and queen-dauphine would unite the crowns of France, Scotland and England, and this dynastic theme of triple monarchy was the leitmotif of the poems and anthems specially composed for the occasion.

Ronsard, the Pléiade's inspiration and Mary's poetry tutor, began with an epic poem in honor of the Guises, eulogizing the Cardinal of Lorraine as a master strategist and explaining how the dauphin, by choosing the most perfect queen alive, had subordinated Scotland and England to France. Mary was the bearer of a magnificent dowry: she was the heir to two kingdoms. By her marriage she had helped to shape a dynasty that would dominate not only the British Isles, but eventually the whole of Europe.

Michel de l'Hôpital, the president of the Chambre des Comptes, or treasury, and an amateur poet, then boasted of a time when the "gallant heirs" of Henry II would each take their place among the crowned heads of Europe — in France, northern Italy and the British Isles. "So shall one house," he foretold, "the world's vast empire share." Only through Mary's marriage could the Valois monarchy subjugate England "without war and murder." Or, to use the metaphor of the pageant, she herself was the Golden Fleece.

Lastly, Joachim du Bellay, Ronsard's chief collaborator in the Pléiade, rhapsodized in an ode to Mary: "Through you, France and England will change the ancient war into a lengthy peace that will be handed down from father to son."

The core assumption of these celebrations was the idea that the British Isles were part of an emerging French empire. Scotland and England were to be the provinces that France had subjugated by the dauphin's union with Mary. This notion was more tangibly expressed in July 1558, when Henry II instructed the Parlement of Paris to register an edict granting French citizenship to all Scots on account of Mary's marriage. It was a contentious demand. To smooth its passage the royal lawyers stressed that Henry intended to safeguard French sovereignty throughout all his provinces and dominions: the purpose of the new edict was to allow him to exercise an imperial monarchy over Scotland modeled on the example of the Roman empire.

Within two years, Henry II's imperial vision would damage the Franco-British project as irreparably as Henry VIII's Rough Wooings had the Anglo-British one. In Scotland, nothing was more important than the idea of national independence. When Henry II had first announced the date of Mary's marriage, the Scottish Parliament sent a delegation of eight commissioners to Paris to discuss terms. The Scots still approved of the dynastic alliance, which they saw as a guarantee of stability and security against England, but asked that their "laws and liberties" be confirmed, to comply with the treaty of Haddington.

Henry referred them to Mary. Now that she had been declared of age, the obligation was hers. She turned for advice to her mother, who in her absence appointed Mary's grandmother Antoinette of Bourbon to represent her. And Antoinette turned to her sons, in particular the Cardinal of Lorraine, even though they had no experience of Scotland and its political and cultural traditions.

Two mutually contradictory sets of undertakings were given. First, Mary promised to observe and keep faithfully "the freedoms, liberties and privileges of this realm and laws of the same, and in the same manner as has been kept and observed in all kings' times of Scotland before." It was an unequivocal declaration of national independence, signed and sealed with her own hand and binding on her successors. A copy was taken back to Edinburgh, where it was kept with the registers of Parliament.

That was nine days before the marriage. But at Fontainebleau on April 4, eleven days earlier, Mary had signed three secret documents to a quite different effect. The first was a conveyance or deed of gift, made, as she noted in its opening clause, "in consideration of the singular and perfect affection that the kings of France had always had to the protection and maintenance of the kingdom of Scotland against the English." In the event of her death without an heir, the king of France and his successors would inherit Scotland, also succeeding to all her rights and title to the throne of England.

Next, Mary acknowledged that if she died without heirs, the king of France would have full rights to all the revenues of Scotland until he was repaid one million pieces of gold, this sum being a necessary reimbursement of his investment in the country's defense and in Mary's education. The sheer size of this debt was yet another guarantee that Scotland would remain a French province for a very long time.

The final document was a letter of renunciation or "protestation" signed jointly by Mary and Francis, in effect a combined oath and prenuptial agreement, whereby she affirmed as Queen of Scots and he confirmed as her fiancé that the dispositions and gifts she had secretly made were valid and effective in law, and would remain so irrespective of her marriage and of any other assurances she had formerly given or might give in the future. The document also nullified any future contract or agreement made by the Scottish Parliament on the strength of the previous contract that she herself made with the commissioners.

It is impossible to believe that in asking her to sign these documents, the Guises acted out of ignorance or naiveté. They knew what they were doing; but did Mary?

The secret documents were extremely clever. They were written in florid and high-flown language to create the illusion that Scotland's national interests *were* indeed protected — but by Henry II and the Valois dynasty. Mary was only fifteen; she was not a constitutional lawyer. Her identity was shaped in France and bound up intimately with Henry II and the Valois dynasty. No one had told her that the secret documents, and especially the third, were illegal by Scots law. Again Mary was inclined to be too trusting. She had already signed thirty-five blank sheets of paper for her mother's convenience. She trusted her other Guise relations in the same way, and might well have been prepared to sign documents presented to her without studying them properly.

Perhaps she should have been more careful, but then, she did not expect her uncles to act illegally. Her adolescence suggests she was uneven in her precocity. She was both older than her years and no more than her years. Diane de Poitiers noticed that she spoke to the Scottish commissioners "not as an inexperienced child, but as a woman of age and knowledge."

What Mary lacked was direct experience of Scottish politics and of the different expectations of the nobles in that country. In most respects, Scotland and France were in parallel universes. The kings of France were almost untrammeled rulers: they got their way even over the Par-

lement and the law courts. In contrast, the Scottish lords saw themselves
less as the ruled and more as co-rulers. Their code of honor was closely
linked to their belief that they were the national "guardians" of the realm
and the "commonwealth." It was on this basis that they justified their
volatile shifts of allegiance. They saw themselves as the protectors of
Scottish interests, the repository of ancient Scottish values. It had taken
Mary of Guise twenty years to come to grips with them, and even then,
d'Oysel, her chief minister, was constantly by her side. Since the Rouen
fête, moreover, her authority had been buttressed by Henry II's seemingly inexhaustible flow of pensions, a gravy train of patronage to keep
the nobles malleable and compliant.

It was almost seven years since Mary had last seen her mother. Despite the strong bond between them, she had become dependent, emo-
tionally and politically, on her Guise relations. This had positive and
negative aspects. Antoinette of Bourbon and Anne d'Este treated her
with genuine love and affection: they were her guardian angels. Her uncles, in contrast, were impassive and detached. Beneath their amiable
and emollient exterior, they could be cold-hearted, even cruel. To them,
Mary was a dynastic asset to be exploited, not an adopted daughter to be
cherished. At this stage, she was mostly susceptible to her uncle Charles,
who although a brilliant courtier and diplomat, was full of grandiose
schemes and prone to overreach himself, his fertile imagination brim-
ming with fresh intrigues.

The true extent of the Guise deception before Mary's marriage was
not proved until the reign of Louis XIV, a century later, when the secret
documents were found. This does not mean that the Scottish commissioners did not smell a rat, because as well as attending Mary's wedding,
they were invited to the state banquet and so witnessed the pageant of
the ships and what that symbolized. Mary's half-brother Lord James
Stuart, now twenty-seven, had been in the crowd outside Notre-Dame
watching his sister's wedding. He disliked everything he saw, partly from
jealousy, partly because many of his friends came from the close circle of
Scottish and English Protestant exiles in Paris, who hated the idea that
there were still French (Catholic) troops in Scotland and were determined to expel them if they could.

The Protestants later accused Mary and her family of a plot to murder
the eight Scottish commissioners on their way home. Since four of them
died of a mysterious illness before they had even returned to their ships,
it was easy to claim that they were poisoned. But there is no evidence

of such a plot: in her letters to her mother, Mary spoke generously of the dead commissioners and mentioned the plague epidemics ravaging France at the time, in particular near Amiens and the channel ports.

More credible is a report that the Cardinal of Lorraine had lobbied the commissioners to surrender the "crown imperial" of Scotland to France (the same one with which Mary had been crowned at Stirling). He wanted the crown kept at the abbey of St.-Denis, the royal mausoleum and the holiest shrine in France. If Francis and Mary had no children, Scotland would be held by each successive dauphin as an *apanage* or duchy of France. A central tenet of the Guise dynastic plan was that every future dauphin would be king of Scotland whether Mary's heirs or not, establishing the country's subordination forever.

No one in Scotland knew about this. Mary's marriage was celebrated in Edinburgh with bonfires and processions, and when the four surviving commissioners returned to submit their report, Parliament of its own accord offered the dauphin the "crown matrimonial," a considerable accolade and Mary of Guise's greatest achievement as regent, because as well as making Francis lawfully king of Scotland with the prospect of a future coronation at Stirling, the grant implied he had the right to retain the throne for the Valois dynasty should Mary have no children.

No sooner had Francis been given this honor than the Hamiltons, the family and followers of the Duke of Châtelherault, the heirs apparent to the Scottish throne if Mary did die childless, joined with the Protestants to oppose it. As a result, the crown was never sent. Parliament's grant was not officially revoked, but neither was it fulfilled. What dictated this reverse was the lords' determination to maintain their own rights and privileges. Hitherto, the French alliance had not threatened these, because the regent's power was limited by the terms of the treaty of Haddington. But when Mary married Francis, the mood began to shift. The lords realized that her position as the future queen of France could mean she would always be an absentee queen of Scots. This was a major change. Scotland might forever become a French province, as indeed Henry II and the Guises intended it should, and unlike the French tributes in honor of the wedding, all those written by Scots stressed their country's equality with France.

Amid this dynastic turmoil, sudden shock waves reverberated around Europe. Mary Tudor, the English queen, fell mortally ill at the age of forty-two. She had long suffered from severe headaches, abdominal pains and heart palpitations. In her final months she was almost blind. Her life had been ruined by the false pregnancy she suffered in 1555, when her womb swelled and her breasts spontaneously lactated. She went excitedly to her confinement chamber, only to emerge more than three months later a broken woman. Philip II had come to resent the fact that his wife was eleven years older than he was. He found her sexually unattractive, joking that she was his "aunt." But he badly wanted a son by her, and decided to put up with her until he knew she would never bear one. In August 1555, he left abruptly for Brussels to resume command of the war effort against France, only to return for a single brief visit in 1557 to bring England into the war.

Mary Tudor died on November 17, 1558, and from this moment a Valois dynastic project that might otherwise have seemed to be little more than a fantasy — a romantic flight of fancy expressed through pageants and poetic allegories, but without any prospect of political success — had real potential.

In London, Elizabeth — Mary Tudor's younger sister and Mary Stuart's cousin — was proclaimed queen. She was twenty-five years old. Her education had closely paralleled her cousin's, but whereas the young Queen of Scots was not noticeably academic, Elizabeth was a natural scholar. She spoke French and Italian fluently and had a smattering of Spanish. She could converse "readily and well" in Latin and "moderately" in Greek. She could keep pace with the ablest of her tutors, and could still denounce an ambassador in Latin when she was over sixty. Her favorite subjects were Scripture, history, classical literature and music, and while Mary could play competently but not brilliantly on the clavichord or the lute, Elizabeth was something of a virtuoso on the virginals.

Elizabeth was attractive with bright eyes and long golden hair. But she was less tall and therefore less striking than Mary, at most about five feet four inches. And her complexion was olive: the Venetian ambassador used the word "olivastra," or sallow. This was something of a disadvantage at a time when a perfect white skin mattered so much to a woman. Unlike Mary's marble-like features, Elizabeth's were said to be "comely," and whereas Mary's delicate breasts and tight slim waist were legendary, her cousin's were merely "well formed."

Elizabeth's first surviving portrait depicted her at the age of thirteen, immaculately dressed in a crimson robe with an embroidered kirtle and sleeves of cloth of gold, her headdress and the strapwork of her gown bedecked with pearls. Her image was always carefully projected, and in this portrait she was holding a book of devotions complete with a marker peeping out to prove she had been reading it. On the lectern beside her, another book, perhaps the Gospels, stood open. None of this was accidental. Elizabeth was a Protestant. But she was a moderate in religion: more a Lutheran than a Calvinist, unlike the man she chose as her chief minister. This was William Cecil, who had first entered politics as Somerset's secretary and then become Elizabeth's steward. In Edward VI's reign, he was promoted to secretary of state, playing a leading role in the Protestant revolution, which brought him into close contact with John Knox, and forming the ideas that were to pit him so vehemently against Mary in later years.

In Paris, Elizabeth's accession was greeted with unconcealed scorn. The Guises swiftly proclaimed their niece to be "queen of England, Scotland and Ireland," challenging Elizabeth's right to succeed her elder sister on the grounds of bastardy and Protestantism. As she had indeed been declared illegitimate by act of Parliament in 1536, and since Mary Tudor's most vaunted policy had been to restore Catholicism, returning England to the papal fold after Henry VIII's break with Rome and burning more than three hundred Protestants at the stake for their faith, the Guise claim was not so outrageous.

In order to reinforce and publicize it, the cardinal ordered the heraldic arms of England to be blazoned with those of France and Scotland on all the plate and furniture belonging to the household of the king-dauphin and queen-dauphine. This was highly provocative, tantamount to establishing a rival monarchy in exile. It was to taint the relationship of Mary and Elizabeth for the rest of their lives.

In the new year of 1559 the Guise campaign intensified. Letters and official documents were sent out to the rulers of Europe styling Francis and Mary as the "king and queen dauphins of Scotland, England and Ireland." At the wedding of Princess Claude, Henry II's second daughter, to the Duke of Lorraine, Mary's heraldic arms were quartered with those of England on the coat armor of her servants.

By the summer, ushers clearing a path for Mary on her way to chapel cried out, "Make way for the queen of England!" Her heraldic arms were further embellished. Above the dynastic symbols of England, Scotland and France was placed a closed or imperial crown to represent Mary's Franco-British empire. Beneath (in French) was the legend:

The arms of Marie, queen-dauphine of France,
The noblest lady in earth. For till* advance
Of Scotland queen. And of England also,
Of Ireland also. God hath provided so.

A young and fiery Protestant called Sir Nicholas Throckmorton, whom Elizabeth had appointed as her special ambassador to France, translated the verse into English, then sent it to Cecil with a drawing of the heraldic arms. Cecil filed it away in his already fast-growing archive on Mary, where it became one of the documents he most often referred to over the next thirty years when preparing memos for Elizabeth about her.

Meanwhile, the Cardinal of Lorraine strenuously lobbied the pope to rule against Elizabeth and in favor of Mary. He was incensed to discover that he was opposed by Philip II's agents. France and Spain were still at war, which made Philip keen to make terms with Elizabeth, to whom he also made a speedy proposal of marriage. Philip wanted to preserve the dynastic alliance of England, Spain and the Netherlands first envisaged by his father, Charles V. It was a guarantee that his territories in Europe and the New World would never be successfully challenged by France.

As a loyal Catholic, Philip saw the logic of Mary's dynastic claim. Spain had never forgotten the insult it had received when Henry VIII had divorced Philip's great-aunt Catherine of Aragon. Philip had no wish to alienate the young Queen of Scots, who as the leading Catholic claimant might be useful to him in the future. But when Mary Tudor unexpectedly died, his instinctive reaction was to try and marry Elizabeth, who was shrewd enough to entertain his overtures until he had committed himself to support her against Mary.

Just when it had seemed that the English throne might be within Mary's reach, it was snatched away. Pope Paul IV refused to declare in her favor, making it impossible to justify the claim that Elizabeth was illegitimate. The pope simply could not afford to offend Philip, whose forces held the balance of power in Italy. Philip's support was also essential to the success of the Council of Trent, the Counter-Reformation assembly that the pope had reconvened to rejuvenate the ideals of Catholicism and take a firm stand against the Protestants.

If this were not enough, a *volte-face* had taken place in France itself. Henry II, taking stock of the likely consequences of the Guise dynastic

* Archaic usage introducing the infinitive, literally "for to advance."

project, decided to draw back. He began to realize the magnitude of what it involved, and had started to resent the ambitions of Mary's uncles. When their archrival Constable Montmorency was finally released from his prison in Brussels, royal policy began to change. The constable had worked on a peace plan with Spain while in captivity. He and his chief ally, Diane de Poitiers, were now vigorously countering a dynastic policy that they both knew could lead to an endless war. Mary's claims and personal prestige would have to be sacrificed for the greater good of France.

Montmorency was reunited with Henry II on October 10, 1558, at Amiens, where he proceeded to lay all the problems since the resumption of the war at the door of the Guises. The king listened, and signed a truce with Philip. The threat of the war spreading to Scotland, England and northern Italy was just too great. France was almost bankrupt. Spain, too, was exhausted and anxious for peace. Elizabeth also wanted peace, which would enable her to move forward and enact a Protestant religious settlement without the fear of a French invasion.

When negotiations opened among the three countries at Cercamp, a consensus quickly emerged. The talks were suspended on December 1, once the news of Mary Tudor's death was confirmed. They were then resumed on the neutral territory of Cateau-Cambrésis, where on April 2 and 3, 1559, two treaties were signed: one between France and England and the other between Spain and France.

The treaties involved dynastic and territorial clauses. Philip II, whose offer of marriage Elizabeth politely declined, would marry Princess Elizabeth, Henry II's eldest daughter, and Henry's sister Marguerite would marry Philip's ally Duke Emmanuel-Philibert. The French abandoned their claims to Savoy and northern Italy, but were rewarded with the return of St.-Quentin and its hinterland. France was to retain Calais for eight years, after which the town was to be returned to England or an indemnity paid in lieu. Lastly, Henry II undertook to "pacify" Scotland.

Such terms struck at the heart of Guise power. Montmorency and Diane had staged a spectacular comeback. They and their clients again dominated the Conseil des Affaires, the nerve center of power-brokering at Henry II's court. When a suit by the Duke of Guise to obtain the office of Grand Master of the King's Household permanently as a reward for his victory at Calais the previous year was rebuffed, he read the signals and left for Joinville to lick his wounds.

While the duke sulked and his brother plotted, a disappointed and bewildered Mary tried her best to appear regal and composed. Although Scotland was not strictly a party to the diplomacy, a subordinate treaty had been made with England at the Berwickshire village of Upsettlington. On May 28, 1559, in a chapel beside the Louvre, Mary and Francis ratified this treaty, closely observed by Henry II and Catherine de Medici.

Since Francis had a stutter, Mary did the talking. She said that "as the queen of England was her cousin and good sister, she and the king her husband were glad of the peace." They would do everything in their power to preserve it. She spoke confidently and with apparent conviction; her words were taken as an olive branch. No claims were made to the English succession. Mary had been forced into a U-turn. The dynastic ambitions of the Guises, previously the main aim of Henry II's foreign policy, now stood in the way of French interests.

At the reception at the palace of the Tournelles that followed the ratification of the treaty of Upsettlington, the Guises were nowhere to be seen. Mary was attended not by her uncles but by the constable. He took the opportunity to reassure the English ambassadors that he had always been a friend of England, and was now more so than ever. Elizabeth, he continued, was "a virtuous and worthy queen." This was the new official line in French foreign policy, and Mary had to stick with it whether she liked it or not. When she wrote to Montmorency afterward, she referred to his "good and happy enterprise."

As Mary gained in experience, she began to see how fickle her allies could be, and how she could quickly become the victim of a change in the balance between the factions. To her uncles in 1559, the peace of Cateau-Cambrésis was a sellout, but if Henry II had approved it, Mary had to take it or leave it. Directly after the ratification ceremony at the Louvre, the king disappeared to the houses of the constable and Diane de Poitiers for a week's holiday, more than satisfied that his policy was once again in safe hands.

When Mary came under pressure, the first thing to suffer was her health. Exactly this happened in the spring of 1559, when she found herself caught between the policy of her uncles and Henry II and the constable.

In March, a report reached Cecil, now firmly entrenched in his position as Elizabeth's chief minister, that Mary was seriously ill. Four days before the ratification ceremony, Throckmorton, one of the English representatives and Cecil's intimate friend and confidant, said Mary looked "very pale and green, and withal short breathed and it is whispered here among them that she cannot long live."

On June 18, she was seen to be "evil at ease" in church, and "they were fain to bring her wine from the altar" to prevent her from fainting. Three days later, she collapsed. A week after that, she was "suffering from a certain incurable malady."

Whether this illness was caused by stress or was perhaps Mary's first attack of acute intermittent porphyria, a worse calamity lay in store. At the end of June, another three-day tournament was held at the palace of the Tournelles. At five o'clock on the 30th, the forty-year-old king returned to the lists. He had insisted on a rematch with the captain of the Garde Écossaise, Gabriel de Lorges, Count of Montgommery, whom he had beaten in their first joust, but who had gained points for striking the king with his lance.

Diane and Catherine did their best to dissuade Henry. But he refused to listen. Montgommery was ordered to rearm. At the first blow, his lance hit the king's chest, then slid up the length of his body armor to his helmet, where it shattered, allowing fragments to slip under the visor. A large fragment hit the king's forehead, lacerating the flesh, and another pierced his left eye. Henry was carried indoors, and the finest surgeons were summoned. They hoped at first to save him, but the brain had been touched. One moment he was coherent and talking normally, the next he was paralyzed or seized by convulsions.

Mary and the dauphin watched day and night with the rest of the royal family at the foot of Henry's bed. He died on July 10 of a massive stroke. The constable stood guard over the corpse while Mary's exultant uncles brought her and Francis from the Tournelles to the Louvre in a carriage. The Guise brothers were suddenly back in power.

Francis was proclaimed king of France, after which the court went into mourning. Then everyone dispersed. The new king was taken by Mary's relatives to Meudon, the Guise château near Paris. Mary went straight to the royal apartments at St.-Germain. She was just five months short of her seventeenth birthday. She was also now queen of France.

7

Betrayed Queen

THE DAUPHIN was crowned Francis II at Rheims, the holiest city of France and the traditional setting for the anointing of kings. The celebrations got off to an inauspicious start. When Francis and his escorts arrived at the gates of the city in readiness for his official welcome on Friday, September 15, 1559, a sudden downpour soaked everyone to the skin. Then the coronation, due to take place at a High Mass on Sunday, had to be postponed for a day. The Duke of Savoy, one of the highest-ranking spectators, caught a fever and everything was put off. Francis II was, in consequence, the only French king to be crowned on a day that was neither a Sunday nor an important feast day. Moreover, the medal struck in advance at Paris, which was to be handed out as a souvenir to those who attended, now bore an incorrect date.

The ceremony was to be held in the vast Gothic cathedral, famous for its magnificent stained glass. On Sunday evening, Mary attended a vigil with her husband, hearing the choir sing vespers and joining in the prayers before retiring to bed in the archbishop's palace. During the service, Francis went to the high altar to lay a gift of a solid-gold statue of his namesake, Saint Francis of Assisi, at the foot of the cross. The Archbishop of Rheims preached a sermon, to which Mary must have paid attention, because the archbishop was none other than her uncle Charles, the Cardinal of Lorraine. It would be his great privilege to anoint and crown the new king next day.

On Monday morning, Francis, dressed in a gown of white damask over a shirt and tunic of white silk in preparation for his anointing, processed in state from the archbishop's palace to the cathedral. As he entered the main door, trumpets sounded and an anthem was sung. He walked up the nave toward the choir, which was decorated with the richest tapestries from the palace of the Louvre. When he reached the high altar, he took his seat opposite the archbishop's throne.

Mary's uncle then conducted the service. Francis was anointed with special chrism from the holy ampulla kept at the ancient abbey of St.-Rémi. There were antiphons and responses, more anthems, then a High Mass followed by a litany. Between the anointing and the coronation, Francis withdrew briefly to a pavilion of crimson and purple velvet set up in the choir, where he changed into his coronation robes. He emerged resplendent, his blue velvet gown lined with crimson taffeta and trimmed with ermine and gold fleurs-de-lis.

Mary's uncle presented the fifteen-year-old king with the scepter, the rod of justice and a ring, which he placed on the fourth finger of his right hand. He then raised the heavy gold crown above his head. There followed more prayers and benedictions, after which Francis was led up some stairs to his throne, on an elevated platform at the entrance to the chancel. He was meant to be wearing his crown, but as it was so heavy, it took at least four nobles to hold it in place and keep it from falling off while the puny Francis took his seat.

When at last the king was enthroned, the archbishop bowed and shouted, "Vivat rex! May the king live forever!" The entire congregation joined in, most of them doubtless relieved that their long ordeal of sitting still — the ceremony had so far lasted more than five hours — was nearly over. They shouted "Vive le roi!" while the organist and other musicians played their instruments at full volume and more or less at random.

When order was restored, the choir sang the Te Deum in a plainsong setting, and as its gentle and evocative strains drew to a close, the keeper of the royal aviaries released some seven hundred goldfinches and other songbirds from their cages hidden in the choir. They chirruped and sang as they flew up toward the roof or perched on the ledges of the triforium, a symbol of peace, tranquillity and the dawn of a new age.

Mary watched everything as a mere spectator. Although the queens of France were also anointed and crowned, the ceremony was held at a time and place different from the king's coronation. Catherine de Medici had been crowned two years after Henry II at the abbey of St.-Denis, on the outskirts of Paris, the usual place for the crowning of a queen. According to the ancient Salic law, women were barred from the throne of France. The queen was a dependent of the king, not his partner. She was a consort, who was excluded from the succession or from exercising the powers of government. Although she might be appointed regent if her husband was sick or abroad, she had no shared rights of sovereignty.

Mary sat with the other female members of the royal family in a special closet to the side of the high altar. Catherine de Medici took precedence after her, followed by Mary's childhood friend Princess Elizabeth, now fourteen and recently married by proxy to Philip II of Spain.

Catherine wore a long black silk dress. The court was still in mourning for the dead king, and black was the color of mourning in her native Italy. She had embraced it right after her husband died, flouting the convention that white was the norm at the French court and continuing to dress in black until her own death in 1589, earning a reputation as something of a somber and brooding presence.

Elizabeth and her younger sisters, Claude and Marguerite, followed their mother's example. Mary alone insisted on wearing a white gown. Since the royal closet was clearly visible to many in the crowd, she must have stuck out like a dove among crows. Her choice of color sparked controversy. Mary wore white to be different. She was asserting her flair for the theatrical, a prospect made all the more attractive by the fact that she knew the color suited her better. And the people loved her for it, nicknaming her "the white queen" as a result.

If Mary's dress caused several in the congregation to mutter, the official banquet did not quite go according to plan either. It was served in the great hall of the archbishop's palace, where an ancient ritual was observed. The nobles and leading guests ate at their usual tables, while the king ate alone at a special table in the middle of the room to symbolize his almost sacred status.

Francis, however, was tired and bored. He kept yawning and wanted to retire to his chamber before the end of the meal. When he decided to leave, the reception broke up early and in some confusion. Overall, it was perhaps the most awkward and least convincing coronation day in French history.

Change was afoot at court even before the ceremony. A palace revolution began the day after Henry II's death. The Cardinal of Lorraine evicted

Constable Montmorency from his suite of apartments and the Duke of Guise usurped his seat at dinner. The brothers took control of the Conseil des Affaires and put their own clients and retainers in many important offices, even replacing the royal chaplains and almoners.

But Catherine was not to be underestimated. Already she perceived Mary to be a threat, and so refused to accept the usual title of Dowager Queen. Instead, she insisted on being called Queen Mother for the rest of her life. The subtle nature of this change was that it helped her keep her place as an active rather than a retired politician. Catherine knew that she had to be cautious in her dealings with the Guise family. Her son regarded both the duke and the cardinal as his principal advisers and as national heroes. She knew she had to be willing to work with them for the time being while she waited patiently for her opportunity to remove them.

Not just the constable and his sons were frozen out, but almost his entire family. Since many of his relations were Huguenots or at least supporters of the Protestants, this action acquired a religious edge. He was replaced as Grand Master of the King's Household by the Duke of Guise, who finally achieved his ambition to hold this office.

Diane de Poitiers, the constable's ally, was scarcely treated better. The magnificent jewels Henry II had showered on her were reassigned to Mary, a not unreasonable decision, since they belonged to the crown and Mary was now queen. But the demand that Diane surrender Chenonceaux, her most prized château, to Catherine in exchange for Chaumont — a pleasant enough building but in a much less desirable location — rankled, and after a brief stay at the lesser property she threw it back at Catherine and withdrew to the château of Anet, one of Henry II's gifts that she was grudgingly allowed to keep. There she lived as a mistress without a king, until she died seven years later, all her former influence evaporated.

As soon as the Guises felt secure, they reinstated their Franco-British project, even though it was discredited by the treaty of Cateau-Cambrésis. They could claim to be reviving Henry II's original foreign policy, but expectations elsewhere had changed. Philip II was flatly opposed to their plans for Scotland and Normandy, which he knew to be linked and designed to win control of the English Channel and the North Sea. Normandy was the province with the largest number of channel ports and the deepest harbors. The Guise family had made themselves the richest and most powerful landowners there, and although they were not inclined to resume an outright war with Philip, he had guessed correctly that they planned to turn the region into their military and naval base.

Philip left the Guises under no illusions that the peace was fragile. When they continued with their dynastic policy regardless, the level of tension rose. Whereas before the treaty the main theaters of European war and diplomacy had been Italy, France and Germany, now the spotlight was on Scotland and England.

The arrogance of the Guises was their undoing. Far from exercising caution or masking their intentions, they broadcast their aims to the widest possible audience. And their niece was at the center of it all. Wherever the French court came to rest and whichever towns it visited, the heraldic arms of Francis and Mary were blazoned with those of England on the gates. Sir Nicholas Throckmorton, Elizabeth's ambassador to France, heard about the design of a new great seal for Scotland. When the seal was produced, it confirmed his worst fears. Francis and Mary were shown seated in imperial majesty above the legend "Francis and Mary, by grace of God, king and queen of France, Scotland, England and Ireland."

After Francis's coronation, these claims were once again inscribed on the gold and silver plate and carved on the furniture with which Mary's household was newly equipped as queen of France. The coup de grâce was delivered when Throckmorton was invited to dinner and then forced to eat his meal off silver dishes bearing the usurped insignia.

From a Guise family perspective, these may have been the correct initial steps to claiming what they believed to be their rightful patrimony, but their behavior toward Mary was cynical. They treated her like a puppet and made important decisions behind her back. As to Francis, they encouraged him to play the man by going hunting. Not a single major initiative can be traced back to Mary or her husband while they were king and queen. The result was that their reign, an interlude of no more than five hundred days, turned into a Greek tragedy in which the main events were played out offstage by actors or unseen forces over which they had little or no control.

Mary's health was poor for more than a year. Her illness and debility may even have worsened. At St.-Germain in August 1559, she was said to be sick after every meal. She fainted in the Spanish ambassador's presence and had to be revived "with acqua composita" or whisky. In

September, when the court moved to Bar-le-Duc, she was at first better, then ill again. She fainted in chapel and was led to her bedchamber, where she fainted once more. Next month, she was said to be suffering from tuberculosis, and at Blois in November she looked very pale and "kept her chamber all the day long."

Stress was a cause, because the reports of Mary's ill health coincided with her pleas of helplessness at the events unfolding in Scotland. Her mother, Mary of Guise, was still sole regent, but even before Henry II was dead, the Cardinal of Lorraine was interfering in the internal affairs of the country. In particular, he urged his sister to crush the Protestants, whom he regarded as political insurgents. He had first advocated this policy in April 1558, when Mary married the dauphin. And his demands soon became shriller and more insistent.

In merging the defense of Catholicism with his Franco-British strategy, the cardinal sought to imitate Mary Tudor. Her attempt to build a pan-European dynastic alliance with Philip II had been closely linked to her persecution of Protestants. The impresario of her much-vaunted campaign against heresy was the papal legate and Archbishop of Canterbury Reginald Pole, who took advice from the Spanish Dominicans. The policy backfired when many important Protestants fled to exile in Geneva and other Swiss and German towns, leaving the poor and less socially influential reformers to be burned at the stake.

The exiles were free to harangue Mary and Pole from the safety of the Continent, using the pulpit and the printing press. And yet despite these drawbacks, the Guises were impressed by what they believed to be the campaign's effect. They wished to see the policy introduced into Scotland, and the cardinal even obtained a copy of the manual used by Pole's inquisitors for his sister, also sending her a delegation of theologians from the Sorbonne who were skilled at rooting out heresy.

Such measures proved to be wholly counterproductive in Scotland, where the threat of religious persecution led to a rapid crescendo of fears about national independence. Whereas patriotism and the pro-French alliance had been mutually compatible at the time of the treaty of Haddington, now the reverse was true.

The impetus for change came from England, where in April 1559 Elizabeth had deftly engineered a Protestant religious settlement. She had played her cards brilliantly. In the opening months of her reign, she had taken a bipartisan approach. She had made soothing noises to Philip II and appealed to Protestants and Catholics evenhandedly. The former supported her because they knew she would reject the pope and the Catholic Mass. The latter took comfort from the fact of her Catholic conformity in Mary Tudor's reign, when Mass was said in her chapel. Only after the peace negotiations had been concluded at Cateau-Cambrésis did Elizabeth start to reveal her true intentions.

Her leading advisers were Protestant. Cecil, her chief minister, had been one of the leading architects of Edward VI's Reformation. Under Mary Tudor, Cecil was a known supporter of the exiles, who included the fiery preacher John Knox. His name had been linked in Edward's reign to an offer to Knox of the bishopric of Rochester, which as a staunch Calvinist Knox had refused. And yet, even Cecil, to save his neck, had attended a special High Mass at his house at Wimbledon, ordered by an indignant Mary Tudor when the extent of his support for the Protestant underground was discovered.

When the English Parliament voted for the religious settlement in 1559, the country became officially Protestant. It was a cue for the pro-English faction in Scotland, which won the support of the Protestants in an outright revolt against a regent who was easily depicted as the instrument of a French, and especially a Guise, tyranny. From this point on, the rebel propaganda would blend Protestantism and nationalism opportunistically.

Up until now, Mary's mother had governed Scotland effectively, but her determination to enforce law and order, to secure higher taxes and, above all, to assimilate the Highlands and border region into a centralized Scottish state had alienated those who believed it was their birthright to rule their own kinship networks and regions. The Scottish lords at heart rejected a centralized monarchy. They wanted to rule themselves as a loose federation of small kings. The regent's policies might still have been acceptable had the flow of French pensions continued, but after the treaty of Cateau-Cambrésis, the tap was turned off. In those altered circumstances, Protestantism offered the perfect excuse to topple an unpopular regent.

Such opportunism, in turn, gave England a unique chance. Protestantism could be made the foundation of a plan to remold the entire British Isles as a single community, so reversing the disasters of the Rough Wooings. France had everything to lose, but the person most likely to be affected was Mary, since as a Catholic queen her crown and reputation were at stake. Cecil was in the vanguard of this policy, which was to color almost everything he did for the next thirty years and more.

Everything depended on whether the English could learn from their mistakes. It turned out that they could, because for the moment they dropped all references to England's claims to imperial overlordship and to the status of Scotland as a satellite. They acted circumspectly, led once again by Cecil, who sought at all costs to avert a full-scale European war: he had become Protector Somerset's secretary just in time for the 1547 Pinkie campaign, when he had witnessed the terrible carnage and narrowly escaped being killed.

On the Scottish side, the nobles in revolt against the regent called themselves the Lords of the Congregation. The most powerful was the Earl of Argyll. The son of the man who had carried the sword of state at Mary's coronation, he was a genuine Protestant who controlled huge tracts of territory in the Gaelic-speaking Highlands, where royal control was weakest, and in the Western Isles (Hebrides) and Argyllshire as far south as the Clyde estuary. His influence even stretched across to Ulster in the north of Ireland, and he was the only Scottish noble with the resources to muster a full-size army independent of the crown. He was joined by the Earl of Glencairn, Lords Ruthven, Boyd, Ochiltree and several others.

By far his most spectacular recruit was Lord James Stuart, Mary's illegitimate half-brother, who quickly emerged as the champion of these lords. Born in 1531 and eleven years older than his sister, he was a big man in every sense. Physically robust, he had a bluff but offensively regal manner and a conviction that as the son of James V, he acted as a man of principle in Scotland's best interests. He was intelligent and well educated, first appearing on the scene when he sailed in Mary's galley on his way to the University of Paris. Destined for a career in the Church, he was appointed Prior of St. Andrews, one of the richest abbeys in Scotland, a position he held as a layman and milked for all it was worth.

Lord James broke with the regent in 1559. His defection was reported in France, where his motive was guessed: he tilted at the regency, and perhaps the crown itself. His methods were cool, calculating and insidious. He advised Mary of Guise that if she would accede to the reasonable demands of the lords, he would support her. Meanwhile, he was continually writing to Cecil, urging him to assist the lords in their campaign to expel the French permanently from Scotland.

Events moved with breathtaking speed in May, when John Knox returned to Scotland from Geneva and joined forces with the lords. His

sermons attacking the pope and the idolatry of the Mass triggered out-
breaks of iconoclasm at Perth and St. Andrews. Church buildings and
ornaments were ransacked in rampant acts of civil disobedience.

In August, the lords appealed to England for military aid against
the regent. Cecil debated the pros and cons in a series of encyclopedic
memos to his colleagues in the English Privy Council. He threw his own
weight unflinchingly behind the lords, but before action could be taken,
he first had to persuade Elizabeth and many of his own colleagues that
an armed intervention to oust a legitimate government in alliance with
its rebels could be justified.

Here lay the seeds of an ideological rift between Elizabeth and Cecil
over Scottish affairs that was to mature over the next thirty years. Despite their ability to work together on almost every other issue, where
Scotland and Mary Queen of Scots were concerned, Elizabeth and her
chief adviser were repeatedly at loggerheads. Whereas Cecil always put
the interests of Protestantism ahead of dynastic considerations, Elizabeth took the opposite approach. Although she was a Protestant, she
kept religion and politics apart, putting the ideal of monarchy and of hereditary descent ahead of religion. When dealing with Mary and her
mother, she found it utterly repugnant that in determining the government of Scotland, legitimate dynastic rights should be overridden by
what amounted to religious preconditions.

In September, Châtelherault and his son, the young Earl of Arran,
climbed aboard the bandwagon. Like many others, they converted to
Protestantism for largely cynical reasons. By October, the vast majority
of the Scottish lords were behind Lord James and Argyll. Only James
Hepburn, Earl of Bothwell, Lord Borthwick and Lord Seton were unwavering in their loyalty to the regent. Bothwell — who had succeeded his
father, Patrick, as earl in 1556 — was the most resourceful in his guerrilla tactics against Lord James, so it was hardly surprising that Mary
would remember him with gratitude when she returned to Scotland.

Between October 19 and 23, the rebel lords did the unthinkable. They
rode to Edinburgh and deposed Mary of Guise from the regency, replacing her with a council of twenty-four nobles voted from among themselves. Their timing was perfect, because the regent was seriously ill. She
was suffering from dropsy, possibly caused by a weak heart. Her legs and
body swelled up, and her French doctor urged her to rest, avoid all stress
and move to a warmer climate. Nothing could have been more impracti-

cal, and the regent ignored his good advice. With characteristic courage she mustered three thousand French troops and took refuge in the port of Leith, which she fortified.

She also appealed to her family, who offered reinforcements under the command of René, Marquis d'Elbeuf, the youngest of the Guise siblings. These troops were to be sent from Normandy, confirming Spanish fears. Until they could arrive, Mary of Guise used her existing forces. When in November the lords refused the offer of a truce, the French garrison made a sortie from Leith and routed them. The Scots attempted to rally but then dispersed, enabling the regent to return to Edinburgh in triumph.

But the respite was brief. Cecil was set on a military intervention, aware that the lords lacked artillery and could never hope to defeat the French on their own. He was starting to hover between his aims of freeing Scotland forever of the Guise threat and remolding the British Isles as a single Protestant community, which meant a virtual annexation of the country. He prepared his ground for a month, then convinced a majority of the English Privy Council on December 27 that the revolt offered the chance of a lifetime to ensure Elizabeth and England's security by effectively turning Scotland into an English dependency.

It was to be the turning point. After the Privy Council meeting, Elizabeth sent a fleet to the Firth of Forth and two thousand troops to Berwick. The fleet arrived in late January 1560 and blockaded Leith. Its orders were to await d'Elbeuf's reinforcements, but these never arrived. Although some French ships left Dieppe, they were scattered by violent storms. They were either forced back to port or wrecked on the Netherlands coast. A mere handful of men landed in Scotland.

Elizabeth swallowed her distaste for the time being and allied with the rebels. The negotiations, directed on the Scottish side by Lord James, led on February 27 to the treaty of Berwick. She committed England to protect the "ancient rights and liberties" of the kingdom of Scotland and defend the "just freedom" of the crown from conquest, and the pact would last for one year longer than Mary was queen of France. Etched into this language was an almost complete contradiction, since "liberty" and "freedom" here meant merely the dislodgement of the Guises by the English: the removal of one occupying foreign power and its replacement by another.

A helpless and increasingly distressed Mary was bypassed in all this. So was her mother, who was mortified by the treaty of Berwick. She pleaded for d'Elbeuf to return, but her requests were ignored. Her brothers sent vague reassurances, but no specific aid. The reasons were partly logistical, reflecting the difficulty of supplying forces over such a long distance, but mainly political. The balance of power in France was once again shifting, this time away from the Guises. Their palace revolution had been too extreme. The death of Henry II had created a power vacuum, enabling the Huguenots to make inroads among the nobles. The Wars of Religion were looming. The Guise brothers were encountering problems all too similar to those already experienced by their sister. The difference was they put their own interests first and left Mary's mother to her fate.

Not everyone shared their sang-froid. Margaret of Parma, the regent in the Netherlands, warned Philip II of a threat to his sea routes. Catherine de Medici had also come to fear the scale of Guise ambition sufficiently to seek support from Spain against her most hated domestic rivals. She had retained much of her independence in politics and diplomacy as Queen Mother, and she now appealed to Philip to intervene as a mediator.

By the beginning of March 1560, a new Spanish move was almost inevitable. The irony is that when d'Elbeuf's ships were lost, the Guises were themselves reduced to seeking Philip's aid in defending their beleaguered sister against her rebels. This played directly into Philip's hands, as he could now step forward as the mediator to whom both sides had appealed for help and advice. In making his plea to Philip, the Cardinal of Lorraine's ambassador railed against the Scottish lords and joked sardonically that Elizabeth's concern for their national independence reminded him of the fable of Reynard the Fox, who used soothing words to coax the chickens down from their high perch only in order to devour them.

Philip appointed two special envoys to intervene before the crisis erupted into war. One was sent to England and the other to France. Their missions made little progress, and in Paris the Cardinal of Lorraine decided on a final roll of the dice. He proposed that France and Spain should unite to subdue Scotland, but only as a prelude to the conquest of England. As an inducement, England would be given as a dowry to the future son of Philip II and Elizabeth of Valois, and this son would in turn marry the daughter of Francis II and Mary. To make this possible, Mary would surrender her own dynastic claim. It was an extraordinary flight of fancy, illustrating not only the delusions of the

Guises, but the extent to which they would trample over Mary to get themselves out of a hole.

Mary was not consulted by her uncle. The mother she adored was dying, and at the same time she was ignored. At Amboise toward the end of April, there was a scene. She confronted the cardinal and "made great lamentations." She wept bitterly and complained that her uncles "had undone her" and "caused her to lose her realm." In reply, the cardinal swore to be avenged of Elizabeth for allying with rebels, but his promises were no more than hot air. Mary burst into tears again, then took to her bed.

In an emotional last letter to her mother, Mary conveyed her deepest love and sympathy, praying "that God will assist you in all your troubles." Catherine de Medici, she knew, had "wept many tears on hearing of your misfortunes." As to the crisis in Scotland, she would insist that her husband "send you sufficient aid," which "he has promised me to do and I will not allow him to forget it."

Mary demanded fresh reinforcements for Scotland, but before they could set sail, Philip II had joined with Catherine de Medici to limit Guise power once and for all. On July 6, after negotiations lasting a month, the treaty of Edinburgh was signed. Purporting to be a coda to the treaty of Cateau-Cambrésis, the accord was in reality a new one. Supposedly between England, Scotland and France, it was actually between England, the Lords of the Congregation and the Guises, with Philip's special envoy in the wings.

The terms were a complete vindication of England and the rebel lords, and a betrayal of Mary and her mother. In a diametric reversal of Guise policy, France recognized Elizabeth to be the rightful queen of England. It was a slap in the face for Mary by her own side. The claim of Francis and Mary to the English throne would be dropped. French troops had to evacuate Scotland without delay; their forts and garrisons were to be razed to the ground. The council of twenty-four nobles would become the official government of Scotland for as long as Mary was an absentee ruler, so enabling it to become the vehicle for the ambitions of Lord James. Moreover, Francis and Mary were to admit that Elizabeth's role in the whole affair had all along been that of an impartial umpire, a disinterested observer who had merely helped to create the right conditions for a negotiated treaty, and not a direct military participant who had allied with their rebels to forward English interests in Scotland.

There was a further sting in the tail. The final clause of the treaty included what might be called a surveillance clause. Despite the humiliating terms of the treaty, Francis and Mary had to promise to ratify and fulfill all of its conditions, failing which England might intervene in Scotland again whenever it thought it necessary to uphold the Protestant religion and to extirpate Catholicism and French influence from the British Isles.

The treaty of Edinburgh, and especially this last clause, was a travesty. It cast a long and inky shadow over Mary's entire career, primarily her relations with Elizabeth and Cecil. It is almost impossible to exaggerate its significance. Mary and Francis had not even been consulted about its terms, nor had Mary even agreed to negotiate with Elizabeth and Cecil as Queen of Scots. A commission, issued in the joint names of Mary and her husband, was admittedly produced in Edinburgh. But this commission was prepared on her behalf by her uncles. It was not Mary's personal act, nor was her mother a party to the treaty. The commission merely illustrates how far their sovereignty had been usurped. Mary's mother was easily bypassed, since she died on June 11, 1560, shortly after the negotiations for the treaty began. Cecil and Lord James were able to stitch everything up without paying any attention to the lawful sovereign power in Scotland.

Mary was devastated by her mother's death. Even this was heartlessly concealed from her: the news first arrived in France on June 18, but was kept secret for ten days. On finally being told, she withdrew to her chamber and wept for a month. When Jane Dormer, one of Mary Tudor's former chief gentlewomen of the privy chamber, who had married Philip II's ambassador in London and was traveling overland to Spain, saw her, she was greatly moved by her distress. The Venetian ambassador, who was also a witness, reported that Mary "loved her mother incredibly." She was so grief-stricken, "she passed from one agony to another."

We know what Mary looked like at this time, because she was first drawn and then painted in her *deuil blanc*.* The drawing is the work of François Clouet, who perhaps also did the accompanying panel portrait. The sittings were completed in or around August 1560, when Throckmorton met Mary at Fontainebleau and the period of official court

* *Deuil*, or "dole," is the French word for mourning clothes. The *deuil blanc* was a white wired hood or cap to which a wide gauze veil was attached, enveloping the body from the top of the neck down. A white streamer attached to the back of the cap hung down loosely.

mourning for her mother had just expired. The portrait shows her as she was approaching the age of eighteen.

It was Mary's idea to send her portrait to England. She was impatient, she said, to find out more about her "sister queen," and hoped to make a fresh start in their relations after the disasters of recent months. Mary wanted to appeal directly to Elizabeth at the level of queen to queen. She had already realized the importance of personal relations in her diplomacy, and offered to send her the portrait if Elizabeth would reciprocate. It was a generous gesture, even if Mary's obvious curiosity about her cousin's true height and appearance partly lay behind it. In the easy, almost bantering style she was beginning to adopt with people when she wanted to get her own way, she made Throckmorton promise Elizabeth would comply, "for I assure you," said Mary, "if I thought she would not send me hers she should not have mine."

When Throckmorton had given his promise, Mary said, "I perceive you like me better when I look sadly than when I look merrily, for it is told me that you desired to have me pictured when I wore the *deuil*." There is no evidence that Throckmorton ever said anything of the sort. The impulse for the exchange of portraits was Mary's, but the ambassador knew what was expected of him.

The portrait shows Mary as a fully mature adult. Her pose is more confident; her face is rounder and fuller, her cheekbones are set higher, and her chin is more fully developed. Her deep-set hazel eyes are wistful and yet reveal her quick intelligence. Her nose is less snubbed and more aquiline than in the earlier chalk drawings. Her compressed lips and slightly pinched mouth convey her great sadness at her mother's death. Her fine auburn hair is, as usual, crimped into ringlets peeping from the edges of her cap.

Everyone who saw Mary remarked on her perfect complexion, and in the portrait her marble-colored skin exactly matched the marbled effect of the semitransparent white gauze veil stretching down to her feet. Beneath the veil she wore a black dress edged with white lace, cut low in the neck and rising in semicircles at the shoulders and across the bosom. It is the portrait of a young woman attempting to cope with her grief. But above all, it is the image of a woman who had grown into the part of a queen.

Throckmorton had already noticed an alteration in her. At his previous interviews with her, she had been flanked by Catherine de Medici or her uncles, who spoke first, leaving Mary to echo their views. But this time they met privately and did not speak French. Although Throck-morton spoke in English and Mary answered in Scots, they understood each other. And Mary seemed much more relaxed. Whereas before she had been tense or arch, now she talked "more graciously and courte-ously." She was more natural and at ease, more willing to follow her in-stincts. She was also a lot more spirited, confidently speaking her own lines and not those scripted by others.

This was Mary's first solo royal audience. It was also her first interview since the making of the treaty of Edinburgh, which she steadfastly re-fused to ratify. The tussle over the treaty was to become a legendary bat-tle of wills, and Mary handled the opening round with aplomb. When asked to confirm it, she dissembled just as her mother had done to Sadler all those years ago. "What the king my husband resolves in that matter," she said, "I will conform myself unto, for his will is mine." It was the first of a series of classic excuses, showing Mary to be Elizabeth's equal in the art of political evasion. She knew very well that she could count on Francis not to do anything by himself.

Mary then changed tack. She produced a killer fact for Throckmorton to report back to his queen. "I am," she said, "the nearest kinswoman she hath, being both of us of one house and stock, the queen my good sister coming of the brother, and I of the sister."

By reminding Elizabeth of their common ancestry as descendants of Henry VII, Mary alluded to her own dynastic rights. "I pray her to judge me by herself," she continued, "for I am sure she could ill bear the usage and disobedience of her subjects which she knows mine have shown unto me."

Mary called for friendship and "amity" between the two queens on the basis of their kinship ties: "We be both of one blood, of one country and in one island." This became a constant refrain of Anglo-Scottish diplo-macy after Mary's return to Scotland.

Finally, she made a pledge: "I will for my part in all my doings make it good, looking for the like at her hands, and that we may strive which of us shall show most kindness to the other." Her offer was as elegant as it was ironic, given that Cecil had only six weeks earlier negotiated the treaty of Edinburgh with her rebels behind her back.

Mary was finding her feet. Throckmorton had no luck with the rati-fication of the treaty, but was fobbed off in such style and with such charm, he hardly seemed to notice. Soon he was forced to report, "As-suredly the queen of Scotland . . . doth carry herself so honorably, advisedly and discreetly, as I can but fear her progress." Her charisma was perhaps in the end more lethal than the dynastic threat she posed. If only, he mused playfully in an aside, "one of these two queens of the isle of Britain were transformed into the shape of a man to make so happy a marriage as thereby there might be a unity of the whole isle."

Back in London, Cecil saw things very differently. "We do all certainly think," he informed Elizabeth in one of his more portentous but revealing memos, "that the Queen of Scots and for her sake her husband and the house of Guise be in their hearts mortal enemies to Your Majesty's person." England was in danger and must defend itself, "and principally the person of Your Majesty." The "malice" of these conspirators was so great, they would never give up "as long as Your Majesty and the Scottish queen liveth."

Cecil's mantra for the rest of his long career was to be Elizabeth's safety. He saw relations between the two British queens less in terms of amity than as an almost cosmic struggle between the forces of good and evil. Between the Rouen fête and her wedding, Mary had been trained to set her sights on the English throne, but after her marriage she was a relative bystander, even a casualty of Guise policy. Cecil never understood or made any allowances for that. Instead, he regarded her as much as her uncles as the instigator and intended beneficiary of an international Catholic conspiracy to depose and kill Elizabeth. He was already her most vehement and determined antagonist.

Cecil's ill will scarcely augured well for the fresh start Mary had proposed, and yet thanks to the arrogance and pretensions of her uncles, it was the assumption on which all her future dealings with England were likely to rest. The Guises had played their poker game and lost. It was now up to Mary herself to see if she could reshuffle their discarded hand.

Linlithgow Palace,
Mary's birthplace

Mary of Guise,
mother of Mary
Queen of Scots

Henry VIII at about the age of fifty-two, a year or so before the first of the Rough Wooings

Antoinette of Bourbon,
grandmother of Mary
Queen of Scots

Prince Edward, Henry VIII's son
and heir, by Hans Holbein the Younger

The Dauphin Francis, Henry II's eldest son, at age eight

Elizabeth, Henry II's eldest daughter, at age fourteen

Mary at age nine

Mary at age fifteen, shortly before her wedding to the dauphin

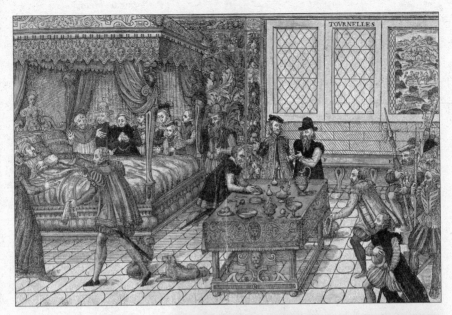

Henry II on his deathbed.
Mary can be seen in profile
with the family at the bedside

Double portrait of
Francis and Mary as king
and queen of France

Catherine de Medici
dressed as a widow

Mary's first letter to a
foreign ruler, sent to Mary
Tudor in January 1554

Ma dame ma bonne seur S'en retournant le sr d'Oysel Lieutenant
du Roy en Ecoséc, ie l'ay prié passant par votre Royaume vous visiter
de ma part, & vous mercier, comme ie fais tresaffectueusement, de
l'affectionnée amitie; dont vous me donies asseurance par votre
derniere lettre, et vous dire que i'ay delibéré de ma part si sincere-
ment y correspondre, qu'il sera si dieu plaist perpetuelle memoire de
deux Roynes auoir esté en vn mesme temps en ceste Isle la ioinctes
d'inuiolée amitie, aussi bien qu'elles le sont de sang & si prochain
lignage. Dont ie vous prie, ma dame ma bonne soeur, le croire, & de
tout ce qu'il vous fera entendre de ma part, tout ainsi que vous
feriés la personne mesme de

Votre bonne seur & cousine

MARIE

8

Return to Scotland

IN THE SUMMER OF 1560, Mary was rumored to be pregnant. The gossip was more than credible, because in a dynastic monarchy the expectation was that rulers should marry and start a family as soon as possible in order to perpetuate their lineage. On the wedding night, a well-established ritual had been followed. Mary and Francis had been tucked in together after their nuptial bed was blessed and sprinkled with holy water. What, if anything, had transpired is unknown, and after that first night the norm for their sleeping arrangements was different. Like all royal families at the time, the king and queen occupied separate suites of rooms in their palaces and slept in separate bedrooms, although Francis was free to arrive unannounced in Mary's room at any time of day or night if he felt disposed to exercise his conjugal rights.

Sex was obligatory for a queen consort of France. Catherine de Medici had been vulnerable to mistresses in the first seven years of her marriage because she disliked it so. Only when Francis was born could she begin to feel more secure. Likewise, the birth of a son and heir would have given Mary an invincible position at the heart of the Valois state. The timing would have been perfect, not least because a pregnant Mary could have instantly overshadowed her dominating mother-in-law. She clearly did think she was pregnant. She put on loose-fitting gowns and insisted that the court depart from Fontainebleau for the cooler air of St.-Germain.

Mary's hopes lasted for six weeks, but by the end of September the pregnancy was proved false and she returned to wearing her usual clothes. Her uncles made light of the affair, joking that a sixteen-year-old king and seventeen-year-old queen had plenty of time for such matters. Then Francis became ill. He had always been a delicate child, nicknamed "le Petit Roi" by his subjects on account of his sickly and runtish appearance.

One evening in mid-November, when he returned from hunting near Orléans, he complained of dizziness and a buzzing in his ear. The following Sunday, he collapsed in church. Soon he was suffering from an acute stabbing pain in his head. Most likely his superficial suffering was from an ear infection — when fluid began to be discharged from his ear, he was immediately confined to bed, but it became rapidly clear he was gravely ill, probably from a brain tumor.

Mary's uncles, the Duke of Guise and the Cardinal of Lorraine, pretended the king merely had a cold exacerbated by catarrh, which they said had affected his ear. This ceased to be convincing when the Spanish ambassador was promised an audience and then fobbed off at the last moment: rumors of a fatal illness or even poisoning started to circulate.

In late November, Francis suffered a series of violent seizures. He was unable to move or speak, gazing impassively for hours at those around his bedside. Mary fought Catherine de Medici for the right to nurse him. Her mother-in-law was becoming insufferable. She mistrusted the Guise family and had dismissed all her son's *gentilshommes* without informing his wife. In the end, the wife and mother nursed Francis together, testing his food to ensure nothing had been tampered with. Every day at dawn, they took up their stations, and despite Mary's continual protests, Catherine also insisted on entering the king's bedroom at night.

The doctors bled Francis, purged him with enemas, and considered performing an operation to bore inside his skull to relieve the pressure of the fluid. Shortly before the operation was due to begin, the discharge from his ear stopped, but just as the doctors were congratulating themselves, a huge eruption occurred and Francis became delirious. By the beginning of December, the pains and discharges were almost continuous and fluids were escaping from his nose and mouth. On Thursday the 5th, he was so debilitated that he lay entirely prostrate, and died late in the evening. No one knew exactly when.

During the night, while Mary kept vigil over the body of her dead husband, Catherine de Medici convened a *conseil secret*. Francis was succeeded by his brother, Charles IX, who was only ten years old. By coincidence, the Estates-General, the most important representative body in France and one that had not met since Charles VIII's reign in 1484, had been summoned to meet at Orléans. Its delegates appointed Catherine to be regent. She was determined to be revenged on the Guises, whose machinations had repeatedly undermined her. She even restored Constable Montmorency to his former positions; hence the palace revolution that had accompanied Francis II's accession only seventeen months before was reversed.

Mary had no place in Catherine's plans. Within weeks of the king's funeral, an unbridgeable gulf separated the young widow and her mother-in-law. The idea that Mary should marry Charles IX was rejected out of hand by Catherine. In response, the Guise brothers attempted to betroth their niece to Don Carlos, the only son and heir of Philip II by his first marriage, to the Infanta Mary of Portugal, who had died giving birth to him fifteen years earlier.

Catherine quickly stepped in to frustrate the Guises. Such a marriage, she believed, would threaten the interests of her other children and risk eclipsing the future progeny of her eldest daughter, Elizabeth, whom Philip had taken as his third wife and who was now living with him in Madrid.

The Guise family's ascendancy was over. In April 1561, the duke decided to withdraw from court. With Montmorency and many of his Huguenot relations back in power, the stage was set for conflict. Within six months, the cardinal and the rest of the Guise faction had followed the duke's example, taking a majority of their dependents and retainers with them. The family and around seven hundred of their adherents gathered at Joinville, later moving to Nancy and Saverne, where they felt secure in their estates on the eastern frontier of France.

Mary read the signals and never risked an outright confrontation with her mother-in-law once Francis II was dead. On her first day as a widow, she voluntarily handed over her jewels as queen of France. A rough inventory was compiled, after which the coffers and bags were sent to Catherine together with the list. Once the jewels had arrived, the Queen Mother ordered a full inventory, listing every item and its value, from the largest pieces such as diamond necklaces and a ruby called "the egg of Naples" down to jeweled embroideries and even individual pearls.

On the same day as the jewels were surrendered, Mary left her royal apartments and went into mourning in a private chamber. During the first fifteen days, she refused to receive any visitors apart from her uncles and a few close friends and family members. Within the space of six months, she had been widowed and orphaned and had lost her standing as queen of France. She sat almost motionless in her white mourning clothes, the room in darkness with the windows blackened and the only light provided by candles.

On the sixteenth day she was willing to receive certain bishops and foreign ambassadors, who offered their condolences. After forty days, she ended her term of seclusion and attended a solemn Requiem for her husband in the Greyfriars church at Orléans.

Although barely eighteen, Mary was able to control her emotions. Her husband's death had followed hard on her mother's. She was at first distraught, but once the initial shock had passed, she came to terms with her loss. A dynastic marriage was not a love match. Mary had done her duty in marrying Francis and had showered him with signs of affection. But she had never loved him. Her mother's death had been far more painful. Moreover, she had been sidelined as queen of France. She was little more than a puppet, the strings pulled by her uncles. But she was still a crowned queen in her own right.

She reviewed her options and decided to return to Scotland. Her mind was made up within a month. "Amongst others," as Throckmorton informed the English Privy Council, "she holds herself sure of the Lord James, and of all the Stuarts." Here Mary showed that despite her experiences with her uncles, she could still be too trusting. Throughout her life, she retained her belief in the importance of family ties. Lord James was her half-brother, the son of her father, James V, even if he was an enemy of the Guises. The reality was that by the terms of the treaty of Edinburgh, he could wield more power in Scotland than she did as long as she stayed away, but she naively believed that if she returned, he would be reconciled to her. Her suspicions turned more toward the ambitions of Châtelherault and his son, the young Earl of Arran, who had made a firm pact with Knox and the Calvinists. Their latest and most outlandish scheme was to send Arran south as a suitor to Elizabeth and so make him king of England.

Mary was learning to speak and act for herself. She was getting her information from her mother's former servants in Scotland. She knew that she could win the hearts of the ordinary people there. Unlike the more selfish nobles, they did not seek war or revolt. "All those who hold themselves neuters," she declared, would support her if she returned, as would "the common people, who now, to have their queen home, she thinks will altogether lean and incline unto her."

But if Mary preferred to take up her own throne rather than to pit her wits against her mother-in-law for years to come, her uncles had other ideas. Their niece was their best hope for making a comeback. They were indefatigable, straining every nerve to engage Spain in the negotiations that they hoped would lead to Mary's betrothal to Philip II's son, Don Carlos.

Mary ignored their intrigues. By acquiescing in the treaty of Edinburgh, her uncles had betrayed her. She did not even begin to forgive them for almost a year. Not one of her letters between Francis II's death and her return to Scotland is on the subject of a second marriage. It is impossible to believe that Mary, who could not resist writing to her family whenever she was genuinely excited, would have said nothing about a new marriage if she had really wanted one then. Instead, her uncles took the initiative, using her aunt Louise, Duchess of Arschot, as an intermediary on account of her many Spanish friends.

Offers came in thick and fast: from the king of Denmark, from the king of Sweden, from the Dukes of Ferrara and Bavaria, and from Ferdinand I, the Holy Roman Emperor and head of the Austrian branch of the Hapsburg family, who was in the marriage market on behalf of his sons. All of these suitors were credible, even if they were not good enough for the Guise family.

Other candidates included the overbearing and ineffectual Arran, whom Mary already held in contempt for his suit to Elizabeth, and Lord Darnley, the fourteen-year-old son of the Earl of Lennox and Lady Margaret Douglas, the couple who had married and lived inconspicuously in Yorkshire after Lennox's hectoring attempt to become Mary's stepfather had failed and he had defected to Henry VIII in 1544.

Court gossip went into overdrive, but the Venetian ambassador, who most feared a marriage to Don Carlos, knew that Mary was not herself a willing partner. It was all the work of her uncles, who sought to manipulate her against her will. In any case, Philip II rejected the suit. He was already closely allied to Catherine de Medici by reason of his recent marriage to her eldest daughter, Elizabeth. His links to Catherine were multiple: through his ambassador in Paris, through her letters to his wife, and through a secret correspondence with Catherine directly, in which Mary was discussed under the code name of "un gentilhomme."

The Guises believed that Philip would sit up and pay attention to them, thinking that Mary could be the key to a future Spanish hegemony in the British Isles if she allied dynastically with his family. At this moment, they could not have been more wrong. Philip's entente with Elizabeth had survived the setback (from a Spanish perspective) of the 1559 Protestant religious settlement: his relations with England were still cordial. He privately knew that he could depend on Cecil and his ally Lord James to keep Mary in her place. Catherine, too, was bound to Elizabeth and Cecil, and for some fifteen of the next twenty years, their rapprochement enabled Cecil to attack the Guises while staying on good terms with Catherine and her children.

When the forty days of official mourning ended, Mary withdrew six miles into the countryside. She used this period of privacy and retreat to collect her thoughts and reconstitute her household from that of a queen to a dowager queen of France. It was a significant change, one she planned with her return to Scotland in mind, because she chose as her new advisers those with recent experience of the country. The most important of her appointments was Henri Cleutin, Seigneur d'Oysel, formerly her mother's chief lieutenant, who was made a knight of honor, the equivalent of a lord-in-waiting. Next in rank was Jacques de la Brosse, the French ambassador to Scotland at the time of Henry VIII's Rough Wooings and a long-serving Guise client who had accompanied her to France in 1548.

Mary's appointments confirm that she was eager to return and wanted to learn much more about her country than she had gleaned from Ptolemy's *Geography* in the schoolroom. D'Oysel had recently married a beautiful young Parisian woman. He had no further desire to live and work in Scotland, but in advising Mary in this interim phase, he made a vital contribution, since only a native Scot knew more about noble factionalism and the habits and idiosyncrasies of the individual lords.

D'Oysel briefed Mary fully on the revolt of 1559–60, urging her to give credence to Lord James despite his prominent role in her mother's deposition. His undoubted treachery aside, he was the most capable of the lords and the only one who could hold them together. So far, he had applied all his wit to his own private ambition, which he had cloaked under the pretense of religion. Lord James was a Protestant, but a pragmatic rather than an ideological one. He was not a hard-line Calvinist, and was no friend of Knox. The trick for Mary would be to turn her illegitimate brother from a virtually autonomous agent into a royal servant. If she could do that, she could succeed in ruling her country as well as any of her royal predecessors.

In February 1561, Catherine de Medici and the court moved from Orléans to Fontainebleau as they usually did about this time of year. Mary ended her retreat a few days later, and on the 16th received Throckmorton and the Earl of Bedford, whom Elizabeth had sent with letters of condolence for her.

It was a successful audience. Mary read Elizabeth's letters, then "answered with a very sorrowful look and speech." Bedford was to thank his mistress "for her gentleness in comforting her [Mary's] woe when she had most need of it." Elizabeth, said Mary, "now shows the part of a good sister, whereof she [Mary] has great need." As she spoke, Mary gained in confidence and reiterated her idea of a fresh start after the debacle of the treaty of Edinburgh, saying she interpreted Elizabeth's letters as a positive gesture and would strive to match her goodwill. She invited the ambassadors to return whenever they wished, asking d'Oysel to escort them back to their lodgings.

They reappeared on the 18th, when Mary repeated her desire for "amity." She and Elizabeth, she started to say, were two queens "in one isle, of one language, the nearest kinswomen that each other had . . ."

But the ambassadors interrupted her, raising the vexed issue of the treaty of Edinburgh, which they insisted be ratified "without delay." It was a peremptory demand, and Mary balked. Perhaps Bedford, Cecil's close ally and one of his most trusted friends, imagined that she could be bludgeoned into submission. If so, he had seriously misjudged her.

She refused to be intimidated, anticipating Elizabeth's own later and more celebrated tactics when browbeaten by her ministers or Parliaments. "She was," Mary said disingenuously of herself, "without counsel." The Cardinal of Lorraine was absent, and none of her Scottish nobles had yet arrived, although some were on their way. Their "counsel and advice" were essential, since "the matter was great for one of her years."

Pressed by Throckmorton, Mary countered that she "was not to be charged" for contracts to which she was not a party. She turned the argument around, accusing Elizabeth of breaking her own agreements, because although she had accepted Mary's portrait in her *deuil blanc* the previous year, she had not sent her own in return. Mary now asked Throckmorton to bring it as he had promised.

It was a shot across the bow, because the exchange of portraits was, in Mary's view, symbolic of her offer to make a fresh start. If the portrait was to be withheld and demands for ratification of the treaty continued, then by implication all subsequent bets were off.

The ambassadors tried again on the 19th, when Mary stonewalled them. She recited the aphorisms on the duties of rulers she had learned from her tutors, saying she must always be "advised" on matters touching her crown and state, turning prudence into an excuse for delay.

She also scored a palpable hit. She cautioned Bedford that if she treated her nobles as contemptuously as he appeared to be suggesting by acting without their advice, she could only expect them to behave as badly in the future as they had done in their revolt against her mother.

If Mary kept her dignity, however, she made no progress beyond this. Elizabeth, if left to her own devices, would almost certainly have offered concessions before very long, since, like Mary, she was well aware of the clandestine nature of the treaty of Edinburgh.

But there was a more formidable obstacle. Mary's most determined opponent was Cecil. To him, Mary's refusal to ratify the treaty was tantamount to a hostile act; it meant in his opinion that she had refused to recognize Elizabeth as the rightful ruler of England, in which case he was more convinced than ever that she was the prime mover of an international Catholic conspiracy to depose and kill his queen.

Mary left Fontainebleau in the middle of March, beginning a three-month tour of her family to say goodbye. It was a gesture typical of her, and yet she received a mixed reception. Her aunts were sympathetic. They could see that she had taken a bold, brave step: she would have no mother to greet her in Scotland and was leaving almost all she knew behind. Her uncles were far from understanding or approving, and perhaps this was why there was so much confusion over the itinerary for these farewell visits.

Mary went first to Rheims, where she was met at the city gates by her uncles and grandmother, but stayed with her aunt, the Abbess Renée, at the convent of St.-Pierre-des-Dames. After three weeks there, she had just set out for the Duke of Lorraine's estates at Nancy when she was overtaken at Vitry-le-François by John Lesley, a young Catholic lawyer who had ridden posthaste from the Netherlands, and at St.-Dizier a few miles down the road by Lord James, who had come from Calais. They had been sent to her by the rival factions in Scotland, as it was expressly said in the case of Lord James, to "grope her mind" before she set sail for home.

Their soundings were the coordinates for her return. Lesley, who arrived first by a whisker, was the emissary of the Catholic contingent. Led by the Earl of Huntly, the murdered Beaton's successor as chancellor and the head of the Gordon family, this group included the Earls of Atholl, Cassillis, Caithness and Crawford. A powerful faction, almost as powerful as the Protestants and far more united, they had convened at Stirling, where they urged Mary to return home unconditionally.

Lesley warned Mary against Lord James, whom he denounced as treacherous and a rebel, a man attacking Catholicism with the intention of overthrowing the monarchy. She ought to have him imprisoned in France, or else she should land in Aberdeen, where her loyal Catholic nobles would meet her with an army of twenty thousand men and march with her to Edinburgh.

Mary sensibly rejected this advice, which was a recipe for civil war. She turned instead to Lord James, the Protestants' representative. His mission was to capitalize on the success of the Lords of the Congregation, who in August 1560 had successfully outflanked the Catholics and convoked a Parliament declaring Scotland to be officially Protestant.

But Lord James was calculating the odds: he had also come to safeguard his own position. For this he was accused by Lesley and even by some of his own side of conspiring "to make himself great." As he talked intently to Mary over the next few days, he came to sense her determination to return to Scotland. Although his own strong preference must have been for her to stay permanently in France, especially given his role in the treaty of Edinburgh, he could see that she had made up her mind.

Throughout their conversations Lord James marketed himself as reliable and incorruptible. Whatever his ambition, he had a keen intelligence, a rare ability to bring people together and predict their reactions. He could even build bridges to the Catholics. And he was, after all, a blood relative, for Mary a strong and important bond.

She talked to him for five days until he steadily won her confidence. They agreed that she would follow his advice and maintain the religious status quo, the existing uneasy balance between Catholics and Protestants, but one in which Protestantism was recognized as Scotland's official religion, even if a majority of people were still privately Catholic and worshiped as such when they could. In exchange for this arrangement, Mary might hear the Catholic Mass in her own chapel at her palace of Holyroodhouse.

In making these terms, Lord James supped with the devil. His compromise was even more abhorrent to the Calvinists than to the Catholics. John Knox, the doyen of Calvinist preachers, and his allies the young Earl of Arran and Lord Lindsay, had already threatened to rebel again if Mary refused to submit to the demands of the reformed Kirk* and if the "idolatry" of the Catholic rite was allowed to persist in any form.

But when French troops evacuated Scotland, in accordance with the conditions of the treaty of Edinburgh, much of the fear of Catholicism evaporated in Scotland and so did most of Arran and Lindsay's support. Many of the former Lords of the Congregation were inclined toward the Reformation, but of these only a small minority were genuine Protestants. And very few were outright Calvinists.

Arran, meanwhile, had become a figure of fun. His suit to marry Elizabeth had been rejected: it had been as forward as it was insolent and presumptuous. As if that were not enough, when the English queen had refused his offer, he set his sights on marrying Mary. He convinced himself he had fallen in love with her, even though he may have seen her only once, briefly, in Paris and from a distance, believing she was a woman who could easily be forced into marriage. It was an arrogant and absurd policy that completely misjudged Mary's character and made Arran look like a fool.

Lord James had little to fear from Arran's quarter. Moreover, to be excoriated by Knox was an attraction where Mary was concerned. She had never seen her brother as a Calvinist. His conversion to Protestantism was largely political, and within two years he would be quarreling with Knox almost as often as she did.

Mary saw eye to eye with her older brother. She even considered appointing him acting governor of Scotland, until she was informed afterward by d'Oysel that he had gone to meet Throckmorton in Paris as soon as he left her, not to mention briefing Cecil in London on both the out-

---

* "Kirk" is the Scottish word for "church," before and after the Reformation. It designates either a building or the institutional church, but is used in this book chiefly to mean the Protestant Reformed Church.

ward and return legs of his journey. When she learned this, she had second thoughts about the regency, but stuck to the rest of their agreement.

Mary had no better option than to make terms with her half-brother. In her final days in France, she was trying to identify those who would be her leading councilors as a reigning queen of Scotland. She had to attempt this blindfolded, and all she could do was trust her instincts. It was not a difficult decision in the end. Lord James already led the council of twenty-four nobles that by the treaty of Edinburgh was the lawful government of Scotland. He had personal charm and outstanding diplomatic skills: he was the adviser most likely to be able to build up a broad consensus.

Mary knew that aligning herself with Lord James would mean accepting his cronies as her advisers. His closest allies were William Maitland of Lethington, the secretary of state to the council of nobles, and James Douglas, Earl of Morton. They formed an axis that — thanks to Lord James's journey to St.-Dizier — would prove to dominate policymaking in the formative years of Mary's rule in Scotland.

Maitland was the cleverest of the three. He had been Mary of Guise's secretary, but defected to the Lords of the Congregation, who sent him to London as their mouthpiece. A genuine and deeply committed Protestant, he had done more than anyone else to steer the Scottish Reformation legislation through Parliament. He formed a close bond with Cecil, at whose London house he stayed. Their friendship was based partly on their shared religious beliefs and partly on their mutual admiration for classical literature. Maitland was known in England as "the Scottish Cecil" and in Scotland would come to be known as "Mekle Wylie" (or Much Wily), a pun on "Machiavelli." Neither nickname was meant to be complimentary, but acknowledged his political suppleness.

Morton was the most dangerous and least complicated. Vindictive, harsh and cruel, he was also a sexual predator who fathered four illegitimate sons and a daughter. Notably rapacious in fiscal matters, he was a technocrat who rose on the strength of his administrative ability and territorial power as the head of the Douglas family. In Edward VI's reign, he was a prisoner in England, where he tasted the Reformation and acquired an English accent. He vacillated in his support for the revolt against Mary of Guise, sitting on the fence until he was sure that the Lords of the Congregation would be victorious.

As soon as Lord James departed home for Scotland, confident of his success, the jockeying for position began. Morton wrote a memo justifying his role in the lords' revolt, which his cousin Archibald Douglas delivered to Mary in person. He blamed others for the "wrong information" she had received about him. Unfortunately he spoiled the effect by asking her to ratify his claim to the estates of the earldom of Angus, which he held in trust for his nephew with a disputed legal title. And Mary ignored the request.

Maitland wrote next, offering his "faithful service" and complaining about "calumniators" and "talebearers." He too was seeking to exonerate himself, in his case against the charge that he was too much Cecil's lackey.

Mary sent an illuminating reply, a remarkably clear-sighted one for an eighteen-year-old. She told him she was willing to forget the past and would judge him solely by his loyalty from now on. She told him candidly that she saw him as the "principal instrument" of all the "practices" attempted against her, and advised him to curtail his "intelligence" with Cecil. "Nothing," she said, "passes among my nobility without your knowledge and advice. I will not conceal from you that if anything goes wrong after I trust you, you are the one that I shall blame first." Her letter does more than anything to explain why Maitland reinvented himself over the next two years as Mary's loyal servant. He was soon working harder for Mary than he was for Cecil, but typically hedged his bets by sending a transcript of Mary's letter directly to Cecil by courier.

Lord James and his allies knew that the treaty of Edinburgh would shortly become a major stumbling block in their relationship with Mary. Its final surveillance clause required her to ratify and fulfill all of its conditions, failing which England might intervene at any time in Scotland — a threat that was now a serious cause for alarm. To deal with it, Maitland drafted a blueprint for a fresh concordat with Elizabeth and Cecil, one in which Mary would renounce her immediate Catholic claim to the English throne and recognize Elizabeth as England's lawful queen during her own lifetime. It was to be a more or less straightforward trade: in exchange for surrendering her immediate dynastic claim, Mary would be recognized in England as "the second person of the realm." She would become heir apparent to the English throne, and would succeed to the throne if Elizabeth died unexpectedly without children of her own.

Maitland was trying to engineer a compromise even while Throckmorton was still badgering Mary to ratify the treaty in its original form.

He argued that the effect would be the same as if she had ratified the treaty. She would have acknowledged Elizabeth to be England's legitimate queen. His idea was that with Cecil's agreement, Mary should travel home to Scotland through England. She would cross from Calais to Dover, where she would be greeted by a Scottish delegation and taken to London. While there, she would be able to meet Elizabeth face to face and conclude the final terms. In Maitland's words, such an interview between the queens "shall breed us quietness for their times."

After visiting her relations at Nancy, Mary returned to Joinville and Rheims, and finally to Paris, where she arrived on June 10, 1561. She was received with honor and lodged at the Louvre. On the 18th, Throckmorton reappeared to try his luck again. Mary replied that she meant to delay her "resolute answer" until she had the advice of her lords, which would not be until after her arrival in Scotland. This would, she assured him, be soon. She meant to take her departure "very shortly." D'Oysel was to be sent to Elizabeth with a message, and to request a diplomatic passport in case through storm or illness Mary's galleys were forced to take shelter in England. Beyond this, she did not propose to commit herself.

Throckmorton was in a quandary. His instructions bound him to persuade Mary to ratify the treaty as a condition of his own recall to London. He therefore called a spade a spade. Mary's insistence on obtaining her lords' approval before the treaty was ratified was, he said, ridiculous. It was these very same lords who had negotiated it. That, retorted Mary, was now less certain. Things, she suggested, were changing: "It will appear when I come amongst them whether they will be of the same mind."

It was not the answer Throckmorton needed. And it only got worse when d'Oysel arrived in London to request the passport. Although such documents were issued routinely for diplomatic travel, Elizabeth flatly refused to grant Mary or her companions any safe-conduct unless she first ratified the treaty. When d'Oysel was asked point-blank when that would be, he declined to answer, thereby causing a furor.

Elizabeth had been advised by Cecil. On July 14, following the clash with d'Oysel, he spoke his mind to Throckmorton. If Mary relinquished her immediate dynastic claim before leaving France, then, providing she also ratified the treaty as the surveillance clause required, she might be recognized as Elizabeth's heir. She might travel to Scotland via Dover, and a personal interview between the two queens could be arranged, which, as d'Oysel was curtly reminded, would be "a friendly meeting" to confirm the "amity."

Cecil had been corresponding with Maitland, but there was a fatal divergence in their approach. Whereas Maitland sought to "allure" Mary toward a compromise with England, believing she might become Elizabeth's heir without ratification of the treaty, Cecil still required her not only to renounce her claim but to ratify as well.

In effect, he vetoed the Scottish scheme. Without ratification, there would be no deal. As if this were not enough, his innermost thoughts were betrayed when, after conceding Mary might be recognized as Elizabeth's heir apparent, he added: "Well, God send our mistress a husband, and by him a son, that we may hope our posterity shall have a masculine succession."

Even if a settlement was reached, Cecil intended to do all in his power to persuade Elizabeth to marry and have children, thereby excluding Mary from the English succession forever. "This matter," he concluded, "is too big for weak folks, and too deep for the simple; the Queen's Majesty knoweth of it, and so I will end."

On July 20, Mary's farewell tour ended at St.-Germain, where a fête was held in her honor lasting four days. Verses by Ronsard celebrated her as "this beauty, honor of our times, who makes both kings and peoples content."

She withdrew from St.-Germain on the 25th and, accompanied by her uncles and a large retinue, headed north for the channel ports. False information about her itinerary was given out to confuse English spies: the tactic succeeded, because although Elizabeth relented at the last moment and issued a safe-conduct, it could not be delivered and Mary traveled without it.

Mary reached Calais on August 10. There she rested for four days before boarding her galley on the morning of the 14th. Her departure was organized by her uncles, one of the last pieces of stage management they would ever undertake for her. They had arranged for the Marquis d'Elbeuf and two of his older brothers to attend her during the crossing. Constable Montmorency's son, Henry, was also with them. He returned to France a month or so after Mary arrived in Edinburgh, and she wrote to thank his father for his courtesy. Her diplomatic escort was Michel de Castelnau, Sieur de Mauvissière, an astute diplomat and intellectual who later played a key role as one of the French ambassadors in Scotland and England and who described Mary in his *Memoirs* as a "naturalized Frenchwoman . . . not just the most beautiful, but the most elegant of all her sex, both in speech and good manners."

Also on board Mary's galley were her loyal companions the four Maries, whom she had teased so mercilessly on their outward journey thirteen years before. A second galley was provided for the rest of her staff, with a flotilla of more than a dozen vessels for the baggage, some of them merchant ships chartered from the Dutch. So many ships were needed because on top of Mary's tapestries, furniture and several hundred coffers stuffed with gowns, gold and silver plate, paintings and other works of art, vast quantities of bed linen and other stores and equipment were transported, including one hundred horses and mules.

She sailed out of the harbor at around noon, watched by one of Throckmorton's servants, who had finally tracked her down. As on her outward voyage, the captain of her galley was Nicolas de Villegagnon. And just as there had been false starts and a smashed rudder when she left Scotland, on this occasion also, as her flotilla attempted to leave the port, there was an accident. At the narrow entrance to the harbor, a badly steered ship collided with another boat, which sank. Mary rushed forward, calling on Villegagnon to rescue the drowning sailors and offering a typically generous reward to anyone who succeeded. But the ship had plunged to the bottom and there were no survivors.

"What an omen is this?" asked Mary as the signal to hoist sail was given again and the journey resumed. She flatly refused to go down to her cabin, and instead the four Maries prepared a bed for her on the poop deck, where she spent her first night at sea.

Since the shock of her double bereavement in 1560, Mary had acted with courage and composure, especially in her dealings with Throckmorton. But as the coast of France disappeared into the haze, she lost her nerve and burst into tears. With her elbows leaning on the stern rail, she sobbed her heart out, her eyes fixed on the speck of land she had just left as it finally slipped out of view. Her last words as her galley sailed into the open sea were "Adieu France. It's all over now. Adieu France. I think I'll never see your shores again."

9

Into the Labyrinth

MARY'S ARRIVAL in the land of her birth took her subjects somewhat by surprise. Her voyage lasted barely five days, almost a record for the crossing and up to a week less than anyone had expected. As soon as Villegagnon's flotilla had entered the open waters of the North Sea after leaving the English Channel, the captain had ordered the two smaller and much faster galleys under his command to sail ahead, leaving the slower transport ships to follow at their own speed.

On Tuesday, August 19, 1561, at six o'clock on what by all accounts was an unusually damp and depressing morning, the galleys sailed up the Firth of Forth and anchored at the entrance to the harbor at Leith. A thick haar, or cold sea mist, cloaked the shore as they arrived, so the sailors were told to wait until it dispersed before tying up at the wooden pier. Mary disembarked shortly before ten, just as the sun was starting to pierce the haze. It ended an absence of thirteen years: she had left a Scots child and returned a French woman.

She washed and rested at a nearby house for an hour or so. Then she rode with her retinue up the steep hill from the port to the town of Edinburgh, and from there down the Canongate to Holyroodhouse, the most magnificent of the palaces of Scotland. The royal apartments were situated there, in the great fortified tower her father, James V, had constructed to eclipse the older buildings on the grounds of the ancient Abbey-Kirk.

It was hardly the homecoming of which Mary had dreamed. She had reached the shores of Scotland before anyone had thought it necessary to organize a welcome party, and her transport ships were left behind. One carrying her horses, including her favorite state palfrey, had been intercepted by an English naval patrol on suspicion of piracy and diverted to Tynemouth, where the horses were detained for a month before being allowed to proceed by land to Edinburgh. Mary was forced to make a lackluster entry into her capital without the familiar trappings of monarchy and on borrowed horses. She was unable to use her own saddles and bridles, which lay in her coffers on the slower chartered vessels.

Not everything was a disappointment. Although Leith was deserted when she arrived, Villegagnon had the galleys fire their guns, which quickly produced a crowd to watch the show. By evening the celebrations had started, and even John Knox had to admit that the people rejoiced with bonfires and music. A company of "honest musicians," he noted with grim satisfaction, "gave their salutations at her chamber window." They were, of course, his fellow Calvinists, who aroused rather different reactions among the members of Mary's entourage. The Catholic Pierre de Bourdeille, who traveled in her suite, complained that when the queen wished to go to bed, several hundred "knaves of the town" congregated under her window, playing fiddles and other stringed instruments and singing psalms "so badly and out of tune that nothing could be worse."

On August 31, almost a fortnight after Mary's return, the provost of Edinburgh finally got his act together. For the queen and her retinue, including her Guise relations, he mounted a civic banquet in Cardinal Beaton's old mansion. The next day, Villegagnon set sail again for home, accompanied by those of Mary's escorts who did not wish to stay longer in Scotland than their official duties required.

When Mary first walked through the largely empty rooms of the royal apartments at Holyrood, she must have felt excitement at taking up her role as queen. She had been trained for such a moment since birth. There would have been sadness too, because these rooms had been unoccupied since her mother's death the year before. The furniture had been stored away, and Mary's own had not yet arrived.

Holyrood was unfamiliar to her. She had never seen it as a child. Now that she had arrived there, she would instantly have noticed that her father's architectural plan was in the French style. The James V tower in which she lived was modeled on the Francis I tower of the château of Chambord, where she had stayed many times. The new west front of the palace with its great expanses of glazing and ornamental carvings was most likely an imitation of the famous terrace at Chambord. It had been built in honor of Madeleine, Francis I's daughter and James V's first wife. Then, when she died and James married again, the vast gardens were laid out by Mary's mother after the patterns she knew so well from her many visits to St.-Germain, Fontainebleau, Amboise and Blois.

Mary found herself to be in more congenial surroundings than she might have expected. She would have noticed very little change in her food and daily routine, because she had brought a skeleton domestic staff with her from France, including *valets de chambre*, ushers, chefs, pantry staff, a tailor, chaplains, doctors and an apothecary. She kept up her French establishment throughout her reign in Scotland, paying for it out of her revenues as dowager queen of France, and it steadily grew to be more than 170 strong.

One thing she discovered to be different was the climate. Scotland was neither Paris nor the Loire Valley. The summer heat lacked the searing intensity of the Paris basin or the Touraine; the winter cold was far more severe than in Picardy or Normandy. Spring came six weeks later, and autumn a month or so earlier.

Her escorts remarked on the poverty of the Scottish people. Bourdeille, with his tendency to exaggerate, predicted it would be easy for Mary to dazzle those who were so poor: the very sight of their penury, he said, had brought tears of sorrow to her eyes when she had first arrived. And Cecil, forever keen to gather news of her reception, heard that "the poverty of her subjects greatly advanceth whatsoever she intendeth."

This was not the whole story. There were more beggars in France than in Scotland. The difference was less the humble, hand-to-mouth existence of the Lowland tenant farmers and Highland cottagers than the absence of a strong urban bourgeoisie outside Edinburgh. The towns and burghs were so small as to be almost inconspicuous. What was really noticeable was the extreme contrast between the cottages and huts of the rural masses and the castles and tower houses of the nobles and lairds — the gulf between rich and poor. Rumors of Mary's vast collections of dresses, diamonds, pearls and jeweled embroideries, her gold and silver plate, her furniture and tapestries, had preceded her return. It was the sheer glamour of a Scottish queen who could rival anything seen at the ruling courts of Europe that most likely overawed her less well-off subjects rather than her wealth as such.

Soon Mary felt sufficiently confident in her new home to stage a triumphal entry into Edinburgh. She wanted to compensate for the lack of spectacle she felt had detracted from her arrival. The *entrée royale* was a core element of the cult of monarchy in France, and she decided it was now the best way to introduce herself to her people.

At eleven o'clock on the morning of September 2, she left Holyrood by a back way, skirting around the town on the north side and reentering through a specially made opening in the town wall. She then rode in state up the hill to the castle flanked by her leading nobility. When they entered the great hall of the castle, a state banquet was served at which Mary was the host and civic dignitaries the guests of honor. As she left, the guns of the citadel thundered a salute.

This completed the formalities of the *entrée*. There followed civic pageants laid on by the Edinburgh town council to welcome her home. When Mary rode down Castle Hill, she was met by fifty young townsmen dressed up as "Moors," clad in yellow taffeta with black hats and masks and bedecked with jeweled rings and gold chains. A procession was formed in which the nobles and "Moors" led the way, after which came Mary, flanked by civic officials who bore a purple canopy aloft above her head. "Moors" and "Turks" were the favorites of the crowds at spectacles of this sort, and companies of "Moors" had appeared in processions designed to honor Mary of Guise when she was regent. The town council had enough of their costumes already in stock, and decided to use what was available.

The procession advanced in stages. When Mary reached the Lawnmarket, she was treated to the first in a series of pageants and moral allegories. A triumphal arch had been built out of wood and painted in rich colors. From a gallery above the arch, a choir of children sang "in the most heavenly wise," and as Mary passed underneath, a mechanical globe constructed to look like a cloud opened, from which emerged a "bonny bairn" who descended on a rope as if he were an angel. He first handed the keys of the town to Mary, then presented her with a Bible and a psalter, bound in fine purple velvet. After handing her the books, he recited some verses:

Welcome our sovereign, welcome our native queen,
Welcome to us your subjects great and small,
Welcome I say even from the very spleen
To Edinburgh your city principal.
Whereof your people with heart both one and all,
Doth here offer to your excellency
Two proper volumes in memorial,
As gifts most worthy for a godly prince . . .

Mary had kept up her Scots, so understood every word. She willingly accepted the gifts, but when the recitation began, she frowned and with some panache passed them to Arthur Erskine, the captain of her guard, who stood nearby. He was a diehard Catholic who had served her in France as her butler. She knew at once that the presentation was more than it seemed: the phrase "godly prince" was code for a Protestant one, and to the Huguenots and other Calvinists, the Bible and psalter symbolized their sermons and psalm-singing. Mary realized that she was being petitioned to adopt a Calvinist religious policy, and she decided to move on.

The procession advanced toward St. Giles Kirk along the High Street. When it reached the Tolbooth, the scene of state trials and Parliament's usual meeting place, a dumb show was performed in which a group of young virgins — their parts played by boys — represented Fortitude, Justice, Temperance and Prudence. These virtues were extolled by a narrator as those that a ruler should have, which was uncontroversial, as this pageant, unlike the previous one, was based on classical literature and was an almost identical repeat of one previously put on for Mary's mother.

Mary stopped next at the High Cross fountain, where a third pageant was enacted. It, too, was accompanied by a commentary, but this time the words were inaudible. Unruly crowds were gathered there, because wine instead of water gushed from the fountain's spouts. Almost nothing could be heard above the noise of breaking glass and drunken revelry. Mary acknowledged the cheers of the drinkers and quickly moved on again.

Farther down the High Street at the Salt Tron, one of several public weigh-houses where commodities sold in the market could be put on the official scales, a fourth tableau was in preparation. Choreographed by the Calvinists in a threatening way to show a Catholic priest burned at the altar in the act of elevating the host, it was stopped by the Earl of

Huntly, who carried the sword of state at the front of the procession and got there well ahead of Mary. In its place, a revised scene was hastily improvised in which effigies of three Israelites were burned for defying Moses, which satisfied the Protestants but also delighted the Catholics, who took it as an allegory against blasphemy.

A fifth and final tableau was staged at the Netherbow Port, the gatehouse at the end of the High Street, which marked the eastern boundary of the old town and the point where the Canongate and the court sector began. Here, after a speech, a dragon made of canvas and papier-mâché was set on fire and a psalm sung while it blazed. This was an apocalyptic theme: the dragon was perhaps an allusion to the pope as antichrist, and it is unlikely that Mary approved.

On arriving at Holyrood, some children who had followed Mary in a cart sang a psalm and chanted yet another attack on the Mass. The provost and civic leaders then produced a coffer packed with gilt plate, which they humbly presented to their queen. She thanked them graciously, studiously ignoring the chanting, after which the procession dispersed and everyone went home.

It had been a stiff, decidedly awkward celebration, an expression of genuine joy and delight on the part of the vast majority of the nobles and people at their queen's return, but also a blatant attempt by a Calvinist minority to dictate her religious policy.

Mary reacted prudently. She could see things in perspective. The Calvinists claimed the Mass was "terrible in all men's eyes." But this was hyperbole. "All men" were not Protestant. The religious divisions in Scotland were no different from those elsewhere in Europe. Mary could judge this from her knowledge of the Huguenots in France. The challenge was to manage things in a way that averted a religious war. Outside Edinburgh and the towns, Protestantism was far from entrenched. Many remnants of the old Catholic system survived as if nothing had ever disrupted it. In Edinburgh, it was easy to find the Mass celebrated openly at Easter: the official Reformation had gone too fast for most people. At other times of the year, Catholics still heard Mass in their own houses and even their local churches, especially in the remoter areas of the country.

As to the former Lords of the Congregation, only Châtelherault and his son Arran had staged a boycott of Mary's *entrée* on religious grounds. The ordinary people of Edinburgh had greeted her warmly as the first adult reigning ruler of Scotland for twenty years. They were not put off by her Catholicism. The Calvinists might dominate the town council, but were of infinitesimal size in relation to the overall population of Scotland. And even the Edinburgh Calvinists had welcomed Mary, if ambiguously. That in itself was something of a triumph. It was certainly against the wishes of their leader John Knox, who had dismissed the day's proceedings as verging on idolatry. "Fain," he wrote, "would fools have counterfeited France."

Mary decided to confront Knox, nipping the threat in the bud. She reckoned that with Lord James and his allies by her side, it was the Calvinists and not a majority of the lords who were most likely to try and oppose her. She was beginning to work out the values and honor systems of the Scottish nobles, which she knew from their treatment of her mother stemmed in most cases from ambition and opportunism more than from religious principle, but which they justified to themselves as protecting Scottish national interests. Lord James was shrewder than most of the nobles, and he was no friend of Knox. If anything, he agreed with his sister's opinion of him.

Since returning to Scotland from Geneva, Knox had been the minister at St. Giles Kirk, the most influential pulpit in the country, where he preached against the pope and the Mass every Sunday for two or three hours at a time. He spoke in the language of prophecy and saw himself as a latter-day Jeremiah. A clash with Mary was inevitable. Lord James was unable to restrain him, since Knox was still smarting from the slight he felt he had received when Mary's half-brother had ignored his demands that she be required to conform to the wishes of the reformed Kirk when discussing the conditions of her return.

Matters came to a head on Mary's very first Sunday at Holyrood. While she was hearing Mass in her private chapel, a fracas erupted in the courtyard outside as Patrick, Master of Lindsay, the eldest son of Knox's ally Lord Lindsay, brought his friends to heckle, shouting for the "idolatrous" priest to be killed. To protect Mary, Lord James stood sentry at the door, but the servant who carried the altar furniture was severely jostled, his candles snatched and "trodden in the mire."

Next day, Mary issued a proclamation declaring her resolve, on the advice of her councilors, to make a final order for "pacifying" the differences between Catholics and Protestants, which she hoped "would be to the contentment of the whole." Meanwhile, "in case any tumult or sedition be raised," she would preserve the status quo. No one should attempt to alter the state of religion "which Her Majesty found publicly and universally standing" upon her arrival. In addition, no one was to harass or molest her servants or any member of her retinue, whether Scots or French, "for any cause whatsoever."

Arran lodged a protest. He complained that Mary's subordinates were being allowed to avail themselves of a concession granted solely to her. If any members of her household attended Mass in her chapel, he held that they had committed "idolatry," an offense "more abominable and odious in the sight of God" than murder or assassination.

Mary ignored this bluster, summoning Knox, the main author of the campaign against "idolatry," to justify himself. It was a risky strategy, but she had been practicing her debating skills on Throckmorton.

Knox arrived on September 4, when Mary received him attended by Lord James. She cut straight to what she believed to be the point. This had little to do with theology and everything to do with armed political resistance. She was well aware that while in exile at Geneva, Knox had written *The First Blast of the Trumpet Against the Monstrous Regiment of Women,* a diatribe against female rulers, published in 1558.

The gist was that "idolatrous" rulers — whether male or female, but especially if female — could be overthrown by force. Knox had asserted that female monarchy was "repugnant" to God and Scripture, and a woman ruler was "a monster in nature." The Old Testament, he claimed, provided the necessary precedents, notably Queens Jezebel and Athalia, both victims of legitimate regicide.

A year after publishing the *First Blast,* Knox had applied his theory to Scotland. He argued that the regent, Mary of Guise, and even the Queen of Scots herself, could lawfully be deposed by the nobles. So far, his dilemma had been Saint Paul's defense of authority: "Let every soul be subject unto the higher powers. For there is no power but of God: the powers that be are ordained of God" (Romans 13:1). Knox was honest enough not to ignore this text, but found it tricky to negate. He at last circumvented it by noting that the word "powers" was in the plural and therefore had to mean more than just the ruler (or "superior power") alone. He claimed that since multiple powers were intended, then the nobility (whom he designated as the "inferior power") had to be included as well as the ruler. Both "powers" were legitimate and both "ordained of God." In which case, concluded Knox, the nobles (or "inferior power") could resist and, when necessary, depose the ruler in a godly cause, because by resisting an "idolatrous" ruler and demolishing "the altars of

Baal," they were in fact "obeying" God's ordinances and "fulfilling" the Ten Commandments.

Knox had made a conceptual leap, turning him from a theologian into a resistance theorist. Mary began the interview by accusing him of inciting her subjects to armed revolt. To this he answered that all he had done was to profess the faith of Jesus Christ.

"You think then," said Mary, "that I have no just authority?"

"Please Your Majesty," protested Knox, "learned men in all ages have had their judgments free." He claimed the right to hold his opinions, adding, "If the realm finds no inconvenience from the rule of a woman, that which they approve shall I not further disallow . . . but shall be as well content to live under Your Grace as St. Paul was to live under Nero."

Mary was stung to be compared to the Roman emperor most berated for his tyranny. She demanded an explanation from Knox, who tried to limit the damage by claiming the *First Blast* had been directed against one particular queen, Mary Tudor of England. She had been a special case, because by repealing the Protestant legislation of her brother Edward VI, she had broken what Knox regarded as the "covenant" made between the English nation and God, and so could be deposed.

Mary saw this as specious. She was discussing her mother and herself, not Mary Tudor. Moreover, the *First Blast* had attacked all women rulers without distinction. As she reminded Knox, "You speak of women in general," proving she had actually read his book. She held her ground and demanded a plain answer. And she returned to her original question. "Think ye," she said, "that subjects having power may resist their princes?"

Finally, Knox answered. "If princes exceed their bounds and do against that wherefore they should be obeyed, it is no doubt but they may be resisted, even by power." He compared rulers to parents and argued that children might lawfully form a confederation to overpower and disarm a father "stricken with a frenzy," and "keep him in prison until his frenzy be overpast . . . It is even so, madam, with princes . . . their blind zeal is nothing but a very mad frenzy." Far from being unlawful, it was a positive act of "obedience" to resist forcibly and imprison rulers who disobeyed God, "until that they be brought to a more sober mind."

Mary was stunned by this speech, which was quite unlike anything she had heard in France. Not even Lord James, a man not normally at a loss for words, could break the silence. After a lengthy pause, she said, "Well then, I perceive that my subjects shall obey you, and not me; and shall do what they like, and not what I command: and so must I be subject to them, and not they to me."

When Knox departed, Mary wept in anger and frustration. She was on the defensive, wishing now that she had not issued the challenge. Yet she had kept her nerve. At least while Knox was in the room, she had shown no visible signs of weakness. In fact, his own impression of Mary was one she might well have taken as a compliment. As he told Cecil, to whom he could not resist sending a verbatim account, "her whole proceedings do declare that the cardinal's lessons are so deeply printed in her heart that the substance and the quality are likely to perish together." In other words, she had proved to be his equal in a quarrel!

The following week, Mary decided to make a royal progress to Stirling, and beyond to Perthshire and Fife. She wanted to see more of her country and her people, and to show herself to them. Far from being homesick for France, she seems to have felt that she had at last stepped into her proper place.

She stayed for two days at Linlithgow, her birthplace, before leaving for a nostalgic visit to Stirling, where her mother had spent so many years. Her sojourn there was brief but eventful. A candle on a table beside her pillow set her bed curtains alight in the middle of the night while she was asleep. Smoke filled the room, and she was lucky to escape before she was suffocated.

On Sunday, the issue of religion came again to the fore. Mary wished to attend a High Mass in the Chapel Royal, where as a baby she was crowned Queen of Scots. Her chaplains were making the arrangements when the Earl of Argyll and Lord James arrived and drove them away. Lord James was honoring his agreement with Mary, but to the letter. He had promised that she might hear Mass in her chapel at Holyrood. It was now clear he did not intend the concession to extend to any of her other palaces. He was hedging his bets after Mary's clash with Knox, keeping the Protestants at bay by appearing to forbid his sister's Masses, whereas in reality he was happy to allow them at Holyrood. That, of course, was still anathema to Knox, who wanted the Mass abolished completely. But it was acceptable to a majority of the other ex–Lords of the Congregation, including the most powerful of them, Argyll.

Mary's next stop was Perth, where another triumphal entry was staged. But although she was acclaimed by the ordinary people and presented with a "heart of gold, full of gold" by the civic authorities, she was once again lobbied by the Calvinists, who had copied Edinburgh's example. The tension was starting to build, and as she rode in the procession, she suddenly felt sick and had to go indoors to recover. The English ambassador, Thomas Randolph, who was accompanying her, described her illness as one of those "sudden passions" to which she was prone "after any great unkindness or grief of mind."

Over the next four and a half years, Randolph got to know Mary extremely well, providing a wealth of information about her. His weekly (sometimes daily) reports to Cecil are invaluable; but he must be seen for what he was: a partisan witness. He was a close ally and former student in Paris of George Buchanan, the brilliant classicist and poet, who was also a Calvinist, a republican and Mary's later vilifier.

Randolph was also a protégé of Cecil, who had used him to smuggle men and bags of untraceable gold coins into Scotland during the revolt of 1559–60, which he did under the code name of "Barnabie." He was then posted officially to Scotland, where he assisted the rebel lords and acted as their liaison with Cecil. He was the right man for the job, as he already knew Lord James and Knox, both of whom he had met in Paris. As a result, he understood the tensions between political pragmatism and religious conviction among the lords, barely disguising his own sympathies, which were closer to those of Knox.

As a resident ambassador in Scotland, Randolph enjoyed the customary diplomatic privileges. He was licensed to attach himself to Mary's court as it made its way from Stirling to Perth, from Perth to Dundee, and from Dundee across the River Tay on the ferry to Fife. After visiting St. Andrews for a week, Mary returned to Holyrood, where her first progress was judged a success. She had taken possession of her country despite the Calvinists' taunts. It was a delicate balancing act, but there was a more compelling reason why her policy was succeeding.

Her concordat with Lord James was paying dividends. Four days after her *entrée* into Edinburgh, Mary named her first Privy Council. A cross section of the nobles, it included the territorial magnates such as Argyll, Châtelherault and the Earl of Huntly. Seven out of twelve were Protestants. And the heavyweights whom Mary placed in her inner cabinet were Lord James and his supporters Maitland and the Earl of Morton. They were the lords who had steered the council of twenty-four nobles after the overthrow of Mary's mother. They had rebuilt their bridges to Mary before she left France, and were to play a decisive role in the tumultuous events of her reign.

When Mary arrived home, the trick was to engineer the transition of power. In 1559–60, the Reformation had combined with the innate factionalism of the lords to create a moment when the monarchy was suddenly vulnerable. When Mary's mother was deposed as regent, the government of Scotland had ceased to be that of the queen and become that of the lords, to the point where Cecil's clerk, filing letters from Scotland in his office in London, endorsed them "Letters from the States of Scotland." The word "States" had strong republican connotations, and the council of twenty-four nobles was to all intents and purposes a quasi-republican institution.

Mary knew she had to restore the monarchy's prestige. Her solution was to choose Lord James as her chief adviser, to preserve continuity, while subordinating him and his cronies to her authority as ministers and servants of the crown. This largely suited them, because their favored relationship to Mary enabled them to maintain their private power and factional interests exactly as before, at the same time indemnifying themselves against any possible reprisals for their part in the revolt against the regent.

Everyone was walking a tightrope. Mary's success was not a foregone conclusion. A fortnight before she left Calais, reality had dawned on the lords. Their queen was coming home and their role was about to change from that of near-autonomous governors to servants of a woman ruler. It was Mary to whom they would now be accountable. For this reason, a quite different relationship toward England would be required, because everyone knew that Mary would never agree to ratify the treaty of Edinburgh.

Cecil had already vetoed Maitland's earlier attempt to break the deadlock over the treaty and its surveillance clause. When Mary returned to Edinburgh by sea and not by land through England, the lords hoped that the issue of the treaty would fall by default. It did not. Lord James and his allies began to panic when Randolph showed them a letter from Cecil demanding his original terms. They felt they had made every effort to bridge the incompatible aims of Mary and Cecil, but had failed. Now "they need look to themselves, for their hazard is great."

But there was time for a final effort. In what looks like a concerted campaign, Lord James tried his luck with Elizabeth, and Maitland with Cecil. Both wrote carefully drafted letters designed to alter English policy and turn it around.

To Elizabeth, Lord James sent an obsequious if highly perceptive letter, regretting Mary had ever "taken in head to pretend interest or claim title" to the English throne. It had been a fatal mistake, caused by the bad advice of her Guise family, but events had moved on. If only a "middle way" could be found, "then it is like we could have a perpetual quietness." He reopened the case for a compromise: in exchange for renouncing her immediate dynastic claim, Mary should be "allured" by recognition as Elizabeth's heir. If the English queen would agree, he would attempt to bring Mary "to some conformity."

Cecil had already rejected this line, but that was while Mary was in France. Now that she was already on her way home, the matter was urgent, and it fell to Maitland to explain why. It would, he argued, be possible for Mary to divide and rule. Her arrival would transform her relationship with her subjects, many of whom would rally to a young and vivacious queen whose merest smiles and frowns would be enough to captivate them. Her Catholicism was unlikely to stop her, since Protestantism had yet to put down roots. She had the unconditional support of the Catholics, and it would be easy to win over many of the Protestants. The lords were unprepared for a fight. Too many were either "inconstant" or "covetous": they could be manipulated or bought off. Maitland's greatest fear was that Mary would pursue a Catholic policy. She was an "enemy to the Religion"; her return "shall not fail to raise wonderful tragedies."

These letters were carefully coordinated. While Maitland's stated the problem, Lord James's offered the solution. It is in this light that Mary's policy must be judged, because on her arrival she chose the "middle way" proposed by Lord James. As soon as she had consulted him, she sent Maitland to London to renegotiate the treaty of Edinburgh in favor of a new accord that maintained the "amity" with England but on terms that were not humiliating to herself or the monarchy.

In seeking to renegotiate the treaty, Mary's immediate dynastic claim was her bargaining chip. By renouncing it and recognizing Elizabeth as England's rightful queen, she could obtain recognition as her heir in return. It would be an honorable exchange. A dynastic claim that was valid in Catholic eyes but almost meaningless without the support of the pope and Philip II would be bartered for something tangible. Who knew how long Elizabeth might live? The result, claimed a jubilant Maitland, who worked out the finer details, would be to bind Mary to a perpetual "amity" with England.

When Maitland reached London, he did not mince words. "I think," he boldly confided to Elizabeth, "the treaty is so prejudicial to Her Majesty [of Scotland] that she will never confirm it." It was "conceived in such form as Her Majesty is not in honor bound to do it." He pointed out (somewhat hypocritically, since he had played a leading role in brokering the treaty) that it assumed Mary herself had authorized the negotiations when in fact she had not even been consulted. It was a point calculated to strike home with Elizabeth, who had never approved of Cecil's clandestine operations over the treaty.

Elizabeth always made it her priority to defend the ideal of monarchy. She now relented, allowing that if Mary would appoint commissioners to review the treaty, she would do the same. A conference would then be convoked and its agenda prepared jointly by Maitland and Cecil, the two secretaries of state. At the time, the decision was a breakthrough. Elizabeth had offered to settle despite Cecil's steadfast resolve that Mary should ratify the original treaty. It was a vindication of the policy of the "middle way" and of Mary's choice of Lord James as her chief adviser.

When Mary returned to Scotland, she was still only eighteen and faced the consequences of decisions made in her absence by others. She needed advisers she could trust, men of proven ability who were able to keep noble factionalism in check. By her concordat with Lord James she could subordinate the lords to her authority, and by preserving the status quo in religion she could secure an enviable degree of stability. In short, she could catch her breath while she learned more about the land she was born to rule. She had refused to ratify the treaty because it was in flagrant derogation of her honor. With the amendments proposed by Lord James and Maitland, a line could be drawn under the episode.

Mary entered a labyrinth on her arrival in Scotland, but so far she had successfully found her way. On his return to Edinburgh, Maitland delivered what previously would have been thought unthinkable: a eulogy of Mary. "The queen my mistress," he informed a mistrustful Cecil, "behaves herself so gently in every behalf as reasonably we can require. If anything be amiss, the fault is rather in ourselves." He even swiped at Knox. "You know the vehemency of Mr. Knox's spirit, which cannot be bridled, and yet doth sometimes utter such sentences as cannot easily be digested by a weak stomach. I would wish he should deal with her more gently, being a young princess unpersuaded."

Before her arrival, Maitland had imagined Mary to be an ideological Catholic whose return "shall not fail to raise wonderful tragedies." He now believed she "doth declare a wisdom far exceeding her age." After Francis II's death, the politics of the British Isles had been dictated by Cecil's agenda. But when Elizabeth conceded that the treaty of Edinburgh was renegotiable, the spotlight fell on Mary. Her proposal of a "fresh start" made shortly before she left France no longer looked naive. Her charisma could yet become her winning card. The benefits were potentially huge. "Surely," concluded Maitland's accolade, "I see in her a good towardness, and think the queen your sovereign shall be able to do much with her in religion, if they once enter in a good familiarity."

By the end of 1561, a solution to Scotland's problems seemed closer than at any point since Henry VIII's death. A young, beautiful and intelligent queen had returned to take up her throne, and within months was well on the road to success. The questions were: Would her charisma be enough, given the inequality between Scotland and England? And would Cecil still get in the way?

10

A Meeting Between Sisters

ON DECEMBER 5, 1561, the first anniversary of Francis II's death, Mary showed her respect for her late husband by putting the court at Holyrood into half mourning for two days. Black velvet was given to her chaplains for use in special Masses to be sung in her private chapel. She herself wore her *deuil blanc*, and at the solemn climax of the Requiem, she lit a great wax candle trimmed with black. The services were thinly attended except by the most loyal of her servants. Even Paul de Foix, the visiting French ambassador, found it prudent to stay away. Many of Mary's household were afraid they would be beaten up by gangs of Edinburgh Calvinists if they were there. None of the Scottish lords would attend, and when Mary asked those who were at court or starting to arrive in Edinburgh for the festive season to wear mourning clothes for a day, they all refused. Despite this, she was undeterred and no detail of the liturgy was omitted.

Then, as Christmas approached, the mood changed. Mary was once more behaving as her Scottish subjects expected her to behave, which meant presiding over a court renowned for its "joyousity." The hospitality and entertainments may have been on a smaller scale than those to which she had grown accustomed in France, but they were almost as glittering. As Thomas Randolph, the English ambassador, reported, "The ladies here be merry, leaping and dancing, lusty and fair." Soon he was almost overcome by it all: "My pen staggereth, my hand faileth farther to write . . . I never found myself so happy, nor never so well treated."

As soon as Mary's transport ships had arrived with her goods and furniture, she restored her palaces to a state of splendor and magnificence unseen in Scotland since her mother was sole regent. Ten cloths of state were for use as indoor canopies over Mary's throne. Five other canopies were for outdoor use: one of crimson satin was effectively a parasol to "make shadow before the queen" on hot sunny days. More than one hundred tapestries were unpacked and used to decorate the walls of the royal apartments and state rooms. Thirty-six Turkish carpets were rolled out on the floors. No fewer than forty-five beds were reassembled, fifteen of them trimmed with gold or silver lace or adorned with richly embroidered valances and bed curtains for the use of Mary and her guests.

Mary's tapestries were the glory of her collection. She had some twenty complete sets, often comprising between seven and a dozen individual pieces, enabling many rooms to be hung. One set depicted the History of Aeneas, another the Judgment of Paris, another the French triumph over the Spanish and papal forces at the battle of Ravenna in 1512. This third set was one of Mary's favorites. It followed her everywhere she went for the rest of her life, and her rooms at Fotheringhay Castle were decorated with it before her execution. A single tapestry was marked in the inventory as "not yet complete." It must have been taken from the factory in a hurry, and as it was impossible to finish it in Scotland, it was cannibalized to make an extra cloth of state.

The queen's gilded throne was high-backed and upholstered with crimson velvet and cloth of gold. The table in her presence chamber was painted and gilded. Low stools were set out for the four Maries and folding stools for important visitors, all of them covered in velvet. As many as eighty-one embroidered cushions were scattered around the rooms, which included thirty-three of cloth of gold, fifteen of cloth of silver and thirteen of satin or brocade. Cupboards or buffets were used to display gold or silver plate, glasses and decorative objects. Side tables had covers of the finest velvet, damask or cloth of gold, one embroidered with the lilies of France in gold thread. For banquets held in the great hall, Mary had two white linen tablecloths, each more than forty feet long.

Lastly, although Mary always preferred to ride on horseback rather than be carried in a litter or a coach, her luggage included a horse-drawn litter covered with velvet and fringed with gold and silk, as well as a coach that looked like a four-poster bed on wheels. Such an equipage was the height of luxury. (Mary of Guise had owned one, as she received a bill for 13 shillings for repairs to it at St. Andrews, but hers was the first seen in Scotland.) Coaches were still rare even in Paris: the only women important enough to have them in Henry II's reign were Catherine de Medici and Diane de Poitiers.

Sumptuousness of this kind was unknown in Scotland. James V's belongings had been more modest. And while Mary of Guise had owned tapestries and personal effects of a similar style and quality, they were fewer in number. There was a separate inventory of her mother's possessions, which Mary inherited and added to her display. Their strength was in the visual and decorative arts. Her mother had amassed twelve sets of tapestries, one on the theme of the Twelve Labors of Hercules, and ten paintings on wood. She had six maps of the world and a pair of globes, one astronomical and the other terrestrial. Ten clocks in jeweled or silk cases took pride of place on her side tables. One of her mother's most treasured items was a panel portrait of herself. It had been secreted away by her executors, and Mary insisted that it be returned to her. This together with her mother's globes were among Mary's own most cherished objects. They too were to be rediscovered in her bedroom at Fotheringhay.

Mary always loved clothes, and her wardrobe was vast. She had dozens of gowns, petticoats, chemises, Spanish farthingales (undergarments of wooden hoops designed to support wide skirts), skirts, bodices, detachable sleeves (often stuffed or on wire supports), jackets, cloaks and mantles. She had drawers crammed with black silk stockings, white crepe stockings, woolen and silk stockings embroidered with gold and silver thread, stockings of the finest Guernsey worsted, and special fine-knitted stockings that showed off the shape of the leg, which she would have worn while dancing. She had silk garters, buttons of pearl, gold buttons, silver buttons and more than fifty embroidered handkerchiefs.

She had wired headpieces, soft linen caps, veils, mufflers, scarves, hats and hatbands. One of her many inventories recorded thirty-six pairs of velvet shoes laced with gold and silver. She had soft leather shoes, leather and buckram shoes, as well as innumerable pairs of gloves, some leather, some of Guernsey worsted and some that were specially perfumed.

Her indoor dresses were made from cloth of gold or silver, silk, satin, velvet or damask, and lined with taffeta or sarcenet. The strapwork and decorations were of gold and silver embroidery or of jeweled embroidery, the buttons of solid gold or silver or else of black or green enamel, the tassels of woven lace. Her outdoor clothes were generally of velvet, damask or Florentine serge, and her riding clothes were of a specially stiffened serge that was decorated with lace and ribbons.

Mary scarcely had occasion enough to wear all these clothes; and this at a time when a reasonably well-off woman might boast of three dresses. A rich noblewoman might have at most two dozen. A poor woman would be lucky to have more than one, perhaps of linen made from home-grown flax, but more usually knitted out of wool that the wearer had carded and spun herself on a spinning wheel.

As a woman ruler exercising a role normally occupied by men, Mary soon showed that she had a passion for frolics and high jinks that inverted sexual or social stereotypes. Almost six feet tall, she could pretend to be a man and liked to roam incognito with her Maries through the streets of Edinburgh wearing men's clothes. Or else she and her four beloved companions would forget their positions and dress up as burgesses' wives. In St. Andrews they amused themselves by playing house, banishing the symbols of royalty and doing their own shopping. At Stirling, they walked in disguise through the streets begging for money, to see who would give and who refuse. At a masque after a banquet in honor of the French ambassador, they appeared dressed as men again, causing shock and consternation. To indulge her wicked sense of humor, Mary employed her own fool or court jester, a woman named Nichola, or "La Jardinière," whom she brought back from France and with whom she loved to banter.

She liked to take regular exercise. She rode one of her horses every day, sometimes alone and for up to three hours. Her favorite sports were hunting, hawking, archery and equestrian events. She even played golf on the links close to the Firth of Forth. When it rained or was too cold to exercise outdoors, she stayed in and played chess, cards or billiards.

Not everyone wanted "joyousity." John Knox complained that Mary "kept herself very grave" in the presence of her advisers, but the moment "her French fillocks" — wanton young women — "fiddlers and others of that band got the house alone, there might be seen skipping not very comely for honest women." He saw himself as a voice crying in the wilderness, because Mary was a gregarious queen who enjoyed the constant cycle of banquets, dancing, masques and dramatic entertainments she had become used to in France.

Her first masque was staged at a banquet in honor of her Guise relations, who were shortly to return to France. The queen, her Maries and
the leading nobles played the main characters. At other times when
they were spectators, the principal roles were taken by professional actors or servants. Sometimes the performance was a dumb show, but usually there were recitations followed by music and dancing, and on special occasions mechanical or clockwork effects. George Buchanan often
scripted these masques. In one, he made Apollo and the Muses march in
procession before Mary while explaining how, being driven from their
homes by war, they flocked to her court. In another, the four Maries
played the parts of nymphs who came to offer their oblations to Mary,
herself depicted as Hygieia, the goddess of health, whose animal was a
serpent drinking from a saucer held in her hand.

Mary employed Buchanan to read to her. Shortly after her arrival,
they were studying Livy's history of Rome for an hour or so each afternoon. How interested Mary really was in Livy is unknown. She had
never much enjoyed history or classical literature, and the true reason
for her afternoons with Buchanan was more likely her love of French
vernacular poetry, because although a native Scot, he was one of the
finest poets writing in Latin and French. Even Ronsard and the Pléiade
had eulogized his work.

Buchanan was soon recognized to be Mary's official court poet. He
scripted most of her masques for the next five years and also spoke the
impromptu verses in honor of Mary Fleming when she was chosen as
"Queen of the Bean" in the revels on Twelfth Night. In accordance with
tradition, a bean was baked in a great "Twelfth cake." Whoever found the
bean in her slice of cake was queen for the day. When Mary Fleming paraded it in triumph, she was given a crown and seated in state on her
throne. Mary would have delighted in this play, which appealed to the
same love of social inversion that had her enjoying the part of a bourgeois housewife at St. Andrews.

Music was indispensable to the masque, and as Mary adored dancing,
she always kept minstrels and musicians on her payroll. She had a consort of five viol players and three lute players. Some of her valets sang
and played the lute. She also liked wind instruments. At first she had
several pipers and a shawm player — a shawm was an early type of oboe.
Later she kept a small orchestra of trumpets, oboes, fifes, drums and ta-
bors. Her domestic staff formed a choir, which sang at her evening functions. When three valets who sang three parts needed a bass to sing the

fourth part, David Rizzio, a young Piedmontese valet and musician who had traveled to Edinburgh in the suite of Signor di Moretta, the Duke of Savoy's ambassador, left his post to enter her service.

One of Mary's sporting and cultural innovations was the equestrian masque, which she introduced within weeks of her arrival. Such events were normally staged outdoors, taking the form of "running at the ring" in costumes or disguises. One team dressed up as Stranger Knights and the other as Female Knights, after which the teams rode in competition on Leith sands. The object of the game was to see which team could score the most points by spearing a ring suspended from a post in a fixed number of turns. But every so often they were indoor entertainments, as when troops of knights rode into the great hall and recited verses eulogizing Mary and pledging themselves as her lieges.

A few days after Mary's inaugural masque, Randolph, the resident English ambassador, was summoned while she was in the council chamber. He found her "where she herself ordinarily sitteth the most part of the time sewing some work or other." When her councilors departed, she asked him to wait. "I will," she said, "talk with you apart in the garden." One of Mary's special skills was the ability to persuade the person with whom she was conversing that he was the only one who really mattered, and Randolph knew the interview was important.

She began in a low key. "How like you this country? You have been in it a good space and know it well enough." Randolph answered, "The country is good and by policy might be made much better." To this she replied, "The absence of a prince hath caused it to be worse, but yet is it not like unto England?" Randolph said there were many countries in far worse condition than Scotland, but few were "better than England," which he trusted she would visit.

He was struggling, because he had little idea what was afoot. But he had said the right thing. A visit to England was exactly what Mary wanted. "I would," she quickly agreed, "be content therewith if my sister your mistress so like."

Mary chose her moment. After Maitland's return from his embassy, there had been a sudden turnabout and his diplomacy had failed. No longer would Elizabeth appoint commissioners to renegotiate the treaty of Edinburgh, to clear the way for Mary to be recognized as heir apparent in England in exchange for renouncing her immediate dynastic claim. As the years went by, Elizabeth developed an almost primordial dread of naming a successor or allowing the succession to be discussed. She had a superstitious fear that to identify her successor would hasten her own death: she was still scarred by the plots and revolts of her brother's and elder sister's reigns. "I know," she said, "the inconstancy of the people of England, how they ever mislike the present government, and have their eyes fixed upon that person that is next to succeed."

The result was that six weeks after Maitland departed, she rescinded her offer and called on Mary to ratify the treaty in its original form. Mary refused to do this, but still believed it would be possible to reach a settlement with her cousin.

She was all the more determined because she did not wish merely to be a figurehead in Scotland. She wanted to rule effectively. With Knox's insubordination so fresh in her mind, she was beginning to fear that her Scottish subjects would never obey her in the manner she thought she should be able to expect unless her right of succession in England was accepted. Her dynastic claim and her prestige in Scotland were connected, because as long as Knox and his supporters were writing to Cecil in England and colluding with him behind her back, her authority was undermined.

If, however, Elizabeth accepted her as the successor, then she would have legitimized her once and for all, making it far riskier for Mary's subjects to appeal to England against her authority, whether she outlived a woman who was only nine years older than herself or not.

Elizabeth had an agenda of her own. She had made it clear to Parliament in 1559 that she did not intend to marry. Her speech was widely reported: "This shall be for me sufficient, that a marble stone shall declare that a queen, having reigned such a time, died a virgin."

Such a statement cannot be taken at face value. Elizabeth was a superb rhetorician who knew that her marriage and the succession were linked. While a marriage, if there ever was to be one, would settle the succession as long as she bore children, it was politically naive for a woman ruler to announce her plans to marry much in advance of her final choice of candidate. Not only would her authority be shaken by factionalism at home, her wedding — assuming she had chosen to marry a foreign prince — would raise the same vexed questions of absentee monarchy that had so humiliated Mary Tudor, when Philip II had left England and his wife after just over a year to attend to more important concerns.

In 1560, Elizabeth had seriously contemplated marriage. She had fallen madly in love with her handsome favorite, Lord Robert Dudley, with whom she had a fling lasting eighteen months. As it was explained by the Spanish ambassador, "Lord Robert has come so much into favor that he does whatever he likes with affairs and it is even said that Her Majesty visits him in his chamber day and night." Their affair seems to have stopped at heavy petting, but was scandalous because Dudley was a married man. His wife was Amy Robsart, whom he had married ten years before.

The gossip ran riot in September 1560, when Amy died at the age of twenty-eight in highly suspicious circumstances. She fell down a flight of eight steps at her home in Cumnor Place, near Oxford, and broke her neck after sending all her servants to enjoy themselves at a nearby fair. A coroner's jury brought in a verdict of accidental death, but whether she died accidentally, committed suicide or was murdered is impossible to judge from the meager evidence. Elizabeth would hear nothing against Dudley, but by the end of the year had decided that marriage to him was too risky. In November, she drew back from giving him an earldom. She picked up a knife at the last moment and slashed the deed of grant.

After Maitland had returned home, convinced that the treaty of Edinburgh was renegotiable, Elizabeth changed her mind. All her life she was prone to such bouts of indecisiveness and vacillation. Sir Walter Raleigh later quipped that "Her Majesty did all by halves." It was a failing she seemed to regard as a virtue, since it gave her more time to weigh her options.

On Cecil's advice, she decided that to replace the treaty would be too dangerous. She knew that she was illegitimate in the eyes of the Catholic Church. She also knew that her father's Parliament had declared her illegitimate when her mother, Anne Boleyn, was executed. And if the treaty was to be picked apart, all that could stand once more in the way of Mary's immediate claim to the English throne was Philip II's decision to recognize and protect Elizabeth.

England's relations with Philip II were still cordial, but that might not last forever. A crack had opened up in the spring of 1561 when Elizabeth refused to send representatives to a new session of the Council of Trent, provoking accusations from Catholics all over Europe that she was schismatic, heretical, illegitimate and immoral. The more Elizabeth thought about it, the more reluctant she was to put a sword into the hands of those who might eventually be in a position to turn it against her.

She had been uncharacteristically frank with Maitland. She usually kept her doubts to herself, but this time she had made them known. She feared a threat to her own security if Mary was named as her successor. However honorable her cousin's intentions, the mere process of naming a successor would stir up a hornet's nest. "Princes," she said, "cannot like their own children. Think you that I could love my own winding-sheet?"

As a result, Mary wanted to meet Elizabeth face to face. She was confident that if she could only talk to her cousin at the level of queen to queen, their differences would quickly melt away and a fresh accord be reached. As soon as she broached the possibility of a visit, Maitland set to work to arrange it.

Nor was Elizabeth unresponsive. The truth is, she was already wavering again. In January 1562, she wrote to her cousin, saying that although she could not yet send her portrait to complete the earlier agreed exchange, this was only because the artist was sick and unable to "set it out." She would, as soon as it was ready, dispatch her picture to her "sister." It was a good example of Elizabeth's genius for public relations, but also an acknowledgment that a settlement between the two queens was still in the cards.

The English Privy Council was deeply divided over Mary's claim to the succession. As long as Elizabeth refused to marry, there would always be those who wished to keep their lines open to someone with as strong a claim to the throne as Mary's. Although she was a Catholic, she had many supporters in England. Her policy of compromise and conciliation with the Protestant lords had been warmly received. And other events worked to her advantage.

Henry VIII's will had set aside the strict rules of hereditary descent. If his children died without heirs, then the throne was to pass to the offspring of the Duchess of Suffolk. By the 1560s, this meant Lady Catherine Grey or her younger sister Mary Grey. They were Protestants, and Catherine's claim was strongly supported by Cecil, who worked tirelessly behind the scenes to promote it.

Elizabeth, however, loathed the Grey sisters. She did what she could to humble them, and when Catherine secretly married the Earl of Hertford and became pregnant, the scandal benefited Mary Stuart. The marriage was discovered in August 1561, when a furious Elizabeth sent both parties to the Tower. A son, called Edward after his father, was born to Catherine. Elizabeth persuaded the Court of Star Chamber to fine the earl £15,000 for seducing a royal virgin and the Archbishop of Canterbury to annul the marriage, thereby denying any resulting children a place in the order of succession.

Whereas Cecil supported Catherine Grey's claim to the succession, Elizabeth favored Mary Stuart's. When looking at it from this viewpoint and not from that of fear of the papacy and Philip II, she believed that the Queen of Scots was undeniably a proper queen and not an upstart. She was a respectable widow, a woman around whom there had never been the slightest whiff of scandal, and although privately a Catholic, in public she had accepted the official Protestant Reformation in Scotland, where her star was rising fast.

The crux was the place of religion. Elizabeth was a Protestant, but not as Protestant as Cecil wanted her to be. The ideological rift between her and Cecil over Mary was fast taking shape. Elizabeth would always be reluctant to settle the succession if that meant identifying a named successor. But in her mind she kept religion and politics apart. Her overriding aim was to defend the ideal of monarchy, and if left to her own devices, she would sooner or later acknowledge the deficiencies of the treaty of Edinburgh and reach an accord with Mary. She would be tempted to recognize her right to be regarded as heir apparent, but without actually naming Mary as her successor.

What terrified Cecil was that Elizabeth might one day overrule him and do exactly this. And what better way was there for Mary to persuade Elizabeth to do so than at a personal interview?

By the spring of 1562, Mary was eagerly making plans for the forthcoming meeting. All suitors for her hand were politely rejected, and if the topic of marriage was broached, she would joke that she would have no one else but Elizabeth.

Mary was so hopeful, she sent another portrait of herself to her cousin. It was a miniature this time, set in a ring behind a large diamond framed like a heart, the sort of token exchanged by royal lovers. The jewel took almost three months to manufacture, but was ready by the middle of June.

Mary wrote verses to accompany it. As with her oration at the Louvre, she needed help, which Buchanan ably provided. The gift had the desired effect. Elizabeth replied two weeks later, sending Mary her own verses in Italian. Mary then reciprocated in French and Italian "with a few more in number written in the best sort she can":

Just one thought gives me pleasure and grief,
My heart fills first with bitterness and then sweetness,
Fluctuating between the doubt and hope that afflict me,
So that peace and sleep flee from me.

Therefore, dear sister, if this verse brings you
The desire to meet that also moves me,
I can only be left in pain and sadness
If the meeting does not happen soon.

These exchanges between the two queens were, in a flirtatious way, the prelude to a rapprochement. Maitland was in correspondence with Cecil, who was attempting to put a damper on the proceedings. He saw where this was heading. He especially disliked the idea of a symbolic marriage, and could never accept that Mary, a Catholic and a Guise, had changed her colors.

On May 25, Maitland was sent again to London, where he stayed until early July. His instructions were to make the final arrangements for the interview.

But Cecil dragged his heels. Maitland complained of his "brief and dark sentences." Then shocking news arrived from France. The Duke of Guise, traveling from Joinville to Paris, had passed through the village of Vassy just as several hundred Huguenots were worshiping in a barn. His retinue tried to break up the congregation, but were repulsed. So his musketeers fired, leaving twenty-three dead and almost one hundred wounded. The duke vehemently denied that he had started the massacre. He always insisted that the violence had erupted because the Huguenots had pelted him and his men with stones. Whatever the truth of this, Cecil's worst fears were confirmed; it seemed that the Guises were set on a religious crusade that would eventually cross the channel and spread to England.

Cecil appealed to Elizabeth to bury the interview with Mary, who knew herself that the massacre would seriously damage her cause. On May 29, Mary summoned Randolph to dissociate herself from her uncles. She "lamenteth their unadvised enterprise, which shall not only bring themselves in danger of their own persons, but also in hatred and disdain of many princes in the world."

Cecil settled down to another round of memos. He was determined to block the meeting and scraped the bottom of the barrel of feeble excuses, claiming it had rained so much, "the great wet" would clog the wheels of the coaches carrying the queens. York or Nottingham had been suggested as the venue, midway between London and Edinburgh. But Cecil argued there were shortages of "wine and fowl" there. He even felt confident enough to draft instructions to Sir Henry Sidney, Lord Robert Dudley's brother-in-law and the president of the Council in Wales, who was to be sent to Mary as a special envoy to say the meeting had been canceled.

Maitland continued to lobby Elizabeth at Greenwich. His efforts seemed doomed, and yet to his delight Elizabeth overruled Cecil. She made her decision on July 6, when a contract was drawn up finalizing the small print for the meeting. It would take place at York in August or September. Mary was to be allowed one thousand attendants as a reflection of the esteem in which she was held. She would be permitted to use "the rites of her religion as at home," a generous privilege. She was not, however, to be a guest, but was to pay her own way. Provision was made for a *bureau de change* where Scottish gold and silver coins could be exchanged for English money to buy provisions and other necessities.

If Cecil was shaken, Mary was jubilant. Maitland was back in Edinburgh by the 15th, when she told Randolph how elated she was and how she could not possibly have received better news. Elizabeth had even sealed the bargain by sending her long-overdue portrait.

Mary showed it to Randolph and quizzed him repeatedly. "How like is it," she asked, "unto the queen your mistress's lively face?" He answered that she herself would shortly be able to judge, when she "would find much more perfection than could be set forth with the art of man."

"That," replied Mary, "is the thing that I have most desired ever since I was in hope thereof." She was almost overcome with joy. "And let God be my witness," she said, "I honor her in my heart and love her as my dear and natural sister." Mary's language shows that she had made a heavy psychological investment in the meeting's success.

Once again she would be foiled. Nine days after Elizabeth first offered the interview, she changed her mind. It was to be postponed until the following year on account of the tragic events in France. Cecil had gotten his way. Sidney, already earmarked to break the news to Mary and whose instructions were ready and waiting, left London on the 16th. He arrived in Edinburgh on the 21st, but Mary was indisposed. Lord James and Maitland had already heard the news. They told Mary, who fell "into such a passion as she did keep her bed all that day," refusing to move or speak to anyone. She received Sir Henry next day, "with great grief . . . as well appeared by divers manifest demonstrations not only in words but in countenance and watery eyes."

Worse was to come. Within three months, the first of the Wars of Religion would have begun in France, and England would have intervened in Normandy on the side of the Huguenots against the Guises. All thoughts of the interview would be sidelined.

The civil war in France broke out less because Catherine de Medici, still the regent there, had opposed the Huguenots than because she had stopped supporting them. After her ten-year-old son Charles IX's accession, her policy had been to eclipse the Guises and deal with the religious question by appeasing the Huguenots. She had little choice, because their rapid advances at court were matched by their expansion in the country as a whole. In the short term, Catherine's policy of allowing them to worship freely in their own homes worked. Their two great leaders, Louis Prince of Condé and Gaspard de Châtillon, Admiral Coligny, were sufficiently powerful to keep their supporters in check. The trouble started when the Huguenots demanded the right of public worship and the Guises successfully detached Anthony, the titular king of Navarre, from his brother Condé.

Anthony of Navarre was lieutenant-general of the kingdom and commanded the royal army. When the Guises won him over to their side, they were seemingly back in power, which pushed their enemies into a revolt. The Huguenots, led by Condé, seized the town of Orléans, followed by Angers, Tours and Blois. When Lyons fell to their forces, Catherine was pressed into a *volte-face*. She turned again to Constable Montmorency and the Duke of Guise, who agreed to bury their differences, and looked for aid to the pope, the Duke of Savoy and Philip II. The Huguenots, for their part, appealed to Elizabeth and Cecil.

Mary was caught squarely in the middle. If she declared her support for her Guise family, she might end up on the side opposite Elizabeth in a war. If she allied with her, she would be accused of betraying her family and her religion, and of assisting heretics to rebel against their lawful sovereign.

Elizabeth was also put on the spot. She saw the extreme danger of alienating Catherine de Medici and Philip II at the same time. She could end up as a Protestant pariah: a heretic queen who seemed always to be making it her business to support her fellow rulers' rebels. She was all too aware of the risks Cecil had taken in Scotland during the revolt of the Lords of the Congregation.

Cecil began to despair of Elizabeth, who was inclined to ignore the plight of the Huguenots. His ally Throckmorton came to the rescue: he knew which card to play. Writing to Elizabeth, he hinted that the Huguenots were likely to be victorious in northern France. If aid was sent to them, the opportunity would arise to recover Calais, which England, to Elizabeth's deep chagrin, had been required to cede to France by the treaty of Cateau-Cambrésis.

It was a golden opportunity to aid the cause of religion and recover a lost territorial prize. But if a war was imminent, the interview between the queens must be postponed.

Elizabeth desperately wanted to recover Calais or another Channel port in its place. Her favorite, Lord Robert Dudley, now took center stage, opening up negotiations with the Prince of Condé and sending Sidney, his brother-in-law, as a special emissary to France. Dudley's diplomacy was at its height in May and June, just when Maitland was lobbying Elizabeth in Mary's favor. Then, on July 17, eleven days after Elizabeth overruled Cecil and agreed to the interview and two days after she had changed her mind again, it all came to a head. Dudley reported to a meeting of the Privy Council at which it was decided to launch a military intervention in France.

On September 20, a treaty was concluded with Condé's delegation at Hampton Court. Elizabeth promised to supply the Huguenots with six thousand troops and a loan of 140,000 crowns, and in return was granted Le Havre as a pledge until Calais was restored.

Mary watched these events with growing alarm and despondency. Since the meeting was postponed, she decided to fulfill a summer plan she had delayed: a royal progress to northeastern Scotland, to see the country and show herself to as many of her subjects as possible there. It was a sensible decision, even if Mary made it chiefly to save face.

She began at Stirling, traveling by way of Perth, Glamis and Edzell. She reached Aberdeen at the end of August, but the journey was difficult. According to Randolph, it was "cumbersome, painful and marvelous long, the weather extreme foul and cold, all victuals marvelous dear, and the corn that is, never like to come to ripeness."

The final stop was Inverness. Randolph thought he could relax when he reached nearby Strathbogie, the principal seat of the Earl of Huntly, the leading Catholic noble and head of the Gordons, whose house was the fairest and "best furnished" in Scotland. Huntly's hospitality was "marvelous great," but Mary refused to go there, even though it was within four miles of her route.

Her summer progress was taking on a sinister aspect. Huntly was in grave disfavor for opposing Mary's proposed interview with Elizabeth. Like many of the Catholic lords, he resented the policy of conciliation toward England, which he saw as a Trojan horse for Lord James and his allies. He was also sulking because he felt he had been cheated. He had been administering the earldoms of Mar and Moray on behalf of the crown for several years, until Mary granted them to Lord James, the first early in 1562 and the second in September, in exchange for the first.

Her anger had been aroused when she first reached Aberdeen. Huntly had turned up to welcome her, but brought fifteen hundred retainers when he was commanded to bring no more than a hundred. A clash was unavoidable when Mary arrived at Inverness only to find that the captain of the castle, apparently on Huntly's orders, refused to open the gates to her.

Huntly, known by the nickname "Cock of the North," was the most important landowner in the northeast, second only to the Earl of Argyll in the west. A staunch adherent of Mary of Guise, he was appointed chancellor after Cardinal Beaton's murder. When discredited for opposing the regent's centralizing policy in the Highlands, he joined the Lords of the Congregation, but was a reluctant, halfhearted recruit. He ended up antagonizing both sides, especially Lord James, to whom he had unwisely boasted that he could restore the Mass in three counties.

Advised by her insinuating half-brother, whom Mary now officially elevated to the earldom of Moray, she decided not to tolerate Huntly's insubordination. When she was denied entry at Inverness, she lodged for the night in the town, but next day returned with a force. She took the castle and hanged the captain from the walls of the battlements.

Huntly was still at Strathbogie, where he was lying low. He feared an attack by Lord James and decided to preempt it by separating Mary from his rival. When she set out back to Aberdeen, he planned to seize her as she forded the River Spey. He mustered his forces under Sir John Gordon, one of his younger sons, who was already wanted for a prison escape. But Mary was forewarned. When she reached the river crossing, she had three thousand men, whereas Sir John had only one thousand. They were hidden in the woods within two miles of the river, but fled as the royal army approached.

Mary was jubilant. She had escaped an attempt to kidnap her. "In all these garboils," wrote Randolph to Cecil, "I assure Your Honor I never saw her merrier, never dismayed, nor never thought that stomach to be in her that I find." Her only regret was that "she was not a man, to know what life it was to lie all night in the fields, or to walk on the causeway with a jack and a knapscall,* a Glasgow buckler, and a broadsword."

Randolph got caught up in the excitement. As an ambassador, he was supposed to be a neutral party, but he confessed, "It may please you to know that in good faith where so many were occupied, I was ashamed to sit still and did as the rest." He was positively disappointed when Huntly's men fled. "What desperate blows would that day have been given, when every man should have fought in the sight of so noble a queen, and so many fair ladies."

In Randolph's mind, it seemed that Mary's ride was part of a medieval chivalric romance. When he reached Aberdeen, the gloss wore off. Mary was royally entertained "as well in spectacles, plays, interludes and other as they could best devise." But lodgings were in short supply, and Randolph had to share a bed with Maitland. To his horror, he later discovered that Huntly had planned to attack the town, burn down the house where he was staying and assassinate Maitland in his sleep.

Mary did not intend to allow Huntly to go unpunished. Lord James had inevitably denounced him as a traitor. She ordered him to surrender, and when he prevaricated, she sent spies to Strathbogie to arrest him, but he saw them coming and escaped through a back door of the castle. By October 12, she had decided to destroy him. Five days later, the earl was outlawed and ordered to yield Strathbogie. Mary's forces were now searching for him, and Huntly was on the run. He had a force of some seven hundred men, and aimed to wear down his opponents in a guerrilla war.

Mary named the newly promoted Earl of Moray as her lieutenant and, together with Morton and the Earl of Atholl, he led an army of two thousand out of Aberdeen. Huntly occupied a hilltop some fifteen miles away at Corrichie, but was forced by Moray's artillery to retreat to the lower

* Jack: a quilted leather tunic worn by foot soldiers, usually plated with iron. Knapscall: a headpiece or helmet.

mossy ground, where he was trapped. Two of his sons were captured, one of them Sir John, who was executed next day at Aberdeen.

Huntly was also taken, but died of a stroke while still mounted on his horse. His corpse was embalmed and sent to Edinburgh, where it was kept until the following May, when it was put on trial in Parliament. As the clerk's report put it, "The coffin was set upright, as if the earl stood on his feet." He was then found guilty of treason, and the family estates were declared forfeit.

Two days before Huntly was outlawed, Elizabeth wrote to Mary, justifying her decision to send an army to France to aid the Huguenots. It was an intimate but awkward letter, full of labored metaphors. She argued that "necessity has no law," that in an emergency "we have no choice but to protect our own houses from destruction when those of our neighbors are on fire," and that she would so conduct herself that Mary's brother-in-law, the young Charles IX, "will think me a good neighbor, one who preserves rather than destroys." (The last of these arguments was taken straight out of one of Cecil's memos, in which he argued that neighboring rulers had a duty to protect each other, especially minors, when their realms were threatened by tyranny.)

Elizabeth was on the defensive. She wanted to avoid a breach with her "sister," but was well aware of the risks. The prospect of a split "gnawed her heart" in case "the old sparks would be fanned by this new fire." That said, Elizabeth used a well-known device of classical rhetoric to say how much she would have preferred not to mention — so enabling her to mention in graphic detail — the innocent victims who had been so barbarously butchered of late by the Catholics. She could not imagine Mary being so infatuated with her uncles that she was able to ignore their terrible crimes.

Mary took this surprisingly well, less perhaps because of the letter's contents than because it was dictated at the height of the attack of smallpox that brought the English queen close to death. "I would write more," Elizabeth had concluded, "but for the burning fever that now holds me completely in its grip."

Her letter arrived on the day of the battle of Corrichie. Mary was so busy, Randolph could not deliver the letter, so he decided to return the next day. He reappeared while Mary was at supper. She eagerly asked for her letter. "Let me see," she said, "what you have for me!"

Randolph, who knew the letter's contents, hesitated, but she insisted.

In the event, she showed no emotion. Her face did not flinch. She put down the letter without saying a single word and returned to her supper "in mirth," just as when Randolph had arrived.

Later, she recalled him. "Now Mr. Randolph, I trust we shall the next year travel as far south as we have done north, with as much ease and more pleasure than we have had of this journey." She was adamant that her meeting with Elizabeth be rearranged. She refused to consider the implications of the letter, that it was canceled for good.

Before retiring for the night, Mary summoned Randolph to her bedroom and asked quietly, "Is my sister sick?" She pointed to the last sentence of the letter she was still clutching, and Randolph explained that Elizabeth was recovering from an attack of smallpox. Mary was genuinely concerned, and shortly afterward wrote the letter in which she referred to her own experience of the disease as a child.

But she was also secretly elated. Smallpox was often fatal. If Elizabeth died, Mary intended to stake her claim to the English succession. And yet the timing could hardly have been worse. She was caught in the middle of a war. She therefore told Randolph that she had decided to be neutral. Her uncles, she said, must surely have acted out of a sense of duty. Beyond this, she was unwilling to get involved.

Mary was struggling with her emotions. So when Randolph, despite promising himself not to say anything "grievous" against her uncles, could not resist the riposte that Elizabeth intervened in France in a godly cause, and that Charles IX, when he was older and wiser, would thank her for it, she smiled broadly and changed the subject.

Almost as soon as Mary had returned to Holyrood from the northeast, she suffered a bout of viral flu. She was in bed for six days, and when she rose was in a more belligerent mood. Lord James and Maitland had given her the news that when Elizabeth's attack of smallpox was at its height, only a single voice had been raised in her favor in the English Privy Council as the successor.

Mary was mortified. Maitland saw the danger and warned Cecil that the war had radically altered the game. Mary, he said, was a "perplexed" queen, boxed in between her uncles and England. To ensure her friendship, she needed a more secure interest in the succession than was provided by Elizabeth's "love." Her religion could no longer be an obstacle when she had so obviously protected the Protestants and destroyed Huntly, the leader of the Catholic nobles.

But Cecil was unmoved. The war in France was going badly for England; the Huguenots were forced to retreat and the English troops were cornered at Le Havre. As he reminded Randolph, there were "two dangers." One was a Catholic victory so overwhelming it would put "us here in danger for our religion"; the other was that the Guises would "build their castles so high," they would attempt to depose Elizabeth.

It was the same old story. Strangely, Randolph's assessment veered to the opposite extreme. However much Mary favored her uncles, he advised Cecil by return of post, "yet she loveth better her own subjects." She understood the need for the amity with England to be greater "than a priest babbling at an altar." Mary, he argued, "is not so affectioned to her Mass that she will leave a kingdom for it." Cecil should be reassured. He should no longer suspect her, because "her desire was never greater to live in peace, nor never more heartily desired the Queen Majesty's kindness and goodwill than now she doth." "Yesterday," said Randolph, "she spoke it and willed me to write the same."

Mary's charisma had worked its magic on Randolph. Cecil, however, was immune. She celebrated Christmas 1562 in fine style, but the joy was hollow. Even before the revelry was over, she was muttering that she had not heard from Elizabeth for two months.

Early in the new year, Maitland wrote to Cecil: "Sir, I cannot think it to be without some hidden mystery that the intercourse of letters (which were wont to go frequently as well betwixt the two queens as us their ministers) is thus ceased on your part."

A full-blown crisis erupted when Mary found out that Elizabeth had summoned Parliament to meet on January 11, and that Cecil aimed to bar her from the succession by an Act of Exclusion. The news had leaked by January 5, when Randolph warned that Maitland was "in great choler." He had reassured Mary that nothing would be "to her discontentment," but the pressure was mounting, and in the last week of January she took to her bed for six days.

On the 31st, a letter arrived from Elizabeth full of specious excuses, which Randolph presented. She "read it quite over twice in my sight" and took it in good part. But she asked to be excused from replying. She had decided to send Maitland back to London to present her case, if necessary directly to Parliament. He was to go first to London, and then on to France.

He left Edinburgh on February 13, when he also carried a letter from

Moray urging how the "love once kindled" between the two queens should be reignited, a task in which he supposed Cecil should never "relent."

It looked very much like more of the same: another attempt to arrange an interview between the queens. But it was nothing of the kind. Cecil was entirely unprepared for what was in Maitland's instructions. Not even Moray now knew his sister's thoughts. She was dissembling, because Elizabeth's prevarication and Cecil's obstinacy were leading her to a different and (from the English viewpoint) far more threatening solution. She would seize the initiative by searching for a husband able to secure her dynastic rights in England. This was to become her policy for the next two years, and to get her own way she would, if necessary, break with her Guise relations.

11

A Search for a Husband

MARY PLAYED by the rules in seeking a second husband. Female monarchy was thought to be an aberration; Knox's account of it in his *First Blast of the Trumpet* was just an extreme version of a well-worn stereotype. The correct "solution" for a woman ruler was to marry and settle the succession. Marriage was a matter of "reputation" and of the "fortification of her estate." Cecil, who continually urged his own queen to marry, used exactly those words. Whether a reigning queen of Scotland would become subordinate to her husband upon marriage was a delicate political conundrum to which there was no textbook answer. Scotland was not France, where the Salic law prevailed. Everything depended on the personalities involved and the opinion of Parliament, but generally the nobles were more settled and less factious when dealing with a man, even if he was merely a king consort.

A complication for Mary was that few kings or princes took widows as their first wives, usually thinking of them as second or subsequent wives in cases where their first marriages had already produced heirs. Against this could be offset the Queen of Scots' transcendent youth and beauty, and the fact that her dowry was a kingdom. She was still only twenty and in her prime. Now that she had decided to marry again, it was a matter of identifying the best candidate in light of her political and dynastic goals.

So far, she had been advised by Lord James, now Earl of Moray, and his allies, whom she appointed to her inner cabinet. Their policy was to maintain the amity with England by seeking a "middle way" in which Mary would be recognized as Elizabeth's successor. In turn, the preservation of the amity gave the Protestant lords a guarantee of the religious status quo.

But the gap between Elizabeth and Cecil over how to deal with Mary was widening. Whereas he grumbled that she sought to win Mary over "by gentleness and benefit" without sufficient regard for her own or the nation's security, Elizabeth joked that her chief minister was more bothered about her "safety" than she was herself. Cecil's belief in an international Catholic and Guise conspiracy turned him into Mary's most ardent and determined opponent, whereas Elizabeth, who repeatedly refused to name a successor out of fear that it would somehow hasten her own death or encourage a dangerous upsurge of factionalism, remained sympathetic to Mary's claim, which she preferred to those of any other candidates.

Cecil's fear when Maitland reappeared in London was that if Mary, frustrated by the setback she had received over the canceled interview, now looked for a European husband, her claim to the English throne would once again become her principal asset over and above her own country. It was likely to attract large numbers of potential suitors, even those who already wore a crown. The greatest danger would be if the pope or Philip II backed her claim, because a papal bull declaring Elizabeth to be illegitimate would be tantamount to inciting English Catholics to revolt. Since even Cecil acknowledged that the official Protestant Reformation had barely scratched the surface in the north of England or Wales, a papal or Spanish intervention was greatly to be feared. A majority of the English people were still Catholics. Moreover, the norms and values of nobles and landowners were tied to the rules of hereditary descent where property rights were concerned, and very few outside Cecil's inner caucus would have agreed that religion should take priority over property rights when considering the succession to the throne.

Mary reshuffled her inner circle of advisers in readiness for her search. It marked the end of Moray's period of ascendancy. Mary was disillusioned with her half-brother. His policy of a middle way had failed, and she began to suspect that he had plotted Huntly's destruction for his own ends. He was the sworn enemy of the Gordons, their rival for the earldom of Moray with its extensive territorial estates. While Elizabeth and Cecil had been delighted by Mary's sudden and unexpected disabling of the leading Catholic family in Scotland, they did nothing to reward her. On the contrary, even as Maitland rode south, Cecil was experimenting with drafts of a parliamentary bill to exclude her from the succession.

In January 1563, Mary took Maitland aside. She told him she was making him her leading councilor. She had almost completely forgiven him for his earlier intrigues with Cecil. Although a genuine Protestant, and thus a man with very different religious ideals from her own, he was so far proving himself to be a friend of the monarchy. His sense of duty made him willing to see himself as a royal servant, which in turn made him more valuable to Mary than was Moray, whose blunt but annoyingly regal manner she was fast coming to resent. She also judged her illegitimate brother to be too manipulative and ambitious to live up to his promises. Moreover, Maitland's friendship over many years with Cecil might now be turned to her advantage.

Mary allowed Moray to drift. She decided to promote the wily and conniving Morton to the chancellorship in place of Huntly. She had not yet plumbed the depths of his villainy. Although a Protestant, Morton was too venal and lascivious to be an ally of Knox and the Calvinists. She knew that he had rebelled against her mother, but his contribution had amounted to little. By advancing him, Mary hoped to be assured of the loyalty of the powerful Douglas clan and at the same time provide a counterweight to Moray, who was likely to resent his demotion.

Mary was starting to assert herself as queen. She was attempting to control the noble factions by creating a broad coalition of advisers, which would enable her to take a tougher line with Elizabeth and Cecil, since she would have wider support throughout the country than before. To help create this coalition, she asked Maitland to recommend some new appointments to the court and Privy Council. They included Catholics like Atholl as well as Protestants like Lord Ruthven, which corresponded to Mary's aim of nurturing ideals of royal service and loyalty to the crown that transcended sectarian divisions.

Mary acted next to raise her stock in Europe. She wrote two letters, both at the end of January. One was to the Cardinal of Lorraine, the other to the pope. She asked her uncle to intercede with the pope on her behalf. Exactly what he was to say was not committed to paper, but the aim was clearly to make Mary more attractive as a prize for Catholic suitors. To the pope she accounted herself "your most devoted daughter," whose uncle would explain "the state of our affairs" and "the need which we have of the assistance and favor of Your Holiness."

On February 13, Maitland set out for London and Paris, armed with parallel sets of instructions. The first related to a position Mary had adopted over the French Wars of Religion once English troops had landed in Normandy. She had offered to act as an independent arbitrator. This was clearly a nonstarter and may well have been a diplomatic blind. Elizabeth was too committed to the Huguenots to consider arbitration at this stage, and the war was going too well for Catherine de Medici to wish to settle.

A second set of instructions concerned the debates in the English Parliament. Should the succession be discussed to Mary's detriment, Maitland was to insist on a right of audience to register a protest in which her claim was set on record. This, although far from a blind, was unrealistic, because Parliament was in no mood for it. Cecil had taken care to ensure that as many Protestants as possible were elected, then lined up his friends to speak. Sir Ralph Sadler, who as Henry VIII's former ambassador had admired Mary as an infant even as he was duped by her mother, delivered what was tantamount to a racial attack: "Now if these proud, beggarly Scots," he said, "did so much disdain to yield to the superiority of England . . . why should we for any respect yield to their Scottish superiority, or consent to establish a Scot in succession to the crown of this realm?"

Maitland's final instructions, delivered to him in March by courier, required him to negotiate first with Alvarez de Quadra, Bishop of Aquila, the Spanish ambassador in London, and then with Mary's relatives in France, to propose a marriage to Don Carlos, Philip II's son and heir. This was what Mary was after. It had been the match so keenly sought by the Guise family after Francis II's death, frustrated by Catherine de Medici's secret diplomacy. Then Mary's interest had been negligible as she made her own plans to return to Scotland. Now she had decided to revive the negotiations, winning support from Maitland, who could see that without exerting pressure on England, there was little hope that Elizabeth would reopen Mary's case. He was also under the illusion that Philip II was a religious pragmatist, taking as his cue Philip's well-known reluctance to support Mary Tudor's burning of the Protestants while he had been married to her. Whether for manipulative reasons or because of a misunderstanding over Philip's stance, Maitland glossed over the problems of religion, arguing that the Calvinists would not re-sist Don Carlos. He chose to believe that Philip, although a Catholic, was not a Catholic ideologue. He and his son would be willing to accept the religious status quo, because he was a "wise politic prince" who governed the many territories under his control "according to their own humor."

Maitland visited de Quadra, who wrote enthusiastically to Philip. "If Your Majesty listened to it," he explained, "not only would you give your son a wife of such excellent qualities . . . but you also give him a power which approached very nearly to [universal] monarchy." To his existing dominions, Philip would add through his son the entire British Isles and Ireland. From the outset, the bait was Mary's dynastic claim, exactly as Cecil feared.

Maitland put his queen's position in a nutshell. Her rebuff over the postponed interview required her to restore her honor and reputation by seeking "such a marriage as would enable her to assert her rights."

De Quadra was flattered and delighted. He had assumed that Mary would try to marry her brother-in-law Charles IX. He was unaware of Catherine de Medici's outright opposition to that idea, and Maitland did nothing to disabuse him. On the contrary, Mary's new chief councilor actively hinted at the prospect of a second Valois marriage to create the illusion of a competition.

Maitland's first report to Mary, sent from London, recounted de Quadra's belief that Don Carlos, who was now almost eighteen, was "very far in love" with her. His second report, written a month later from Chenonceaux, where he was following the French court, was more gloomy. He had heard from the cardinal, who had sent a strongly worded letter meant for Mary. The signs were not good. Her Guise rela-tives, noted Maitland, had little regard for her declared wishes or feel-ings. They paid attention to her only because of her "grandeur" as a reigning queen, from which they derived their own "advancement and surety." Or, to put it simply, they would help their niece only to the extent that they first helped themselves.

As to Catherine de Medici and the nobles, they "care not greatly of your marriage or with whom it be, provided it bring with it no peril to this crown." It was a reaction as cold as it was cutting, reflecting Cathe-rine's unrelenting concern after Francis II's death to keep Mary at a dis-tance. A marriage to Don Carlos, said Maitland, was opposed as vigor-ously as before by the Queen Mother. Her own daughter Elizabeth was still childless as queen of Spain, and however much she longed for Philip

II's support against the Huguenots, Catherine did not intend to hand him a claim to the English throne on a plate.

In England, Maitland continued to Mary, there were "three factions": the Catholics, the Protestants, and the queen. Elizabeth, while remaining single, wanted Mary subordinated to a man in such a way that she herself "had least cause to stand in fear." The Catholics, for whom Lady Margaret Douglas, the Countess of Lennox, was fast becoming a mouthpiece, insisted on a Catholic marriage, one that would put Elizabeth under maximum pressure. The Protestants also wanted Mary to marry for religion, but to someone who would defend the Protestant cause.

Anticipating the reaction of the lords in Scotland, Maitland believed that "albeit the best part" would support Mary, there would be "divers malaperts," notably Knox's allies, who would vehemently speak out against Don Carlos.

The Cardinal of Lorraine had demanded an answer to his letter within six weeks, which Maitland thought unrealistic. As he had discovered, the Guise family were in no position to antagonize Catherine de Medici while both were engaged in a bitter war against the Huguenots. The Guises had performed a turnabout and decided that Mary should not marry Don Carlos. They proposed instead an alliance with Archduke Charles of Austria, with whom Mary's uncle was already negotiating behind her back.

It was an unwelcome twist. The archduke was the third son of Ferdinand I, the Holy Roman Emperor. In principle highly suitable for Scotland — he had a reputation for being a moderate rather than an extreme Catholic — his main disadvantage was that he was already a candidate for Elizabeth's hand, for which reason Mary rejected him, as she did all other suitors except Don Carlos.

The cardinal decided that Mary was doing too much thinking for herself. He was apprehensive about his niece's growing independence, so sent Philibert du Croc to Edinburgh as his special envoy in May. Du Croc had arrived by the 15th, and Mary was closeted with him for days. Their talks were private and were accompanied by a flurry of dispatches to Maitland and the Guises. Randolph saw du Croc as a threat: "She useth no man's counsel but only this man's . . . and assuredly until Maitland's return, she will do what she can to keep it secret."

Du Croc, however, left empty-handed. Mary would decide nothing until Maitland returned, and despite the envoy's best endeavors, he merely obtained her thanks and a request for further information about the archduke's personality, income and the dowry he would offer.

This was largely diplomatic froth, and when Maitland reached Edinburgh on June 24, he found more urgent matters in his in box. Some reports of Mary's marriage plans had leaked, and Knox was on the prowl. As Maitland had predicted, the influential preacher would do all he could to stir up resistance to Mary's betrothal to a Catholic.

Knox's imagination was in full flight, leading to a spectacular showdown. Preaching as the representative of the Kirk before the same session of Parliament that declared Huntly's embalmed corpse guilty of treason, Knox attacked Mary's proposed marriage.

She summoned him to Holyrood the same afternoon. "I have," she exclaimed, "borne with you in all your rigorous manner of speaking, both against myself and against my uncles; yea, I have sought your favors by all possible means. I offered unto you presence and audience whensoever it pleased you to admonish me; and yet I cannot be quit of you." Her voice quivering with rage, she threatened, "I shall be once revenged," and burst into tears of self-pity. Knox began to justify himself, but Mary jumped straight to the point: "What have ye to do with my marriage?"

He replied that since so many of her nobles were flatterers, neither God nor the commonwealth were "rightly regarded." At this, she once more demanded: "What have ye to do with my marriage?" And "what are ye within this commonwealth?"

Knox's bile was up. "A subject born within the same, madam. And albeit I neither be earl, lord nor baron within it, yet has God made me (how abject that ever I be in your eyes) a profitable member within the same."

He then proceeded to repeat everything he had said in his sermon, prophesying that if Mary married a Catholic, the realm would be "betrayed" and she would end her days in sorrow. One doubts whether any other ruler in the sixteenth century was so roundly rebuked.

Mary was speechless. No longer was she able to measure up to Knox in a quarrel. He had broken every convention of political speech, every rule of courtesy, because although she was an anointed queen, he addressed her as an equal and even a moral inferior. She always found it hard to suppress her emotions. This time she was overwhelmed and simply "howled."

Knox waited while she dried her eyes, then compounded his offense. He declared he had never much liked weeping and "can scarcely well abide the tears of my own boys" when he beat them. Since, however, he had spoken the truth and Mary had "no just occasion to be offended," he must endure her tears rather than offend his conscience or "betray my commonwealth through my silence."

At this final affront, Mary ordered him out of the room. She had reached a crossroads. For the moment, she had no choice but to bide her time. She saw clearly what was at stake. She had read the *First Blast of the Trumpet* in which Knox had explained that a Catholic "idolatress" was by simple definition a woman guided by hot and uncontrolled out-pourings of passion instead of cool reason.

Knox was a fanatic, a misogynist and a prude. He had devised his own rudimentary theory of the psychology of female Catholic rulers. Because they were idolatresses like Jezebel or Athalia, they ruled from the heart and not the head. Their idolatry was inflamed by their sex. They had set reason aside, since if they had been governed by reason, they would have long ago converted to Protestantism. It followed that Knox equated Catholicism in a woman with unbridled sexual lust. It would not be long before he would be accusing Mary of the crimes to which she must automatically be prone merely *because* she was a Catholic and not a Protestant.

This was the foundation of the stereotype that Mary ruled from her heart, unlike her cousin Elizabeth, who always ruled from the head. Since Elizabeth was a Protestant, Knox exempted her from his vicious attacks on female rulers. Unlike Mary, she was a queen "by a miraculous dispensation of God." It is clear where some of these ideas came from, since on this point Knox followed Calvin's opinion that female monarchy deviated from the "proper order of nature," with the exception of those special women who would be "raised up by divine authority" to be the "nursing mothers" of the Protestants.

No impartial witness of Elizabeth's cavorting late at night in her bed-room with Lord Robert Dudley while his wife was still alive could have compared her favorably to Mary at that stage. Knox had embarked on a strange sectarian fantasy. Starting from his belief that Mary was an idolatress who attended her Mass in secret, he came to malign her as a femme fatale: a manipulative siren whose moral defects and unfitness to rule were evident from her dancing, banquets and flaunted sexuality.

Soon he was convinced he had unearthed a scandal. It was centered

on Pierre de Bocosel, Seigneur de Chastelard, a poet on the fringes of the
Pléiade. Mary had admitted Chastelard to her service, and he had written some poems for her to which she had unwisely responded in the tradition of courtly love. It was an innocent gesture, but the poet fell madly
in love with her. The night before Maitland left for London, Chastelard
hid under her bed, where he was discovered with a sword and dagger.
He was banished, but followed Mary to Fife, entering her bedroom
again, only two days later, while she was undressing. This time he was
tried for treason. On the scaffold, he delivered a fine rendition of Ronsard's "Hymn to Death," after which he cried out, "Adieu, the most beautiful and the most cruel princess in the world."

Chastelard had an obsession, but Knox blamed Mary for leading him
on. The charge was leveled by the time of Maitland's return and linked
to Mary's love of dancing late into the night. "In dancing," he claimed,
"the queen chose Chastelard, and Chastelard took the queen." She would
rest her head on his shoulder, "and sometimes privily she would steal a
kiss of his neck."

Knox's charge of sexual transgression was a travesty. In reality, Mary
had been so terrified by the poet's appearance in her bedroom, she refused for months to sleep alone and would not retire for the night unless
Mary Fleming, the chief of her four Maries, slept in the same room.

Once Maitland was back home, he sought to neutralize Knox by denying plans for Mary's marriage. This was strictly true: there were negotiations but no wedding plans. It was a splitting of hairs worthy of Knox
himself, but the preacher continued to subvert Mary, exploiting the jealousy that had arisen between Moray and Maitland, and attempting to
bind himself to Moray, who was out to court as much popular support as
he could.

The events of 1563 put a considerable strain on Mary. She was already
becoming estranged from her uncles when suddenly she was hit by a report of the Duke of Guise's assassination. The duke, returning from a
routine inspection of his army near Orléans on February 18, was shot
three times in the shoulder and died a few days later. His assassin was a
Huguenot, who under torture implicated the Protestants in his crime.
Mary was devastated. Although the Cardinal of Lorraine had been her
mentor, the duke was her favorite uncle.

The news reached her in St. Andrews on March 15. She was "marvelous sad, her ladies shedding tears like showers of rain." She decided

to console herself by riding for days on end across the fields, hawking and hunting as she passed from place to place. Randolph accompanied her to Falkland, her favorite hunting palace, beside the gently sloping Lomond hills. As the ambassador took his leave, she suddenly regaled him with an account of "all her griefs and the great adventures that have fallen unto her since the death of her husband, and how she was now destitute of all friendship."

Mary felt very much alone. In some ways she was too gregarious to be a queen. She could be naive, impulsive and impatient, sending Maitland to negotiate her marriage to the most eligible bachelor in Europe and then expecting quick results. Although she could work with her advisers, one doubts whether she liked any of them. They were too self-interested, too concerned with their private quarrels for anyone not brought up in Scotland to understand. Mary was a charismatic queen who showed her emotions, but she also needed emotional support. With her mother dead, this she could rarely obtain. Scotland was still in many ways a foreign country. There were few people she could really talk to apart from the four Maries, who knew Scotland no better than she did. She loved her family, but often felt badly let down by them. And now the Duke of Guise was dead, which brought the terrible dangers of religious war home to her.

A week or so later, Randolph did exactly the right thing. While Mary hunted near Pitlessie, he brought her a letter of condolence from Elizabeth, written in her own hand. She stopped her horse and read the letter, "not without some tears that fell from her eyes."

"Monsieur Randolph," she said, "I have now received no small comfort, and the greatest that I can, coming from such a one as my dear sister, so tender a cousin and friend as she is to me. And though I can neither speak nor read but with tears, yet think you not but that I have received more comfort of this letter than I have of all that hath been said unto me since I heard first word of my uncle's death."

Mary tucked the letter into her bosom. At dinner when she seemed to be alone, she took it out and read it again, saying aloud, "God will not leave me destitute. I have received the best letter from the queen my good sister of England that ever I had, and I do assure you it comforteth me much."

Mary was genuinely grateful for the letter, but this latest scene was a piece of theater from beginning to end. Mary knew she was being watched by her servants, who she knew would be suborned by Randolph, who in turn would inform Cecil. She was learning to put on a show. Despite her comfort at Elizabeth's letter, she was under no illusions that the interview between the queens would ever take place, but rather than show her disappointment, she would keep Elizabeth guessing.

Mary did not intend to be snubbed again. She meant to be fully occupied during the summer months, when a rescheduled interview might have been expected. To avoid looking like a suppliant, she threw herself into plans for another royal progress, this time to Ayrshire, and from there into Argyllshire and the western Highlands, the main Gaelic-speaking territory.

Knox, meanwhile, notched up another scandal. A few days after Huntly's attainder, Parliament passed an act against adultery, the penalty for which was now death. That very same night, one of Mary's French chaplains was discovered in bed with another man's wife, providing further "proof" of the moral deficiencies that sprang from saying or hearing Mass. Knox could scarcely contain his glee, attempting to inflame his congregation at St. Giles Kirk against Mary and the licentiousness of her court. His strictures were always the same:

> O Lord, if thy pleasure be, purge the heart of the queen's majesty from the venom of idolatry, and deliver her from the bondage and thralldom of Satan . . . The queen's idolatry, the queen's Mass, will provoke God's vengeance . . . Her house, whither her subjects must resort, is become the haunt of dancing and carnal concupiscence . . . Get thee to the prophets, I say, for Elijah saith: "The wrath of God shall not spare rulers and magistrates . . . The dogs shall lick the blood of Ahab and eat the flesh of Jezebel" . . . The devil takes the depraved at his will. Dancing is the vanity of the unfaithful, which shall cause the people to be set in bondage to a tyrant.

When the summer progress finally began, Mary's attention turned to archery, hunting and hawking, the outdoor sports she loved so much. But her European diplomacy was not forgotten. In late July, before she left Argyllshire, Luis de Paz, an agent of de Quadra, came all the way from London to visit her. Mary eagerly received him.

Elizabeth's response was to warn Maitland that such secret diplomacy with Spain must cease. Her bluster would not have mattered had Mary been supported by her Guise family. This was her best opportunity to escape from the straitjacket into which the English were now steadily trying to force her. Only through an alliance with a major European power could she make the sort of marriage to which she aspired, and since Catherine de Medici refused to help her, that ally, she believed, must be Philip II. Her case against the archduke was that, although a Hapsburg and Philip's cousin, he lacked sufficient weight to force England to recognize her claim to the succession, not least because he was already Cecil's favored candidate for Elizabeth's hand.

Even as Mary was closeted with de Paz, her uncle was bargaining for the archduke. The cardinal was so unmoved by Mary's objections to his diplomacy and so supremely confident of his ability to talk her around, he traveled to Innsbruck without her knowledge to sign a treaty.

It was a wasted journey. Mary always intended to reject the suit, and when she understood the full extent of her uncle's double dealing, she rebelled. As she later told her aunt the Duchess of Arschot, "Not that I don't consider it great and honorable, but less useful to the advancement of my interest, as well in this country as in that to which I claim some right."

A sudden crisis blew up when Sir Thomas Smith, the new English ambassador to France, reported what he believed to be the terms of Mary's offer to Don Carlos. He was misled by inaccurate French intelligence, wrongly supposing that her uncle had brokered a deal with the pope to give England as a dowry to Don Carlos. If only he had known that Philip II had entertained Maitland's overtures in the first place solely to counter a possible marriage to Charles IX, and would end them as soon as he knew for certain that Catherine had vetoed that idea, there would have been no reason for the English to panic.

But if Smith was misled, his report was taken at face value. In March 1563, three months after the Huguenots had been resoundingly defeated at Dreux and their leader Condé captured by the Catholics, the two sides made peace at Amboise. Then, to Elizabeth's bewilderment and chagrin, their forces had united in a joint campaign to expel the English army of occupation from Normandy.

As the combined French armies advanced on Le Havre, the English dug in, but bad weather and plague compounded their woes. By late July, they had no option but to surrender. It was an ignominious defeat, although not a complete disaster, because the Guises had failed to achieve the mastery in Normandy to which they had aspired.

It was bad enough, however, to persuade Elizabeth not to get involved in a war for the next twenty years. She became an isolationist in foreign policy, which made her all the more determined to dictate the terms of

Mary's marriage. If England was to become a fortress, it was essential to box in Mary and coerce her to cut her own European links; otherwise, the perennial problem of Scotland as the back door into England would arise again, just as it had repeatedly under Henry VIII.

Elizabeth's policy toward Scotland went into overdrive after her retreat from Normandy, and came to focus on a right to veto any husband Mary might actually seek to choose. When Elizabeth and Cecil heard of the cardinal's visit to Innsbruck, Randolph was told to inform Mary that if she married the archduke or anyone else from the emperor's family, the amity with England would be at an end.

Randolph delivered this message on September 1 at Craigmillar Castle, one of Mary's favorite retreats just south of Edinburgh. She greatly disliked what she heard. As he spoke, she constantly interrupted him, asking so many questions "that scarce in one hour could I utter what might have been spoken in one quarter." She demanded that his message be submitted in writing so that she could reflect on it.

Then Smith's report arrived on Cecil's desk and Randolph was recalled to London. Mary angrily insisted on knowing what marriages were "sortable" for her. As Elizabeth complained, her royal cousin was threatening a showdown. She wanted to know "whom we can allow and whom not; secondly what way we intend to proceed to the declaration of her title."

It was turning into a battle of wills. In an ideal world Mary would have ignored Elizabeth and gone her own way. By demanding, whether mockingly or sarcastically, to be told whom she might marry, she was giving the English queen a clearly defined hold over her. But then she had precious little choice. Scotland was a weak country lacking the troops or cash reserves to fight a war without foreign aid. Even the richer Lowland territories were reliant for their trade and commerce on the northern counties of England. Since Catherine de Medici refused to help her, and as Philip II was still protecting Elizabeth for the moment, Mary was in a vise. At the same time, she felt she needed to be recognized as Elizabeth's successor if she was to bolster the legitimacy of her reign and curtail the insubordination of Knox and her more turbulent lords.

By the end of 1563, Mary had brushed aside her uncle's advice to marry the archduke, but this did not mean she had to cut her diplomatic links to the Continent. The extent of her breach with her Guise family was exposed when she bitingly remarked, "Truly I am beholden to my uncle: so that it be well with him, he careth not what becometh of me."

But if support from France was for now beyond her reach, she might yet outmaneuver Elizabeth and Cecil.

On November 17, Randolph was given new instructions spelling out the limits of Mary's choice of husband. They had proved to be a drafting nightmare. The main gist remained constant. Mary was to be asked to marry someone, preferably an English nobleman, who was fully committed to the amity. If no one met those qualifications, she might seek English permission to marry a foreigner as long as he was prepared to live in Scotland after he was married. He must be "naturally born to love this isle" and be "not unmeet" (Elizabeth was fond of double negatives), but "no one" from Spain, France or Austria would ever be acceptable.

That was the first draft. Elizabeth then softened her line. Once again, she followed her instincts as a queen and a woman and laid down the basis of a compromise. Should Mary accept this English advice, her dynastic claim could be reinstated, in which case "we will not be behind on our part to satisfy her as far forth as if she were our only natural sister or dear only daughter."

Cecil took one look at this and balked. First he crossed out "dear only" in the last sentence. Then he deleted Elizabeth's entire paragraph. "We do," he wrote in its place, "promise her, that if she will give us just cause to think that she will in the choice of her marriage show herself conformable," then "we will thereupon forthwith proceed to the inquisition of her right by all good means in her furtherance." Mary might submit evidence in support of her right to the succession. A legal adjudication would then follow, declaring whether or not her claim was upheld. Only then, "if we shall find the matter to fall out on her behalf," could Mary expect to be treated as Elizabeth's "natural sister or daughter."

Cecil's amendments prevailed. They were not only insulting to Mary, suggesting that an anointed queen and sovereign of an independent nation should submit herself to the jurisdiction of an English court, they introduced the novel element into the equation that her right should be put on trial.

It was an extraordinary requirement. Simply by seeking to marry and so fulfill the universally accepted obligations of a woman ruler, as Cecil could only wish his own queen would do, a course had been set by which Mary would be turned into an English suppliant. Already regarded as the chief antagonist, she had been imagined as the defendant in a court case.

It was surely a drastic overreaction that exposed the increasing paranoia in England over the threat of Guise and papal conspiracy. If Mary had as yet failed to find a husband who was able to secure her rights, she might still renew the attempt. But first she would have to deal with the demands of Elizabeth and Cecil, whose efforts to assert their superiority over her and her country were becoming as blatant as they were threatening. Conciliation had been replaced by confrontation. The trick would be to appear compliant, then to pull off a fait accompli.

12

❧✦❧

"My Heart Is My Own"

I N NOVEMBER 1563, Elizabeth had asserted a right of veto over Mary's choice of husband and advised her to marry an English nobleman. If Mary refused, she might consider a foreigner as long as he was not from Spain, France or Austria, but the conditions laid down were so narrowly restrictive they would be impossible to fulfill with honor. No names had yet been mentioned. Did Elizabeth all along have one of her own favorites in mind? It was impossible to be sure.

Cecil then went one step further, demanding a trial of Mary's dynastic claim. Elizabeth gave in to Cecil's nagging, but softened the blow by sending Mary a token of her affection. It was a jewel, a diamond ring that she "marvelously esteemed." Not the least of her challenges, in the year before she finally broke free of English intimidation over the terms of her marriage, was making sense of such mixed messages.

When Randolph presented the ring, Mary was delighted. It symbolized Elizabeth's role as her "lover" and seemingly turned the clock back to the spring of 1562, when the agenda was a meeting between the two "sister queens" as the prelude to their symbolic union. As if to emphasize the point, Mary "often looked upon and many times kissed" the ring.

But Randolph sent it to her ahead of the stipulations for her marriage. When Mary read these, she showed great presence of mind. Not wishing to protest outright, but keen to set a different tone in her relations with England, she indulged her mischievous sense of humor, pointing to two rings on her fingers, one Elizabeth's, the other the gift of her late husband, Francis II. "Well," she said with a smile, "two jewels I have that must die with me and willingly shall never [be] out of my sight." By this she made it clear she was keeping her options open.

She then laid on a charade, poking fun at Randolph. Everyone joined in, and the scenes may have been loosely scripted. She would open with the line "Randolph would have me marry in England!" Argyll or someone else would then call out, "Is the queen of England become a man?" Mary would ask, "Who is it there in that country whom you would wish me to marry?" To which Randolph would have to answer lamely, "Whom you could like best. Maybe there is so noble a man there as you could like?"

Or else Mary would pose as an innocent bystander, eager to oblige the English queen if only Randolph could be less "obscure" about whom Elizabeth wanted her to marry.

It was all, at least on the surface, innocuous. It also caused a lot of merriment in the court. Randolph, whose instructions expressly forbade him from naming any particular candidate, could only squirm or suggest that a delegation be sent to London to quiz Elizabeth in person.

Despite Mary's sang-froid, the pressure was taking its toll. For two months she had been sick. Her affliction, as Randolph reported, was "divers melancholies." She disguised these as best she could, but "often weeps when there is little apparent occasion." The day after her twenty-first birthday, she was in bed all day after dancing late into the night. With Knox as vigilant as ever, the official explanation was a cold "being so long that day at her divine service."

But the sickness persisted. As Christmas approached, Randolph knew that "her disease . . . daily increaseth." This is the first mention of symptoms that would later suggest a gastric ulcer rather than porphyria. "Her pain," he continued, "is in her right side." Here he was mistaken. He was soon correcting himself, reporting that the abdominal pain was on her left side. She had taken "divers medicines, but hitherto findeth herself little the better. Upon Saturday she was out of her bed, but took no great pleasure in company nor to have talk with any."

Knox had partly caused Mary's distress. She had bided her time until the lords gathered in Edinburgh for Christmas, when she had finally decided to strike against the man she believed to be her scourge. During her last summer progress, a priest who had celebrated Mass in her private chapel in her absence was threatened by two Calvinists. When they were imprisoned for defying Mary's proclamation on religion, Knox summoned "a convocation of the brethren" to free them. His letter, an implicit incitement to acts of violence, was shown to Mary, who consulted the Privy Council, a majority of whom thought it treasonable. Mary was elated by their response. It looked as if Knox had gone too far this time, and despite Maitland's deep reservations, he was put on trial.

The court was packed. Mary was flanked by her lords, and when she saw Knox standing at the other side of the table with his cap off, she first smiled and then "guffawed." She was in high spirits, saying, "This is a good beginning . . . Yon man made me weep and shed never tear himself. I will see if I can make him weep."

A worried Maitland whispered in her ear, advising her to be quiet. He believed the proceedings to be ill-advised and wanted to minimize the damage, but Mary wanted her revenge on a man she saw as her implacable enemy. Knox was charged with conspiracy to "raise a tumult" against her. By attempting to "convoke" the Protestants to free the offenders, he had summoned Mary's subjects to arms and so threatened armed resistance against his lawful queen.

Knox conducted his own defense. He argued that there was a difference between a legal and an illegal assembly. In this he was supported from the table by Lord Ruthven, a Calvinist, who pointed out that Knox "makes convocation of the people to hear prayer and sermon almost daily." To this, Knox himself added that all his actions were authorized by the Kirk, and therefore he had acted lawfully as a minister of the Gospel.

In the end, it came down to a vote, taken in Mary's absence. Knox was acquitted, and when she returned to the court to hear the verdict, she demanded it be taken again. The result was exactly the same. Her illegitimate half-brother, the Earl of Moray, who was jealous of Maitland's promotion over him, was the cause. He had spoken in favor of Knox's prosecution in the Privy Council, but then used his influence in the court to secure an acquittal. It was his first open act of betrayal. Mary had been humiliated. She was resentful and exasperated, incredulous that victory could have been snatched away from her like this.

But she was willing to learn from the experience. In an attempt to avoid new conflicts, she reined in Maitland somewhat, so that he and Moray were once more treated as equals. Her aim was still to balance the factions by creating a broad coalition of advisers willing to subordinate their private quarrels to royal interests, and to a large extent she was succeeding. Her working Privy Council, with whom she met almost every day at Holyrood, embraced a wide cross section of lords.

Heartened by their support and for the moment taking no discernible offense at Moray's duplicity, Mary decided to pick herself up and begin again after her suit for Don Carlos failed. She was still looking for a husband, but this time the focus of her diplomacy was England. She intended to test Elizabeth by teasing out the full extent of her unreasonable demands. It was becoming clear that Mary had a deep-seated need to secure recognition of her "grandeur" not just from her own subjects, but from the other rulers with whom she came into contact.

Mary signaled the start of her latest round of diplomacy at a Shrovetide masque at Holyrood in February 1564. It was performed at a banquet said to be the most sumptuous and extravagant in living memory. There were three courses, each comprising not a single dish but a choice of some forty or fifty dishes of every conceivable kind of fish, fowl, game and meat, followed by jellies, pies, cakes, baked pudding, tart and fruit, all carried into the great hall by servants dressed in costumes of black and white. Mary, her attendant lords and the four Maries also wore these colors.

A boy dressed as Cupid led the procession that served the first course to the strains of an Italian madrigal sung by Mary's choir. After a suitable break, while the diners washed their hands and faces in silver bowls and dried them on white linen towels, the second course was served. At the head of the new procession was a beautiful young girl representing Chastity. As the waiters served the food, Latin verses were recited: a eulogy of the pure mind and radiant beauty that the child symbolized and which were iconic of Mary herself. Lastly, a boy in the character of Time ushered in the serving of the third course. The choir sang another divertissement, this time a setting of verses foretelling that as long as heaven and earth might endure, the mutual love and affection of Mary and Elizabeth would prevail.

Randolph, the English ambassador, attended the feast and sent a report of it to Cecil. He also shuttled between Mary and her advisers, seeing first the one and then the others. He was becoming increasingly apprehensive, since almost three months had elapsed since he had delivered his message and Mary had so far declined to make any answer to the conditions he had conveyed about her marriage. He had no way of knowing what was in her mind, and believed that only Maitland really knew what she was thinking.

The symbolism of the masque seemed unequivocal. But the answers given directly to Randolph by Mary and her councilors were more quixotic and far less easy to interpret.

Maitland began the parley. He made "great protestation" to Randolph of Mary's love for Elizabeth. And yet, how much better it would be, he said, if so difficult and confidential a topic as her marriage could be discussed face to face between the queens alone, without the intercession of their subjects. He clearly had Cecil firmly in his sights.

Moray and Argyll then joined in. They coordinated with Maitland to inform Randolph that because his stipulations for Mary's marriage had been "only general," her reply could only be "uncertain." It was a delphic response, a masterly piece of rhetorical obfuscation akin to what Elizabeth, when giving an equally unacceptable answer to her Parliaments on the subject of Mary, would later call an "answer answerless."

But there was a clear method to this. Mary, closely advised by Maitland and Moray, was inching toward a position in which Elizabeth would be forced to break cover and give the name of her favored candidate for Mary's hand. Randolph was summoned to see Mary in a private interview a week after the Shrovetide banquet to receive his answer to the terms he had delivered before Christmas. He had no choice but to write a letter afterward to Elizabeth, rather than simply sending in a report to Cecil as he was usually expected to do.

Once again, Mary had taken a leaf out of Elizabeth's own textbook. One of the English queen's time-honored ploys was to muddy the waters when imparting unwelcome news, saying one thing herself while getting her councilors to say something rather different. Now Mary did something similar. Her councilors had been noncommittal. They had not bound their queen to pay attention to Elizabeth's conditions.

But Mary did not want to leave it like that. She was still striving as hard as she could for a settlement in which her dynastic rights in England were recognized. Her councilors, she said, had given the mere "words" of her answer, but she wanted Randolph to explain to Elizabeth the full extent of their "meaning."

"Princes," she confided, "at all times have not their wills, but my heart being my own is immutable." Mary paused as she spoke, to emphasize that she was "without evil meaning" to Elizabeth. When she continued, she said that she longed for "nothing more" than her fellow sovereign's lifelong love and goodwill.

Mary had momentarily exposed her inner self. All her resolve and determination, her strength of will and sense of "grandeur," were packed into these words. They were disarmingly honest. They recognized that her sheer determination and force of character would not always allow her to get her own way even if she had set her mind on something. And yet they were layered to the point of being enigmatic. They implied that she always intended to keep something of her own in her heart: to retain a part of herself for herself, no matter what politics and policy forced on her. To this extent, she would always remain elusive. Her goodwill toward Elizabeth was genuine, but not to be presumed upon.

What most struck Randolph at the time was Mary's sincerity. She spoke straightforwardly, wholeheartedly and without guile. She was to be trusted. "The word of a prince," he added reassuringly, was of far greater worth than "the mutable mind of inconstant people."

Randolph had appealed to a well-known maxim of the Athenian rhetorician Isocrates, Elizabeth's favorite classical author. In a passage she had learned by heart as a child, he had advised rulers: "Throughout all your life show that you value truth so highly that your word is more to be trusted than other people's oaths." For Elizabeth, it was a lifelong moral axiom. She claimed time and time again in her own letters and speeches that her word alone was sufficient, because the words of rulers were the badges or symbols of their authority.

Mary's tactics worked. Honesty brought forth honesty, and Elizabeth sent Randolph an answer explaining that her preferred candidate for her cousin's hand was none other than her own favorite, Lord Robert Dudley.

It was a breathtaking reply. Dudley had first been mentioned as a possible husband for Mary during the 1563 session of Parliament. In what at the time was an off-the-cuff remark and not a considered proposition, Elizabeth urged Maitland, then preparing to advance Mary's suit for Don Carlos, to recommend Dudley on account of his qualities and graces, a hint Maitland brilliantly deflected with the riposte that if Lord Robert was so desirable as a husband, Elizabeth had better snap him up herself!

Mary was likely to be gravely affronted at the prospect of inheriting her cousin's castoff lover. Randolph was beside himself at the thought of transmitting such advice. He broke the news in a lengthy audience at

Perth. Mary could scarcely believe her ears. She listened implacably. At length she said, "Monsieur Randolph, you have taken me at a disadvantage." But she quickly recovered her wits, her mood switching in a matter of seconds from incredulity and bemusement to anger.

"Do you think that it may stand with my honor to marry my sister's subject?" The whole idea, she said, was insulting, whereupon Randolph only made things worse, replying that there could surely be no greater honor than to match herself with a nobleman "by means of whom she may perchance inherit such a kingdom as England is."

Mary bridled. Why did Dudley make any difference to her dynastic prospects in England when she was already the strongest claimant by hereditary right?

"I look not for the kingdom," she said, "for my sister may marry and is like to live longer than myself. My respect [i.e., concern] is what may presently be for my commodity, and for the contentment of my friends, who I believe would hardly agree that I should abase my state so far as that!"

Mary laid out her position clearly. She countered those of her critics who had been unable to understand why, once her initial policy of conciliation failed, she continued so relentlessly to set her sights on recognition as the heir to a woman who was only nine years older than she was and perfectly capable of bearing children if only she chose to marry. Despite the apparent illogic of such a policy, it was in fact the only possible course of action open to her. The "commodity" of princes was their honor and reputation. It was vital for Mary to safeguard them, and given her emphasis on family ties, the "contentment" of her relatives also mattered to her, despite her recent disappointment with them.

Still more important after her latest and most embarrassing clash so far with Knox was her continued belief that her Scottish lords and the Calvinists would come to obey her in the way she expected only when her dynastic rights were recognized. She most of all needed Elizabeth's friendship to arm herself against the volatility of her lords and to bolster the legitimacy of her reign in her own country. Since Knox had only just finished yet another round of correspondence with Cecil, using Randolph as his intermediary, it was a burning issue for her.

Up until now, Mary had conducted this interview with Randolph in the presence of her advisers, but she now ordered them to leave, her Maries excepted.

"Now Monsieur Randolph," she began again in her most winning and confidential tone. "Doth your mistress in good earnest wish me to marry my Lord Robert?" Randolph assured her that it was so. "Is that," she said, "conforming to her promise to use me as her sister or daughter, to marry her subject?" Randolph said he thought it might be.

"If I were," she said, "either of them both, and at her disposition, were it not better to match me where some alliance and friendship might ensue, than to marry me where neither of them could be increased?"

Randolph hesitated. He then replied, "The chief alliance my sovereign desires is to live in amity with Scotland."

"The queen your mistress," said Mary, "being assured of me, might let me marry where it may best like me, and I always remain friend to her as I do."

Mary paused, thinking carefully. Finally she said, "These things are uncertain." There were many risks and no guarantees. She did not intend to marry Dudley simply on her cousin's say-so. She would not rule out the suggestion completely. But she needed more time to think it through.

While Randolph was pondering this, the lords reappeared and everyone went to supper. Afterward, the discussion resumed by candlelight in Mary's bedroom. Tempers were beginning to fray, and Moray was unable to restrain himself. He took a dig at Randolph, but one that was also a swipe at Mary, exposing the rising tension as it began to sink in that whoever married her was likely to become king and could make or break their careers.

"Why," Moray asked Randolph, "do you not persuade your own queen to marry, but trouble our queen with marriage that yet never had more thought thereof than she hath of her dinner when she is hungry?" This was vintage Moray: bluff, shrewd, cynical, always calculating the odds. He was a master of the thinly disguised insult. Mary laughed and walked away.

From Elizabeth's viewpoint, of course, Dudley's candidacy was logical. By marrying him, Mary would be subordinated to a Protestant male on whom the English queen knew she could always rely, a man she still loved and trusted never to betray her. More maddeningly eccentric was her idea of how it might play out. Of two minds about allowing Dudley to leave her sight, she came up with the almost ludicrous proposal that there would be a ménage à trois or extended royal family. Mary, Dudley and Elizabeth would all live together at Elizabeth's court after Mary was married, where the English queen would bear the costs of the "family," which, she said, was merely the right way for "one sister" to behave toward "another."

This was not a ploy to drag out negotiations and so stall plans for any marriage Mary might seek to make on the Continent; Elizabeth was serious. Of all her designs, it comes closest to fantasy and makes nonsense of the traditional interpretation that she was always astute and never prone to making decisions on the basis of her emotions. As Randolph ruefully observed, Elizabeth's idea of happy families turned "this comedy" of Mary's marriage "altogether liker to a tragedy."

Mary spent the rest of the spring and early summer of 1564 shoring up her position. She adopted two tactics. Her first was to capitalize on potentially her greatest asset: her popularity with the ordinary people of Scotland, so visible to her when she first returned from France and on her summer progresses. A popular queen was likely to be a strong one. She was the "fountain of justice" responsible for the impartial administration of the legal system, and she took her opportunity to score points off Knox and the Calvinists when a group of poor people petitioned her against the judges of the Court of Session, many of them Knox's friends, whom they accused of favoring the rich and powerful and sitting in judgment on each other's cases and those of their friends and kinsfolk.

The Lords of Session had been established by Mary's grandfather, James IV, to sit regularly in Edinburgh and hear legal cases as expeditiously and fairly as they could. They were the highest court of justice, independent of the crown, later incorporated as a special College of Justice, charged to act professionally and set an example for all the other judges in Scotland. Despite this, they were ignoring their responsibilities to poorer litigants who could not afford to pay for justice.

The time had come for Mary to intervene. She issued a reforming ordinance that required the Court of Session to sit more frequently to hear the cases of the poor. All the judges were to sit at least three days a week, "as well after noon as afore noon." For their extra work, Mary generously increased their salaries. But she also made it clear that she expected justice to be done without fear or favor. She even arrived unannounced in the courtroom one Friday afternoon to watch some poor people's cases being heard. All this was something of a publicity coup, and went a long way to undermine Knox's claims that she was a queen who was interested only in dancing and courtly frippery and not in her subjects' welfare.

Mary's second tactic in the spring of 1564 was to hit back quietly at Elizabeth for attempting to destabilize her country. The previous year, Elizabeth had asked for a favor. She had written to Mary to see if she would give the Earl of Lennox, who had caused so much trouble in Scotland during Henry VIII's reign, a passport to return home from his long exile in England. It was a totally disingenuous move. Elizabeth was no friend of Lennox, whose wife, Lady Margaret Douglas, had only recently been released from the Tower. Elizabeth interceded for Lennox only at the height of Mary's suit for the hand of Don Carlos, when she wanted to stir up trouble.

In making her request, Elizabeth had played with fire. The son and heir of Lennox and his countess was Henry, Lord Darnley, now seventeen, stunningly handsome and the claimant with the best hereditary right to the English throne after Mary herself. When Elizabeth made her bid for the passport, Darnley was not expected to accompany his father to Scotland. On the contrary, he was living at the English court, where he could be closely watched, effectively under house arrest. He was allowed to wait on Elizabeth, to whom he sang and played on the lute in the evenings. But if his father returned to Scotland, would Darnley follow? And if so, could he conceivably become a suitor for Mary?

It seemed a fantastic idea. Randolph dismissed it out of hand. Then, in the last week of April 1564, Mary decided she would take Elizabeth at her word, grant the passport for Lennox and allow him to return to his ancestral home. A close friend of Knox, William Kirkcaldy of Grange, wrote from Perth: "The Earl of Lennox will obtain license to come home and speak [to] the queen. Her meaning therein is not known, but some suspect she shall at length be persuaded to favor his son."

From this moment on, the smart money backed Lord Darnley as a likely husband for Mary. He was probably the last person Elizabeth had in mind. If Mary married him, her claim to the succession would be greatly strengthened, because unlike Mary, he was male and born in England, which countered the two overriding objections to her claim in the Parliament of 1563: that she was a woman and a foreigner.

Darnley was not a Protestant, but neither was he an orthodox Catholic. He did not take his religion very seriously, and was able to attend a Catholic Mass in the morning and a Protestant sermon in the afternoon unfazed by any sense of inconsistency. This made him seem less threatening in Scotland than Don Carlos or Archduke Charles. Beyond this, he was physically attractive. He was four years younger than Mary, stood fully six feet tall and looked as svelte and lissome as had Lennox in his youth, when he had followed Mary of Guise from Stirling to Edinburgh and St. Andrews. He was more effeminate and baby-faced than his father, but the implications of that were not yet talked about.

Once Mary had set in train the mischief of agreeing to readmit Lennox to Scotland, she proposed to Randolph that both queens should appoint commissioners to meet on the border at Berwick-upon-Tweed to discuss terms for the possible Dudley marriage. Elizabeth immediately took fright. Mary was dueling with her. Already she regretted ever asking Mary to repatriate Lennox. But the damage was done, and she could hardly refuse to condone the very course of action she herself had proposed. She appealed, clandestinely as she believed, to Maitland and Moray to block Lennox's return, but her overtures were rebuffed and reported with glee to an increasingly determined Mary.

In the summer of 1564, Mary put herself deliberately out of reach. She let Elizabeth stew and went on another progress, this time to the far north, starting at Inverness and moving on to Gartly in Aberdeenshire and then to Easter Ross. She was attempting to bind together the Gaelic-speaking parts of the country to the Scots-speaking Lowlands by their shared allegiance to herself. And the plan was working, in that she was well received and lavishly entertained by her hosts at every stop. Her willingness to order the court to put on Highland dress helped, as well as her liking for the harp and bardic poetry. She was even willing to listen to bagpipes, which, more than the harp, were coming to be regarded as symbolic of Highland identity and culture.

By the time Mary was back in Edinburgh, there was yet another twist. Catherine de Medici had sent Michel de Castelnau, Sieur de Mauvissière, on a mission to both Elizabeth and Mary, uncertain about the endgame in Mary's suit for Don Carlos and anxious to prevent a future crisis.

Catherine was assiduously promoting a peace policy after the combined Catholic and Huguenot victory over the English at Le Havre. She wanted peace within France and between France and its neighbors; she especially wanted peace in the British Isles to guarantee the Anglo-French entente. She therefore proposed a new double alliance: Elizabeth should marry Charles IX, and his younger brother and heir, Henry Duke of Anjou, should be given to Mary.

When Castelnau put the offer to Elizabeth, she sidestepped it with a joke. "The king," she said, laughing, "is both too big and too small!" She meant that France was too powerful to match with England, whereas Charles IX, who was by now fourteen to Elizabeth's thirty, was too young. She had no desire at all to marry Charles. If she left the country, she would become an absentee ruler, and the king of France could hardly be expected to live in England. But she asked Castelnau to thank Catherine for the honor she had done her and reiterated her support for their entente.

Mary was distinctly frostier. She would soon be twenty-two and the Duke of Anjou was barely thirteen. That, however, was not the reason for her disapproval. She rejected Catherine's proposal less because of the age disparity than because she held the offer to be unworthy. It was a consolation prize. As Castelnau put in his report, Mary "had as big and restless a spirit as her uncle the Cardinal of Lorraine." The offer was insufficient for her "grandeur." Despite this, he found her absolutely "enchanting": a woman "in the flower of her youth" who was "esteemed and adored by her subjects."

Castelnau knew Mary already, having watched her grow up at Henry II's court. He brought her letters from her Guise relations, and as she gradually thawed, she gave him better reasons for rejecting the offer. "Of all the kingdoms and countries of the world," she said, "none touches my heart more than France, where I was nurtured and had the honor of being a queen . . . But I cannot really imagine how I could return there in a lesser role, and in any case if I left my realm of Scotland unattended, I might even be in danger of losing it."

Mary flatly refused to consider a marriage to a prince who she believed (wrongly, as it turned out) had no prospect of ever inheriting a throne. Her next move was to send a gentleman of her bedchamber, Sir James Melville, as her ambassador to Elizabeth. He left Edinburgh in late September, instructed first to play down the effect of an angry letter Mary had written protesting against her cousin's clandestine efforts to block Lennox's repatriation, and then to stand ready to act on her dynastic claim should Parliament, currently prorogued until October, reassemble. Lastly, Melville was charged to negotiate secretly with the Countess of Lennox to obtain a passport for Darnley to travel north.

Mary was still pretending to consider Dudley as a possible husband while quietly investigating the prospect of marrying Darnley, whose dynastic assets, if united with her own, would give her an almost invincible claim to the English succession, provided the rules of hereditary right were followed.

When Melville reached the palace of Whitehall, Elizabeth rolled out the red carpet. She smothered him with blandishments, seemingly taking him into her confidence and granting him up to three interviews a day. He was flattered, even as he realized it was all part of a game. In fact, his entire visit was orchestrated to bring the topic of conversation continually around to Dudley.

One of Elizabeth's ploys was to summon Melville to her bedroom to view her collection of portrait miniatures. The pretext was that she "delighted often" to look at Mary's portrait; but when the hapless envoy arrived, the miniature at the top of the pile was none other than Dudley's, lovingly wrapped in tissue paper with the sitter's name inscribed on the outside in Elizabeth's own hand.

The inscription read "My lord's picture," wording that suggested an outright declaration of love and the more baffling to Melville: to show this inscription to him while attempting to advocate Dudley's merits as a husband for Mary only served to expose Elizabeth's own indecision about the course of action on which she had embarked.

Melville was then closely interrogated about Mary. What did she wear? What color was her hair? How beautiful was she? Was she tall? What were her favorite pastimes? Could she play the virginals? The questioning was endless, and it was almost more than the patriotic and loyal Melville could bear to describe his queen with enough faint praise to avoid offending the vain and jealous Elizabeth.

At the end of a grueling nine days, Melville looked forward to his departure. He was exhausted, but was obliged to stay on to attend Dudley's creation as Earl of Leicester in honor of his suit for Mary, an event at which he and the French ambassador clearly spotted Elizabeth tickling her kneeling favorite's neck.

It seemed to be just another, if more elaborate, royal charade. More perilous was that portion of Melville's mission that concerned Darnley. Elizabeth was becoming wary of him. His stock had risen relative to the gossip about him, which Elizabeth curiously sought to counter by promoting him. At the ceremony for Dudley's ennoblement, Darnley bore the sword of state and was invited to the official reception.

Asked point-blank about Mary's opinion of him, Melville dissembled. "No woman of spirit," he said, "would make choice of such a man that was more like a woman than a man, for he was very lusty, beardless and lady-faced." It was just about a credible answer, because while Darnley was polished and urbane, his character was tainted by recklessness, sexual excess, pride and stupidity. He was almost certainly bisexual, as was the vogue of young hedonistic courtiers in France. Contemporaries had ways of making sexual excess known, and when Darnley was described as a "great cock chick," the pun was intentional. The Cardinal of Lorraine was more polite, merely dismissing him as "a polished trifler."

When Melville returned to Edinburgh, he found Lennox had preceded him. The earl was riding high in Mary's esteem and occupying some of the best rooms at Holyrood. His estates, traditionally concentrated in the region of the Clyde around Glasgow, had been forfeited on his defection to Henry VIII twenty years before, but on October 16 were fully restored. Mary had to speak "very comely" to her councilors to pull this off. The lords in Parliament sanctioned the grant to please her, but Argyll predicted that Lennox had enemies who would stir up trouble if he was allowed to lord it over them.

Mary was unwilling to listen to such objections. She relished too much the irony of proclaiming how his rehabilitation was "at the request of her dearest sister Elizabeth."

Lennox, meanwhile, was greasing the wheels. His wife used Melville as a courier to send luxurious gifts for Mary and her lords. Mary's presents included "a marvelous fair and rich jewel," a clock and a jewel-encrusted mirror. Maitland and Atholl were each given a ring set with a large diamond, while the four Maries received such "pretty things" as Lennox "thought fittest." Moray, the sworn enemy of the Lennoxes, significantly received nothing — a deliberate slap in the face.

Mary's gift may well have been the famous Lennox Jewel. According to Randolph, everyone who saw it thought it stunning, and it bore a coded message that Mary would have appreciated. Designed to be worn around the neck on a gold chain or ribbon, the jewel is in the shape of a golden heart. On the outer face is a crown surmounted with three white fleurs-de-lis on an azure background, set with three rubies and an emerald. Beneath it is a winged heart with a huge sapphire as its centerpiece, above which is a crown set with rubies and an emerald. These emblems are supported by the classical-style figures of Faith, Hope, Truth and Victory. Around the border is the motto "Who hopes still constantly with patience shall obtain victory in their pretense."

The jeweled crown opens, and inside the lid are two hearts and a golden true-love knot, pierced with Cupid's arrows and over them the motto "What we resolve." Below, in a cavity within the crown, is the monogram *MSL* (for Matthew and Margaret Stuart Lennox) inscribed in white, blue and red enamel. Inside the lid of the winged heart are two joined hands holding a green hunting horn by red cords, with the motto "Death shall dissolve." Within another cavity is a skull and crossbones.

The inner portions of the jewel are a celebration of the marriage of Lennox and Lady Margaret Douglas. Its fulfillment is, however, to be the victory of the "pretense" proclaimed by the motto around the classical figures. The word "pretense" had no pejorative connotation in the sixteenth century. Someone who "pretended" merely asserted a claim to something to which he believed he had a right. This inscription most likely applied to Mary's claim to the English throne, and then to Darnley's, and the winged-heart centerpiece may have hinted at a unification of their claims.

Whether or not this was the "marvelous fair and rich jewel" presented to Mary, she had decided by the end of 1564 to prise Darnley away from Elizabeth. Her plan slowly unfolded after the failure of the meeting between the English and Scottish commissioners that was held at Berwick in November.

Maitland had opened the bidding. He proposed that Mary be allowed to marry the husband of her choice, with certain specific exceptions, in exchange for which she would be recognized as Elizabeth's heir apparent. This went nowhere. But his subsequent offer, that she be recognized as heir and confirmed as "second person of the realm" in return for which she would renounce her immediate Catholic claim and marry Dudley, was little more than everyone had been led to think would be acceptable.

Although a Dudley marriage was not ideal, it was a serious offer on the Scottish side. If Mary had been named as Elizabeth's heir, she would have married him as the price of a dynastic settlement.

But when the report of the English commissioners arrived in London, everything went back to square one. Elizabeth retreated at the prospect of actually identifying a named person as her successor. It was not that she objected to Mary as such; it was the same old story as before.

Cecil, who had always opposed a settlement with Mary, made no effort to persuade Elizabeth to honor her commitment. Instead, he wrote a serpentine letter to Maitland and Moray in which he attempted to dilute the clear and binding commitment given on the English side into a nonbinding pledge, urging the Scots to move forward by way of "friendship" and not by way of "contracting." The "ticklish" matter of "princes determining their successors" was cited: the primordial problem of Elizabeth's "winding-sheet." Faced with identifying her heir, her nerve had failed.

On Christmas Eve, Maitland and Moray wrote an anguished reply. They were incredulous, both that the settlement was in jeopardy and that Cecil had taken this opportunity to revive his earlier proposal of a trial of Mary's claim.

The Scots were quite candid. Without a guarantee of Mary's place in the succession, Cecil should know that Dudley was "no fit match" for their queen, even if he was made a duke, a higher rank than the earldom he now held.

And this was the rub. Dudley was a nobleman of Elizabeth's own creation. He owed everything he possessed to her. He lacked sufficient patrimonial estates of his own and a claim to a throne. In addition, he was the son of a convicted and executed traitor, since his father had been the same Duke of Northumberland who had engineered the Protestant Lady Jane Grey's failed coup before Mary Tudor's accession in 1553.

Darnley, by comparison, was the genuine article: a scion of the royal house of Tudor whose dynastic pedigree would be unassailable if it was annexed to Mary's own. Mary had exposed her thinking to Randolph: "Princes at all times have not their wills, but my heart being my own is immutable." She had intended to keep something of herself in reserve. Now she would finally break free of English coercion. Although Darnley was still Elizabeth's subject, he would shortly be across the Scottish border. She had not yet definitely made up her mind to marry him. Snatching him away and marrying him were two very different things. But he was her insurance policy. She would reintegrate him into Scottish politics, where he would become a foil to attempts by Elizabeth and Cecil to dictate her choice of husband.

Mary did not quite know how she would manage this. To her astonishment, she had in the end to do nothing. A fiasco at the English court, one for which Elizabeth would forever rebuke herself, was about to catapult Darnley into Scotland and from there into Mary's arms.

13

❧❦❧

A Marriage of Convenience

ENGLISH POLICY toward Mary was beginning to fall apart. Elizabeth was losing her nerve. On September 23, 1564, she had written to Cecil, then ill at home, to ask for his advice. It was the scrappiest, most faltering note she ever wrote. Typically it was in Latin, the language into which she retreated when lost for words: "I am in such a labyrinth that I do not know how I shall be able to reply to the Queen of Scots after so long a delay. I am at a loss to know how to satisfy her, and have no idea as to what I now ought to say." Cecil filed it away with a cryptic comment: "The Q[ueen] writing to me being sick."

The Dudley marriage plan had been a disaster. No one had ever bothered to ask Lord Robert for his opinion, and he had no desire whatever to marry Mary and live in Scotland. He was dropping frantic hints and doing everything within the bounds of discretion to evade the nomination, which, since he could hardly refuse to marry Mary if asked to do so by Elizabeth, meant finding an alternative husband. In Dudley's mind, the ideal surrogate was the English-born and supposedly loyal Darnley, whose candidacy he and his close friends supported.

One of Dudley's mentors was Throckmorton. Although a staunch Protestant, he did not agree with Cecil's reading of the threat Mary posed to Elizabeth's security. He had been arguing since before she left France that she should be recognized as the heir apparent to the English throne on the condition that she ally permanently with England.

Throckmorton lobbied for Darnley to get Dudley off the hook. By the end of December, Cecil could write of "a device" to steer Elizabeth in this direction. She had, he believed, "no disposition thereto"; but she proved him wrong. She was far from always living up to her image as a politically astute, dispassionate ruler. That, of course, was the Protestant stereotype. As Knox had proclaimed, Elizabeth ruled from the head and Mary from the heart, because in his eyes a Protestant queen was an "exceptional" personality able to overcome the frailty of her gender, whereas Catholic queens were not.

Over the winter of 1564–65, Elizabeth ruled from the heart. She allowed her emotions to dictate policy. Although she had herself named Dudley as the preferred suitor for Mary, she now pulled back, because if he really went to Scotland, she would lose the only man she had ever loved to a rival — for it seemed to her, in her jaundiced state after the Berwick conference, that Mary was actively seeking to steal Dudley from her!

At the start of February 1565, Darnley was given a passport to visit his father. For a few days, but just long enough to allow him to cross the border, Elizabeth decided it would, after all, be better if he and not Dudley married Mary. Mary had only to sit back and wait for Darnley to arrive. When he was safely in Scotland, she had Elizabeth exactly where she wanted her.

No longer did anyone suppose that Mary would marry Dudley. Just this was certain: she did intend to remarry. The pace quickened during a visit to St. Andrews in early February. Randolph had accompanied her, and she chatted with him after supper, sitting beside the fire.

"Not to marry," she said, "you know it cannot be for me. To defer it long, many incommodities ensue." She had made up her mind. But she also had regrets. She gently chided the ambassador that it could all have been so different if Elizabeth had lived up to her own rhetoric and treated her in a womanly, sisterly way.

"How much better were it," she said, "that we being two queens so near of kin, neighbors and living in one isle, should be friends and live together like sisters, than by strange means divide ourselves to the hurt of us both."

Randolph protested Elizabeth's friendship, but Mary knew the score. It was all just words. "To say that we may for all that live friends," she said, "we may say and promise what we will, but it will pass both our powers!"

Mary deplored the lost opportunity. Two women rulers, working together for the benefit of the British Isles, could have achieved "notable things." Now there were too many obstacles, not least that of marriage.

It was the classic dilemma for women rulers. Should they marry and have children, fulfilling the expectations of their councilors and subjects and settling the succession in their countries? Or should they stay single and keep their independence? It was difficult enough, but Mary's own choice had a further complication. Even if Elizabeth had already made her decision not to marry, she refused to allow Mary to trump her in marriage. If her cousin married, then the dueling between them would get steadily worse.

Mary began to prepare herself. A marriage to Darnley must by now have been firmly in her sights. If a settlement could not be reached with England, then should she not consider allying with the man whose hereditary rights, if united to hers, would make their dynastic claim invincible?

She certainly meant to do something. To damp down speculation that she intended to marry a Catholic, she reissued her proclamation of 1561 confirming the religious status quo. She also sent away her confidential secretary and decipherer, Raulet. He was a Guise retainer, the only person other than Mary to have a key to the black box containing her secret papers. She replaced him with David Rizzio, the young Piedmontese valet and musician who had arrived in the suite of the ambassador of the Duke of Savoy and stayed on as a bass in Mary's choir.

By sending Raulet home to France, Mary signaled her concern over security. She intended to prevent copies of her private letters from reaching the Cardinal of Lorraine. For some unknown reason, Raulet had fallen under suspicion. A month later, when he was preparing to board his ship at Leith, his trunk was seized with all his books and papers.

Mary first met Darnley in the depths of winter. The date was Saturday, February 17, 1565; the place, Wemyss, a tiny coastal village in Fife. Randolph wrote from Edinburgh to advise Dudley that his surrogate had safely arrived. Sir James Melville, an eyewitness, reported that Mary "took well" with Darnley, saying that "he was the lustiest and best proportioned lang [i.e., tall] man that she had seen."

Had there been an instant physical attraction? Darnley flattered himself that it was so. He stayed in the same house as Mary for two nights, then went to greet his father, Lennox, at Dunkeld. On arriving there, his first thought was to write a letter of thanks to Dudley, whom he offered to satisfy as if he were his own brother.

After lodging for five nights at Dunkeld, Darnley returned in time to cross the Firth of Forth on the same ferry as Mary. On the 26th, he decided to make a grand gesture. He attended Knox's sermon at St. Giles Kirk in Edinburgh, dining afterward with Moray and Randolph, who went there every week. In the evening at Holyrood, he danced a galliard with Mary. All eyes were watching them, but it was impossible to judge her reaction. Whatever was going through her mind, she kept her thoughts to herself.

The weather took a sudden turn for the worse. Violent snowstorms were followed by a frost of such intensity, nothing had been seen like it since the winter of Mary's birth. By the first week of March, pathways in the Lowlands were still closed. Darnley, however, was flourishing. Since travel was out of the question, the court settled down to a series of lavish banquets and masques. Darnley attended them all, and his courtly manners were "very well liked." As Randolph briefed Cecil, he "governs himself that there is great praise of him."

Mary as yet paid Darnley no unusual attention, which is why Randolph could praise him. She treated him courteously, but no more. She was biding her time. She had asked Randolph to give her a final answer about Elizabeth's intentions. By March 15, she was getting irritable and impatient, asking him every day when a reply was likely to be received.

Elizabeth's answer was ready on the 5th, but because of the snowdrifts, it reached Edinburgh on the 14th. When Randolph opened the packet, he knew there would be trouble. It took him two days to pluck up the courage to deliver the message. Elizabeth had dug in her heels. Where Mary's dynastic claim was concerned, "nothing shall be done until Her Majesty [Elizabeth] shall be married, or shall notify her determination never to marry."

This was truly a bombshell. It repudiated everything on which the Anglo-Scottish amity had depended and made a mockery of Mary's policy of conciliation toward Elizabeth. When the English queen had claimed the right of veto over Mary's marriage and named Dudley as the most favored suitor for her hand, it was clearly understood that compliance would be rewarded by recognition as Elizabeth's heir. Now the goalpost had been moved and everything depended on Elizabeth's decision on her own marriage, something that would likely never happen.

Mary listened in silence, then became angry, claiming Elizabeth had played a game of cheat and retreat. She had misled her and made her waste her time. "To answer me with nothing," she said, "I find great fault and fear it shall turn to her discredit more than my loss." "I would that I might have been most bound to my sister your mistress; seeing that cannot be, I will not fail in any good offices toward her, but to rely or trust much from henceforth in her for that matter I will not."

She then walked out and went hunting. Randolph tried to soothe Moray, who was "almost stark mad" with rage. Mary, Moray believed, would strike out on her own and Scotland would be in danger. For this, he was the "sorrowfullest" man alive. Maitland, to whom Randolph went next, was equally exasperated, although — always more thoughtful and insightful — he had privately expected a setback.

When Mary left Randolph, she "wept her fill." Later he tried to see her again, but she ignored him and retired to her privy chamber. He returned to Maitland and Moray, urging them not to act too hastily, but they cut him off, one saying, "The die is cast."

Next day, Mary rode to the beach beside Leith, where she watched Darnley and his companions "running at the ring." She approached Randolph, and "with the tears standing in her eyes" declared her love for her "sister queen," to whom she said she owed "such obedience as to her own dear mother." It was as big a compliment as Mary could be expected to pay, but far from being the prelude to a fresh attempt at conciliation, it marked the start of Mary's move to break free. She ended the conversation by requesting a diplomatic passport for Maitland. She was sending him through England on an urgent mission to France. Randolph saw the danger, but utterly failed to dissuade her. No one except Mary knew the purpose of Maitland's mission, but Randolph inferred that he was to consult her family about her marriage plans.

Elizabeth's policy was in disarray. She had miscalculated the impact of her message. It had caused a rift that seemed to be the more final in that it had apparently closed the gap that had existed between Cecil's position and her own. Whereas Cecil always wished to impose religious preconditions on the settlement of the succession, Elizabeth had been less dogmatic, keeping religion and politics apart and preferring Mary's claim to that of the disgraced Lady Catherine Grey. But if a settlement was put on indefinite hold, the effect of such distinctions would be immaterial.

Beyond this lay a more pressing danger. The return of Lennox and

Darnley to Scotland had triggered a realignment of the noble factions. Moray, Argyll and Châtelherault took the first step, signing a bond offering one another mutual support and promising to oppose a Catholic marriage. They cloaked their spite in claims about religion, but really they suspected Lennox of plotting to make his son king and so oust them from power.

Lennox had not been slow to look for allies. He had reintegrated himself in Scotland with remarkable speed. His supporters included the Earls of Atholl and Caithness, Lords Seton, Ruthven, Home and others. The first three were Catholics, the rest Protestants. What united them was not religion but ambition. Atholl saw an opportunity to displace Argyll from his preeminence in the Highlands, and all were keen to oust Moray and annex his lands. There would be rich pickings for the Lennoxes and their friends if Darnley married the queen. Already Darnley was known to have consulted a map of Scotland, when he was heard to say loudly that Moray's estates were far too extensive for his needs.

If, therefore, one effect of Elizabeth's message was to strengthen Mary's resolve to seek a marriage that would take her closer to the English succession, another was the resurgence of factionalism. Everything was once again on a knife edge. Moray realized this, and when Randolph tried to speak to him, he shooed him away. "The devil take you," he said. "Our queen does nothing but weep and write!"

On March 31, Randolph wrote candidly to his friend Sir Henry Sidney. He put the blame on Dudley for not taking more interest in Mary, whom he had not even met. She had grown into a woman of "perfect beauty." "How many countries, realms, cities and towns," he suddenly waxed lyrical, "have been destroyed" to satisfy the lusts of men for such women, and yet Dudley, who had been offered a kingdom and the opportunity to lie with Mary "in his naked arms," had spurned both, causing Darnley to arrive.

Randolph's apprehension was justified. By the first week in April, Mary and Darnley were in the early stages of a courtship. They were staying at Stirling Castle, where they spent most of their time together. Mary showed him off as her partner in a game of billiards played for high stakes against Randolph and Mary Beaton. It was neatly arranged, as it was an open secret that Randolph and Beaton, one of the four Maries, were lovers. And so it seemed to everyone watching that two young couples were playing against each other.

It had been settled that whoever won, the women would share the kitty. When Randolph and Beaton emerged the victors, Darnley was seen to present Mary with a ring and a valuable brooch set with two agates.

When Darnley then fell sick, Mary nursed him herself. She sent him food from her table and visited him at almost all hours, sometimes after midnight, a bold thing for her to do. He was "very evil at ease," and for his comfort and convenience was lodged in the royal apartments in the castle, an extraordinary honor, where he remained for over a month. He succumbed first to a cold and then to skin eruptions. His symptoms were a measles-like rash, "marvelous thick," accompanied by "sharp pangs, his pains holding him in his stomach and his head" — almost certainly the onset of syphilis, caught in England.

When Maitland reached London, purportedly on his way to France, the true nature of his mission was clear. He delivered an ultimatum demanding Elizabeth's consent to Mary's marriage to Darnley. The result was chaos and confusion.

On April 23, a frantic Elizabeth signed letters recalling Lennox and Darnley to the English court, only to countermand them at the last moment. Instead, Throckmorton was to be sent to Edinburgh. His first instructions were issued on the 24th, ordering him to explain to Mary how much her proposed marriage was "misliked," but a week later they were replaced by others advising her to marry Dudley or else choose another English nobleman. If Mary agreed to this, a string of concessions would be offered. Only if Dudley was chosen, however, would Elizabeth ever agree to have Mary's title to the English succession "either published, endorsed or enquired of."

Cecil was a worried man. On May 1, the first of a series of extended debates was held by the Privy Council to decide how best to "disallow" Mary's marriage to Darnley.

After two years of bandying words, Mary felt she had nothing to lose. She had already more or less decided to bypass England. She was more cautious about burning her bridges to the Continent. She first wrote to Spain. In reply, the Duke of Alba, answering on behalf of Philip II, reassured her that "no alliance would be more advantageous to her for assuring the success of her claims and the quiet of her country than one with the Lennox family."

Mary then approached Castelnau to see if he could secure French backing, a delicate request given her poor standing with her former mother-in-law. But to his surprise, Catherine de Medici agreed. Her response — that Darnley was at least preferable to Don Carlos or Archduke Charles because such a marriage would encourage a return to the "auld alliance" — was strictly a non sequitur. What she really meant was that it would strengthen her own bargaining position with Elizabeth, who would need to rely even more on a French entente if the amity with Scotland collapsed.

Catherine was singularly duplicitous. Castelnau was ordered to inform Elizabeth that France opposed the Darnley marriage, with the result that much confusion and resentment resulted from what shortly became a pantomime of ambassadors rushing to and fro between three capitals, trying to work out who had said what to whom and when.

This worked to Mary's advantage, as it left her free to deal with the domestic opposition to her marriage. When Moray, summoned to appear before his sister in Darnley's sickroom, refused to sign a document pledging his support, claiming that her marriage was too hasty and that "he misliked" it "because he feared that the Lord Darnley would be an enemy to true religion," he was forced to withdraw from the court in disgrace.

Mary sought to isolate Moray. Cautious support for her marriage straddled the religious divide, and there were other factors working in her favor. One was so incongruous it was almost whimsical. Maitland was in love with Mary Fleming, the chief of the four Maries. His wife had died, and he wanted to marry a woman eighteen years younger than himself. It was a difficult courtship: the running joke was that Maitland was as well suited to Fleming as a Calvinist was to be pope.

Maitland's wedding was delayed for almost two years, but as Randolph knew, if Mary wanted to marry Darnley, Maitland would support her "from the love he bears to Mary Fleming." There was nothing Fleming would not do for Mary, and she had Maitland wrapped around her little finger. It is true that his loyalty was severely tested, but in the end he was (as the English ambassador saw it) "blinded to further and prosecute this marriage."

The key factor in Mary's favor was that Morton had attached himself to Darnley. So far, he had been closely allied to Moray, taking a prominent position as the head of the Douglas clan. But as a Douglas, he was the Countess of Lennox's cousin. His defection from Moray was triggered when the countess surrendered her claim to the contested estates of the Earl of Angus, Morton's nephew, for whom he was acting as guardian. This was the equivalent for Morton of a lottery win, since he could now strip his nephew's estates for his own profit. He had previously asked Mary to confirm his claim when he had sent his cousin Archibald to visit her before she left France. She was unwilling to act then, claiming it was a family dispute. The issue had smoldered, but now Morton was bought off.

Throckmorton arrived to see Mary at Stirling on May 15 while Darnley was still a convalescent. He had traveled to Scotland with Maitland, who had never intended to go to France. The talk of his mission there had been a blind.

Throckmorton found the gates of the castle shut firmly against him, and was obliged to seek lodging in the town. When he finally obtained an audience with Mary, he handed her a letter from Elizabeth and a "Determination" from Cecil: a formal document signed by a majority of the Privy Council advising her to put Darnley aside and marry either Dudley or another English nobleman. The document had an impressive list of signatories, but as Dudley was for some reason not among them, its impact was greatly reduced.

Mary was unimpressed. Throckmorton warned Cecil she was "so far past in this matter with my Lord Darnley as it is irrevocable, and no place left to dissolve the same by persuasion and reasonable means." In reality, she gave a spirited justification of herself. She pointed out that she had advised Elizabeth of her intention to marry as soon as she had made up her own mind. As her cousin had informed her that she might choose her own husband as long as she rejected anyone from the Continent in favor of an English nobleman, she felt she had acted honorably. In Darnley she had lighted on both an English nobleman and Elizabeth's "near kinsman," a choice with which she could only suppose the English queen would be delighted.

Mary did not waste time in negotiations with Throckmorton. Right after the audience, she raised Darnley to a sufficient standing among the Scottish nobles so that she could marry him. She knighted him, created him a baron, then made him Earl of Ross, all on the same afternoon. He in turn created fourteen knights on his own account, but his true colors were glimpsed within a week. Mary had promised to make him Duke of Albany, a title reserved for Scottish royalty, but deferred the ceremony because she wished to see how Elizabeth would react to her answer to Throckmorton. When Darnley learned of the delay, he drew his dagger on Lord Ruthven, who had brought the message. He also threatened Châtelherault, even though the duke had said (admittedly treacherously) he would support the marriage and signed a paper to that effect.

Darnley's defects of character were beginning to emerge. He had managed to behave himself for three months, but was unable to keep it up. The prospect of marrying Mary had gone to his head. Spoiled as a child by his mother, he was overconfident, arrogant and willful. Far too handsome for his own good, he was a narcissist and a natural conspirator. Already he was described as "proud, disdainful and suspicious." He was soon regularly getting drunk, and his sexual license was suspected when he was found to be so intimate with David Rizzio, Mary's new confidential secretary, "they would lie sometime in one bed together."

But as yet these defects were just specks on the horizon. The bulk of our information comes from Randolph, who was a hostile witness. His task was to prevent Mary's marriage to Darnley, and so he found as many reasons as he could to discredit the man he saw as the greatest threat to England since Don Carlos.

By the third week of May, Mary believed she was in love with Darnley. She doted on him. Even Randolph had to admit it. But it was a brief infatuation, brought on by Darnley's sexual attractiveness rather than true love.

By June 3, the relationship was already turning sour. Mary's marriage to Darnley would in the end become purely one of convenience. She had trapped herself, because even when she began to realize what Darnley was really like, she had no choice but to go ahead if she was to maintain her independence and not seem to be Elizabeth's pawn.

If Randolph is to be believed, Darnley's behavior became so "intolerable," Mary suffered a severe attack of melancholy. She was not just depressed, her whole appearance changed. "Her majesty is laid aside," he confided to Dudley, "her wits not what they were, her beauty another than it was, her cheer and countenance changed into I wot not what." Mary had become "a woman more to be pitied than any that I ever saw." The change was so sudden, necromancy was suspected.

And yet on the very same day, Randolph could tell Cecil how Mary was excitedly keeping abreast of her French and Spanish diplomacy, and doing all she could to bind her lords to herself "by gentle letters and fair words." She was carefully building up support for her marriage.

A juggernaut had started to roll. Cecil convened the English Privy Council on the 4th to discuss Mary's answer to Throckmorton. The debate lasted all day. The entire life story of the Queen of Scots was retold, from the treaty of Haddington and her marriage to the dauphin to Henry II's claim that she was the heir to a triple monarchy, and with special reference to the blazoning of the heraldic arms of England on her escutcheons and the cries of her ushers to "make way for the queen of England!" Nothing was omitted or allowed to be forgotten.

Cecil led the debates and wrote the minutes, explaining why Mary was so dangerous. By announcing her intention to remarry and settle the succession in Scotland, she would benefit from a surge of new support. A majority of Elizabeth's subjects — he spoke repeatedly of "the people of England" — favored Mary. They would flock to a queen who, unlike their own, was prepared to marry and have children. The sixteenth century was an age of gender stereotypes. By marrying, Mary would do what Cecil and every other male councilor and head of household wanted a female monarch to do. She would put a man at the head of the royal family. She would recreate a truly regal monarchy and prove that it would be her heirs, and not Elizabeth's, who would eventually unite the thrones of England and Scotland.

The "people of England," Cecil said, whether Protestants or Catholics, would be so won over by this, they would be "drawn away from their allegiance" to Elizabeth and transfer it to Mary.

It was a fascinating argument, opening for us a window into a lost world in which royal marriages had the power to alter people's lives. Cecil was Mary's most single-minded opponent. Any marriage that she contracted while Elizabeth was unmarried posed a threat to England's security. To speak of "drawing away" the "allegiance" of the people was uniquely resonant, as it was the language of the law of treason. Cecil implied that if Mary married without Elizabeth's consent, this in itself was a hostile act.

No one challenged Cecil. In fact, some privy councilors argued that if Mary married Darnley, the threat was even greater than if she had allied herself to Don Carlos or the archduke. Their fear was that Darnley, an English subject of royal blood, had the potential to raise an army in England in support of the Catholic cause and so begin a civil war.

With hindsight, it seems like a tremendous overreaction, and yet this is what Cecil and his allies believed. Their starting point was Throckmorton's report, which had argued that Mary was so intent on marrying Darnley, "the matter is irrevocable otherwise than by violence." The main purpose of the debate was to decide whether England was willing to go to war. Hence the meeting turned into a full-scale strategic review, ending with an action plan that included asking Elizabeth to look more favorably on the plight of Lady Catherine Grey.

Cecil gave Elizabeth a copy of these minutes. She was not usually disposed to listen to his fears about her "safety," but this time she did. The Earl of Bedford, the most senior English border official and the governor of Berwick-upon-Tweed, was ordered to keep a strict watch and to assist Randolph in planting spies. The Countess of Lennox was sent to the Tower, and the letters recalling Lennox and Darnley were reissued. Randolph received them in his diplomatic bag on July 2 and delivered them at once. As he reported, their recipients were "marvelously abashed."

When Mary heard that Lennox and Darnley were recalled, she burst into tears. Up until now, her weeping had tended to stem from disparate groups of emotions: grief at her bereavements, sadness or depression during her attacks of melancholy, anger and frustration with Knox or her factious nobles, or helplessness and self-pity at the betrayals of others. This time, she felt them all. A great storm of emotion overtook her, and she did nothing for a fortnight. Then the skies cleared and she made her decision. She ordered Lennox and Darnley to remain in Scotland and defy Elizabeth.

This was the point of no return. Mary rallied herself and began the arrangements for her wedding. She had made her choice and would live with it. "I know," she told Randolph, "that your mistress went about but to abuse me, and so was I warned out of England, France and other parts, and when I found it so indeed, I thought I would no longer stay upon her fair words, but being [as] free as she is, I would stand to my own choice."

Just as Cecil had given the English Privy Council a lecture that went back to the treaty of Haddington, now Mary also harangued Randolph with the history of her diplomacy with Elizabeth, starting with her desire for a "fresh start" and an exchange of portraits, continuing with her efforts to arrange a meeting between the queens at York or Nottingham, and concluding with the innumerable zigzags and weasel words produced by English attempts to dictate her choice of husband.

It was now, she said, "too late." What could have brought great mutual benefits had come to nothing. And it was entirely Elizabeth's fault. "For if your mistress would have used me as I trusted she would have done, she can not have [had] a daughter of her own that would have been more obedient to her than I would have been."

Mary could not quite let go of the vocabulary of kinship and dependency that had colored her relationship with Elizabeth since the death of Francis II. But she vigorously defended her right as a queen and a woman to choose her own husband. This was not Elizabeth's business, even if Mary had hoped that Elizabeth might approve of her choice. "Let not her be offended with my marriage, no more than I am with hers," she said, "and for the rest I will abide such fortune as God will send me."

Randolph then made a serious mistake. He suggested almost as an afterthought that Mary might do well to convert to Protestantism. Elizabeth might treat her better if she did so. Mary hit back instantly.

"What would that do?" she asked.

"Peradventure," replied Randolph, it would "somewhat move Her Majesty to allow the sooner of your marriage."

"What!" Mary exclaimed. "Would you that I should make merchandise of my religion, or frame myself to your ministers' wills? It cannot be so."

This suggestion was humiliating and degrading. It made Mary even more determined to do as she pleased.

"You can never persuade me" — she almost spat out the words — "that I have failed to your mistress, but rather she to me; and some incommodity it will be as well for her to lose my amity as hers will be to me." With that, she got up and left.

On the morning of Sunday, July 22, 1565, the banns for Mary's marriage were read in St. Giles Kirk, and in the afternoon Darnley was created Duke of Albany. At nine o'clock on the following Saturday evening, heralds appeared at the Market Cross in Edinburgh to proclaim that the couple would be married the next day. And when the celebrations were ended, Darnley would be made king of Scotland.

Darnley himself had insisted on this, even though Parliament was usually consulted before royal titles were bestowed. The Privy Council had debated the proposal for much of the day. Even as Mary was fast winning the support of those lords who were wavering over the marriage, Darnley was losing it by boasting that he cared more for the English Catholics than the Scottish Protestants, a typically insouciant and foolish remark that canceled out the goodwill he had earned by attending Knox's sermons.

Mary was unwise to yield to Darnley's influence over the title. She acted naively and impulsively, and would have done better to insist that Parliament be asked, using it as an excuse to delay a decision until she could see whether he was going to be able to handle this sort of power.

But he would not take no for an answer. The force of gender stereotypes had come into play. He was the man and Mary was the woman, and for the moment she felt that she had to indulge him. She needed the wedding to go ahead if she was not to lose face. She even attended the Privy Council, talking around a sufficient number of lords to win the vote. When the decision was finally agreed, the heralds had been kept waiting for over six hours to make their proclamation.

The wedding took place on Sunday the 29th. Shortly before six o'clock in the morning, Mary was led to the altar of her private chapel at Holyrood by the Earls of Lennox and Atholl. She wore her *deuil blanc* to signify that she was a widow, dressed exactly as when Throckmorton had described her after the death of Francis II.

When Darnley arrived, the wedding vows were exchanged and the bridegroom put three rings on the fingers of Mary's right hand, the middle one set with a rich diamond and enameled in red. (The wedding ring was then worn on the right hand.) They knelt for the prayers, after which Darnley abruptly left. He wished to avoid the charge of "idolatry" and so went straight to the royal apartments, refusing to attend his own nuptial Mass.

The service finished, Mary returned to her bedroom. She was elated and determined to make a success of the marriage in spite of Darnley's behavior. She particularly wanted the ceremony to be inclusive. All the nobles had been invited, and in a ritual staged to mark her passage from widowhood to wifehood, she allowed each of those attending to remove one of the pins holding her veil to her gown. Everyone except Darnley then left while Mary changed her clothes. As Randolph primly noted, they "went not to bed, to signify unto the world that it was no lust [that] moved them to marry, but only the necessity of her country, not long to leave it destitute of an heir."

There followed great "cheer and dancing." Or as Knox sourly put it, "During the space of three or four days there was nothing but balling, dancing and banqueting." Mary and Darnley walked in procession to a state banquet attended by their lords. As they entered the great hall, trumpets sounded. And in an echo of Mary's wedding to the dauphin at the cathedral of Notre-Dame, heralds cried out "Largesse," showering money on the guests. The afternoon was spent in dancing and revelry, after which there was a second banquet to which others were invited. The dancing went on late into the night, "and so they go to bed."

Randolph was invited to the second banquet, but was obliged to refuse. As Elizabeth's ambassador he could do nothing that seemed to recognize the marriage. Already Elizabeth and Cecil were determined never to acknowledge Darnley as king or even as Mary's husband. Mary understood this and wanted Randolph to attend the banquet so badly, she kept Mary Beaton, his mistress, away from him for a fortnight beforehand, promising that they could dance together if he attended, but to no avail.

On Monday at midday, heralds proclaimed Darnley's title as King of Scotland. A dual monarchy was to be established in which sovereign power would be "conjointly" exercised by the king and queen. The official style for all letters and state papers would be "Henry and Marie, King and Queen of Scotland."

The proclamation was heard in a resentful silence. Of all the nobles, no one so much as said "Amen," until the lone voice of Lennox cried out, "God save His Grace."

The wedding had been on a sumptuous scale, yet the celebrations were largely hollow. At the first banquet, the royal couple had been served by Atholl and Morton, the only nobles to offer their unreserved support for the new king. No one else was willing to wait on Darnley or serve his food. Moray, Argyll and Châtelherault had refused to attend. All three were in disgrace and building up their armies. They intended to rebel, to force Mary to separate herself from a man they feared would try to destroy them. They wrote to Cecil seeking troops and artillery, and sent a messenger to London to press their case. Next they approached Randolph and the Earl of Bedford, appealing for money. They demanded £3000, the same amount Cecil had given them six years before when they had begun their revolt against Mary's mother.

On July 30, Elizabeth sent John Thomworth, a gentleman of her privy chamber, as a special ambassador to Scotland. He was to see Mary, while at all times ignoring Darnley, and remonstrate over her "very strange" and "unneighborly" conduct. He was to tell her she was beguiled by "sinister advice."

Elizabeth decided to set Mary straight. She must know that "she forgets herself marvelously to raise up such factions as is understood among her nobility." A civil war was in prospect, or so Elizabeth supposed. Mary should reconcile herself to Moray, a man "who has so well served her."

But Mary was having none of it. Before Thomworth reached Edinburgh, Moray was declared a rebel. And when the ambassador arrived, Mary gave him a lecture. A queen, she said, had every right to marry without rendering an account to other princes. It had not been her practice "to enquire what order of government her good sister observed within her realm," nor did she believe it was the custom of princes to interfere in the internal affairs of neighboring states. Princes were "subject immediately to God, and to owe account or reckoning of their doings to none other." No one should know that better than Elizabeth herself.

As to Moray, Mary advised Elizabeth "to meddle no further with private causes concerning him or any other subjects of Scotland." She should follow this advice unless she wished Mary to reciprocate by taking up the cause of her mother-in-law, the Countess of Lennox, whom Elizabeth had unworthily imprisoned.

Mary was undaunted. Whatever Darnley's many failings, she believed she had the upper hand. Whether she had actually improved her position is arguable, but she had gained the confidence to act strongly. For the first time in their relationship, she and not Elizabeth drew up an offer of terms. According to these, Mary and her husband would do nothing to enforce their immediate dynastic claim, nor to countenance English rebels. They would not ally with foreign princes against Elizabeth, nor seek to alter the religion, laws or liberties of England. In return, they expected Elizabeth not to ally with foreign princes or Mary's rebels, and in particular to settle the English succession by act of Parliament in Mary and Darnley's favor.

For the first time in her life, Mary was setting the agenda and reveling in it. After a brief infatuation in which she had been sexually attracted to Darnley and had believed she found love, she had come to realize that her marriage was purely one of convenience. Darnley, she knew, was unpopular. But her own popularity could counter this, and as a married queen with every likelihood of producing heirs, her position was more secure than at any time since her return from France. Only the jealous, perfidious Moray and his allies stood between her and a long and happy reign. And she intended to deal with them next.

14

Enter Bothwell

MARY COULD NOT have been more focused or astute in dealing with Moray's latest revolt. She first isolated her opponents politically, then took to the field to rout them. Her skill and courage were more than sufficient to disprove Knox's stereotype that a Catholic woman ruler would be too busy dancing or pursuing a life of material and sexual indulgence to rule her country properly.

She began by exposing the flaws in the rebels' propaganda. They were claiming that her marriage to Darnley would mark the end of the religious status quo. From the outset they had sought to ignite sectarian hatred between Protestants and Catholics for their own political ends.

Their campaign had begun on Palm Sunday. A Catholic priest celebrating Mass in Edinburgh was abducted in his vestments by a Calvinist gang and frog-marched to the Market Cross, where he was tied up and pelted with ten thousand eggs for four hours. Eggs, the Catholic symbol of Easter, were deliberately chosen for this act of ritual humiliation. Youths armed with cudgels then arrived to beat the victim senseless; he was freed only after a last-minute intervention by the provost. The fact that so many free eggs were conveniently supplied shows that the attack must have been powerfully backed, although nothing can be proved to link it to Moray.

Mary had countered this shocking act of violence by confirming the religious status quo. A majority of the Protestants did not think their re-

ligion was under threat. Her policy of creating a broad coalition — in the
Privy Council and the country at large — had so far worked. Even Knox
did not believe there was a genuine threat to the Kirk this time, and he
declined to speak out in Moray's favor, a major prize for Mary and one
she greatly relished.

To make sure that she kept bipartisan support, she put out further reassuring signals to the Protestants. She ate meat for the first time during
Lent, defying the Catholic obligation to avoid it between Shrovetide and
Easter. She did not hear Mass as often as she usually did. She attended a
Protestant baptism, and volunteered to go to sermons if preachers could
be found of whom she approved. She dropped heavy hints to Randolph
and others that her subjects should "live as they list," following their own
beliefs.

A fortnight before she married Darnley, Mary wrote to the leading
Protestants, reassuring them that they would always be free to worship
according to their consciences. A month after the wedding, her religious
proclamation of 1561 was confirmed for a second time, this time in the
joint names of the king and queen.

Mary, meanwhile, was mustering a royal army. By July 19, ten days before the wedding, she had around six thousand troops within easy reach
of Edinburgh. And in his abstracts of Scottish papers for the same day,
Cecil made an entry that was brief but portentous: "The Earl Bothwell is
sent for."

No more formidable adversary, no more deadly rival of Moray, could be
imagined. Like his father, Patrick, who had died in 1556, Bothwell was
hereditary Lord Admiral and the Sheriff of Edinburgh. In the Lowlands,
the family held Crichton Castle, just south of Edinburgh, and Hailes
Castle in East Lothian. Their main stronghold was the Hermitage, an almost impregnable fortress in the border region of Liddesdale, not far
from the Debatable Land.

Bothwell was seven years older than Mary. He cut a dashing figure: he
was stocky like his father, although his hair was darker and his complexion ruddier. He had a military bearing and sported a mustache in the
French manner. His eyes were darting and his gaze restless. The only
known portrait of him is a miniature, one of a pair depicting him and his
wife, Lady Jean Gordon, and made to celebrate their marriage in 1566.
The work is of modest quality, but Throckmorton, who met Bothwell at
Orléans in late 1560, caught some of his traits. In a dispatch to Eliza-

beth, he described him as "a vainglorious, rash and hazardous young man." His enemies should "have an eye to him and keep him short."

Bothwell's military and chivalric ethos is the key to his character. Like his father and grandfather, he saw himself as a "man of honor," who favored trial by combat and fought at least a dozen duels. He regarded violence as a valid method of settling disputes and an alternative to a judgment in a court of law.

It is not exactly that he held civilian values in contempt. He had a rough and a smooth side. He could discipline soldiers and pirates, but like Mary was educated in France. He had attended the University of Paris, where he learned to write in a stylish italic hand neater than Mary's own. He had dipped a toe into the world of learning, but always preferred action to words. He regarded honor and nobility as virtues conferred by birth and on the battlefield rather than acquired in libraries or the council chamber. He could swear profusely and profanely. And yet he was far from being a philistine; he knew some classical history. But the books he collected were on mathematics, military strategy and chivalry, not moral philosophy, poetry or literature.

Bothwell, unlike his father, was a Protestant. Or, to be more precise, he was a nonideological Protestant. He was a skilled swordsman but not an advocate of the spiritual weapons of prayer and repentance. That is why men like Throckmorton, whose civilian values caused them to loathe everything for which Bothwell stood, could describe him as "a man of no religion." He was not, however, a Catholic sympathizer. If he did not take his Protestant beliefs too seriously, he refused to attend Mass even when begged to do so by Mary.

Somewhat unusual for a Protestant, Bothwell was virulently anti-English. More consistently than his father, who had flirted with England when it suited him, he had always been a nationalist and unwaveringly loyal to Mary and her mother. During the revolt of the Lords of the Congregation, he was the regent's most stalwart supporter. When Cecil sent £3000 to assist the rebels in October 1559, it was Bothwell who intercepted the courier. This man was one of Randolph's secret agents: Captain Cockburn, Laird of Ormiston. He was smuggling the first instalment of £1000 in untraceable gold coins across the border when Bothwell waylaid him in East Lothian, taking the money to Crichton. Although Moray and Arran promptly went there with a force, Bothwell had posted lookouts, and half an hour before they arrived he fled with the money. The lords occupied the castle and stripped it of its furniture.

Bothwell challenged Arran to a duel, but Arran, an inveterate coward, refused to fight. From then on, Bothwell was excoriated by the lords for putting their cause in jeopardy.

Shortly afterward, Mary's mother had sent Bothwell to command a special force of eight hundred French and Scots troops at Stirling. In April 1560, he and the Catholic Lord Seton ambushed the commander of the English forces attacking Leith. As a border lord used to dealing with the so-called Scottish "riders" whose depredations were at their worst in his own region of Liddesdale, Bothwell was an expert in guerrilla warfare.

A few weeks later, he was sent to France carrying letters to a worried Mary. As he wrote to reassure her mother before he left, "I have made the greater haste to the effect I may return again with the army." He was planning to travel by way of Denmark, where he hoped to persuade Frederick II to lend Mary of Guise his fleet to transport five thousand German mercenaries to Scotland. Then the regent died and Bothwell's commission expired. He dallied for several months in Denmark, where he had an affair with a beautiful dark-haired Norwegian girl, Anna Throndsen, the daughter of a retired nobleman and admiral, who later sued him for breach of promise.

Presenting himself to Mary at Paris and Orléans, Bothwell was barely in time to deliver his letters and receive his reward before Francis II died. He was given 600 crowns of the sun, a typically generous amount, the fee of a *gentilhomme* of the *chambre du roi* and essentially a pension. Lord Seton received a slightly higher amount, and these rewards were for their loyalty to Mary's mother, not for services to Francis. In Bothwell's own words, "Amongst others she [Mary] rewarded me much more liberally and graciously than I had deserved, which angered my enemies to the greatest degree."

Bothwell's mortal enemy was Moray, who never forgave him for thwarting the lords in 1559–60. While the council of twenty-four nobles ruled in Scotland, Bothwell avoided them by crisscrossing to and from France. He was in Scotland in February 1561, but was at Calais in the summer to assist in bringing Mary home. As he was Lord Admiral, it has been assumed that his role was to escort Mary. This is mistaken, because her official escorts on the voyage were French. Her galleys were commissioned by her uncles, and Bothwell's role as Admiral was to organize the flotilla of chartered transport ships that carried the queen's baggage and animals.

After Mary had returned home, Bothwell's feud with Moray intensified. As a leading lord, he was appointed to the Privy Council but rarely
took up his seat, as Moray gave him no respite. In a series of incidents,
he was attacked or set up by Moray. Then, in December 1561, Bothwell
gave his rival a golden opportunity. He went out on what amounted to a
stag night with Lord John of Coldingham. Lord John was about to
marry Bothwell's sister, Jane Hepburn,* and they decided to pay a late-
night visit to Arran's mistress in Edinburgh, a woman named Alison
Craik. On a previous occasion, they had been admitted and enjoyed
themselves, but this time they were refused entry. They broke down the
door, leading next day to a confrontation with Arran and his friends that
sparked a riot. Just when things were hotting up, Moray appeared on the
scene to restore order.

Bothwell was expelled from Edinburgh, but fresh conflicts arose with
Arran, whose delusion had then been to marry Mary. Bothwell, the
swashbuckling military officer who had loyally defended her mother,
ridiculed Arran and insulted him for his presumption. The result was
another round of feuding, because even if Mary regarded Arran with the
utmost suspicion, he was still an ally of Moray.

At Mary's insistence, the two men were cursorily reconciled, but
Arran behaved like a man possessed. He was mentally disturbed, once
calling for a saw to cut off his legs and for a knife to slash his wrists. Now
he accused Bothwell of treason. In a series of increasingly incoherent depositions he claimed that Bothwell had advised him to abduct Mary as a
prelude to marrying her. These were serious charges even if they sprang
from a delusion. An investigation had to be conducted into this alleged
conspiracy.

Bothwell had nothing to hide. He willingly surrendered and, when
Arran accused him of treason, challenged him to a duel. In the end,
every one of Arran's charges was dismissed. But Moray would not free
Bothwell. Instead, he used his power in the Privy Council to imprison
him in the dungeons of Edinburgh Castle. After three months, Bothwell
decided to escape. Legend says that he broke the bars of his cell and
scaled down the Castle Rock without a rope, but the more likely if less
romantic explanation is that he bribed someone to let him out by the
gate.

Bothwell fled to his mother's house in Haddington and then took ref-

---

* Bothwell's sister is variously called Jane, Jean and Janet in the sources.

uge at the Hermitage. He wrote to Mary offering to submit, but she was furious with him for breaking out of prison. Despite her own wicked sense of humor, any amusement or gratification she might have felt over his clumsy pranks at Arran's expense was wiped out by his cavalier approach to justice. Banishment and exile were the penalties for aristocratic lawlessness, and this is what she had in mind for Bothwell. "Anything that he can do or say can little prevail," wrote Randolph to Cecil, because "her purpose is at the least to put him out of the country."

This was shortly before the Gordons were destroyed by Moray at the battle of Corrichie. Mary's brother was at the height of his power. In Bothwell's view, he was deceiving his sister by pursuing his personal and political vendettas under the cloak of enforcing law and order. Even Châtelherault had lost out, his stronghold at Dumbarton seized by Mary on her manipulative brother's advice. Worried that he could be next, Bothwell decided to leave of his own accord, and in late December 1562 took ship for the Continent. When this was reported to Mary, she lost her temper and insisted that Bothwell would be punished.

It looked like the end for Bothwell. Moray appeared to have won hands down. But within a month, Mary had begun her marriage suit for Don Carlos. Moray was allowed to drift, then Maitland was sent on his mission to the Spanish embassy in London. Mary was taking a tougher line against English attempts to intimidate her. In that context, she started to view the nationalist and Anglophobe Bothwell more favorably.

Randolph had been the first to spot the change. Mary was asking where Bothwell was. She wanted him back, to be "reserved, though it were in prison, in store to be employed in any kind of mischief that any occasion may move." It is an illuminating insight that holds the key to Mary's relationship with Bothwell over the next two years. As an irritant, both to England and to Moray, he could prove invaluable. His lawlessness in this respect was likely to be a positive advantage.

Matters are complicated by confusion over Bothwell's movements. The usual story is that he stayed in exile until March 1565, when he briefly returned, only to be forced back to France. He was actually in and out of Scotland, especially in 1564, when he met Mary secretly at Dunbar. It would be tempting for the historian to try and mold this evidence to prove a romantic liaison between Mary and Bothwell even before Mary married Darnley, but this would be to fall into a trap.

Bothwell never reached France at the beginning of 1563, because he was shipwrecked. He was driven by a violent storm onto Holy Island, off the coast of Northumberland, and captured. Held first at Tynemouth Castle, he was taken to London and imprisoned in the Tower. He was in a confident, brash mood, informing his captors (somewhat contradictorily) that he was sufficiently in favor with Mary to be on his way to visit her relations, but was on no account to be sent back to Scotland to face the wrath of his enemies. This was typical Bothwell, always bluffing his way out of trouble.

Mary asked for news of him just as Maitland was headed for London. "Whatsoever they say against him," she said, "it is rather from hate of his person and love that they bear otherwise than that he hath deserved." She asked Randolph to write to Elizabeth to demand Bothwell's return. "I do desire," she said, "that he may be sent hither again into Scotland. So shall the pleasure be great and I will gladly requite the same." But if Bothwell had returned, she would have put him back in prison for breaking out of Edinburgh Castle. A summons had already been issued against him for this offense, which simply awaited his reappearance.

Bothwell had risen in Mary's opinion because he was the perfect foil to Moray and his English friends. Randolph was well aware of this. "One thing I thought not to omit," he told Cecil, is "that I know him as mortal an enemy to our whole nation as any man alive, despiteful out of measure, false and untrue as a devil."

By April, Mary was again demanding to know why Bothwell was in the Tower. "If he were here," said Randolph from the vantage point of Edinburgh, "he would be reserved for an evil instrument." A reply came that Bothwell was in custody for a legal dispute and that a trial was imminent. This seemed to satisfy Mary. By the end of June, Bothwell was released on parole. "Lock up your wives and daughters!" was Randolph's advice to Cecil.

As soon as Bothwell was freed, Mary forgot about him again. She was too busy negotiating for a match to Don Carlos and against one to Archduke Charles of Austria to worry about him. Just as before, Bothwell was in limbo. In desperation, he decided to petition Mary, because he had no money and under the terms of his parole could not leave England. In December, he met Randolph near the border and suggested a deal. If Randolph would use his influence with Mary (proving of course that Bothwell as yet had none) to persuade Elizabeth to allow him to leave England, then in return he would go voluntarily into exile in France.

Randolph agreed. But it was at precisely this moment that Mary's duel with Elizabeth over her marriage was nearing its climax. She had just received Elizabeth's advice to marry an English nobleman and Cecil's even more insulting demand that her dynastic rights be tried in an English court of law. She did not intend to allow Elizabeth to keep any Scottish nobles in England against their wills, and therefore wrote to her, not once but twice, for Bothwell. So too did Mary Fleming, the chief of the Maries, who persuaded the lovesick Maitland to write as well. Even Randolph was induced to write, possibly at the intercession of his own lover, Mary Beaton.

By February 1564, Randolph knew he had been tricked. "Such as have written, and I amongst the rest, in favor of my Lord Bothwell," he advised Cecil, "saving the queen and Mary Fleming, repent their haste. It is found out that this way it is purposed to bring him home." And sure enough, a week later Bothwell slipped across the border and saw Mary at Dunbar. When the news got out, everyone thought Moray was in danger. "Bothwell," said Randolph, "was come secretly to speak with the queen with many horses," and Moray, who would normally have accompanied her, was ordered to stay at home.

Then Bothwell was off again to London on his way to France, carrying a packet of letters. It sounds conspiratorial until one realizes that this was one of Mary's theatrical ploys to irritate Elizabeth. The rendezvous at Dunbar was so "secret" that Randolph knew all about it. And the letters Bothwell carried were innocent. We know to whom they were addressed: Elizabeth, Cecil and Dudley! When Bothwell had delivered them, he was finally allowed to board his ship for France. He had been a convenient go-between on his way south to the channel ports.

One purpose of the rendezvous was to prove to Elizabeth that Mary had lords who would champion her refusal to have the terms of her marriage dictated to her by England. Mary, said Knox's friend Kirkcaldy of Grange, wanted to be sure that Bothwell was "at all times ready to shake out of her pocket."

There was also a less melodramatic explanation. Mary placed a high premium on family ties, and one little-known fact about Bothwell is that when his sister, Jane Hepburn, married Lord John of Coldingham, she became Mary's sister-in-law. Lord John, another of James V's illegitimate sons, was one of Mary's most spirited and sophisticated courtiers, noted for his love of dancing. He was a great favorite of hers, and she granted him Dunbar Castle, on the edge of the cliffs of East Lothian. When Bothwell went to Dunbar, he had merely visited his sister, with whom Mary happened to be staying.

The main purpose of his visit was probably even more straightforward. Lord John had unexpectedly died a few weeks before. Bothwell's sister was a grieving widow, and Mary had gone to comfort her. What could be more natural than that Bothwell, waiting close by on the other side of the border for his passport to travel to France, should want to see his sister before he embarked?

And there was a final connection. Bothwell's grandmother had been James IV's mistress. The rumor, almost certainly unfounded, that his father, Patrick, was the king's son persisted. If it was true, then Mary and Bothwell were cousins. This explains something that perplexed Randolph after Mary's return to Scotland. "The Earl Bothwell," he wrote, "hath given unto him old lands of his father's in Teviotdale and the abbey of Melrose. Some say that he is near sib unto Her Grace." Randolph could not understand why people were hinting that Mary and Bothwell were near siblings, but the old rumor about Bothwell's grandmother was the reason.

After almost a year in France, Bothwell was restless again. In late February 1565, he addressed a petition to Mary asking her to allow him home. He was suitably humble in his request, offering to accept conditions as long as they were appropriate to "his calling and birth." Mary's reaction was mixed. "Of herself," said Randolph, "she is not evil affected toward him, but there are many causes why he is not so looked upon as some others are."

Bothwell, however, rarely stood on ceremony. He returned to Scotland without waiting for a reply. Moray called for him to be outlawed. "It is said the queen hath granted it," wrote the Earl of Bedford from Berwick, "but whether she will suffer it to be performed, some doubt."

Bothwell went first to see his mother at Haddington, then quickly moved on. He feared for his safety and was right to do so: Moray was out to get him. Bothwell, in Moray's mind, was forever a running sore. Moray would never forgive him for stealing that all-important first installment of English gold from Captain Cockburn. "He followeth the matter so earnestly," continued Bedford in his next report, that "Scotland shall not hold them both."

Soon Bothwell was back at the Hermitage. He was in yet more trouble, because his servants had reported "divers words" he was said to have spoken in France. These included threatening to kill Moray, Maitland and Cecil. He had also (allegedly) joked that Mary and Elizabeth between them "could not make one honest woman." More dubiously (since Bothwell's servants hoped to receive rewards for their information), he was said to have slandered Mary, calling her the "whore" of her uncle the Cardinal of Lorraine. Not deigning to reply to such charges, Bothwell counterattacked by accusing Maitland of attempting to suborn his barber to poison him.

Mary was weary of this squabbling. The allegations and counterallegations were leveled within days of Elizabeth's bombshell that she would do nothing further on the subject of Mary's dynastic claim until she herself had married or resolved to remain single. Darnley was now within Mary's sights; Bothwell was a tiresome distraction. "She hath sworn," said Randolph, "upon her honor that he shall never receive favor at her hands." His return was "altogether misliked," and he was ordered to appear in court to answer the charge of breaking out of prison. Despite this, Mary set light terms. Moray wanted Bothwell arrested under pain of treason, but Mary intervened to make sure that he was given a modest bail of £200.

This made the outcome inevitable. Bothwell's trial date was set for May 2, 1565. Moray arrived in Edinburgh with a large force in an attempt to overawe the jury and secure a guilty verdict. But Bothwell absconded, possibly with Mary's connivance. Once more she was keeping him up her sleeve. Maybe he had his uses after all. But until she was married to Darnley, he was to keep out of the way. "It is believed," wrote one of Randolph's spies, "that the Queen's Majesty would [do] him good, but I trust Her Grace will not declare the same at this present."

Bothwell returned to France, where he stayed for just a few weeks. He was found guilty by a jury in his absence, but Mary intervened to prevent the justice clerk from sentencing him. Bothwell escaped without forfeiting his lands, but the verdict was left on the record in case he misbehaved in the future.

His position was transformed by Moray's revolt against the Darnley marriage. Mary recalled Bothwell ten days before the wedding. She was determined to be revenged on her brother for his treachery, and could act with impunity because she was no longer dependent on his allies for support in her Privy Council. She knew that Bothwell would be ideal for the role. Their enmity ran deep, and to complete the package, she also

unleashed George, Lord Gordon, son and heir of the defeated Earl of
Huntly, whose family had been virtually annihilated by Moray at the
battle of Corrichie. When the Gordons were attacked and their lands
and goods seized, Lord Gordon, a Protestant, was saved by the advocacy
of his father-in-law, Châtelherault. Instead of being sent to the gallows,
he was put under house arrest at Dunbar, where he stayed for over two
years.

A fortnight after Mary recalled Bothwell, she freed Lord Gordon on
parole. His discharge was proclaimed in Edinburgh, and two days later
he presented himself to Mary and Darnley at Holyrood. He returned to
loyal service, and was restored first to the lordship of Gordon and then
to the earldom of Huntly and to all the lands and dignities that had belonged to his father. His reinstatement still required an act of Parliament to confirm it, but Mary had to all intents and purposes restored
Huntly to his birthright.

And Huntly reciprocated. Not only did he help pursue Moray and his
allies until they fled, but he also allied with Bothwell. He became Bothwell's closest friend and supporter, a friendship sealed by the ties of kinship the following year when Bothwell married Huntly's sister, Lady
Jean Gordon.

When Moray and his allies rebelled, time was of the essence. Although Mary had recalled Bothwell, she did not wait for his return before launching her reprisals. The revolt had to be nipped in the bud, and
since relatively few lords rallied to the rebel cause, Mary seized her opportunity. Her musters were completed by August 26, and she pledged
her jewels to pay her soldiers. She had great success in raising a royal
army. She rode out of Edinburgh at the head of between eight and ten
thousand men, outnumbering her opponents by five to one. This time
Moray would discover that he did not face a middle-aged dowager queen
who suffered from dropsy and had to rely on unpopular French troops,
but an energetic, charismatic and infuriated queen who had the loyal,
unstinting support of her own native Scots.

Since Moray had last been seen in Ayrshire, Mary rode in that direction, across the central belt of the Lowlands. With Darnley at her side,
she set a brisk pace in spite of the wind and driving rain. She sported a
pistol in her saddle holster and a steel cap on her head, while Darnley
wore a gilt breastplate. She was at ease and in her element. In a campaign aptly known as the Chase-about Raid, so called because Mary
drove the rebels before her without stopping, she retained the initiative

from the start. As Randolph reported, "The queen followeth them so near with such forces — and so much the stronger by reason of her musketeers — that she giveth them no time to rest in any place." Even Knox was forced to admit her gallantry. "Albeit the most part waxed weary," he wrote, "yet the queen's courage increased man-like so much that she was ever with the foremost."

Moray was joined in Ayr by all the rebel lords except Argyll. They stuck together, and when they heard that Mary was on the move, they went past her in the opposite direction. They entered Edinburgh on the 31st, hoping to capture the town while Mary's troops were in pursuit. But the citizens were loyal and trained the castle guns on them. The rebels withdrew some fifty miles to the west, and were in Dumfries by September 5. Mary almost intercepted them on the 4th, but when they slipped past her again, she let them go and returned to Edinburgh to consolidate her forces and allow them to rest.

Argyll had not merged his men with Moray's. He ran his own campaign in the Highlands, where he was attacked by the Earl of Atholl and the Lennoxes. When they finally boxed him in, Mary left him there, sealing him off within his own enclave. She turned next to Moray's base in Fife, cutting his supply lines. On the 13th, she issued a manifesto at St. Andrews, attacking those who "under pretense of religion" had raised "this uproar" so that they might "be kings themselves." Moray and his allies countered from Dumfries, denying they were traitors or rebels and claiming they acted in ways true to God, their queen and the "commonwealth." Their complaint was that Mary, by marrying Darnley and proclaiming him king without the consent of Parliament, had trampled on the legitimate rights of her nobles.

On the 17th, Bothwell landed at Eyemouth from France. Within a quarter of an hour, he was riding in search of Mary, whom he found at Holyrood on the 20th. Bedford, under orders from Cecil to arrest him at sea, had commissioned one Wilson, a notorious pirate, to seize him. Wilson had successfully tracked down Bothwell's ship, but, daring as ever, Bothwell escaped with six or eight men in rowboats. His equipment was lost, but he still had some armor and a few boxes of pistols.

Mary was delighted to see him, later giving him the credit for her success. On the 28th, she reinstated him on the Privy Council. After Lennox and Atholl, Bothwell would become her most trusted adviser. Randolph greatly feared him. "His power," he said, "is to do more mischief than ever he was minded to do good in all his life."

But his return delighted Randolph in that it provoked the first of many rows between Mary and Darnley over who should be lieutenant-general of the royal army. Mary wanted Bothwell in command; Darnley wanted his father, Lennox, in the post. In the event, they split the role. Lennox was appointed to lead the vanguard, and Bothwell was to join Darnley at the head of the main battle army. In Argyllshire, the Earl of Atholl was named as Mary's chief lieutenant.

By October 4, Mary's army had risen to between ten and twelve thousand men. Huntly, said Randolph, had brought "a great force out of the north." He, too, wanted his revenge. He "imputeth the overthrow of his father only to my Lord of Moray, which is approved by the queen's self." Bothwell, meanwhile, "taketh great things upon him and promiseth much." Mary's "chief trust" was in Bothwell, Huntly and Atholl.

They did not disappoint her. On the 8th, the main battle army left Edinburgh and marched toward Dumfries. Mary rode beside it, still wearing her steel cap.

But there was no battle. By the time the royal forces arrived, Moray and his allies had fled. They simply slipped across the border to Carlisle. After several days' rest, they reached Newcastle-upon-Tyne on the 16th. Moray's appeals for English reinforcements had fallen on deaf ears. The Privy Council had debated the pros and cons, but in the end decided that a military intervention in Scotland could lead to an open-ended commitment. Such aid as was received was purely financial. Randolph first slipped 3000 crowns to Lady Moray, after which Bedford handed over £1500.

Argyll, meanwhile, was trapped in Argyllshire. He expected Mary to turn on him next, but nothing happened. With Moray and the rest of the rebel lords on the run in England, she was content to disband her forces. Her victory was complete. She was more powerful and popular than ever before. She had married Darnley and thereby unified their claims to the English throne. She had routed her enemies in flight, with no bloodshed. And she had two loyal and devoted advisers in Bothwell and Huntly.

Mary succeeded because she had snatched the initiative. While Elizabeth dithered and English policy was in disarray, she had acted decisively. She had shown great courage and untiring energy. Her assurances over religion had been masterful. She had stood aloof from England, and had seen Elizabeth and Cecil back away from assisting Moray to rebel a second time.

Elizabeth wanted another round of diplomacy, but Mary could not see the point. She would never allow another ruler to intervene in the internal affairs of her country or in her struggle with her rebels, whom she meant to punish severely. She had told the English queen just that in a strongly worded message handed to Randolph on September 7. Her cousin had never been angrier and more frustrated with Mary than when this note arrived.

Elizabeth had expected Mary to defer to her wishes almost as a client queen, unaware that Mary had now found the strength to stand up for herself in a way she had not been able to do before. England, and especially Cecil, imagined Scotland to be a satellite state. But Mary was hardly going to revert to a policy of conciliation when she was winning. Moray's overthrow was nearly as complete as the elder Earl of Huntly's three years before, and she did not intend to relent. She made just one exception. When Châtelherault humbly apologized for his role in the revolt and went into voluntary exile, he and his family were pardoned. But that was the extent of Mary's forbearance.

On December 18 and 19, heralds appeared in full armor at the Market Cross in Edinburgh to summon Moray and his allies to appear in Parliament on March 12, 1566, "to hear and see the doom of forfeiture orderly led against them."

Mary's summoning of Parliament and this advance warning of her intention to confiscate the lands and goods of the rebel lords would prove to be another watershed. It was also the prelude to one of the two most dramatic assassinations in Scottish and British history.

15

A Marriage in Trouble

THANKS TO A stroke of luck in the archives, we know what Mary was thinking and what drove her to engage the rebel lords in a military showdown after Bothwell's return from exile. In a series of interviews at Holyroodhouse, she poured out her heart to Castelnau, the special ambassador whom Catherine de Medici and Charles IX had sent to Scotland. He first saw Mary and Darnley in their presence chamber and then privately in the palace garden, and wrote down everything they said in a "Discourse" sent to Charles, which is now in the Bibliothèque Nationale in Paris.

After Castelnau had presented his diplomatic credentials and congratulated the king and queen on their marriage, Mary declared her intention to crush the rebels. They had abandoned their lawful allegiance in the name of religion and wished to depose her. For this reason, her case touched every other ruler, since if her rebels were allowed to behave in this way and seek the collusion of England in their crimes, there could be no stability or order in the world.

Mary was unflinching in her convictions. Moray and his allies were not just rebels with political grievances or an ax to grind; they were outright "republicans" — she used the word herself — set on destroying the "ancient monarchy." The rebels would depose and kill her and Darnley, and then create a "republic" in which sovereignty was vested in the nobles. They had already deposed her mother. She was next on the list. Her view of republicanism was an early prototype of the domino theory.

Once Scotland had fallen, the subversion would spread to England, to the Netherlands and to France.

As she talked, tears welled from her eyes. She spoke with deep feeling, her outlook shaped by her memory of the events of 1559–60, when the Lords of the Congregation deposed her mother and Elizabeth legitimized the government of the rebels by making the treaty of Edinburgh.

Mary reminded Castelnau of the long tradition of the "auld alliance" and of her own Guise and Valois connections. She then demanded military reinforcements. "All my hope is in France," she said. Castelnau was appalled, because his charge was to persuade her to settle with Moray. A new French intervention could spark a protracted civil war that might spread to the whole of the British Isles. He urged Mary to compromise, but she flatly refused. "It is incompatible with my honor and with the safety of my person and that of the king my husband, because these rebellious subjects of their bad faith and evil will have decided to kill us both."

Mary had redefined the issues. This struggle was no longer about her marriage, but had taken on an ideological dimension. She saw a clear antithesis between monarchy and republicanism, between divine-right rule by an anointed queen and anarchy, between French and English influence in Scotland. She refused to listen to Castelnau's pleas that "utility," "prudence" and "expediency" obliged her to make concessions. Her rebels must be punished as a point of principle; otherwise, it would not just be her own authority but the institution of monarchy itself that would be undermined. Castelnau noted that Darnley, whose French was fluent, was even more insistent in his defense of the ideal of monarchy than Mary. The discussion lasted for four hours and ended up going around in circles.

Next day, Castelnau was summoned to the palace garden, where Mary strolled with Darnley amid the ornamental borders and fruit trees planted by her mother. She looked radiant and relaxed, and when Castelnau joined them, she turned all her charm on him, saying that one day she would ask her council to consider everything he had suggested. Meanwhile, the time for talking was over and she intended to fight.

When Castelnau asked if she would risk everything in a battle, Mary did not flinch. "Yes," she said. "Because to play for time is no longer to be a queen." He tried to argue that the amity with England was essential both for Mary's authority in Scotland and to secure her claim to the English succession. Mary did not dispute this. "Yes," she said. She was willing to reinstate the amity and offer Elizabeth her affection and goodwill, "but only when she declares me to be her successor if she dies without children."

Castelnau sensed a change in Mary. On his last visit, he had been struck by her sense of "grandeur." His comment then was that she "had as big and restless a spirit as her uncle the Cardinal of Lorraine." This time, it was not so much that she was proud or overconfident. She had become defiant, tending to dramatize her problems and generalize from them. Moray's revolt, an uprising of a serious but not unfamiliar type in the age of the Wars of Religion, posed in her mind a general threat to European monarchy and so required some form of extraordinary aid. Mary saw Moray as a republican revolutionary, and was unable to comprehend Castelnau's inability to agree with her.

Writing privately to Paul de Foix, the resident French ambassador in London, Castelnau said he found Mary "intractable." Her courage and willfulness were such that she had adopted an "all or nothing" approach. She equated Protestantism with political revolution and would prefer to abandon her throne rather than negotiate with Moray.

Castelnau overstated the case. The changes in Mary were straws in the wind rather than an outright shift in her policy. A swing toward Catholicism was in the offing, but it had not yet begun, and there is a better explanation for its appearance. Darnley, as Castelnau also observed, was throwing his weight around. He saw himself as king of Scotland, with his own ideas and "enterprise." He wished to visit France to impress Charles IX with his Catholic credentials. He was eager to be integrated into the European dynastic system and awarded the Order of Saint Michael, the highest badge of honor in France, which he asked Castelnau to obtain for him. He was engaged in a policy of courting the Catholic powers while attempting to persuade Mary to outlaw the Protestants after Moray's revolt.

Darnley was not a devout Catholic. He was certainly not a Protestant, despite his well-timed visits to Knox's sermons, but was cynically exploiting religion for his own political purposes, chiefly as a way of drawing attention to himself. His efforts to secure European influence and recognition were just another aspect of his narcissism.

His ambition was truly overweening. Mary had allowed him to be styled king to appease his vanity. She was prepared to allow him to take an equal share in governing Scotland, but he expected her to cede all her power as a reigning queen to him. He really believed that their marriage had made her his subordinate. He was now the King of Scots, and his primitive view of the sexual act led him to think that his authority was most clearly asserted in bed, where he would do what he liked when he liked. The fact that Mary was at first submissive in the bedroom was merely because she badly wanted a son and heir.

Now Darnley quizzed Castelnau about the recent meeting of the ruling families of France and Spain at Bayonne. In a spectacular gathering of the two courts in May and June 1565, Catherine de Medici had been reunited with her eldest daughter, Elizabeth, Mary's childhood friend in Henry II's nursery and for some years queen of Spain. The meeting was attended by Charles IX and the Duke of Alba, one of Philip II's trusted advisers; its purpose was to restore the close family links established by the treaty of Cateau-Cambrésis. But the meeting was a flop, because Philip never bothered to turn up. All over Europe, however, the Protestants were talking about the event. They feared that in a secret session, a Catholic League had been organized to crush them. It was this that obviously interested Darnley.

From the day of his marriage, he had been plotting to alter Scotland's religion again. His confidants were Sir James Balfour, David Rizzio, John Lesley, David Chalmers and Francis Yaxley — an ill-assorted but highly intriguing quintet.

Balfour was Darnley's right-hand man. An unscrupulous lawyer who had attended Castelnau's second interview in the garden at Holyrood, he had been an accessory to the plot to murder Cardinal Beaton all those years ago. He had been captured by the French at St. Andrews Castle, and rowed in the galleys with John Knox. Now a judge in the Court of Session in Edinburgh, he was notoriously unprincipled. His career was all that mattered to him, and he switched sides with bewildering rapidity.

Rizzio, Mary's new confidential secretary, was already said to be Darnley's "only governor" and the man who "works all" in his counsels. As noted earlier, the two men were found in bed together, which in view of Darnley's swing toward Catholicism led inevitably to the accusation that Rizzio was a papal spy.

Lesley and Chalmers were leading Catholic lawyers. Lesley (later Bishop of Ross) was to become one of Mary's most active and prolific defenders after her flight to England. His name was linked to Balfour's soon after the Darnley marriage. Chalmers was connected to Balfour in the Court of Session, and both of them to Bothwell as former advocates in the Admiralty Court, over which he presided as Lord Admiral.

Yaxley was an oddball. He was an Englishman and an ardent Catholic who had served Philip II and Mary Tudor and was a protégé of the Countess of Lennox. He arrived in Scotland shortly after Darnley's marriage and proclaimed himself a Scot. Regarded by the countess as a bridge between her son and the English Catholics, he quickly won Darnley's trust. He was employed as a special agent and sent to Philip II to say that Mary lacked confidence in her French relations and so wished to commit herself and her country to Philip's protection. Naturally it did not occur to Darnley to ask Mary's permission first.

Yaxley was a menacing character: he had once been Cecil's servant, knew the identities of his foreign agents and had diplomatic accreditation throughout Europe. He claimed to know many Catholics "of good power" in England who were ready to declare themselves for Spain if Mary displaced Elizabeth as queen.

Darnley also dabbled alarmingly in Ireland, where Spain wanted to extend its influence. A vibrant trade existed between Spain and the south of Ireland. The country was overwhelmingly Catholic. The Gaelic chiefs were in revolt, and many would be willing to consider Mary as queen of Ireland in place of Elizabeth. Philip II was starting to wonder whether his policy toward the English queen had become just a little too cordial. And if so, Ireland was far more likely to be the back door into England than Scotland was. If Darnley stirred up trouble there, it would be of even greater interest to Philip than Mary's claim to the throne of England, which remained ineffective if not almost entirely useless without the pope's backing.

Yaxley's mission to Spain was successful. He obtained letters from Philip to Darnley, but was shipwrecked and drowned in the North Sea on his return journey. His body washed up on Holy Island along with his document case and a vast sum of money in crowns and ducats. The letters he was carrying were addressed to the "King of Scots" alone: no mention was made of Mary, who was invisible in Darnley's diplomacy. Although Darnley had been Mary's husband for less than six months, he was flexing his muscles and attempting to make his name as one of Europe's greatest Catholic kings.

At Christmas 1565, Darnley showed just how Catholic he meant to be. He attended midnight Mass and then matins, followed by High Mass, where he prayed "devotedly upon his knees." No longer did he attend

Knox's sermons at St. Giles Kirk. He talked of restoring the Mass in Scotland and granting "liberty of conscience" to Catholics at the next Parliament, due to assemble in the spring to attaint Moray and his allies. Darnley was creating dangerous waves. His actions put in jeopardy the religious compromise that Mary had worked so shrewdly over the past four years to establish. She tried at first to distance herself from his campaign. At Christmas, when he attended Mass so conspicuously, she stayed up all night playing cards and went to bed at dawn, thereby missing the services. But he persisted, almost daring her to prove that she really was a true Catholic.

Darnley believed that as king he was above the law, because unlike Mary, he had no explicit permission to worship privately as a Catholic in the Chapel Royal at Holyrood. It did not occur to him to ask whether his actions would be seen as divisive, and he was so arrogant and insensitive, he gloried in what he took to be a crusade against the Protestants.

He planned to bring his campaign to its climax during the week of the Catholic festival of Candlemas, in early February 1566. He was then to be invested with the Order of Saint Michael, which Castelnau had secured. He eagerly looked forward to the ceremony, which he regarded as a vote of confidence by the Catholic powers.

A grand delegation arrived in Edinburgh for the investiture. Charles IX's ambassador was accompanied by another from Mary's uncle the Cardinal of Lorraine, who brought an extraordinary letter from the pope congratulating Mary and Darnley on the official restoration of Catholicism. Of course nothing of the sort had yet been attempted, and we know from a series of documents in the archives at Simancas that the pope's letter was the result of Yaxley's visit to Spain, where Darnley had been marketed as the man who would overturn the Scottish Reformation.

The day before his investiture, Darnley summoned as many of the Scottish lords as he could find and invited them to attend Mass. Only the Catholics accepted. The Protestants, in particular Bothwell and Huntly, refused. Darnley lost his temper. He stalked out of the ground-floor room, locking the door and threatening to throw away the key. Mary came downstairs from her apartments to see why her husband had been shouting, anxious to avoid a scene. She did her best to pacify Darnley, who was becoming distressingly abusive. She even took Bothwell and Huntly, her most loyal supporters, by the hand and tried to lead them to Mass. Still they refused. Darnley then said loudly that he intended to restore High Mass at St. Giles Kirk in Edinburgh, an idea that would surely have driven Knox to apoplexy if he had heard it.

Mary had finally become enmeshed. Although Darnley had championed the Catholic cause only because he felt it gave him international prestige, he had pricked her conscience. In effect, he had capitalized on her own devout Catholicism to plant in her mind the idea that she had failed her Church and fellow Catholics by accepting a religious compromise that had denied them the right of public worship, which must have rankled when her husband was flaunting his faux Catholicism in front of so many distinguished foreign diplomats.

Could there have been other reasons why Mary was persuaded to veer toward Catholicism? The influence of her Guise family was surely one. When her uncle's ambassador, still under a misapprehension about her policy, warmly congratulated her on restoring the Catholic faith to Scotland, she would have known in her heart that she had let down her family as well as her coreligionists.

The second reason is that she was five months pregnant. Whatever Darnley's personal failings, he had quickly fathered a child. Castelnau had been the first to pick up the rumors. He noticed that Mary was draping herself in a mantle even when indoors. She stopped riding her horse and was seen traveling in a litter. She had been suffering from sickness and abdominal pains, which Randolph called a recurrence of "her old disease," but was this time the initial stages of pregnancy. She kept to her chamber for five days and stayed in bed. Even Randolph then guessed that the rumors might be true. He quickly began taking an inordinate interest in Mary's menstrual cycle, bribing his contacts among her female bedchamber staff for up-to-date information.

At the High Mass on Candlemas Day, Mary and Darnley bore candles to the altar in the Chapel Royal at Holyrood, accompanied by Lennox, Atholl and a congregation of some three hundred more. Darnley had gotten his way. Mary even promised that "she will have the Mass free for all men that will hear it." It was a significant statement, instantly seized on by the Protestants and causing consternation. A week after the investiture, Darnley and his friends swaggered up the High Street in Edinburgh, boasting that they had overturned the Scottish Reformation at a stroke.

But Mary was not in thrall to Darnley. What Randolph did not yet realize was that her marriage was in trouble. "Jars," or quarrels, had first arisen between the newlyweds over the rivalry between Lennox and

Bothwell to command the royal army against Moray. Lennox afterward claimed that everything had come to a head "about November" 1565, when Mary had "suddenly altered" in her affection for his son. She knew then that she was pregnant. And it must have occurred to her that she would not need to indulge her husband's every whim for much longer, now that he had served his sexual function.

Things had worsened in December, when Mary pardoned Châtelherault and his family, the ancient enemies of the Lennoxes, for their part in Moray's revolt. The Lennoxes were furious about the pardon. Darnley told her bluntly that as her husband and superior, he forbade any further remissions. No one had ever talked to Mary like that and got away with it. Her reaction was predictable. She would not be dictated to by a man she had raised up from nothing.

Mary's estrangement from Darnley was apparent at Christmas 1565, when there had been several spectacular rows. Although the quarrels took place in their private apartments, the news soon leaked. Whereas, said Randolph, "a while [ago] there was nothing but 'King and Queen, His Majesty and Hers,' now 'the Queen's husband' is the most common word." Quite simply, Mary had decided to demote him. In her proclamation after her marriage, she had conceded a dual monarchy in which power was exercised "conjointly." In state papers and on recently minted coins, Darnley's name had taken precedence.

Now this arrangement was canceled. Where in official documents Darnley's name had previously appeared first, it was placed second. And where the legend on the coins had read (in Latin) "Henry and Marie, by the grace of God, king and queen of Scotland," it was altered to say "Marie and Henry . . . queen and king . . ." By a profound irony, a motto on the coins had read: "Those whom God hath joined together, let no man put asunder." This, too, was expunged, replaced by a text from Psalms, "Let God arise and let his enemies be scattered," to celebrate Mary's victory over Moray.

At Darnley's investiture ceremony at Candlemas, Mary even denied him the right to bear the royal arms. Three days later, Randolph advised Dudley, "I know now for certain that this queen repenteth her marriage: that she hateth him and all his kin."

Darnley's drunkenness had become a flashpoint. Sir William Drury, who was marshal of Berwick and the Earl of Bedford's deputy there, told how when Mary had asked her husband to moderate his drinking at a private dinner party in Edinburgh, he had snarled back at her and she had left the table in tears.

As quickly as Mary had granted Darnley a royal title, she decided to strip him of it. She could not prevent him from signing his letters "Henry R" (i.e., Henry Rex) if he chose to do so, but she could deny him the "crown matrimonial." Fortunately for her, that could be granted only in Parliament. And if the crown matrimonial was withheld, then Darnley could never be crowned. He would enjoy no legal status as king, and could make no claim to the succession should Mary die childless.

Mary meant business, because when she refused Darnley the right to bear the royal arms, she made it clear to him that the crown matrimonial would be denied too. So the Lennoxes decided to wrest it from her without her consent. When Randolph excitedly described the "mislikings" between the royal couple in his reports to Cecil, he knew that the chief cause was Darnley's ambition for the crown, "which she is loath hastily to grant, but willing to keep somewhat in store, until she know how well he is worthy to enjoy such a sovereignty."

Mary's mood lightened in the week of Darnley's investiture. Suddenly she was happy again. It could not have been Charles IX's ambassador who was the cause of her delight. He had strongly advised her to pardon Moray and the exiled lords on the condition that they promise to live "like good subjects" — the last thing she wanted to hear. She had already said, with an uncharacteristic degree of vitriol for her, that she hoped Moray would die in exile.

The clue is provided by two dispatches from Randolph to Throckmorton, newly discovered in the archives in Edinburgh, describing the background to the ceremony. One was written on February 7 and the other on the 10th, the same day as the investiture.

Mary, as Randolph sensationally reported, was preparing to renew her immediate Catholic claim to the English throne. The third and final reason for her shift toward Catholicism was her conviction that her supporters in England were "never so great." Perhaps Yaxley had told her this. Or perhaps her uncle's ambassador had recycled some information that Yaxley had spread while in Spain. Or again, maybe Cecil had all along been correct in his prediction that the "people of England" — whether Protestants or Catholics — would flock to Mary's cause if she married and produced an heir.

Mary's pregnancy must have greatly reinforced her self-esteem. With

Elizabeth still unmarried, it would have made her think that she could vindicate her claim one way or the other. Despite the fragile state of her relationship with Darnley, their union had been a triumphant success in a dynastic sense, which was all that royal marriages were then intended to be, and she could begin planning a new campaign to assert her rights in England.

A few days before the investiture ceremony, a sumptuous banquet was held in the great hall at Holyrood in honor of the visiting ambassadors. Catching sight of a portrait of Elizabeth that had no doubt been deliberately set in position for the purpose, Mary rose and declared in the full glare of publicity that there was "no other queen of England but herself."

This was a dramatic change. Ever since Pope Paul IV had failed to endorse her dynastic claim at the time of Elizabeth's accession in November 1558, Mary had reluctantly accepted that her only viable option was to secure recognition as the English queen's successor. But then Paul IV had died, as had his immediate successor. A new pope, Pius V, had recently been elected. One of his first public acts had been to write the letter to Mary that was delivered by her uncle's ambassador at the request of Philip II, after Yaxley's mission to Spain.

Mary's claim at the banquet that she was the rightful queen of England must have caused a furor. Randolph predicted disaster, saying, "This court is so divided that we look daily when things will grow to a new mischief."

Whereas three years before Randolph had assured Cecil that Mary was "not so affectioned to her Mass that she will leave a kingdom for it," now he found her "bent to the overthrow of religion." She would stop at nothing in her desire to restore Catholicism. All her efforts were aimed at this, and linked directly to her dynastic claim. Since her victory over Moray, she had never been more powerful. She would now, as Randolph feared, seek to further her success through "her most idolatrous Mass." He concluded, "I pray you burn this letter."

All this made for a confusing situation. When Mary denied Darnley the use of the royal insignia at his investiture and at the same time tried to lead Bothwell and Huntly by the hand to Mass, a huge contradiction arose in her policy. She was unwilling to be bullied by her dissolute and conspiratorial husband, yet she had become embroiled in his "enterprise" to restore Catholicism, not (as he wished) to impress a putative

Catholic League, but because after Pius V's election, she believed she could use Catholicism to achieve a final recognition of her dynastic claim.

Meanwhile, Darnley was furiously plotting against his wife. Randolph knew that there were "practices in hand," contrived between Lennox and Darnley, "to come to the crown against her will." The new session of Parliament was imminent. It had been set to begin on March 12, 1566, when the leaders of Moray's revolt (other than the pardoned Châtelherault) would be punished and their properties forfeited to the crown.

The Scottish lords had an overwhelming motive for wishing to disrupt, or preferably to cancel, the new session. If Parliament met as planned, Moray and his allies would be stripped of their lands and titles. They also feared that as the next stage of Mary's agenda, the Mass would be restored.

A plot was inevitable, not least because opposition to Mary's aims extended more widely. The Catholic lords intended to join with the Protestants to resist the forfeiture of the rebels on the grounds that such action might become a precedent for noble forfeitures generally. They were not affected at the moment, but might well be in the future.

All this rebounded on Mary. When she had struck out on her own, marrying Darnley as a fait accompli and abandoning her policy of conciliation, she had triumphed. But with the exiled lords, their allies and the Lennoxes all scheming against her, the balance of the factions tilted back against her. Mary's position reverted from almost indomitable strength to dangerous isolation. Warned of a plot by her ever-loyal servant Sir James Melville, she dismissed his fears. "She had," she said, "also some advertisements of the like bruits, but that our countrymen were well worthy." If this was indeed Mary's response, her confidence was ill-advised.

The plot was shaped, according to one of its leading participants, "about the 10th day of February," the very same day as the investiture. Darnley lay at its heart. But he was pointed in the right direction by Maitland, whom Mary had marginalized in favor of David Rizzio, and who was determined to get Moray and his allies pardoned and recalled from their dishonorable exile as the prelude to his own rehabilitation.

To forward the plot, Lennox went deep into Argyllshire for a secret rendezvous with the Earl of Argyll. He made Argyll what he hoped would be an irresistible offer. He was to contact Moray and the exiled lords in England, and if they would agree to grant Darnley the "crown matrimonial" in the next Parliament and so make him lawfully King of Scots, then Darnley would switch sides, recall the exiles home, pardon them and forbid the confiscation of their estates. Finally, he would perform the ultimate U-turn and reestablish the religious status quo as it had existed at the time of Mary's return from France. By this route, everyone would get what they wanted at Mary's expense. Darnley would become king with full parliamentary sanction, Moray and his allies would be reinstated as if they had never rebelled, and the Protestant Reformation settlement would be restored.

The plot took off when Moray accepted these terms. Its logic was crude, if devastatingly effective. But there had to be a scapegoat, someone to blame for misleading Darnley and orchestrating the recent swing toward Catholicism. The ideal candidate was Rizzio, once Darnley's lover, but who Maitland, the supreme insinuator, had falsely informed Darnley was sleeping with Mary and toward whom Darnley, with his patchy grasp of reality, developed murderous intentions.

Now everything started to fall into place. Rizzio, as Darnley became convinced, had not just betrayed him; he had done so by seducing his wife. And since Rizzio was already said to be a papal agent thanks to his association with Darnley, it followed that he must be responsible for the swing toward Catholicism.

On February 9, Maitland had written a letter to Cecil that has all the elements of a smoking gun. "Nothing," he said, "is on either part so far past but all may be reduced to the former estate, if the right way be taken . . . I see no certain way unless we chop at the very root — you know where it lieth, and so far as my judgment can reach, the sooner all things be packed up the less danger there is of any inconveniences." If Maitland's syntax was convoluted, his meaning was crystal clear. The clock had to be put back to where it had been before Mary married Darnley. The way to do that was to "chop at the very root" — i.e., assassinate Rizzio. The sooner it was done, the better.

Just as Cardinal Beaton's assassins had once sought Sir Ralph Sadler's approval and indemnity before carrying out their plan, so Maitland courted Cecil. His letter, he assured Elizabeth's chief minister, was "a gage of my correspondence to your disposition." In other words, he asked Cecil to speak now if he disapproved, and if not, then forever to hold his peace.

Almost overnight and by a masterly propaganda exercise, the unfortunate Rizzio was transformed into the queen's illicit lover and "evil counselor." He was the man everyone could agree to hate: excoriated as foreign, lowborn, proud, ambitious, a papal agent, a spy, a sycophant and a voluptuary. It was a perfect cover story, because the greater the marital estrangement between Mary and Darnley, the more Darnley suspected her of adultery; and the more she became politically isolated, the more she found herself relying on her private secretary, since Moray was in exile and Maitland eclipsed and spending time plotting with Darnley.

If Maitland was the insinuator, the Earl of Morton, also avidly seeking Moray's rehabilitation as well as a powerful position for himself when Darnley obtained the crown matrimonial, was the technocrat. Morton was the leader of the Douglas clan, and it was the Douglases and Lord Ruthven, Morton's relative by marriage, who planned the assassination.

On March 6, 1566, Bedford and Randolph jointly wrote two letters to Elizabeth and Cecil. A "great attempt" was to be made in Scotland. Darnley had persuaded the Protestant lords to promise to make him king by the consent of Parliament. Bonds had been signed to this effect. He would obtain a grant of the crown matrimonial in return for the restoration of the exiled lords and of the Protestant settlement. "The time of execution and performance of these matters is before the Parliament, as near as it is."

Randolph left Edinburgh a week before the assassination. Mary had belatedly discovered that he had covertly bankrolled Moray during his revolt. Two witnesses were produced who confessed to acting as intermediaries, smuggling bags of untraceable gold coins to Lady Moray, and although Randolph barefacedly denied the charges, he was ordered to leave the country. He was first accused on February 19 and offered a safe-conduct across the border, which he refused to accept. He later left under threat of forcible expulsion, but held out long enough to send Cecil his final reports on the plot.

Everything had so far gone like clockwork. Cecil was fully briefed about Rizzio's assassination and did nothing to prevent it. The timing was ideal, as it would soon give Mary a great deal more to think about than reasserting her claim to the throne of England.

Others also knew of the plot. Randolph wrote to Dudley: "I know that if that take effect which is intended, David [Rizzio], with the consent of the king, shall have his throat cut within these ten days."

One of the few left completely in the dark was Mary. Others were Bothwell and Huntly, the closest of friends since Bothwell had married

Huntly's sister. They were staying at Holyrood on the night of the assassination and found themselves caught in the middle of it.

The plot was planned for Saturday, March 9. The previous evening, Randolph and Bedford wrote to Cecil and Dudley from Berwick to confirm that Morton was already in position and Argyll would soon be there. Moray and the exiled lords would arrive at Berwick next day from Newcastle, and would be in Edinburgh on Sunday morning. The murder of "him whom you know" would have been carried out by the time they rode through the main gates at Holyrood.

The night of March 9 was to be one of the longest and most terrifying of Mary's life. And this was to be only the first of a kaleidoscopic sequence of murderous events in her country.

16

Assassination One

AT EIGHT O'CLOCK on the fatal Saturday evening, Darnley led Lord Ruthven and an accomplice through his private apartments in the James V Tower at the palace of Holyroodhouse. Partly feeling their way in the darkness, each man with only a candle to see by, they climbed a secret stairway within the walls. Through a door concealed in the paneling, they entered Mary's bedroom on the floor above. Leading off the bedroom was a smaller adjoining chamber, or supper room, about twelve by nine feet, where Mary was eating with a group of friends, including Rizzio.

Beyond the bedroom and through a door on the left was a larger room, the queen's outer, or presence, chamber. At the far end of the outer room, a locked door opened onto the main staircase, which led down to the ground-floor lobby. Morton and the rest of the conspirators, around eighty in all, came up the main staircase to the door of the outer chamber, where they waited until it was unlocked from the inside by Ruthven's man, who had passed through Mary's bedroom.

Darnley was the first to appear. He entered the supper room and spoke to Mary. She was surprised to see him, but not unduly perturbed. He offered her soothing words and put his hand on her waist. This was to give Ruthven — an ungainly man suffering from liver and kidney failure, from which he would die only a few weeks later — time to reach the supper room, and for his accomplice to unlock the door. The delay must have been at least five minutes, because Ruthven clanked up the stairs in heavy armor and yet could barely walk (as he said himself) twice the length of his own chamber.

Ruthven finally staggered into the supper room looking ghastly. "It would please Your Majesty," he spluttered, "to let yonder man Davie come forth of your presence." Mary kept her wits and turned to Darnley. "What do you know about this?" she demanded. Darnley lamely professed ignorance.

Mary turned to Ruthven. "What is his offense?" She was thinking on her feet, instantly guessing why Darnley and Ruthven had burst in, but equally aware that Darnley would be irresolute and therefore potentially manipulable.

Ruthven said that Rizzio had done great offense to Mary, to the king, to the nobility and the country. "And how?" she asked.

"If it please Your Majesty," said Ruthven, "he hath offended your honor, which I dare not be so bold to speak of. As to the king your husband's honor, he hath hindered him of the crown matrimonial, which Your Grace promised him . . . And as to the nobility, he hath caused Your Majesty to banish a great part of them, and to forfeit them at this present Parliament."

"Leave our presence under pain of treason," ordered Mary. "If he hath offended in any sort, we shall exhibit the said David before the lords of Parliament to be punished."

Ruthven ignored her. Turning to Darnley, he said, "Sir, take the queen your sovereign and wife to you." When Mary rose in anger, Rizzio hid behind her back, clutching the pleats of her skirt. Mary's guests and chamber servants tried to grab Ruthven, who lurched forward and drew his dagger. There was a scuffle during which Morton and his company charged into the room. In the commotion, the dining table and all its contents were overturned, and only the quick thinking of the estranged wife of the Earl of Argyll, Mary's half-sister and one of the guests, avoided a disaster when one of the candelabra fell to the floor next to the tapestries on the adjacent wall: a blaze would have started had she not promptly snatched it up.

Ruthven and another conspirator struck at Rizzio with their daggers behind Mary's back. She later said that the blows had been so close to her, "she felt the coldness of the iron." Darnley was distraught. It is sometimes alleged that one of the plot's aims, if not to assassinate Mary, was at least to trigger a fatal miscarriage, as she was six months pregnant. This is implausible, because if anything had happened to Mary,

Darnley's plan would fail. He had not yet secured the crown matrimonial and was entitled to claim it only as long as he was married to Mary. If she was killed, his status as king consort would die with her. The throne would descend to the heir apparent, the exiled Duke of Châtelherault, the sworn rival of Lennox and Darnley. Only if the Rizzio plot had taken place after Darnley had been crowned, or after Mary's child was born, would her death have made sense. Darnley could then have remained as king or been appointed governor, and so maintained his position. Otherwise, he and his father would not have survived for more than six months. The factions would have turned against them, murdering them or driving them into exile.

Ruthven grabbed Mary and gave her to Darnley, telling her not to be afraid. "All that is done," he tried vainly to reassure her, "is the king's own deed and action." Rizzio, cowering in an alcove, was then hauled from the supper room, watched by as many of Morton's men as could fit into the small space, some of whom carried guns. One man pointed a cocked pistol at Mary to prevent her interposing herself between the hapless Rizzio and his captors while he was dragged out. Ruthven shouted an order to carry Rizzio down to Darnley's apartments by the secret stairs, but his voice was drowned in the melee. The crush was intense. Rizzio was propelled into the outer chamber, where the rest of the plotters were waiting. They surged forward and stabbed him in a frenzy, just short of the door beside the main staircase. Darnley refused to join in the butchery, so one of the conspirators seized his dagger and used it to deliver the final blow. Darnley's dagger was left in the corpse, to signify his connivance in the plot.

Mary, meanwhile, was paralyzed by fear. She really believed she would be killed too. We have her own account, sent to Paris. The assassins, she said, burst in, grabbed Rizzio and "most cruelly took him forth of our cabinet, and at the entry of our chamber gave him fifty-six strokes with whiniards [daggers] and swords."

When Rizzio was dead, the assassins fled and Morton went downstairs to seal the gates and doors of the palace. Guards were posted outside certain rooms, and Mary's black box containing her ciphers and secret correspondence was retrieved from Rizzio's room and returned to her.

Where Mary's own account differs from the rest is in the later plans of the conspirators. She was convinced that the plot was meant to have two stages: first Rizzio's assassination, then a palace coup in which

Bothwell, Huntly and the Earl of Atholl, the leaders of her forces during the Chase-about Raid, would be killed. Sir James Balfour, Darnley's
erstwhile protégé, was also said to be on the death list. He was to be
"hanged in cords" to stop him from exposing the true extent of Darnley's
treachery.

Bothwell and Huntly were lodging elsewhere in the palace. They
heard the uproar in the James V Tower, guessed that their lives might be
in danger and escaped out a back window by climbing down a rope.
Atholl and Balfour also managed to slip away or talk their way out of
trouble.

Back in the supper room, Mary was given a lecture. Ruthven told her
exactly what he and his supporters thought of her. According to Mary's
version, he said they "were highly offended with our proceedings and
tyranny, which was not to them tolerable; how we were abused by the
said David . . . in taking his counsel for the maintenance of the ancient
[Catholic] religion, debarring of the lords which were fugitive, and entertaining of amity with foreign [Catholic] princes and nations with
whom we were confederate; putting also upon [the Privy] Council the
Lords Bothwell and Huntly."

Ruthven's account of the speech is slightly different. In his narrative,
the main charge was that Mary had ruled "contrary to the advice of your
nobility and counsel, and especially against those noblemen who were
banished." To this Mary retorted with justified sarcasm that Ruthven
had himself been one of her privy councilors since the overthrow of the
elder Earl of Huntly three years before!

Whichever account is correct, Mary was harangued, and her own
summary proves that she understood the plotters' agenda and what she
was up against. She burst into tears, but refused to be dealt with in this
manner. Turning to Darnley, she demanded, "Why have you caused to
do this wicked deed to me, considering I took you from a low estate and
made you my husband? What offense have I made you that you should
have done me such shame?"

Darnley was consumed by jealousy. He indignantly assumed that he
had been made a cuckold and so turned into an object of scorn. He complained that Mary had not "entertained" him since Rizzio came into her
favor. Whereas before their marriage she used to visit him in his apartments, now she played cards with Rizzio until one or two in the morning. "And this is all the entertainment that I have had of you this long
time." Darnley became increasingly explicit. He had been denied sex by

his wife. On the occasions he had come to visit her, "she either would not or made herself sick." He had noticed the change last Christmas, and he wanted to know why.

Mary fought back her tears. "It is not," she said, "a gentlewoman's duty to come to her husband's chamber, but rather the husband to come to the wife's." Royal protocol required the husband to initiate all sexual advances, which is why, in the palaces of the sixteenth century, a private passage linked the king's bedroom to the queen's which was meant for the king's use and rarely, if ever, hers. According to the courtly handbooks, a queen's duty was to produce children and otherwise be "chaste, loyal and obedient."

Maitland had done his work well. Darnley was beset by his belief that Rizzio was Mary's lover. He also harbored suspicions that Mary found him sexually inadequate. His ego was bruised. "Am I failed in any sort?" "What disdain have you of me?" "What offenses have I done you . . . seeing I am willing to do all things that becometh a good husband?"

Now Darnley's vicious streak came to the fore. He was stung by the reference to his social inferiority. "Suppose I be of the baser degree, yet am I your husband and your head, and you promised me obedience at the day of our marriage and that I should be participant and equal with you in all things."

Mary would never be told by anyone except a reigning monarch that she was someone's equal, let alone inferior. "For all the offense that is done to me, my lord, you have the weight thereof, for the which I shall be your wife no longer nor sleep with you any more, and shall never like well until I have caused you to have as sorrowful a heart as I have at this present."

Ruthven, a witness to this battle of words, interposed. His illness was so far advanced, his armor so heavy, he was desperate for a drink and a seat. He brought a note of almost grotesque comedy to the scene. Asking Mary's permission, he "called for a drink for God's sake!"

He was served a cup of wine, which he downed in a single gulp. Mary looked at him in revulsion. "If I or my child die," she threatened, "you will have the blame thereof."

Ruthven began to answer, but a messenger knocked at the door. Bothwell, Huntly and Atholl had escaped. Ruthven and Darnley departed, leaving Mary to pace up and down in her bedroom for several hours. Barely able to take in what had happened, she was further dismayed that

Morton had stationed a sentry at the entrance to her outer chamber and that she was denied the comfort of her four Maries and domestic staff.

Darnley at last returned to supervise the removal of Rizzio's corpse. It was dragged through the door of the outer chamber and unceremoniously hurled down the main staircase. From there, it was carried to the porter's lodge at the palace entrance, where it was stripped by the porter's servant and Darnley's dagger removed.

For the whole of that long night Mary did not sleep. When Darnley returned at eight o'clock the next morning, she vented her rage on him in a quarrel that lasted two hours. Darnley then left in a fury. Before allowing him to depart, however, Mary persuaded him to let her gentlewomen return.

Morton and Ruthven were ruffled by this news. They knew that Mary could be a skillful operator and an unflinching adversary. She would not wait to be carried along by events, but would seek to dominate them. She could act naively and impulsively, but in a crisis would always try to hold the initiative. She had kept her head at the start of Moray's revolt. Her technique, then as now, was to appeal to men such as Bothwell and Huntly, on whose loyalty she knew she could rely. And in her present predicament, the way to contact them was through her gentlewomen.

Mary frantically scribbled her instructions. She then sent her ladies away to turn them into letters for her loyal supporters. She also had a letter sent to Argyll, who despite his aid for Moray and his allies was someone she still largely trusted. Whatever his religious views, he was a royalist and anti-republican, who later opposed Mary's forced abdication and was prepared to fight as her lieutenant. His catchphrase was "God first and then our prince in God, under God and by God's laws." The repudiation of an anointed queen, not to mention regicide, was something he found alien and abhorrent. Whatever her alleged offenses, Argyll held Mary to be a legitimate ruler, and for this reason she knew she could depend on him now.

Her plan shaped in her mind, Mary turned next to Darnley. She decided to stage a piece of theater for his benefit. When in the afternoon he returned, she pretended that she was about to miscarry. Morton and Ruthven had reimposed the ban on her gentlewomen. Mary was determined to get them back. So she played her part and the midwife was summoned. Mary had already primed this woman to confirm that she was about to go into labor. With his child's life and therefore his own position at stake, Darnley rushed a message to Morton and Ruthven, who reluctantly withdrew the ban, and Mary's ladies returned.

But the lords were suspicious. They told Darnley to beware lest Mary change clothes with one of her Maries and flee, leaving her stand-in behind. Darnley took the point. The guard at the entrance to the outer chamber was redoubled. Orders were issued that no one would be allowed to leave the queen's apartments who wore a muffler over her face or whose identity was suspect.

By eight in the evening, Moray and the exiled lords, with the important exception of Argyll, had come back to Holyrood. Darnley welcomed them, after which they retired to supper at Morton's house in the Canongate. When Mary learned of Moray's return, she sent one of her ushers to bring her half-brother to see her. When he arrived, she received him with open arms. She embraced him, saying that if he would be content to be reconciled, she would gladly accede to his request. According to Sir James Melville, she also said that if he had been there the previous night, he would not have allowed her to be so roughly treated as she had been, a sentiment causing the duplicitous Moray to weep.

The chief concern of the exiled lords of the Chase-about Raid, however, was their forfeitures. Darnley had to deal with this right away, as Parliament was due to meet in two days' time. His first public act after Rizzio's murder was to discharge Parliament. He ordered the heralds to proclaim at the Market Cross that everyone summoned to the new session should leave the town of Edinburgh within three hours under penalty of treason. When, therefore, Moray and the exiled lords duly appeared at the Tolbooth on the 12th "to hear and see the doom of forfeiture" against them, the building was almost deserted.

Mary knew of Darnley's proclamation. She believed it to be a bad mistake on his part, which played into her hands. She would split the lords in two, pardoning Moray and the exiles but striking against Morton, Ruthven and the Douglases. Morton's role in the Rizzio plot was clear. She would deprive him of his post of chancellor, which she would give as a reward to Huntly. Maitland she could not quite pin down, but she meant to punish him. She had warned him when he had sought her favor in France, in the months before her return to Scotland, that she would hold him accountable as the "principal instrument" of any "practices" against her. She would give his lands to Bothwell, then leave him to stew in his own juice.

That left just Darnley to account for. Mary knew him to be stupid, cowardly, vain, drunken, dissolute and narcissistic. He was also violent, vindictive and an inveterate liar. His plot could hardly have been clumsier. And yet it might still succeed if she did not act quickly.

Mary was now twenty-three, still young but no longer on such a steep learning curve. She had been back in Scotland for over four years and was able to confront the challenge of controlling the noble factions. She had learned that the way to deal with the lords was to divide and rule. To do this now, she needed to keep the feckless and murderous Darnley in her power, turning him into a weapon against his former allies, who in return for his treachery would become his mortal enemies.

Her next move was brilliantly attuned to Darnley's primitive psychology. She waited until Moray had returned to his supper at Morton's house. When she and Darnley were alone, Mary offered to have sex with him to prove that her affection for him had not changed as he had claimed. She said that he could come to her later in the evening, and she would sleep with him all night. She had a pretty good idea that he would be incapably drunk by bedtime. She meant to buy time so that she could find a way to deal with him and make her escape from Holyrood.

Darnley accepted the offer. The assignation was fixed and he withdrew to his apartments to ready himself. He met Morton and Ruthven, to whom he boasted of his forthcoming tryst. This, he declared, was the way to handle women. The lords were skeptical. Darnley, they said, "grew effeminate again." The fear of a reconciliation between Mary and her husband was the plotters' nightmare, but as they still had Darnley's bond consenting to Rizzio's murder and bearing his signature, they were not yet unduly anxious. Darnley had persuaded them that he would conquer Mary, making victory and the crown matrimonial almost a certainty. Morton and Ruthven knew, however, that their lives would be in jeopardy if Mary held her ground.

As Mary had predicted, Darnley failed to turn up for the assignation. It was something Randolph and Bedford could not understand. In their report to Cecil, they explained: "We know not how he foreslow himself, but [he] came not at her, and excused himself to his friends that he was so sleepy that he could not wake in due time."

Ruthven's version was more or less the same. Ruthven said that he had waited in Darnley's dressing room for so long, he finally went to bed himself. George Douglas, Morton's cousin and a half-brother of Ruthven's wife, then came to tell Ruthven that, despite repeated attempts to wake Darnley from his drunken stupor, he was too far gone to be roused.

At dawn, Ruthven reproved him. "You did not keep your promise to the Queen's Majesty to lie with her all that night." Crestfallen, Darnley replied, "I was fallen on such a dead sleep, I could not be awakened." Naturally he blamed someone else: his faithful bedchamber servant William Taylor.

"But," said Darnley, "I will take my nightgown and go up to the queen." Ruthven, who despite the gravity of his illness still had a sense of humor, said, "I trust she shall serve you in the morning as you did her at night!"

And so it was. Darnley climbed the secret stairway for the second time in three days to emerge at Mary's bedside. She was asleep, or pretending to be asleep, so he was forced to sit and wait for an hour until she was ready to speak to him.

"Why did you not come up yesternight?" she finally asked. Darnley said that he was so deeply asleep, he had not stirred before six o'clock. "Now," he continued, "am I come, and offer myself to have lyen down by you."

But Mary said she felt too unwell to make love. She now wanted to get out of bed.

Darnley did not argue. Instead, he changed the subject, asking when Mary would pardon the exiled lords. He also said that he expected a full pardon for everyone involved in the Rizzio plot. Mary beguiled him with soothing words. She was winning him over, and at eight o'clock he returned downstairs "very merrily." He began to swagger, boasting to Morton and Ruthven that everything was in hand. They urged him not to underestimate Mary. She had been trained in France by her Guise uncles, both masters of political deception. "Trust me," said Darnley. "Let me alone, and I will promise to bring all to a good end."

Later that morning, Darnley attended a secret meeting with Moray, Morton, Ruthven and the leaders of the Rizzio plot. The Faustian pact was sealed. Darnley would get Mary to grant full pardons to the lords, he would confirm the religious status quo, and they would offer him the crown matrimonial in the next session of Parliament. Darnley was utterly confident that he had pulled off a coup. But the lords wanted a guarantee. They knew Mary better than to think she would agree without a fight. They asked for written proof, signed by Mary, confirming her intention to grant their pardons.

Darnley returned to Mary, who had carefully prepared her lines. She told him the facts of political life in Scotland. As soon as the lords had obtained their pardons and their forfeitures were dropped, they would ditch Darnley, who would never secure the crown matrimonial at their hands. Why should they give it to him when they had already gotten what they wanted? And when would he receive it? Had he not himself just discharged Parliament without setting a date for its recall?

Mary had never been more independent or self-reliant, the more so because, despite all the stress and anxiety, her health held up. If she really suffered from acute intermittent porphyria, it mercifully left her alone. True, she was pregnant and often felt sick, but her mind was as nimble as ever. She explained to Darnley how badly he would be treated if the exiled lords and his co-conspirators were allowed to get their way. Were they not all Protestants? If he failed to convert to Protestantism, he would be toppled. And yet if he did convert, he would be reviled by the very European rulers he had so recently been eagerly courting.

Step by step, Mary persuaded Darnley to retreat. The way to rule Scotland, she said, was to rise above and then balance the rival factions, not to ally with any one of them "and so become enslaved." "By this persuasion, he was induced to condescend [i.e., agree] to the purpose taken by us." He agreed to escape that night with Mary to Dunbar, the nearest impregnable royal fortress and the home of Bothwell's widowed sister. Already Mary had received replies from Bothwell and Huntly to the letters she had sent through her gentlewomen. They advised her to flee, over the walls of the palace if necessary by means of ropes and chairs,* and then ride through the night.

The question was how to get past the guard at the entrance to Mary's apartments. To this she also had the answer. She and Darnley should not disguise their intention to leave Holyrood, but should mislead the lords as to the date of their departure. They would be gone before the lords realized they had been duped.

On cue, the midwife reappeared to inform Darnley that if Mary was not quickly allowed a change of air, she would miscarry. Her Maries and other gentlewomen dutifully confirmed this. Darnley hastened downstairs to tell the lords and to seek permission for Mary to depart next

---

* It seems that using ropes and chairs was the standard way to escape from a building in case of fire.

day. Her French doctor then arrived to emphasize the urgency of the request. Mary, he confidently claimed, would miscarry if she did not leave Holyrood soon.

The lords reluctantly agreed, provided Mary sign a paper confirming their pardons. Shortly after four o'clock, Moray, Morton and Ruthven went upstairs with Darnley to visit Mary. The lords waited until Darnley fetched her from her bedchamber. They all knelt and Morton made their petition, after which each spoke individually. Mary heard them out, then gave a gentle answer: "I was never bloodthirsty nor greedy upon your lands and goods since my coming into Scotland, nor will I be upon you."

Mary invited them to draw up whatever document they liked, and she would sign it. She then took Darnley in one hand and Moray in the other and walked about the room for an hour in conversation. Afterward, she returned to her bedchamber. She sent for Maitland, who arranged to remove the guards. It was also agreed that the lords would voluntarily leave Holyrood after supper that night. Even as they spoke, the lords were drafting their articles of pardon, and the document had only to be signed by Mary for the formalities to be completed.

The guards disappeared, and at six o'clock the lords handed Darnley their articles. They agreed to leave Holyrood and withdraw to Morton's house as soon as they were assured that Mary had actually signed.

Darnley, now won back to Mary's side, saw no reason to hurry. He first ate a leisurely supper, and by the time Archibald Douglas, Morton's cousin, returned to collect the signed articles, it was already late. Darnley lied that he had shown the document to Mary, who had pronounced it "very good." But she was feeling sick, he said, and not up to dealing with paperwork. She was turning in early and would sign the articles next morning.

Shortly after midnight, when all was quiet, Mary and Darnley slipped out of Holyroodhouse through a subterranean passage leading off a wine cellar, where they met half a dozen of her most trusted servants, who had fetched her horses from the stables. Without delay they rode hard through the night to Dunbar, a twenty-five-mile journey that took them five hours, which the pregnant Mary found grueling. Several times on the way she had to dismount and vomit.

At dawn, the lords were appalled to find Mary had flown the coop. By then it was no secret where she had gone. They dispatched a messenger to Dunbar to ask her to fulfill her pledge by signing their articles of pardon. But Mary made no immediate reply. She kept the messenger waiting for three days, by which time the exiled Earls of Glencairn and Rothes had separately made their peace with her and obtained their pardons. This opened the floodgates: Argyll had been joined by many of the exiled lords of the Chase-about Raid at Linlithgow, where they debated the terms on which they would settle with Mary and decided to start negotiations. Only Moray held back, hedging his bets and waiting for something to happen.

As the lords' party began to collapse, so Mary's increased. Bothwell and Huntly had already shaped the nucleus of an army. This increased when Mary ordered the landowners of the Lothians and the adjoining counties to muster their troops in her defense at Haddington and Musselburgh.

Early on the morning of Sunday, March 17, Moray left Edinburgh to confer with Argyll and his allies at Linlithgow. Mary — with Darnley safely in tow — left Dunbar the same day, entering Edinburgh with her forces on the 18th. She had between three and five thousand troops, more than enough to occupy the town. Rather than return to Holyrood, in case any of the lords were still there, she lodged in a house in the High Street, later moving to a larger one closer to Edinburgh Castle. A fortnight later, after suitable improvements had been made and her clothes and personal effects delivered, she moved into the castle itself, where she could finally feel secure and await the birth of her baby.

The day after Mary returned to Edinburgh, she sent Sir James Balfour to Linlithgow to offer terms to the rebels of the Chase-about Raid. They would be pardoned and their estates returned to them, provided they withdrew temporarily to their own houses and made no attempt to intercede for Darnley's co-conspirators in the Rizzio plot.

Argyll accepted on the spot, while Moray decided that it was not worth risking another revolt. Once they had received their pardons and waited for ten days, they were allowed to return to court. By the end of April, they were restored to the Privy Council and a grand ceremony of reconciliation was staged. Atholl, Bothwell and Huntly on the one side, and Moray, Argyll and Glencairn on the other, stood before Mary and joined hands. Maitland was excluded from this ceremony. He was still firmly under a cloud, living at Dunkeld under house arrest.

Bothwell gained most from his role in organizing Mary's escape. For his unflinching loyalty, she rewarded him with the captaincy of Dunbar, granting him both the castle and its surrounding estates in succession to her late half-brother, Lord John of Coldingham.

Morton and the remaining conspirators lost the most from the Rizzio plot. All eighty of them were denounced as rebels and their goods forfeited to the crown. Their houses were stripped bare, and Morton was forced to surrender Tantallon Castle, his fortress on the cliffs at the entrance to the Firth of Forth.

Mary took great satisfaction at his fall. She intended to be unremitting to the Douglases. She saw Morton as the principal villain after Darnley, not least because it was Andrew Ker of Fawdonside, a noted Douglas client, who had leveled the pistol at her as Rizzio was dragged to his death.

Morton, Ruthven, Ker and the rest fled to England. Morton and Ruthven wrote a groveling letter to Cecil, protesting the justice of their cause and assuring him that they had only acted on Darnley's orders and for the "preservation of the state and the Protestant religion."

But their position was undercut. A week earlier, Darnley shamelessly denied his role in the Rizzio plot "upon his honor, fidelity and the word of a prince." He had not, he said, even known of the conspiracy, "whereof he is slanderously and sakelessly traduced." He confessed to exceeding his powers by inviting the exiled lords of the Chase-about Raid to return home without Mary's knowledge, but that was the full extent of his crime.

By issuing this barefaced denial, Darnley dug his own grave. The first act of revenge by the Rizzio plotters was to post to Mary the bond he had signed that committed him to the assassination of those "who abused the kindness of the queen," and especially "one stranger Italian called David." The bond had even stated that the deed might "chance to be done" in Mary's private apartments or elsewhere in Holyrood.

If this were not enough, Moray showed his sister the bond that the exiled lords signed at Newcastle, which promised Darnley the crown matrimonial without her prior consent. Mary now had all the proof she needed of her husband's treachery, but the motive of those who denounced Darnley was to publicize their own blood feud with him, not to admit their guilt.

Mary had shown extraordinary daring and presence of mind during the Rizzio plot. At the height of the crisis she had kept her nerve. Her enemies were the first to concede that she had shown amazing coolness.

If she was a winner, the main losers were Darnley, whom she forever afterward despised, and the Douglases, who were in dire straits. Elizabeth was especially furious when the Douglases fled across the border. Their plot had failed, and she disowned them. She advised Morton to find "some place out of our realm" to hide until Mary's wrath had eased or he was acquitted in a legal trial. He duly left for the Netherlands. But Mary's letters had preceded him. He was denied entry and slipped back into England within a month. He was then ordered to "convey himself to some secret place, or else to leave the kingdom."

As to Lord Ruthven, he died at Newcastle six weeks after he was outlawed. On his deathbed, he exclaimed "that he saw Paradise opened, and a great company of angels coming to take him." As his days had been numbered before the plot, his personal sacrifice was limited, but his family lost all their property when he was outlawed.

By the end of April 1566, Mary was back in control. The theme of her policy would be reconciliation. But the effect of the Rizzio plot would prove to be corrosive. Darnley's instability and folly had shown that with friends like him, no one lacked for enemies. Lennox was furious with his son, who had brought all his plans of the past twenty years to the verge of catastrophe.

Darnley might still survive as long as Morton stayed in exile and could not get his hands on the man who had so brazenly double-crossed him. His best hope was Mary's baby. It was due in June, and once he was the father of the heir to the throne, he might think about staging a comeback.

If, however, the leader of the Douglases ever returned to Scotland, Darnley's life would be in peril. By assassinating Rizzio, he had let the genie out of the bottle. Nothing he could do would ever make it possible to put it back again.

17

❧❖❧

Reconciliation

MARY'S EFFORTS to reconcile her lords, with the sole exception of Darnley's co-conspirators, were genuine. In arranging this new phase of détente, she was assisted by Castelnau, who reappeared at Holyrood a month after Rizzio's assassination and shuttled to and from Paris until a resident French ambassador was sent to replace him.

Castelnau had won everyone's respect for his impartiality during the Chase-about Raid, which made him the perfect intermediary between the noble factions after the Rizzio plot, when the difficulty was not reconciling the lords to Mary, but reconciling the factions to each other.

Highest in Mary's esteem were Bothwell and Huntly. Their credit soared because of their unquestioning loyalty to her. At this critical moment in her reign, she needed advisers she could trust, who could act as a foil to the rest of the lords and keep an eye on Darnley.

Bothwell and Huntly agreed to work together. A fortnight before Rizzio's murder, Bothwell had married Huntly's sister, Lady Jean Gordon, by Protestant rites at Holyrood. Mary herself attended the reception, to honor her two most loyal supporters and out of deep respect for Bothwell's sister, the widow of her favorite half-brother, Lord John of Coldingham. Mary loved weddings and always came to those of the people she liked or wished to honor, usually bringing an expensive gift or paying for the bride's dress or the wedding banquet. She gave Jean Gordon a gorgeous wedding dress of cloth of silver lined with white taffeta, and also paid for the banquet, a sumptuous feast that was followed by jousting and "running at the ring."

Since his recall from France, Bothwell had made a remarkable political recovery. He had moved from the sidelines to the center of events, incurring the jealousy of those he had eclipsed. He was Moray's sworn enemy. Maitland also hated him. Relations reached their lowest point when Bothwell threatened to hunt down the disgraced secretary of state and kill him because of his involvement in the Rizzio plot.

Under Castelnau's auspices, such feuds were eased or appeased. The one person who consistently thwarted Mary's attempts to pacify her country was her husband. When she learned the full extent of his and Lennox's treachery to obtain the crown matrimonial behind her back, she banished Lennox from her court. Darnley would be much harder to manage, because she was pregnant and he was the baby's father. She could barely believe that the man she had married could have acted in this way, conspiring with the lords to murder her confidential secretary and then dissolving Parliament without consulting her. Darnley even had the cheek to continue denying his involvement after his exiled co-conspirators had sent Mary his own signed bond approving the assassination.

Darnley just carried on plotting. Shortly after Mary's grand ceremony of reconciliation, he wrote letters to Charles IX and Catherine de Medici in which he called himself "king of Scotland" and used a signet seal emblazoned with the royal arms. He protested his innocence of Rizzio's murder and sought to ingratiate himself to Charles as his "dear brother."

Not content with this, he continued to pursue his Catholic policy despite his promises to his co-conspirators to restore the Protestant settlement. He wrote to Philip II and the pope, complaining of the state of the country, which he claimed was "out of order" because the Mass and the Catholic religion had not yet been restored. This, he said, was entirely Mary's fault.

More bizarre were his schemes to capture Scarborough and the Scilly Isles, both belonging to England but hundreds of miles apart. He plotted to take Scarborough Castle, a partially ruined fortress on the coast of Yorkshire some two hundred miles equidistant from London and Edinburgh, where ships could unload men and supplies in the adjacent harbor. He also studied maps of the Scilly Isles, off the southernmost tip of Cornwall, to which he staked a ridiculous claim encouraged by a small group of malcontent islanders and Cornish Catholic gentry. The Scillies were six hundred miles from Edinburgh. They had no connection with Scotland, but Darnley knew that they had been fortified during the reign of Edward VI, because he noted on his maps the positions of several disused blockhouses and a fort.

Darnley was cooking up some madcap scheme to invade England by landing Catholic armies from the Continent at what he believed to be strategic locations in the British Isles. But he had given no thought as to how, even if a landing could be accomplished in the Scilly Isles, these troops were then to be transported to the mainland. And if troops were landed at Scarborough, how would they be supplied with food and munitions once they started their march to wherever Darnley supposed they were going?

Cecil's spies knew of all these plans. However foolish and implausible, they had to be taken seriously. Mary's popularity had soared in England on the news of her pregnancy, as Cecil had always predicted. He was never shy of resorting to underhand methods where Mary was concerned. He next took an agent provocateur into his service, one Christopher Rokesby, whom he sent without Elizabeth's knowledge — she detested such men — to try and inveigle Mary into a plot, and so discover how far she was likely to be implicated in Darnley's schemes.

Rokesby slipped into Scotland in May, securing access to Mary by posing as a loyal Catholic fleeing persecution in England and presenting her with an ivory carving of Christ's crucifixion. Having gotten her attention, he told her that many of the leading Catholic nobles were weary of Elizabeth and willing to depose her, provided Mary would consent to a coup or an assassination attempt. He urged her to give him some token, and dropped the names of those leading English landowners who would join the revolt if shown proof of her support.

Mary was not fooled by this. She was by now eight months pregnant. No one knew exactly when she would enter her confinement chamber and the child would be born. As medical understanding of gynecology and human reproduction was sketchy, even qualified doctors supposed that female fetuses spent longer in the womb than male ones because they were the weaker sex, and so the length of a pregnancy was thought to vary accordingly.

But Mary was clearly approaching her time, staying in bed for much of the day and traveling only short distances in a horse litter. And with the hopes and fears of childbirth at the front of her mind, she was far too busy to get involved in plotting. She was also in too strong a position with an heir on the way to taint her claim to the English throne with dubious activities. She could be naive and impulsive, but was rarely vindictive: in spite of her dueling and angry clashes with Elizabeth over the years, she had never yet conspired against her. She dismissed Rokesby, who returned to England, where he set about building up a dossier that seemed to prove that he was employed by the Catholic nobles in precisely the way that he had claimed. Armed with these documents, he crossed the border to try again.

Mary was one step ahead. Rokesby had barely finished renting rooms in Edinburgh on the day of his return when he was arrested and his papers were seized. Mary was positively gleeful when a highly incriminating letter from Cecil in code was found in his possession. When deciphered, it established Rokesby's guilt and showed that Cecil had offered him a generous reward if he succeeded. The new English ambassador to Scotland, Henry Killigrew — Cecil's brother-in-law, who had been sent to replace Randolph after his expulsion for covertly bankrolling Moray's second revolt in the Chase-about Raid — was the one person to whom Elizabeth's chief minister had confided his true relationship to Rokesby. And Killigrew was appalled at the letter's discovery. He wrote instantly to Cecil, to warn him to prepare for Mary's reprisals.

It was an extraordinary, rare lapse of security on Cecil's part. But Mary, preoccupied with the birth of her child, bided her time, waiting until it suited her to counterattack. She then sent one of the very few letters that she ever wrote to Cecil personally.

"Since our first arrival within our realm of Scotland," she began, "we ever had a good opinion of you, that you at all times had done the office of a faithful minister." Mary said that she had never doubted Cecil's motives until her good opinion was shaken "by the strange dealings of an Englishman named Rokesby." Since then, she "began a little to suspend our judgment, until we receive further trial therein."

Mary's rebuke was as dignified as it was measured. She explained how she had asked Robert Melville, her agent in London, to discuss the Rokesby affair with Cecil. He had reported that Cecil was "nothing altered" from his former "good inclination" toward Mary "of the which we were not a little rejoiced." She urged Cecil to "persevere in nourishing of peace and amity." In this way, she felt confident that he would "do acceptable service" to Elizabeth.

This was more than artful. It was a thinly veiled threat, because Mary

— after Melville's discreet inquiries — knew quite well that Elizabeth would have been furious to hear of Cecil's use of an agent provocateur.

On June 3, 1566, Mary went into her confinement chamber at Edinburgh Castle. She still feared for her safety after the Rizzio plot and insisted that the Earl of Argyll, whom she trusted to defend the Scottish monarchy with his last drop of blood despite his quarrels with her or her mother, move into the room outside and stay there night and day. Her son, Prince James, was born between ten and eleven on the morning of Wednesday the 19th. It was a long and difficult labor. And yet Mary did not entirely lose her sense of humor, crying out halfway through that if she had known how painful it was going to be, she would never have got married in the first place.

The baby, when he finally arrived, was in excellent health. Mary was exhausted but triumphant. She had produced a legitimate male heir and settled the succession to the throne in her country, exactly what women rulers were supposed to do. All this greatly strengthened her hand with Elizabeth, who pouted and sulked when she heard the news. The guns of Edinburgh Castle were fired to salute the birth, and there was spontaneous rejoicing. Some five hundred bonfires were lit in the town and its suburbs alone.

Killigrew visited Mary in her confinement chamber two days later. After he had congratulated her on a safe delivery, she thanked him, but politely asked to be excused a lengthier interview. She was still in bed and troubled by pain in her breasts. She spoke faintly on account of her physical weakness, her voice interrupted by a hollow cough.

Killigrew was taken to the nursery to see the baby. He was, the ambassador said, in a phrase resonant of his predecessor Sadler's when he first admired the infant Mary, "a very goodly child." Killigrew watched him "sucking of his nurse," Margaret Little, and afterward saw him "as good as naked, I mean his head, feet and hands, all to my judgment well proportioned and like to prove a goodly prince."

Childbirth was a risky, life-threatening prospect in the sixteenth century. Women prepared for the worst. On June 9, Mary had summoned her lords to hear her will. Three copies were kept: one she sent to her Guise relations, one she retained herself, a third she signed and sealed and gave to those who would become regents if she died. No copy of the document now survives. Our only information about it is that the regency was vested in a committee. Either Darnley was not included, or else (and more likely) he was there, but counterbalanced by two independent regents. One was Lord Erskine, a moderate Protestant and the captain of Edinburgh Castle, whom Mary promoted to the earldom of Mar. The other was the Earl of Argyll, already completely rehabilitated.

Later Mary's enemies claimed that she had nominated Bothwell to be regent, which was quite untrue. However much the swashbuckling border earl had risen in her estimation, he was still a controversial figure whose violent temper and love of dueling made him less qualified than Mar and Argyll to protect the interests of her family and the monarchy.

But Mar and Argyll were equally offensive to Darnley. He was furious: the terms of Mary's will caused a smoldering resentment among the Lennoxes. Killigrew supposed that their taste for plots could only be further aroused.

There was a codicil to the will, a testamentary inventory of Mary's jewels, which does survive. It has sixteen rubbed and water-damaged folios listing more than 250 lots, beside which are marginal notes indicating the names of those to whom particular lots were to be given. The lists were compiled by Mary Livingston, one of the four Maries with responsibility for the queen's jewels, and Margaret Carwood, Mary's favorite bedchamber woman. The marginal annotations are Mary's own. Her handwriting is untidy, even for her. She several times complained of cramps during the final weeks of her pregnancy, although her handwriting since her adolescence regularly descended into scribbling.

Mary scrawled a note that her bequests should take effect if she and her baby died. She knew she might be close to death; her handwriting could be an indication of her mental as well as her physical state. The inventory certainly shows where her thoughts were, because it was not the Scottish nobles or her husband who dominated the lists of beneficiaries, but her Guise family. Though she had been back in Scotland for almost five years and must often have felt very alone, she would never forget her French links. Even a quick note from France brought her pleasure in later life, and she received letters from her family that brought tears of joy to her eyes.

Out of fewer than sixty people named as beneficiaries, fourteen were members of the Guise family. They were uppermost in her mind, listed first after her bequests to the Scottish crown and taking the lion's share of precious and showy items. One lot, a magnificent collection of rubies, pearls, brooches, collars, gold chains, earrings and a belt with a gold buckle and studded with precious stones, was to be handed down in the family in perpetuity.

After the Guises came Mary's Scottish relatives and the four Maries and their families. Half a dozen of her relations were listed, with a preference for the women and children: first the Countess of Argyll, then Moray's wife and eldest daughter, then Francis the orphaned son of Lord John of Coldingham, to whom Mary was godmother. The four Maries were to receive less costly gifts, but more intimate ones that were a sign of their former playmate's love.

Mary made typically generous bequests to her ladies-in-waiting, to the surviving parents of the four Maries and to her bedchamber servants. Carwood was left one of Mary's portrait miniatures framed with diamonds, and "une petite boîte d'argent" ("a small silver box"), perhaps the celebrated casket, marked with the monogram of Francis II, in which the queen's enemies later said they found her most incriminating letters.

Darnley appeared on the lists, but was not especially favored. He was to receive up to twenty-six items, ostensibly a tenth of the inventory, depending on how Mary's annotations are interpreted. But the reality was far less. His legacies tended to be specific objects and not the whole collections of jewels left to her Guise family; for example, a watch decorated with diamonds and rubies, or a diamond ring enameled in red. Of the ring, Mary wrote, "It was with this that I was married; I leave it to the king who gave it to me." Her terse comment suggests her overwhelming sense of disappointment with her husband. A close examination of Darnley's gifts reveals that very few items, if indeed any, were bequests over and above the return of gifts that he and his father, Lennox, had made to her.

The nobles, headed by Moray, Argyll, Atholl and Mar, were to have lesser gifts, confirming the degree to which Mary saw her will as a family affair. Maitland's name was omitted — he was still in disgrace — but Bothwell's and Huntly's were included. Bothwell was to get an ornamental jewel containing a diamond in a black enameled setting and a badge or brooch with the figure of a deer set with eleven diamonds and a ruby. Much was made of these putative bequests after Mary's forced abdication, but in 1566 they were thought unremarkable. Far from gaining special treatment, Bothwell had ranked in his usual place in the list of privy councilors. Moray and Mar were both ahead of him, and no undue

favor was shown. In fact, in view of his steadfast loyalty to Mary and her
mother, the marvel is not how much but how little he would receive
from Mary's will if she had died in childbirth.

Mary's health was fragile after the birth. She needed to rest, and in July
and early August she went on holiday. Her refuge was Alloa in Clack-
mannanshire, a quiet and picturesque spot that was only a short trip by
boat along the Firth of Forth from Edinburgh. She was a guest of the
Earl of Mar and was accompanied by Moray and Bothwell. Castelnau,
whose instructions from Catherine de Medici were to reconcile Darnley
to Mary, also arrived. He had his work cut out. Relations between the
royal couple were plummeting. There was a rumor of a split at the end of
April, and Darnley had twice threatened to go and live in the Netherlands.

When Mary had been preparing to go into labor, the couple had
agreed to a purely nominal truce. This now imploded. Darnley arrived at
Alloa independently, staying for only a few hours. Castelnau barely got
to speak to him. Compiling a report for Cecil on August 3, the Earl of
Bedford wrote, "The queen and her husband agree after the old manner,
or rather worse." Mary seldom ate with Darnley anymore, and never
slept "nor keepeth no company with him" — the euphemism for sexual
intercourse — "nor loveth any such as love him." Already the king and
queen led separate lives.

At Alloa, their quarrel had intensified. Mary even swore at him. She
used words, said Bedford fastidiously, that "cannot for modesty nor with
the honor of a queen be reported."

On his way through England on one of his return journeys to Paris,
Castelnau blithely assured Bedford that Mary and Darnley were reconciled. This was either wishful thinking or a ploy to deceive the English.
Bedford had his own spies and was well aware of Darnley's jealousy and
paranoia. "He cannot bear," wrote one of these sources, "that the queen
should use familiarity either with men or women, and especially the ladies of Argyll, Moray and Mar." All her attention had to be on him for
every second or else he stormed out in a tantrum.

It was the little things that caused the most trouble. Mary loved dogs,
and when Sir James Melville gave Darnley a fine water spaniel he was
sent from England, Mary had an attack of pique. She berated her loyal
Melville, calling him a "dissembler" and a "flatterer." She too adored
spaniels, and was jealous that the dog had not been offered to her first.

"How can I trust you if you can give something like this to someone I hate?" she asked.

By the middle of August, Mary had moved on to Peebleshire, where she went stag hunting in the hills close to the Water of Megget. Moray, Mar and Bothwell were by her side, and Darnley came and went.

Then, on the 22nd, she interrupted her holiday. Something had happened that immediately made her sense danger and brought back all the fears she had suffered since the Rizzio plot. She abruptly returned to Holyrood for two days. Up until now, Prince James had stayed with his nurse at Edinburgh Castle in the care of Mar and his wife. Mary believed he was about to be kidnapped and decided to move him to the greater security of Stirling Castle, the fortress on the rock to which her own mother had once taken her.

She feared that Darnley might try to kidnap James, because she knew the birth of their son had shifted the balance of power again. It strengthened her hand significantly in the short term. She had produced a male heir and settled the succession in Scotland. The arrival of an heir also made her less dependent than before on the nobles. But paradoxically it also made her more vulnerable in the longer term, because if Darnley or the lords chose to attack her, they could seize the heir to the throne and appoint a regent to rule during his minority, thereby giving themselves power for up to fifteen or twenty years.

Taking no chances, Mary raised a force of five hundred musketeers, who surrounded Prince James's litter as it was brought from Edinburgh to Stirling. She again left her son in the care of Mar and his wife. When they had safely arrived at Stirling, Mary set out for Perthshire, where she resumed her holiday.

Now that James was well protected, she allowed herself to spend the rest of it hunting and hawking with her perfidious husband. She was keeping up appearances. In the last week of August they were in Glenartney, a lush red-deer forest in the vicinity of Loch Earn. From there they moved on to Drummond Castle near Crieff, and afterward returned to Stirling.

By the start of September, Mary felt more relaxed. She even yielded to the pleas of Moray and Atholl and allowed Maitland to reoccupy his former position as her secretary. He arrived at Stirling on the 4th, and the next day she dined with him alone. He made his humble submission, after which Mary went back to Edinburgh. Maitland was ordered to appear there on the 11th, where a short time later he was reconciled to Bothwell. How deep this reconciliation went is open to doubt. But as Maitland wrote jubilantly to Cecil, he was back at court and once more pulling the levers.

Darnley, however, was furious. Mary, he believed, was reconciling the lords in order to build a consensus against him. There was no reason for him to think this. All she was doing was trying to restore order and harmony between the feuding factions after the upheaval of the Rizzio plot. But Darnley was recalcitrant. When she returned to Stirling to fetch him back to Edinburgh, he refused to leave. He was misbehaving shamelessly, even in front of the new resident French ambassador, Philibert du Croc, who had now been sent to Scotland in place of Castelnau.

Darnley made a shocking announcement. He intended to separate himself from Mary and go and live abroad. He avowed this in what du Croc called "a fit of desperation." His tirade brought matters to a head. Mary "took him by the hand, and besought him for God's sake to declare if she had given him any occasion for this resolution; and entreated he might deal plainly, and not spare her." Darnley pushed her aside. Even the lords were appalled at this. They too wanted to keep him in Scotland, since apart from the dishonor that he would do to Mary by leaving her, it was clear that Darnley plotting abroad would be more dangerous than Darnley plotting at home.

Du Croc was bewildered by this public fight. He had never seen anything like it, especially when Darnley conceded that he had no specific grievance beyond a sense of outrage that he was not adequately recognized as king. What rankled was Mary's refusal to crown him. Despite her innumerable attempts to calm him down, he refused to be silenced. At last he stalked out of the room, saying to Mary, "Adieu, madame, you shall not see my face for a long space," and to the lords, "Gentlemen, adieu."

Mary returned to Edinburgh. When Darnley arrived at the gates of Holyroodhouse a week later, insisting that her councilors must be evicted before he would deign to enter, he was personally hauled inside by his wife. She spent most of the night trying to drum some sense into him, and when she failed, the Privy Council was summoned with du Croc as an independent witness. Darnley was then asked to explain exactly what it was that he complained of, and when he was unable to give any credible answer, but continued to ask for a separation from Mary, the Privy Council wrote officially to Catherine de Medici, setting down a record of his insanity and seeking French cooperation should he attempt to establish a royal court in exile.

Mary's health then collapsed. It happened while she was staying at Jedburgh, close to the English border. It had been her intention to go there and preside at her Justice Ayre, a circuit or traveling court that dealt with criminal cases and spent a week or so in each location. The circuit, which included Teviotdale and Liddesdale, had been delayed on account of the late harvest, but was due to begin on October 8. Bothwell, within whose jurisdiction these areas fell as lieutenant of the borders, left Edinburgh on the 6th to prepare for the queen's arrival.

Darnley, who was still threatening to go and live abroad, refused to accompany Mary. But Moray, Argyll, Maitland, Atholl and Huntly were among the forty or so in her train. Scarcely had she passed Borthwick, eleven miles southeast of Edinburgh, when she heard that Bothwell had been violently ambushed by his old enemies the Elliots of Liddesdale, who were notorious for their brigandage throughout this region of rough border terrain.

At first Bothwell was said to be dead; then it was confirmed that he was alive but severely injured, with sword wounds to his body, head and hand. He was dragged on a sledge to the Hermitage, his nearby citadel in the valley of the Hermitage Water, where he lay critically ill.

Mary came to visit him, but not for another week. The facts emerge from the reports to Cecil of Lord Scrope and Sir John Forster, the English officials who were closely monitoring these events from their respective vantage points of Carlisle and Berwick. Since, like all the English, they loathed Bothwell, they are unlikely to have drawn a veil over any of his transgressions.

According to their accounts, Mary did go to the Hermitage, but not as soon as she arrived at Jedburgh. She first conducted the Justice Ayre in the usual way, which lasted a week. Only after the court adjourned on October 15 did she ride to visit Bothwell. She was concerned about his injuries: he was, after all, one of her most loyal privy councilors.

And there was perhaps another reason. Bothwell, her border lieutenant with a commission to root out the "riders" and bandits of the region, was due to begin a new session of the Justice Ayre in Liddesdale on the 16th. She may have planned to attend the new session. Nor did she ride to Bothwell's home alone. Contrary to later calumnies, she was accompanied by Moray and all her leading courtiers. Perhaps they had intended to stay overnight at the Hermitage. If so, they changed their minds. They stayed for only two hours and then returned to Jedburgh the same day. The journey was up to thirty miles each way across rough country using the most direct paths. This seems astonishing until one realizes that forty miles was then considered a normal day's riding. A round trip of fifty miles was above the norm but not out of the ordinary. Sixty miles was pushing it, but feasible in good weather.

Mary's ride was on the 15th or 16th, according to whether Scrope's or Forster's date is accepted. Then, on the 17th, she fell dangerously ill. A few days earlier, she had complained of "spleen." When she finally collapsed, she was in agony from the pain in her left side. She vomited blood several times and then lost consciousness. Within two days she was suffering convulsions and had lost the power of speech. Next day she lost her sight. By the 24th, she had improved, but suffered a relapse the next day. At the height of the crisis she lay apparently dead for half an hour: "eyes closed, mouth fast, and feet and arms stiff and cold."

She was saved by her French surgeon, Charles Nau, who was said to be "a perfect man of his craft." He tightly bandaged her big toes, her legs from the ankles up, then her arms. He massaged all her limbs. After this, he forced open her mouth and poured wine down her throat. He also administered a clyster, or enema. This caused vomiting and diarrhea, enabling her to discharge a large residue of "corrupt" (old) blood. Within three hours, she had recovered her sight and speech and begun sweating. It was a bravura performance by her surgeon: no better treatment could have been given in the absence of a blood transfusion.

Whatever Mary's precise illness, it could not have been acute intermittent porphyria or any other type of porphyria. She had suffered severe internal bleeding and hemorrhagic shock. These symptoms have no connection with porphyria. It is possible that she had more than one disease, but the obvious diagnosis is a gastric ulcer that burst after the exertion of the long ride and the anxiety associated with Darnley's treachery.*

The lords had no interest in the medical causes of Mary's illness, but were alarmed by her condition. With Darnley on the loose, they had no appetite for a change of regime. They foresaw anarchy if their queen

---

* Mary had only just recovered after childbirth, but her resolute actions to protect her son from kidnapping suggest that she was not suffering from postpartum complications.

died and for the moment wanted to keep her on the throne. While the factions in Scotland were interested in the potential benefits of a long royal minority, they were not yet ready for a regency that would be hotly disputed until the ambiguity of Darnley's position as an uncrowned king consort was resolved.

Such fears concentrated their minds on the advantages of Mary's rule while she was alive. Studiously ignoring their own conduct, they blamed Darnley for her breakdown. His behavior had pushed her to the brink. Explaining their dilemma to the Scottish ambassador in Paris, Maitland put the case in a nutshell: "She has done him so great honor . . . and he on the other part has recompensed her with such ingratitude, and misuses himself so far toward her, that it is a heartbreak for her to think that he should be her husband, and how to be free of him she sees no way out."

Just as at the outset of the Rizzio plot, Maitland was hinting at the solution to a problem that Mary herself had not yet defined. Up to now, she was trying to live with the consequences of her marriage, keeping up appearances and balancing the advantages of a legitimate male heir against the disadvantages of Darnley. But the lords, including Bothwell, saw things differently. By removing Darnley, they would be doing everyone a favor.

As the year 1566 moved to its close, it seemed as if Darnley, with his manic obsession to be crowned king and his longing to make a name for himself in Catholic Europe by restoring the Mass, was an intractable problem for everyone except his own family. It would be very convenient to lose him. He had served his purpose by fathering a male heir. His conduct was intolerable; he was politically expendable. He did not bother to visit Mary at Jedburgh until she had almost recovered. Even then, he stayed for only one night, returning to Lennox's stronghold at Glasgow the next day.

Bothwell was in Jedburgh a week before him, carried in a horse litter back to his lodgings. He was well enough to sit on the Privy Council, and within a week had "convalesced well." He was in Mary's retinue on November 9, when she left Jedburgh to begin a royal progress through Berwickshire and East Lothian. He performed his job of lieutenant of the borders without mishap or misadventure. The progress ended on the 20th, when Mary and her privy councilors arrived at Craigmillar Castle, three miles south of Edinburgh, where they stayed for almost a fortnight.

Mary had scarcely been at Craigmillar a week when she was ill again. It was the second time in her life when she said she had been close to death or wished she really were dead. The first was when she was struck by the viral disease known as "the sweat" at the age of thirteen and a half. At Jedburgh, she had certainly been close to death, but her collapse was so sudden, she had been unable to think much about it. Now the narrowness of her escape was dawning on her.

According to du Croc, who was an eyewitness, "she is in the hands of the physicians, and I do assure you is not at all well. I do believe the principal part of her disease to consist of a deep grief and sorrow." She was unable to shake off her mood. "Still she repeats these words, 'I could wish to be dead.'"

"You know very well," continued du Croc, "that the injury she has received is excessively great, and Her Majesty will never forget it." Slowly but surely, the truth slipped out. There was more to this depression than Mary's brush with death. Darnley had been to visit her. There had been further rows. "Things are going from bad to worse . . . I do not expect upon several accounts any good understanding between them." There were two overwhelming obstacles to a reconciliation. "The first is, the king will never humble himself as he ought. The other is, the queen can't countenance any of the lords speaking with the king without immediately suspecting a plot between them."

But if du Croc was brooding over Darnley's deficiencies, Mary's collapse at Jedburgh put a reconciliation of a quite different order within her grasp. When mortality was staring her in the face, she decided that if she failed to recover, then "the special care of the protection of our son" was to be given to Elizabeth, who should come to regard Prince James as her own child. It was fairly obvious that Elizabeth did not intend to marry, in which case Mary could best protect her son's life and dynastic rights in Scotland and England by making this extraordinary gesture. She knew that despite their earlier dueling, Elizabeth would always respect the ideal of monarchy and give precedence to hereditary rights over religious differences. In England, James would be brought up as a Protestant, in which case his prospects in both countries would be unrivaled.

A message was sent to England, and Elizabeth reciprocated. Her reply, conveyed to Mary through Robert Melville, does not survive, but on November 18, during the final stages of the progress in East Lothian, Mary quoted from it. She wrote a letter to the English Privy Council, expressing her thanks for the "good offers" she had received from her "dearest sister," which she proposed to follow up without delay.

Although made on the spur of the moment, Mary's offer to name Elizabeth as her son's "protector" was a masterstroke. It was flattering enough to appeal to the English queen and yet enigmatic enough not to pose a direct threat to the Scottish lords, since a protector, in the sense meant by Mary, was not the same as a governor or regent, for which she had made provision in her will. And it kept Darnley out of the picture.

The result was the prospect of a new accord giving genuine substance to the kinship ties between the two British queens. Such ties had always underpinned their rhetoric, but had so far not amounted to much. As Mary herself once exclaimed, they were all just empty words! Her latest gesture enabled her to play the part of a "natural sister" or "daughter" to Elizabeth with real conviction. She had entrusted her son to Elizabeth's protection. Although she did not die at Jedburgh, it did not mean she could not hope to benefit from the sudden thaw in their relations. When her cousin responded with her own "good offers," the way was open for a fresh round of diplomacy in which Mary hoped to secure recognition of her claim to the English succession.

Elizabeth was willing to compromise. For once, she made her own decision untrammeled by Cecil's intervention. Cecil had for some time been pressing her to marry Archduke Charles, and the English Parliament had been summoned after a gap of three and a half years. In a series of heated debates and backroom deals, Parliament was vainly petitioning her to marry and settle the succession in her own country, just as Mary had done so successfully in hers. Advised by Robert Melville in London, Mary knew how and when to play her cards.

Elizabeth bitterly resented the lobbying of her councilors. She turned her rage first on a delegation of lords and next on a committee of both houses of Parliament for discussing what she regarded as her personal affairs. Her "good offers" to Mary came shortly after the crisis over the parliamentary debates.

In a fractious minute, perhaps read only by Cecil, Elizabeth railed against the "lewd practices" of those who had been lobbying. She turned to her own solution, negotiating on the level of queen to queen.

Her terms were breathtakingly simple. She would retract her demand that Mary ratify the original treaty of Edinburgh. Instead, a new "treaty of perpetual amity" would be negotiated. This would leave the peace intact but remove all clauses detrimental to Mary's honor. "Our meaning," said Elizabeth, "is to require nothing to be confirmed in that treaty but that which directly appertains to us and our children, omitting anything in that treaty that may be prejudicial to her title as next heir after us and our children, all of which may be secured to her by a new treaty betwixt us."

Elizabeth was willing to acknowledge Mary's rights as heir apparent. In return, those rights would be rigorously defined and narrowed down. If Elizabeth married and had children, Mary's claim would lapse. And to remove forever the threat of an attempted usurpation, an "engagement" or "reciprocal contract" would be signed. This would provide mutual guarantees whereby each party recognized the other to be a lawful ruling queen, and neither would do anything to harm the other. As Elizabeth explained it, "This manner of proceeding is the way to avoid all jealousies and difficulties betwixt us, and the only way to secure the amity."

Elizabeth would certainly have gained by these terms. They meant that Mary would ratify the substance if not the form of the treaty of Edinburgh, thereby tying up the loose ends of the past five years and guaranteeing Elizabeth's security.

But for Mary, the offer was still a breakthrough. She must have felt elated. True, the concessions were almost identical (as she was not shy to point out) to those of the "middle way" first proposed five years before. A settlement on such lines could have been agreed on Mary's side at almost any point since her return from France.

What mattered was that the terms were now on offer. But Mary had read the fine print. To validate fully her dynastic claim, the obstacle of Henry VIII's will remained. The will had excluded the Stuart line of succession, specifying that if Elizabeth died childless, then the offspring of Henry VIII's younger sister, Mary Duchess of Suffolk, should inherit the throne.

Mary recalled Robert Melville to advise her. She knew that her trump card was to secure a judicial examination of Henry VIII's will. Its validity had been contested. By the Third Act of Succession, passed in 1544, Parliament had empowered Henry to settle the succession by his "last will and testament signed with the king's own hand." Whether he had signed it was disputed. Although the will was "signed," it was probably not with the king's own hand but with a stamp, a device used by Henry in the last months of his reign. When this stamp was used to sign documents, an impression of the king's signature was made on the paper and the signature inked in later by a clerk. The procedure was meant to spare an increasingly restless Henry from the trouble of signing state papers. But the will was not a normal document: it was a unique legal instrument, which by the terms of the Act of Succession should have been signed by the king in person.

Mary's argument was plausible. Henry VIII's will is listed to this day in the official register of documents signed "by stamp." And her claim that the witnesses and the stamp itself were "feigned," meaning that they were affixed when the king was already dead or unconscious, is also credible, if unprovable.

Elizabeth was not unsympathetic. She had never placed much reliance on her father's will, which is why she had always preferred Mary's claim to that of the remaining Grey sisters, each of whom made clandestine marriages.

On January 3, 1567, Mary wrote to say that she accepted the offer of her "dearest sister," subject to a judicial examination of Henry VIII's will. Since Elizabeth was known to be ready to dissolve Parliament and had in fact done so on the 2nd, this review would take place quickly.

After one more frustrating exchange of letters, everything was agreed. This was to be the settlement of which Mary had always dreamed. On February 8, she ordered Melville to return to London. Her health was recovering and she was happy and excited at the prospect of the new treaty.

Then, at two o'clock in the early morning of February 10, while Melville was still packing his bags, Darnley was assassinated. From the moment the news reached London, Mary's reconciliation with Elizabeth was a dead letter. When Melville reached London on the 19th, Cecil refused to let him into his house. No further talks took place between Mary and Elizabeth. There would be no judicial examination of Henry VIII's will.

The theme of the autumn and winter of 1566–67 was reconciliation, and yet in a shocking act of terrorism, the king of Scotland and two of his personal servants were suddenly murdered. The prospect of a dynastic accord between the two British queens is not in itself proof that Mary played no part in or had no foreknowledge of the assassination. But it makes her complicity improbable. Nothing can be proved by circumstantial evidence. The imminent dynastic accord does, however, create a compelling new context for a reinvestigation of Darnley's murder, forcing us to consider afresh the true facts of the first British gunpowder plot.

18

※

Plot and Counterplot

DARNLEY'S MURDER involved three distinct elements: a conspiracy, a crime and a cover-up. Part of the fascination of this murder mystery has always been to unravel the mass of almost bewildering evidence surrounding it. There are as many accusations or disclaimers as there are actors in the drama. In a debate lasting over four hundred years, no one has given a wholly satisfactory explanation of what happened on that night or why. The fundamental facts are still contested, such as who killed Darnley, precisely where he died and why it was decided to kill him in a gunpowder plot at the house where he was staying — and yet he was apparently strangled in a nearby garden after vaulting a wall at two o'clock in the morning clutching a chair.

Since there *was* no unbiased report, all accounts of Darnley's death must to some extent be hypothetical. Beyond this, the who, where and why in the equation have been confused. The details of the murder, as reinvented afterward by those eager to cast the spotlight on others, have been muddled with those of the conspiracy and crime itself. To orchestrate the cover-up, fact was mingled with fiction, creating fresh stories, each with their own internal logic. Such stories are not the true facts of the events they purport to describe, though they will be important to us, because they tell us about the motives of those who attempted to conceal their deeds or blame them on others. In the end, the stories about Darnley's death took on a life of their own.

First we have the conspiracy to consider. We need to understand how it came about that Mary's husband ended up staying not in his usual apartments at Holyrood, but in a borrowed house on the southern outskirts of Edinburgh, its cellars mined with gunpowder. It sounds simple, but there are layers to this conundrum.

The most basic is archival. The key documents for Darnley's murder are all resolutely English, presenting problems of bias and selectivity. But in this case such problems are greatly magnified, because when the archives were catalogued and bound into large leather volumes in the nineteenth century, they were jumbled up. The idea was to select the most important papers and rearrange them in chronological sequence, but this was easier said than done.

The papers are kept in London, split between different collections. One, itself subdivided between the British Library and the Public Record Office, contains the bulk of Cecil's working papers as Elizabeth's chief minister. The other consists of the reports of the Earl of Bedford, the most senior English border official and the governor of Berwick, and those of his deputy and military adviser, Sir William Drury, and their subordinates.

But when these papers were catalogued in the nineteenth century, documents were pulled out of Bedford's and Drury's papers and added to Cecil's to fill in gaps. Others were shifted about to suit the dictates of the new bound volumes, causing their provenance to be obliterated. So an enclosure sent with a covering letter may be separated from the letter itself, or vice versa. Worst of all, many documents were rearranged under incorrect dates, in which mistaken order they were edited for publication, thereby misleading generations of otherwise responsible historians.

Lastly, whereas most of Cecil's collections were printed at some length, those from Berwick were dealt with in a cursory fashion, their potential largely overlooked. Up until now, no historian has fully digested their contents, which means deciphering the often intractable handwriting folio by folio. But the task is rewarding: it is in this portion of the archives that the most exciting new facts about Darnley's death will be discovered.

The conspiracy began after Mary's breakdown at Jedburgh, for which the lords blamed Darnley. When Maitland wrote of her "heartbreak . . . that he should be her husband," he had taken the first step. The lords brooded over the problem while Mary lay ill at Craigmillar Castle. They had followed her there on November 20, 1566, staying for almost a fortnight and then accompanying her to Holyrood for a few days before she set out for Stirling to celebrate the baptism of Prince James.

At Craigmillar the outlines of a plot were shaped. The lords hated Darnley, but they also feared the Lennoxes. They wanted Mary to pardon Morton and the exiled Rizzio conspirators and recall them to Scotland, where their forces, combined with those of the other lords, would keep the Lennoxes in their place. As at the start of the Rizzio plot, it was the supremely devious Maitland who set events in motion. The question was how to induce Mary to grant the pardons.

Maitland began by talking to Moray and Argyll. He said the first step would be to obtain a divorce for Mary, freeing her of her greatest liability while reducing the threat of an attempted coup by Darnley. She would be so grateful, she would reward those who had helped her by pardoning the exiles.

Moray was unpersuaded, as was Huntly, whom Maitland sounded out next. We do not know why, but most likely they made the point that a divorce would still leave Darnley dangerously on the loose and that something more sinister would be required to silence him.

Maitland pressed ahead. Argyll was willing to join the conspiracy, and Huntly dropped his objections. Naturally they named their price: they wanted their tenure of their ancestral lands to be confirmed by act of Parliament, making all previous forfeitures null and void in law and preventing a challenge to their territorial rights in the future.

Bothwell was approached last. He expressed similar doubts to Moray and Huntly, but agreed to support the plan. Everyone then went to see Mary in her private room to urge her to separate herself from Darnley.

The lords said afterward that she had agreed to a divorce in principle, as long as it was legal and "not prejudicial to her son." This is entirely possible. As a devout Catholic, she would have wanted an annulment of her marriage rather than a divorce, but provided it could be obtained in a way that guaranteed her son's legitimacy and rights of inheritance, it was not unthinkable.

Maitland said that if Moray still dissented, "I am assured he will look through his fingers thereto and will behold our doings saying nothing to the same."

At this Mary became agitated. What did Maitland mean? She quickly replied: "I will that ye do nothing whereto any spot may be laid to my honor or conscience, and therefore I pray you rather let the matter be in the state as it is."

Obviously, all along there had been more to Maitland's plan than he had disclosed. At this stage, Maitland intended Parliament to play a crucial role in exonerating the conspirators, and the most credible interpretation is that he envisaged a divorce for Mary in conjunction with a trial or legislation in Parliament in which the lords declared Darnley guilty of treason on trumped-up charges.

Mary forbade this. She preferred to leave things the way they were than get involved in anything disreputable. Although frustrated and distracted by Darnley's increasingly reckless, unpredictable behavior, her own priority was Elizabeth's latest offer of a final dynastic accord. She was negotiating with Elizabeth at the level of queen to queen through her ambassador, and wanted nothing to interfere with the diplomacy that would at last recognize her claim to the English succession. She could handle Darnley in her own way in her own time. She was, after all, his wife and the queen.

Maitland ended artfully, promising Mary, "You shall see nothing but good and approved by Parliament." And there the matter was left.

After a few days' rest at Holyrood, Mary and the lords rode to Stirling on December 10. The royal baptism took place a week later. Prince James was almost six months old, much older than the usual age for a Catholic baptism then, because the ceremony had been delayed by Mary's illness and by the travel arrangements of the Duke of Savoy's ambassador. The event was choreographed as a glittering three-day fête, modeled on the festivities on the theme of reconciliation staged at Bayonne by Catherine de Medici the previous year. This was not a coincidence. The sequence of entertainments, masques, a mock siege and banquets, ending with a spectacular fireworks display on the last day, was to be the culmination of the process whereby Mary reconciled her lords and salved the wounds caused by the Rizzio plot.

Nothing like this had ever been seen before in Scotland. The cost was well beyond Mary's private means, far exceeding the annual receipts of her income as dowager queen of France. To pay for the special effects, she raised taxes and borrowed £12,000 from the merchants of Edinburgh.

The expenses of the royal household soared to more than £5500 per month from the time the baptism was planned, an increase of fifty percent over previous levels. Costumes had to be paid for and craftsmen employed to build the stages and paint the scenery for the masques. Vast quantities of fine foods and wines had to be requisitioned and transported to Stirling. The comptroller of the royal artillery spent six weeks preparing for the fireworks display. Cannons, gunpowder, saltpeter and much more were brought to Stirling from Edinburgh, Dundee and elsewhere. They were hauled from the river up the side of the steep Castle Rock by night so the final spectacle would be a surprise. This fête would rival anything Mary had known, even the one when she was seated beside her mother and future first husband in Henry II's pavilion at Rouen fifteen years before.

The baptism was performed according to Catholic rites. Charles IX's representative, the Count of Brienne, and the resident French ambassador, du Croc, were much impressed, but the Protestant lords boycotted the service. Moray, Bothwell and Huntly stood outside the chapel door, "because it was done against the points of their religion." Elizabeth acted as the child's godmother, choosing the Countess of Argyll as her proxy. The English queen was so serious about her role, she presented Mary with a font of solid gold weighing 333 ounces. Bedford, sent to Stirling by Elizabeth as her ambassador, was instructed to play down its size. He was to speak of it modestly, pretending that it would be too small for a child of six months, but commending it to Mary for her next baby.

The Countess of Argyll played her part to perfection. Although a Protestant, she took her place beside the font, for which she was later rebuked by John Knox. Bedford, a staunch Protestant, refused to attend the service and stood outside talking to Moray and Bothwell. This was embarrassing, but nothing in comparison to Darnley's absence. Although in residence at Stirling, he spurned everyone and everything, his pride still pricked by the fact that he had not been crowned king. "His bad deportment," wrote du Croc to Mary's agent in Paris, "is incurable, nor can there be ever any good expected from him . . . I can't pretend to foretell how all may turn; but I will say that matters can't subsist long as they are without being accompanied by several bad consequences."

With Darnley sulking, it fell to Bothwell to welcome the ambassadors. "All things for the christening," wrote Sir John Forster sneeringly to Cecil, "are at his appointment and the same scarcely well liked of with the rest of the nobility as it is said." Bothwell's elevation was a reflection of Mary's great favor and of the fête's unifying theme. His own reconciliation to Moray and Argyll had been one of the most remarkable events of 1566, because it ostensibly ended his bitter feud with the two former leaders of the Lords of the Congregation, whose gold he had stolen in his ambush seven years before.

The diplomats of Europe had been invited to the baptism; Mary was unwise to allow Bothwell so much prominence. Alongside Moray and Huntly, he stood behind her chair at the state banquets, which gave him ideas above his station. Sir James Melville said that afterward Bothwell "had a mark of his own that he shot at," taking his metaphor from archery: a "mark" is a target. This was a shrewd insight.

Perception can be as important as reality, and Bothwell's role caused a good deal of talk as he strutted about in his new clothes. Mary had given Moray a suit of green, Argyll a suit of red, Bothwell one of blue. These were just three of the costly outfits she distributed to her courtiers, but the greatest expense was lavished on the banquets. At the supper on the last day, the food was not carried into the great hall as usual by waiters in costume, but served from an extraordinary mechanical "engine." No one knows today how it worked, but the first two courses, comprising fifty or so separate dishes, were brought in on a moving platform operated by twelve satyrs and six nymphs, and the third course was delivered by a "conduit," some kind of mechanical belt. Before the fourth and last course, a child dressed as an angel recited verses after being lowered from the ceiling in a golden globe.

All was in place for the serving of the final course when the moving platform collapsed, bringing the banquet to a premature close. Perhaps it was just as well. Bastian Pages, one of Mary's favorite valets who had helped Buchanan to stage the masques, had been taunting the English delegation. Dressed as satyrs with long tails and whips in their hands, he and his fellow Frenchmen had been wiggling their tails in gestures said to be obscene. A fracas ensued when one of the Englishmen told Bastian it was only out of respect for Mary that he did not stab him through the heart. On hearing the hullabaloo, everyone turned around. By then tempers had flared and order was not restored until Mary and Bedford intervened.

If Darnley had so far gotten away (literally) with murder, he soon had reason to fear for his life. On Christmas Eve, Mary pardoned Morton and over seventy more of the exiled Rizzio plotters. She even pardoned Andrew Ker of Fawdonside, who had leveled a pistol at her as Rizzio was hauled from her supper room. More remissions followed on January 12,

1567, the last day of her stay at Stirling before she returned to Edinburgh with her son.

Bedford had managed to coax Mary into issuing the pardons after the baptism. He attributed his success to the mediation of the lords, especially Moray and Bothwell. They had joined forces to persuade her. Finally, she was browbeaten into agreeing. She had still not fully recovered from her illness at Jedburgh, when her gastric ulcer had burst. She was mentally and physically exhausted, making it easier to talk her around. Two days before she granted the pardons, the French ambassador du Croc found her in bed again, "weeping sore" and complaining "of a grievous pain in her side."

Although Rizzio's murder nine months before had struck at her sense of personal security and loyalty to servants, Mary had put her revulsion aside in the interest of reconciling her volatile and factious lords to the crown and to each other. Her policy was magnanimous, but she had never fully grasped the limitations of the honor culture in Scotland, where a stance of loyalty to the crown was an attractive option to lords wishing to advance their own ambitions, but was primarily an option chosen for self-interested reasons. Mary's lords wanted to serve her solely on their own terms: they were incapable of settling their grudges even when they got what they wanted. Now she had pardoned Morton and the Douglases at a time when her disillusionment with her husband and her abdominal pains caused her to choose the path of least resistance. It was a crucial error.

Yet there was more to this than met the eye. Mary had performed an astonishing *volte-face*. She had always refused to pardon the Rizzio plotters. How had this change come about?

Part of the explanation lay in her policy of reconciliation, which took on its own momentum. But the key was her anticipation of a dynastic settlement with Elizabeth, which would lead to a new treaty to replace the offensive and dishonorable clauses of the treaty of Edinburgh. So far, the negotiations had been conducted at the level of queen to queen. Elizabeth had taken charge of her own diplomacy toward Mary, and Cecil was left out until the perfect opportunity to intervene presented itself, when Elizabeth named Bedford, of all possible ambassadors, to attend the baptism.

Bedford was one of Cecil's closest allies and particularly susceptible to his views. He was to talk privately to Mary during the baptism celebrations and persuade her to pardon Morton and the rest, allowing them to come home. In return, Cecil would not hinder the reconciliation of the queens.

This does not mean that Cecil had changed his attitude toward Mary. His move was deeply cynical. He was double-crossing Mary, because he was well aware that as soon as the pardoned Douglases returned to Scotland, they would demand their revenge on Darnley, and the resulting feud would put more on Mary's plate than she could handle and so doom any chances of an understanding with Elizabeth.

Cecil's intuition was correct. On January 9, Morton wrote to him from Berwick as he prepared to cross the border. He said he was so grateful for Cecil's intervention to obtain the pardon, he would do him "such honor and pleasure as lies in my power." Nothing would be too much trouble, whether it was to be performed in Scotland or elsewhere.

As if in concert, the very same day, Bedford, whose route crossed Morton's traveling in the opposite direction, reassured Cecil that Morton was wholly beholden to him.

It was doubtless for this reason that the editor of the enlarged edition of Holinshed's *Chronicles,* published in January 1587 and barely two weeks before Mary's execution at Fotheringhay, wrote that the "cause, the contriving and the execution" of Darnley's murder was at least in part the responsibility of "great persons now living." The term "the great personage" was well known at that time to mean Cecil. This statement was suggestively attributed to Morton in the *Chronicles,* where it was ingeniously interpolated into a quotation to evade censorship.

It is not, of course, that Cecil conspired to assassinate Darnley; he was far too clever for that. But from the beginning, his policy toward Mary had relied on attempts to destabilize her rule by causing mayhem at critical moments. He could, after all, have dissociated himself from Maitland and Knox if he had ever wanted to, by breaking off their secret correspondence. He could definitely have given Mary an early warning of the plot to murder Rizzio.

Morton was triumphant. The impossible had happened. The Douglases and their allies were heading back to Scotland, and Morton was set on a blood feud with Darnley for his treachery in denying his bond to kill Rizzio and letting his co-conspirators shoulder the blame.

The consequences were inevitable. Almost as soon as Morton arrived home, he arranged a rendezvous with Maitland and Bothwell at Whittingham Castle in East Lothian. The castle was a few hours' ride from

Edinburgh, and it was there, on or about January 14, that the conspiracy against Darnley turned into an assassination plot.*

We have Morton's own account of it, one that is admittedly biased and incomplete. It was doctored to suit the circumstances in which it was given, hence it cast all the blame on Bothwell and minimized the role of others, not even mentioning Maitland. According to this *ex parte* testimony, as soon as Morton rode into the yard at Whittingham, Bothwell proposed that the lords deal with Darnley once and for all by killing him.

Fortunately Drury, Bedford's deputy and right-hand man at Berwick, had his spies in Scotland and was watching events closely. He too was alert to the rendezvous. His reports, previously known only from brief printed extracts, prove that Morton and Bothwell were the joint ringleaders, not Bothwell on his own. When those responsible for the cover-up rewrote history to exonerate themselves, Morton's role was airbrushed out of the picture. Always the invisible man where assassinations were concerned, he was agile enough to cast the blame on others. But the handwritten originals of Drury's reports show that he was always there.

Drury knew that Morton and Maitland had attended the rendezvous with Bothwell. So had Morton's cousin and chief henchman, Archibald Douglas, whose servants later admitted that he was at the scene of the murder, where he managed to leave one of his slippers. (He was said to have worn slippers to muffle his footsteps.) Sir James Balfour was another conspirator. Although one of Darnley's protégés on whom he relied, he had changed sides and now supported Morton and Bothwell. He was no stranger to assassination, as he had been an accessory to Cardinal Beaton's murder all those years ago. He now procured the bulk of the gunpowder, for which he was remembered, because instead of paying

* One of the greatest confusions about the plot would be the later claim that the lords had signed a bond at Craigmillar in which they had promised to murder Darnley. According to this legend, the bond was in Bothwell's possession on the day of his final confrontation with his enemies and given to Mary as an insurance policy before Bothwell fled for his life. Although confidently discussed in innumerable works of history, an authentic copy of this bond has never been found. It was only in hindsight that it was ever said to have existed, and when Mary demanded to see a copy of it while in exile in England — which she would hardly have needed to do if the original bond had been presented to her by Bothwell, as the legend claims — a "copy" was obtained, but this was forged by the unscrupulous Sir James Balfour, who took the opportunity to denounce his own enemies as Darnley's murderers while omitting the names of those he wished to protect.

cash, he bartered it for olive oil. Balfour cannot be pinned down to the meeting at Whittingham, but later sought a pardon for his role in the conspiracy.

So at the heart of the plot lay Morton and Bothwell, who were assisted by Maitland, Balfour and Archibald Douglas. Others in full support included Argyll and Huntly. Moray prudently stood aloof. He was a conniver, because he knew when the explosion would take place and made sure to be away at his house in Fife that night. He had foreknowledge of the murder, but decided to "look through his fingers." He had always loathed Darnley, whom he blamed for his exile in England after the Chase-about Raid. It was typical of his self-serving ambition that Moray made no attempt to warn his sister of her peril.

Darnley, meanwhile, had left Stirling and gone to Lennox's heartland at Glasgow. He stayed there, as Bedford briefed Cecil, "full of the smallpox." In fact, his complaint was secondary syphilis. He had massive skin eruptions, his whole body covered with evil-smelling pustules, for which he was prescribed a remedy known as "salivation of mercury" lasting six to eight weeks. This had been the standard treatment for syphilis for the past thirty years. It involved sweating the patient while administering large doses of mercury either orally or topically as ointment. The doses were given until the gums ached, the teeth loosened and saliva flowed copiously. The patient's breath began to stink as the gum tissue died from the toxic mercury. Despite this, the treatment could be effective. It usually ended with a series of medicated (usually sulfurous) baths.

Mary decided to confront Darnley in his bolthole and persuade him to return with her to Craigmillar Castle. We can see exactly how this came about. On January 20, she alerted her ambassador in Paris to her intentions. She had been warned by one of the ambassador's servants, William Walker, that just as she had feared before, Darnley was plotting with the Lennoxes to kidnap Prince James and crown him. He then planned to rule as regent for the next twenty years, keeping Mary forever in prison.

Walker had identified his source. It was William Hiegate of Glasgow, another of the ambassador's men. He had flaunted his information to Walker, claiming that Darnley was so jealous of the lords who advised Mary, either they or he had to go. If necessary, he would kill Mary's councilors to get them out of his way.

In Mary's judgment, this report of Darnley's threats rang horribly true. He had talked this way the previous autumn, when du Croc had witnessed his tirades. She summoned Hiegate, who denied his words. He would admit only to hearing a rumor that Darnley himself was in danger, which he had reported to Lennox, thereby putting Darnley on his guard.

Mary felt deeply vulnerable. She was unsure of her ground, but she cursed Darnley. "And for the king our husband," she exclaimed, "always we perceive him occupied and busy enough to have inquisition of our doings." She knew that his spies were observing her while he and his father carried on with their cabals.

Fearing that Darnley was plotting a coup d'état, Mary went to Glasgow on her own initiative. But she was deceitfully encouraged by Bothwell and Huntly, who escorted her and provided her bodyguards on the first stage of her journey. They supported her plan to confront Darnley and bring him back to Edinburgh. Of course they did, because Mary had unwittingly played into the hands of her conspiring lords. She was about to become the instrument whereby Morton would be revenged on her husband for his treachery in the Rizzio plot.

Glasgow was swarming with Darnley's armed retainers. Mary's visit there was so fraught with danger, she needed bodyguards the whole time. Bothwell and Huntly had taken her as far as Callander House, the home of Lord Livingston, one of her former guardians and the father of one of her Maries. Thereafter, the Hamiltons, the family and followers of the exiled Duke of Châtelherault, provided her escort. As the most hated rivals and enemies of the Lennoxes, they were the ideal foil to a kidnapping attempt.

Mary left Edinburgh on January 20 or 21, arriving at Glasgow on the 22nd. She went straight to Darnley's sickroom. Syphilis was known to be contagious, but as Mary did not propose to touch him, she was willing to take the risk. She did not visit him out of love or concern for his health. To preempt his suspected coup, she was determined to settle him at Craigmillar, a few miles from Holyrood, where the Catholic laird of the castle, Sir Simon Preston, was fiercely loyal to Mary and could be trusted to keep her errant husband safely under house arrest if she asked him to.

Mary's actions had nothing to do with the rendezvous at Whittingham Castle. She was ignorant of what her lords were plotting. She was concerned by the reports of Darnley's intention to kidnap James and imprison her, which she had heard independently.

When Mary saw Darnley, she played on his character defects. She challenged him over the charges of Walker and Hiegate, and when he repeatedly denied them, she professed to be satisfied in order to win him over. He was so narcissistic, he refused to accept any blame for his plotting and cast himself as the victim instead.

Mary sat at his bedside, returning several times over the next couple of days. But he was becoming paranoid and refused to accompany her back to Edinburgh.

At last she did what had to be done. It is sometimes said that Mary, being the stronger of the two personalities, was able to dominate her spineless husband when they were talking face to face. While her charisma was always her best asset, it was not enough on this occasion. To win over Darnley, she had to prove her affection for him in the only way his carnal and degenerate nature understood. This meant offering to have sex with him again as soon as he was cured, as long as he first returned with her to Craigmillar.

Mary wanted to get Darnley back to Edinburgh, where she could watch his every move. If he was planning a coup, she had to detach him from the Lennox clan in their strongholds of Glasgow and Dumbarton. She had her own agenda, centered on her efforts to secure a final dynastic accord with Elizabeth. In comparison to that, everything else was secondary. She had a motive to keep Darnley under house arrest, but certainly not to kill him. If Darnley was to be murdered, everything Mary had yearned for since her return from France would be lost, because Elizabeth would instantly end their talks and even demand reprisals. Darnley, for all his failings, was her kinsman too.

The state of affairs after the baptism was one of exceptional complexity. There were three independent intrigues. Darnley was conspiring with the Lennoxes to imprison Mary and rule in the name of Prince James. She, encouraged by Bothwell and his allies, was determined to bring Darnley to Craigmillar, where he would be under her thumb. This would give Mary what she really needed: breathing space until her reconciliation with Elizabeth was complete. Finally, there was the assassination plot hatched at Whittingham Castle. Morton and the Douglases thirsted for revenge on Darnley, while Bothwell had seen an opportunity to step into his shoes. They planned to kill him, and whether they did it at Craigmillar or elsewhere was almost immaterial. The only place in the whole of Scotland that the assassination could not be attempted was in Glasgow, where the Lennoxes and their retainers held sway.

On January 31 or February 1, Mary arrived back in Edinburgh, bringing Darnley in a horse litter. But whereas she returned to her usual apartments at Holyrood, he was lodged in the relative isolation of a hastily furnished house on the outskirts of town, where he was to finish his medical treatment with a course of sulfurous baths.

It was afterward said that this house was Mary's choice. This is false. The house was the Old Provost's Lodging, one of a group of dwellings leading off a quadrangle belonging to the old and partly ruined collegiate church of Kirk o'Field, on a site occupying high ground on the southern fringes of the capital. It was Darnley's decision to stay there, and there is evidence to prove it. One of his own servants, later employed by Lennox, testified that "it was devised in Glasgow that the king should have lain first at Craigmillar, *but because he had no will thereof,* the purpose was altered and conclusion taken that he should lie beside the Kirk o'Field" (italics added).

Since a Lennox retainer (if committing perjury) would be expected to say the exact opposite, this testimony is compelling. John Hepburn, Bothwell's cousin, stated about the same thing. Claude Nau, the brother of Mary's surgeon and her secretary after 1575, also recorded that Kirk o'Field was Darnley's choice, although he was not an eyewitness and wrote several years later.

Darnley was not stupid. He refused to lodge at Craigmillar because he loathed and feared Sir Simon Preston and had already heard whispers of a plot. But until Darnley was cured, he did not want to return to Holyrood. He was prodigiously vain and could not bear the thought of people seeing his pocks and pustules. Until his treatment was completed and his skin back to normal, he wished to remain in seclusion. Even in the privacy of his upstairs bedroom at Kirk o'Field, he wore a taffeta mask over his face.

He was willing to return to Edinburgh, but to a house of his own choosing. The one he selected was far enough away from the hurly-burly of the High Street and the Canongate, yet close enough for Mary to visit, which suited his sense of self-importance. It was vacant because it had recently come into the possession of Robert Balfour, brother of Sir

James. It was situated near the top of a steep hill beside the fields, in an area known for its salubrious air. (Holyrood was on low ground and known for its winter smog from the burning of wood and coal fires.) This house was ideal for a convalescent. Where Darnley miscalculated was in the fact that Sir James had changed sides and was allied with Morton and Bothwell.

Darnley was installed at the Old Provost's Lodging by Saturday, February 1, and was expected to remain there until at least the 10th, the earliest date his treatment was likely to finish. When cured, he planned to return to his usual apartments at Holyrood.

The layout of the building is important. The living quarters were in the older part of the house. Built on two stories with a cellar, it had two large bedchambers doubling as reception rooms, one above the other on each floor, two smaller galleries or withdrawing chambers for the servants, and an outside kitchen. A newer one-story extension comprised the hall or presence chamber, equipped with a leather chair of state beneath a black velvet canopy, but otherwise sparsely furnished.

The main house had three doors. The front door faced the ancient quadrangle. There was a side door into a private garden, where Mary liked to walk and sing when she visited Darnley. There was also a postern gate, which led out of a passage from the cellar at the rear of the house. At the rear was a backstreet known as Thieves Row, on the opposite side of the town wall from the quadrangle. The town wall marked the boundary line of Edinburgh, and the back of the house abutted it.

A bathtub had been set beside Darnley's bed on the second floor to enable him to take his medicated baths. The room had been made as comfortable as possible. The walls were hung with six tapestries and there was a Turkish carpet on the floor. There were two or three cushions of red velvet, a high-backed chair covered with purple velvet and a small table covered in green velvet. The bed, which Mary had given Darnley as his traveling bed only six months earlier — it had originally been her mother's — was hung with violet-brown velvet, richly decorated with cloth of gold and silver and embroidered with flowers.

Directly below on the ground floor was Mary's room. It contained a bed of yellow and green damask in which she slept when she stayed overnight, which she did twice. It is sometimes claimed that until the last moment, she had also intended to sleep there on the night of the murder. This is another story invented for the cover-up, because if Mary had planned to spend the night at Kirk o'Field, she would not have kept her horses waiting outside to take her back to Holyrood. When she announced her intention to leave and cried "To horse," the equerries were ready and she left immediately. In fact, some who claimed Mary made a last-minute change of plan to avoid the explosion were forced to say that she walked back to Holyrood, which on a cold and pitch-black February night, in unsuitable shoes and a long dress, is improbable and was certainly not Mary's style, quite apart from the fact that the night watchmen who patrolled the streets after dark would have spotted her.

It was, however, even as Mary and Darnley laughed and talked on the evening of Sunday, February 9, while Bothwell played dice with Argyll and Huntly nearby at the table with the green velvet cover, that a group of men, working surreptitiously below them, mined the house with gunpowder. The explosive was supplied by Sir James Balfour. A dozen or so barrels were brought in the side door or the postern gate or through an underground passage cut into the cellar from the house next door, which was also owned by the Balfours. The explosive was packed into the foundation, with the residue poured into sacks that were heaped on the floor of Mary's bedroom.

In the charges later filed by Moray, it was claimed that two days before, Mary had ordered a bed and some tapestries to be taken from Darnley's chamber and replaced with less valuable ones, returning the more costly objects to Holyrood. One of Darnley's servants testified that a brand-new bed of black velvet was replaced with an old purple one.

Unfortunately for Moray, these facts can be checked. The inventory of items destroyed in the blast proves that neither the bed nor the tapestries were changed. Darnley's bed at Kirk o'Field on the night of the murder was all along the one he would be expected to have used.

By Sunday the 9th, Darnley was celebrating because his pustules had disappeared. His recovery, as he supposed, was complete. The very next day, he planned to return to Holyrood. On the 7th, he had written to Lennox in an optimistic vein. He also spoke of his reconciliation and "love" for Mary, whom "I assure you hath all this while and yet doth use herself like a natural and loving wife." When fresh rumors of plots had reached him, he had reported them to Mary. She had immediately investigated them, but nothing could be discovered. Darnley's was hardly a trusting nature, but Mary had boosted his confidence.

Darnley's vanity had been appeased. His thoughts are characteristically luminous. The following night, he planned to climb the secret stairway at Holyrood and enjoy his conjugal rights once more, thereby reestablishing himself as king of Scotland in the way he knew best. And yet, even as Bothwell rolled the dice with Argyll and Huntly, Sir James Balfour's men were at work.

Back in London, Cecil watched and waited. Unlike Mary, he was entirely prepared for what was about to happen in her country.

An artist's impression of Francis II's coronation at Rheims

Mary in her *deuil blanc,* or white mourning clothes

Sir Nicholas Throckmorton, Elizabeth I's special ambassador to France and Scotland

James Stuart, Earl of Moray, Mary's half-brother

James Douglas,
4th Earl of Morton

At age forty-two, William
Cecil chose this design for
his funeral monument at
Stamford, in Lincolnshire,
where he depicted himself as
a Protestant chivalric knight

ABOVE: Charles,
Cardinal of Lorraine,
Mary's uncle

RIGHT: John Knox
haranguing Mary
with his theories of
lawful resistance
to rulers

BELOW: The palace
of Holyroodhouse,
showing the James V
Tower at left

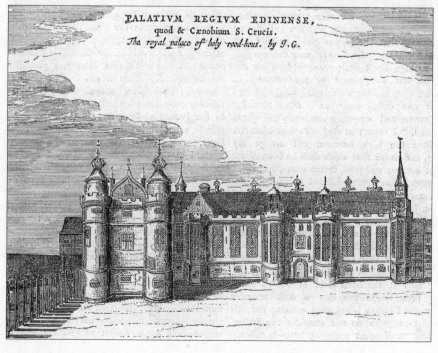

PALATIVM REGIVM EDINENSE,
quod & Cænobium S. Crucis.
The royal palace of holy rood-hous. by J. G.

Elizabeth I holding a sieve, a symbol of chastity and her refusal to marry, by Federico Zuccaro

Lord Darnley, about age nine, by Hans Eworth

Lord Darnley at age seventeen or eighteen

The Lennox Jewel

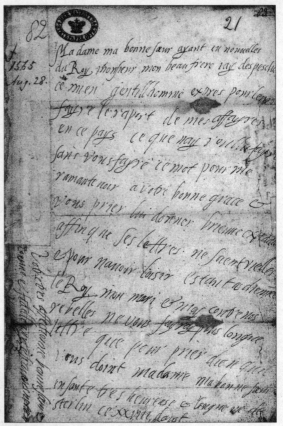

Mary's letter to Elizabeth, dated August 28, 1565, asking for a speedy passage through England for the courier bearing her letters to Charles IX of France requesting aid against her Scottish rebels

James Hepburn, Earl of Bothwell, at age thirty

19

❧❖❧

Assassination Two

THE DAY BEFORE Darnley's assassination was one of the happiest Mary could remember. It was the feast of Quinquagesima: the Sunday before Ash Wednesday and the beginning of Lent, one of the last opportunities before Easter for the banqueting and dancing that gave her so much pleasure. This particular Sunday had a carnival atmosphere. At midday, she attended the wedding reception of Bastian Pages, her favorite valet and stage designer, whose satyrs at the masque for Prince James's baptism had offended the English by wiggling their tails. He had married Christina (or Christily) Hogg, one of Mary's gentlewomen and another of her favorites, that morning at Holyrood in the Chapel Royal. Mary presented the bridal gown, which was richly embroidered and expensive. The celebrations were expected to last until midnight.

Mary left the reception in midafternoon, promising to join in the dancing before the end of the evening. She then changed clothes for an official banquet at four o'clock in honor of the Duke of Savoy's ambassador, who was returning home. The host was the Bishop of Argyll, John Carswell, who occupied one of the larger houses in the Canongate. When she left around seven in the evening, she was accompanied by Bothwell, Argyll and Huntly, but not Moray, who had slipped away to Fife, claiming that his wife was ill and likely to suffer a miscarriage. Maitland also found it prudent to absent himself, and Morton was still bound by a curfew. Under the terms of his pardon, he was barred from coming within seven miles of Mary or the court.

Around eight o'clock, Mary and her train of lords and ladies rode to Kirk o'Field for a party to mark the end of Darnley's convalescence. There was perhaps music and some dancing, at the very least wine and conversation. As the evening drew late, Darnley became increasingly amorous. He wanted Mary to stay the night. He started touching her, but she had already promised Bastian and Christina that she would attend their wedding masque.

Mary always kept her promises to her favorite servants. Shortly before eleven, she rose to leave and called for the horses. Darnley tried to dissuade her, and to fend him off, she drew a ring from her finger as a token, saying that on the following night she would sleep with him.

Moray, later reporting this to Guzman de Silva, the Spanish ambassador in London, said that Mary "had done an extraordinary and unexampled thing on the night of the murder in giving her husband a ring, petting and fondling him after plotting his murder." This, said Moray, who was himself in Fife and saw none of the events he so boldly claimed to be describing, had been "the worst thing" about this cold-blooded deed.

In appeasing Darnley, Mary was playing for time. When he had foreseen the prospect of house arrest and refused to lodge at Craigmillar Castle, her objective was to get him back to Edinburgh under her control and away from the influence of Lennox and his retainers. How to deal with his sexual urges when he was at close quarters was something that time and adequate supplies of whisky — for which he had obviously acquired a taste — might handle. She had, after all, played this game after the Rizzio plot. One step at a time would do for now.

As Mary left Kirk o'Field, she passed the entrance of her own bedroom. It was a tense moment for Sir James Balfour's men, two of whom were hidden inside. In the quadrangle she saw Nicholas Hubert, a valet nicknamed "French Paris" (previously Bothwell's servant), who by the light of the torches she noticed was unusually dirty. While everyone had been at the party upstairs, he had used his duplicate keys to give Balfour's men the run of the lower floor. He had also helped them fill the sacks of gunpowder that were piled in a heap where Mary's bed had once stood. She looked at him and exclaimed, "Jesu, Paris, how begrimed you are!" He blushed.

Mary reached Holyrood around half past eleven. She did not stay long at the masque: the dancing had almost finished and it was time for the ladies to "put the bride to bed," a courtly ritual of laughter and fun in which Mary played the leading part. She went to bed herself half an hour or so after midnight. Darnley continued drinking after she had left, then ordered his horses to be ready for early the next morning to take him back to Holyrood. He too retired around midnight, attended by William Taylor, his loyal bedchamber servant, who slept on a pallet in the same room. Two more of his chamber servants, Thomas Nelson and Edward Simmons, lay nearby. They slept in the gallery adjoining Darnley's bedchamber, which overlooked the town wall and Thieves Row. They, in turn, were attended by Taylor's page, Andrew McCaig. Another half-dozen servants and perhaps three or four grooms were sleeping downstairs.

A little after two in the morning, a bright flash lit the sky, followed by a huge explosion. The noise, a tremendous "crack" resembling a volley of twenty-five or thirty cannons, startled everyone for miles around. Windows were flung open and candles hastily lit. Dogs barked and raced around in a frenzy. People rushed to their front doors, and a small crowd congregated in the streets nearest to the blast. The night was black; it was only when dawn broke that the scale of the devastation could be seen. Everything within a radius of a hundred yards or more of Kirk o'Field was covered in a thick layer of dust. Large splinters and chunks of timber peppered the ground. The Old Provost's Lodging had been razed to its foundation. All that remained was a pile of rubble.

In a garden some forty feet away, on the other side of the town wall and on the far side of Thieves Row, Darnley and William Taylor were found dead under a tree in their nightshirts. The extraordinary thing was, there was not a mark on their bodies. Close by were a chair, a rope and Darnley's furred cloak. A dagger was also found in the garden, adding to the mystery, since neither man had been stabbed.

Of Darnley's remaining servants, Nelson and Simmons survived the blast. Nelson was found clinging to the town wall. Simmons also escaped unscathed, unlike poor Andrew McCaig. He was found dead, but whether his body was extricated from the rubble or found in the garden next to those of Darnley and Taylor is disputed. Of the downstairs servants, at least one was killed. The others appear to have survived the blast.

How had Darnley and his servant not only left the house but got across Thieves Row into a garden on the other side of the town wall? And why did men fleeing for their lives carry furniture? Had the victims made an improvised escape, or were they killed in the house and their bodies dragged outside? If they had indeed escaped, where were they killed, and by whom? And why was the explosion needed?

The explosion woke Mary, who sent Bothwell and the captain of the guard to investigate. She was stunned by their report. As she scribbled to her ambassador in Paris later that same day, "The matter is horrible and so strange as we believe the like was never heard of in any country." The house had been "blown in the air . . . with such a vehemency that of the whole lodging, walls and other, there is nothing remaining, no, not a stone above another, but all either carried far away or dung in dross to the very groundstone. It must be done by force of powder, and appears to have been a mine. By whom it has been done, or in what manner, it appears not as yet."

Mary thought the mine was detonated while Darnley "lay sleeping in his bed." Her assumption, like everyone else's at first, was that Darnley and Taylor were catapulted over the wall by the force of the explosion and landed in the garden. Only later did it emerge that this was impossible. If they had been hurled into the air, they would have been burned or scorched by the explosion and their bodies crushed or bruised from the impact of their fall.

Mary's mind was racing. She had herself at one time regarded Darnley as the "King of Scots," and was well aware that the murder would be interpreted in Europe as regicide. A more scandalous crime could hardly be imagined. While she could not wholly have mourned his passing, she was devastated that this had happened. The timing could not have been worse. Less than forty-eight hours earlier, she had ordered Robert Melville back to London to conclude the longed-for settlement with Elizabeth. She had even swallowed her pride and written a conciliatory note to Cecil. In sentences replete with irony, she wished him to accept her good opinion of him in spite of their differences. He would, she hoped, become a "well willer of all our good causes" — as she had every right to expect after pardoning the Rizzio plotters. This was the letter that Melville was not at first allowed to deliver, because when he arrived at Cecil's house, the news of the explosion had preceded him and he was forced to grovel for an audience.

As with the Rizzio plot, Mary believed this one had been aimed against her. She was convinced that whoever was guilty, the crime "was dressed as well for us as for the king; for we lay the most part of all the last week in that same lodging, and was there accompanied with the most part of the lords that are in this town that same night." Had it not been for Bastian's wedding, she would be dead too.

Mary was a good actress, but not this good. She took a traditional view of the role of Providence and believed God had personally intervened to save her from this appalling crime. Her thoughts were quite clear on the matter. She was an anointed queen and God was on her side. Casting her mind back to what seemed in hindsight to be her almost miraculous decision to leave Kirk o'Field shortly before eleven o'clock in order to attend the wedding masque, she said, "It was not chance, but God that put it in our head."

The true facts of the crime are more difficult to unravel than in the case of the Rizzio plot, but independent sources do exist. Darnley's end is described by the Duke of Savoy's ambassador, Signor di Moretta, and by the Cardinal of Lorraine's agent, Monsieur de Clernault. They had both attended Prince James's baptism and were about to return home. Thereafter, fresh evidence was collected by Drury, whose handwritten reports recorded the exact position of the house at Kirk o'Field and of the garden where the bodies were found, and who clearly relished the challenge of attempting to solve this astonishing murder mystery.

There are two versions of Moretta's description. According to the first, Darnley heard a group of men trying to enter the house with duplicate keys. Since it was a routine security measure to block a keyhole at night by leaving the key in the lock on the inside, some noise would be made until the genuine keys were dislodged. The doors could then be opened using the duplicates. Darnley peered out the window of his bedchamber, which overlooked the side of the house. He saw armed men, realized his predicament, ran downstairs and tried to escape. But the house was surrounded. He was captured and strangled. A slow fuse to the gunpowder was then lit.

The trouble with this theory is that it wrongly assumes that Darnley ended up in the garden of the Old Provost's Lodging inside the town wall, whereas his body was found beyond Thieves Row outside the wall. It also seems odd that anyone would try to break *into* a house that was about to explode. And why blow up the house at all if Darnley was already dead?

Moretta's first account is not very plausible, and is contradicted by the testimony of Bothwell's cousin, John Hepburn. When interrogated by

Moray, Hepburn confessed his role in the explosion, but denied touching Darnley or even seeing him. He said that he and his accomplices lit the slow fuse, then locked all the doors of the house before going a safe distance away to wait for the bang. Darnley would have heard the noise of their keys jangling — not to break into the house, but to lock the doors from the outside. If the duplicate keys were left in the locks, the exits would be blocked, forcing him to escape through a window rather than waste precious time trying to open the doors.

Moretta himself arrived independently at a similar conclusion. He replaced his first version of events with another. "The king," he said, "heard a great disturbance, at least so certain women who live in the neighborhood declare, and from a window they perceived many armed men round about the house." (We will hear more later about these women.) "So he, suspecting what might befall him, let himself down from another window looking on the garden, but he had not proceeded far before he was surrounded by certain persons." His captors quickly strangled him, using the sleeves of his nightshirt. His body was then dragged to the garden, where it was found.

This seems a lot more credible. It has the makings of a hypothesis, except that the second-floor window from which Darnley escaped could not have been the one overlooking the garden of the Old Provost's Lodging, but must have been the one in the adjoining servants' gallery, which overlooked the town wall and Thieves Row.

In that case, Darnley heard the noise, grabbed his dagger, threw his cloak over his shoulders against the cold night air and then used the rope to climb down onto the chair below, or Taylor tied the chair to the rope and lowered Darnley, who was sitting in the chair. Either way, the drop was at most sixteen feet, which with a rope would be feasible.

It was this same method of ropes and chairs that Bothwell and Huntly had recommended to Mary as her way of escape from Holyrood after the Rizzio plot — it was a standard expedient in case of fire. Once Darnley was in Thieves Row, he was intercepted and killed by a second, as yet unidentified group of assassins.

In escaping through the window of the adjoining gallery, Darnley and Taylor must have awakened Nelson and Simmons, but since Nelson ended up stranded on the town wall, it is possible that he and Simmons were climbing through the window at the moment of the explosion, or perhaps they had looked down and witnessed the deaths of Darnley and

Taylor, and so preferred to cling to the wall rather than risk the same fate. We do not know.

Clernault adds a few details. He said that Darnley was found "mort et étendu," meaning his corpse was not crumpled or in a heap, but laid out. He did not think Darnley was killed where his body was discovered. He also said that McCaig's body was found not in the rubble but with those of Darnley and Taylor in the garden, some sixty or eighty paces from the house, suggesting that he too had made his escape using the rope and chair.

The Frenchman believed Darnley was suffocated, which is the best explanation of why no marks were visible on the body. But he thought asphyxiation was the result of smoke inhalation, which is unconvincing. Only if the gunpowder in the cellar had failed to ignite properly might this have happened, since even a long fuse was unlikely to create enough smoke to penetrate the upstairs floors.

Drury then takes up the story. Although based at Berwick, some sixty miles south of Edinburgh, his spies were quickly on the scene and he managed to send his first report to Cecil on the day after the murder. Darnley's body, he said, "was found in the field and strangled as it should seem. His lodging after the death was blown up with powder." Drury knew within a matter of hours that Darnley had been strangled and not catapulted into the air by the force of the blast. He claimed that Lennox's body was found beside his son's. This was a mistake, since Lennox was actually in Glasgow. Moretta had drawn the same conclusion at first, saying that as well as Darnley and Taylor, "the father of the king" was killed. It was ten days before it was confirmed that Lennox was still alive.

Drury then started his detective work in earnest. By the end of February, he knew that John Hepburn was one of the murderers. Hepburn had watched the house after Mary's departure for Bastian's wedding masque, waiting until the candles were snuffed out and everything was quiet before deciding it was time to light the slow fuse.

Drury also discovered that Balfour had supplied the gunpowder. His men had concealed the barrels in Edinburgh a week before the explosion. By this time, Bothwell, Balfour and David Chalmers had been denounced as Darnley's murderers on placards affixed to the Tolbooth. The rumor mill was in overdrive. Everyone was talking about the gunpowder plot and had their own theories. Chalmers, already linked to

Balfour and Bothwell in the tight-knit circle around the Court of Session and the Admiralty Court, was unpopular, but there is no proof he was involved in killing Darnley. The cover-up had started: the finger was pointing only at Bothwell and his known associates.

Drury was skeptical. His particular interest was in Morton's activities and the extent of his pact with Bothwell. He was determined to follow this up, and his tenacity and persistence would be rewarded. In late April or early May, he wrote triumphantly to Cecil: "Morton is noted to have assured friendship to Bothwell, which, to be the thankfuller now for his favor showed him in his absence and trouble, he intendeth to continue."

Despite the furor caused by the explosion, Morton was so grateful to Bothwell for his role in obtaining his pardon that he was standing by him. This gave substance to Drury's claim that Bothwell had all along been working in concert with the powerful Douglas clan.

Sir James Melville corroborated this assessment. He wrote in his memoirs that after the baptism, "the Earl Bothwell ruled all in court, and brought home some of the banished lords, and 'packed' up a quiet friendship with the Earl Morton." To "pack" in this sense, which is also used by Shakespeare, means to bring someone into a conspiracy, or to plot, scheme or intrigue using secret or underhand methods.

Drury was convinced that Morton and Bothwell stood shoulder to shoulder before, during and after the murder, only falling out later. He got increasingly excited, believing he had almost cracked the case. He told Cecil, "It was Captain Cullen's persuasion" — i.e., his idea — "for more surety to have the king strangled and not only to trust to the train of powder."

This largely adds up. If a decision to be doubly sure of the victim's demise was made at the scene, it could explain why the Old Provost's Lodging was blown up, even though Darnley had been strangled. A last-minute change of plan would have caused confusion, especially when Darnley complicated everything by escaping out a window. At the very least, this is a better explanation than the alternative: that the house was blown up to throw the investigators off the scent.

James Cullen, an explosives expert, was the man in question. He had served as a mercenary in France, Denmark and possibly Poland before returning to the garrison at Edinburgh Castle. He was Balfour's hench-man, denounced to the Privy Council as his accessory and interrogated. Although he was said to have confessed, mysteriously no charges were ever brought. He was released and allowed to seek refuge in the Orkney Islands. But he was plainly meant to stay there, since when he returned to the Lowlands after an absence of four years, he was summarily arrested by Morton and hanged.

"Sir Andrew Ker with others," continued Drury, were "on horseback near unto the place for aid to the cruel enterprise if need had been." This was Drury's pièce de résistance. He had found out that Ker of Fawdonside, the man who had leveled a pistol at Mary during the Rizzio plot, was waiting in the wings at Kirk o'Field with a force. Hitherto a man who hedged his bets between the Lennoxes and the Douglases, Ker had switched his allegiance to Morton when Darnley betrayed the Rizzio conspirators. Apart from Morton, he was the man who had most openly avowed his intention to take revenge on Darnley.

That Ker and his men were lurking in the alley has always been a well-known fact. What no one knew was precisely why he was there and for whom he was working. This has led to various conjectures, of which the least probable is that there may have been several plots to murder Darnley. It has even been claimed that the explosion at Kirk o'Field was organized by Darnley to kill Mary, and that Ker and his men had been lined up by the Lennoxes to whisk Darnley back to Glasgow as soon as the deed was done.*

Such speculation can finally be set aside. The handwritten originals of Drury's reports prove that Ker was working for Bothwell, for whom he was running errands. He was seen as a "great carrier of intelligences and letters" for him. And later, at the final showdown between Mary and her rebel lords, Ker and his men would line up on Bothwell's side.

Drury concluded, "The king was long of dying and to his strengths made debate for his life." In simple terms, he had tried to reason with his captors. This seems likely, but it is also possible that the information was planted by the Lennoxes. Their propaganda machine was already in gear, its aim being to turn Darnley into a martyr. To that end, he was said to have been at prayer before his death, reciting Psalm 55:

Hear my prayer, O Lord, and hide not thyself from my petition . . . My heart is disquieted within me, and the fear of death is fallen upon me. Fear and trembling are come upon me, and a horrible death has overwhelmed me . . . It is not an open enemy that has done me this dishonor, for then I could have borne it.

* R. H. Mahon, *The Tragedy of Kirk o'Field* (Cambridge, 1930).

This was disinformation. The idea that Darnley foresaw his fate is incredible, entirely of a piece with a remark attributed to Mary by the Lennoxes and allegedly spoken as she left Kirk o'Field for the last time. "It was about this time last year," she was supposed to have said, "that David Rizzio was slain." (It was in fact eleven months.)

One of Drury's finest contributions was to send a colored drawing of the assassination to Cecil. It is magnificently detailed, almost a visual narrative of events. It marks the exact positions of the church of Kirk o'Field, the quadrangle, the Old Provost's Lodging, the postern gate, the town wall, Thieves Row and the garden where the corpses were laid out. It also shows Ker of Fawdonside and his men. In what seems to be a cul-de-sac on the far side of the garden where the bodies were found, armed horsemen can be seen. They are not in Thieves Row itself. A closer look shows that the cul-de-sac lacks an entrance as well as an exit. The cul-de-sac is a fiction: there was no passage at the place where these mounted men appear. This is a clue of a type that is well understood by art historians, who are used to dealing with narrative history paintings. The clue is there to indicate the secret presence of Ker's men in the vicinity, not to indicate their precise location, which Drury did not know.

We can now begin to refine our hypothesis. Darnley went to bed, but shortly afterward heard noises outside and the rattling of keys. He looked out a side window, where he caught sight of a group of men. Realizing his life could be in danger, he decided to escape. He grabbed his cloak and a dagger, then he and Taylor used the rope and chair to descend from the window of the adjacent gallery. This would enable them to reach Thieves Row on the opposite side of the town wall. But when they turned to flee, they were cut off by Ker's men, who emerged from the shadows and strangled them. Their bodies and the things they had been carrying were then left in the nearest garden.

Further evidence from Moretta fleshes out the hypothesis. He said that when Darnley had escaped through the window only to be captured in the darkness below, women whose cottages were within earshot heard him cry out, "Oh, my kinsmen [*Eh! fratelli miei*], have mercy on me, for the love of Him who had mercy on all the world!" And Drury's drawing shows how this might indeed have been possible. Built into the wall of the garden on the opposite side of Thieves Row from Darnley's lodging are several cottages.

It is easy to explain Darnley's last words. The Douglases, hiding with

Ker of Fawdonside, were Darnley's kinsmen, because his mother, the Countess of Lennox, was born Lady Margaret Douglas. Moretta's evidence makes it almost certain that Darnley was killed by the Douglases, whom at first he assumed to be his allies and willing to protect him, unaware that they had changed sides. He attempted to reason with them, which fits with Drury's information. But their exile and forfeiture after the Rizzio plot had turned them into mortal enemies. They were in the wings at Kirk o'Field on Morton's instructions to make sure that Darnley did not escape alive.

Moretta twice referred to "certain women" in the neighboring cottages. Up until now, no one has managed to identify them. Amazingly, their original sworn depositions survived, tucked away in the archives. Barbara Martin testified that before the blast, she was looking out her window and saw thirteen men go past. After the "crack," eleven came back the other way, two of whom had "clear things" on them as they went by. She shouted at them, calling them traitors and saying they were up to no good. If only she had gone on to explain what these "clear things" were, we might be able to say for sure that these men were the murderers.

Meg Crokat was in bed with her two children when she heard the bang. She ran to the door stark naked, and looking out saw eleven men running past. She harbored suspicions about their identity, because one of them wore silk. They were clearly important people, not ordinary Edinburgh citizens. She called out to ask what the "crack" was, but they ignored her. Seven headed off in one direction, and the rest in another. Crokat was present when her neighbor Martin shouted to the eleven returning men that they were traitors.

There is a discrepancy between the locations of the cottages as shown in the drawing and as given in the women's depositions. Drury's drawing situates the cottages in Thieves Row, whereas the depositions say they were in the Friar Wynd, which is closer to the middle of the town. In one respect, it does not matter. Regardless of precisely where the women lived, everyone agreed that they were the nearest neighbors to Kirk o'Field, and they knew something. During the cover-up, it slipped out that they had been "in doubt whether it were better for them to tell or hold their peace. Although they daintily tempered their speech, yet when they had blabbed out something more than the judges looked for, they were dismissed as fools."

The women had identified someone close to the center of power. But the key fact about their evidence is not that it raises new questions, but that it was suppressed. Although carefully filed away by the clerk of the Privy Council, it was never used. It was even brought to England in 1568 and shown to Cecil, but was quietly buried. By then, the cover-up story was more important than the truth to everyone involved in this drama, except to Mary herself.

The women's depositions are dated "11 February 1567," the day after the blast, which is hardly slow progress. Mary had called for a full investigation, demanding speedy and draconian retribution for the murderers. Unfortunately for her reputation, the criminal justice system in Scotland relied on the oversight of the Privy Council. It was unlikely that the council would act efficiently or impartially when up to half of its members were either the instigators of the very crime they were supposed to be investigating or the men who had privately condoned it. In the case of the women from the cottages, it was these same lords who suppressed the crucial evidence.

Mary was terrified. She feared she had been the target and would be the next to be assassinated. Since security at Holyrood was fairly lax, she decided to move back into Edinburgh Castle, just as she had done after Rizzio's murder. This move, which involved fetching a hundred mules and carts to transport her clothes, bed linen and furniture, to say nothing of her papers and personal effects, would have taken the best part of a day.

Only two days after the explosion, Mary ordered a proclamation to be read at the Market Cross in Edinburgh, offering a reward of £2000 to anyone prepared to inform against the murderers. On top of this, a free pardon was promised to the first guilty person willing to confess and turn queen's evidence. This followed a lengthy debate in the Privy Council at which Mary herself insisted on the offer of a pardon in the hope of solving the murder mystery. But the offer, generous as it was, elicited no further information. If anything, it backfired by raising public expectations and stirring up gossip and feverish speculation about the explosion and its perpetrators.

And much of this conjecture turned toward Bothwell. In a short time, he had risen high in Mary's favor. He was her most loyal and energetic champion, but for all that he was arrogant and unpopular. It was soon whispered that he was one of Darnley's assassins, the talk sparked by the actions of his servants, who had been spotted rolling barrels about. Such rumors were inflamed by the fact that when he had arrived at Kirk o'Field in his capacity as sheriff of Edinburgh, he ordered everyone to return home and refused to allow onlookers to pick through the rubble in search of clues. He had also refused to allow anyone to make a detailed examination of Darnley's corpse.

Bothwell's vilification began on the night of February 16. This was done anonymously, in a series of placards and notices that were stuck to the doors and walls of churches and public buildings. To conjure an atmosphere of fear and expectation, strange voices were heard in the night, "crying penitently and lamentably" for vengeance. A ghostly figure prowled Edinburgh's streets calling out that Bothwell had murdered the king. The Lennox faction was behind this. It was the beginning of yet another blood feud, but the Lennoxes had tapped into a popular mood. To make their point as explicitly as possible, likenesses of Bothwell were posted on gateways and scattered through the backstreets and alleyways at night with the legend "Here is the murderer of the king." These were followed by more incendiary slogans, of which the worst was "Farewell gentle Henry, but a vengeance on Mary."

Mary urged her ambassador in Paris to secure the goodwill of Catherine de Medici at all costs. This was easier said than done. The news of the murder reached Paris as early as February 19, and for a fortnight or so Mary was said to be innocent. On the 21st, a Venetian source described the explosion as a Protestant plot to kill Mary as well as Darnley. Its purpose was to trigger a long royal minority, so that the rebel lords could rule in the name of Prince James, whom they would bring up as a Calvinist.

But opinion quickly swung around. The explosion was such an outrage, it seemed hard to credit that Mary, in her position of power, did not have foreknowledge of it. The public denunciations of Bothwell made matters worse, especially in the eyes of diplomats who had seen him strutting at the baptism. Moretta, who had now reached Paris by way of London, was one of those who began to succumb to the force of the rumors, not yet condemning Mary, but failing to exonerate her.

Du Croc, who had left Scotland three weeks before the murder and was now ordered back, was also suspicious. Hindsight had given him insights he had never had when he was in Scotland, and he started to put a more menacing gloss on what he so vividly remembered of Mary's rows with Darnley.

Soon Mary herself was in the court of public opinion. The explosion, she was warned by her ambassador in Paris, had shocked and astonished all of Europe. The story was spreading that she was "the motive principal of the whole of all, and all done by your command." Her role was keenly debated in France, and "for the most part interpreted sinisterly." Catherine de Medici was no friend to Mary, and even the Guise family were disowning her.

It was essential, the ambassador urged, to show now "the great virtue, magnanimity, and constancy that God has granted you . . . [and] that you do such justice as to the whole world may declare your innocence . . . without fear of God or man." Otherwise, it would have been better if Mary herself had been murdered.

These were harsh words to speak to a queen, but worse was to come. Catherine and her son, Charles IX, delivered an ultimatum. If Mary failed to avenge the murder, they would consider her as utterly disgraced. All the old enmity between Catherine and Mary had returned.

If this were not enough, Mary was betrayed by her own family, who failed to give her the benefit of the doubt. Already she had gently scolded the Duke of Nemours, who had attended her wedding to the dauphin and recently married her widowed aunt Anne d'Este, that her relatives no longer wrote to her as often as they should. She must have felt stabbed in the back when her uncle the Cardinal of Lorraine wrote a conciliatory letter to Moray. He offered to set aside their differences, suggesting they unite to restore order and decency in Scotland. The cardinal sent his agent, Clernault, back to Edinburgh to deliver the letter personally. According to Drury, Clernault brought a secret message to Moray. In all of this, Mary was ignored by her uncle, just as she had been bypassed so many times before.

Even the Catholics deserted Mary in her hour of need. Encouraged by Darnley's zeal at the time of his investiture with the Order of Saint Michael, the pope had named the Bishop of Mondovi as the papal nuncio in Scotland. On her Privy Council's advice, Mary had refused to receive him. The bishop traveled no farther than Paris, where he was in limbo.

On hearing of the murder, he sprang into action, convinced that Mary's failure to restore the Mass and to dismiss her Protestant advisers were the causes of her downfall. He had predicted it all, he said. If only Mary had joined the Catholic League agreed between France and Spain at Bayonne, "she would have found herself now completely mistress of her realm, with authority enabling her to restore entirely the holy Catholic faith. But she never had the will to listen to it." Everything that had happened, the bishop steadfastly believed, was Mary's own fault for throwing in her lot with the heretics.

Elizabeth delivered the coup de grâce. Mary's world collapsed when, a month after Darnley's death, she read a letter brought by Henry Killigrew, Cecil's brother-in-law. He was received by Mary in her mourning clothes in a dark room in Edinburgh Castle. "I could not see her face," he later wrote to Cecil, "but by her words she seemed very doleful and did accept my sovereign's letters [*sic*] and message in very thankful manner as I trust will appear by her answer which I hope to receive within these two days."

Either Killigrew was indulging in diplomatic newspeak or the room was so dark he could not gauge Mary's true reaction. She was fighting back tears. Elizabeth's letter, for once written by herself and not on her behalf by Cecil, did not mince words:

> My ears have been so astounded, my mind so disturbed and my heart so appalled at hearing the horrible report of the abominable murder of your late husband and my slaughtered cousin, that I can scarcely as yet summon the spirit to write about it. And as much as my nature forces me to grieve for his death, so near to me in blood as he was, so it is that I must tell you boldly what I think about it, as I cannot hide the fact that I grieve more for you than for him. Oh madam! I should neither perform the office of a faithful cousin nor an affectionate friend if I studied more to please your ears than to preserve your honor. Therefore I will not conceal from you that people for the most part are saying that you will look through your fingers at this deed instead of avenging it, and that you don't care to take action against those who have done you this pleasure . . . I exhort you, I counsel you, and I beg you to take this thing so far to heart that you will not fear to touch even him whom you have nearest to you if he was involved . . .

A more stinging or vehement rebuke of one reigning queen by another could scarcely have been imagined. Mary reeled from this latest blow, offended not least by the remark that she would look through her fingers, coincidentally the very phrase that Maitland had applied to Moray during the conversations at Craigmillar, when she had answered: "I will that ye do nothing whereto any spot may be laid to my honor or conscience." Was the use of this phrase deliberate?

Worse still, Elizabeth advised Mary to arraign Bothwell ("him whom you have nearest to you") if the Lennoxes were to accuse him of the murder. And then, most crushingly of all, she dictated her revised terms for the political settlement that Mary so eagerly desired. There would be no more talk of reconciliation, of a "new treaty of perpetual amity" to replace the offending clauses of the treaty of Edinburgh, or of Mary's claim to the English succession. All that was now forgotten. Instead, Elizabeth insisted Mary ratify the original version of the treaty. This matter, said Elizabeth caustically, "has gone undone for six or seven years." It was time to end it once and for all.

If Killigrew expected to have an answer within two days, he was sadly disappointed. Mary found Elizabeth's letter so insulting, she refused to reply at all. Killigrew left Edinburgh a week later, empty-handed.

What must have especially rankled was Elizabeth's references to Bothwell. How had she learned that he "whom you have nearest to you" had already been accused of the crime? It was easy enough. Cecil had obtained full transcripts of the placards posted on the walls of Edinburgh. He had even gotten hold of the "sayings" of the prowler who called nightly for vengeance on the murderers. He was corresponding with Lennox, who was furiously rebuilding his bridges.

When Killigrew returned to London, he may not have brought a letter from Mary to Elizabeth, but he brought one from Lennox to Cecil, offering to collaborate with him in avenging the murder of his son.

Suddenly, history was to be rewritten. On the last occasion that Darnley had spoken to an English diplomat, he had repudiated his allegiance to the English queen. Now, as Lennox reassured Cecil, his son had all along been her most loyal subject and his own particular "acquaintance" and good friend.

Morton also looked to England. This most villainous of the Scottish lords, who had written to Cecil in obsequious terms as he had crossed the border on his way to rendezvous with his allies at Whittingham Castle, now sought Cecil's protection against the reprisals he knew would be sought by Lennox. He called himself "your assured friend" and wrote to reiterate his offer to do "anything in my power to gratify you."

At the beginning of 1567, Mary had been at the height of her powers and about to reach a final political accord with Elizabeth on mutually agreed terms. Two months later, she seemed more vulnerable than ever before. This had not been the aim of the lords, but was a byproduct of their shortsighted lust for revenge. What Mary urgently needed to do was to track down the murderers and bring them to justice.

Cecil, meanwhile, was unrelenting, preparing the case against her and seeking to move in for the kill. "I fear," wrote Mary's ambassador in Paris, "this to be only the beginning and first act of the tragedy, and all to run from evil to worse, which I pray God of his infinite goodness to avoid."

It was an accurate prognosis, and yet not a single piece of uncontaminated evidence has ever been found to show that Mary had foreknowledge of Darnley's murder.* Everything depends on the assumption that she was already engaged in an adulterous affair with Bothwell. What wrecked her reputation was that, instead of throwing Bothwell to the wolves, she decided to defy the world and throw in her lot with him.

---

* Two letters Mary was later said to have written from Glasgow will be the subject of chapter 25.

20

A Love Match?

FOR MARY, the weeks and months after the explosion were the most critical of her life. Her integrity was on the line. What happened then tarnished her reputation forever, and rightly so. She made no serious effort to bring Darnley's murderers to justice. The worst that could be proved before the murder is that she wanted the depraved and dangerously conspiratorial Darnley safely under house arrest at Craigmillar Castle and that she had gone to Glasgow, the stronghold of the Lennoxes, to fetch him. It is not what happened before the murder that precipitated her downfall, but the astonishing events that followed it.

Mary really was alone after the explosion. Even the Guises found her to be such a liability that her uncle made terms with Moray behind her back, provoking an angry response. Mary's usually prolific correspondence with her family abruptly stopped. She felt she had no one to trust. Looking back on the months since her illness at Jedburgh, she must have guessed that most if not all her lords had known of the plot to kill her husband.

When Killigrew had arrived to present Mary with Elizabeth's letter of rebuke, he was entertained to dinner by Moray and the lords and assured that every effort would be made to arrest the guilty parties. Mary herself promised Killigrew that the assassins would be unmasked. But how was she to give substance to her promise when the very same lords who dined with Killigrew were the leading conspirators? They stuck together and had already suppressed the testimony of the women in the cottages beside Thieves Row.

Moray was the most farsighted of the bunch: he would choose to go voluntarily into exile until the dust settled, which only served to reinforce Mary's suspicions of his guilt. He realized that Darnley's murder was not the end but just the beginning of a catastrophic downward spiral in which violence and retribution would reach a frenzy.

Bothwell, with his typical bluster and lack of subtlety, saw it very differently. He had a "mark of his own that he shot at." He planned to step into Darnley's shoes and Mary's bed. For this he would reap the vengeance of the lords for seeking to overawe them. He would be the scapegoat for Darnley's murder, but as yet this dramatic turn of events was still several months away. Until then, the pact between Morton and Bothwell (as Drury had noticed) held fast, which was long enough for an isolated and confused Mary to make her own mistakes.

Mary's psychology is crucial. She had been brought up in the luxury and safety of Henry II's court and never felt completely secure after she had left the shores of France. The factionalism of the lords was relentless and on a scale beyond anything she could have imagined. Violence was endemic in Scotland. Politics were tribal, based on organized revenge and the blood feud. An anointed queen she might be, but the monarchy lacked the financial resources and centralized institutions of France.

As for her recent ordeals, she had been publicly insulted by Knox, who had compared her to Nero, the worst of the Roman tyrants, but when she had called on him to explain himself, she had been forced to back down. She had discovered Chastelard under her bed, armed with a sword and dagger. Her uncle Francis had been assassinated by the Huguenots. Her secretary had been dragged from her in her apartments and brutally murdered in the next room. A loaded pistol had been pointed at her by Ker of Fawdonside, and now she had escaped death in a gunpowder plot by what seemed like a hairsbreadth.

Mary saw a common thread linking all these events. Her conversations with Castelnau eighteen months before had shown that she had an ideological understanding of politics. She had claimed then that the rebel lords sought to depose and kill her in order to create a "republic." She had long believed that her mother's deposition was the beginning of a trend, and the murderous events at Kirk o'Field proved just how right she had been.

When she picked herself up after the initial shock of the murder, she shaped her own destiny. In what she saw as desperate circumstances, she took a breathtaking political gamble. She wanted Bothwell to protect her by controlling the noble factions. A poacher was to be turned into a gamekeeper. She did this because, rightly or wrongly, she saw him as the monarchy's champion and the only man who could save her from a fate similar to Darnley's.

As she went over and over everything in her mind, her priorities switched from identifying the murderers to deciding what they were likely to do next — this time to her and perhaps her child as well. Her chief suspect in the explosion was always Moray, her illegitimate half-brother. It was in this mistaken belief that she would make a tragic, pragmatic decision to protect herself and her son, believing that whatever else she did, she would be surrounded by treachery and deceit.

She had another reason to think this way. She was approaching her twenty-fifth birthday, the age when Scottish rulers by tradition revoked the grants and rewards they had improvidently made in their youth. Mary would be expected to reassert the power of the monarchy against the nobles, and yet Darnley's murder had crippled her ability to do so. With the dynastic settlement with Elizabeth a dead letter, she decided to claw back her power by supporting Bothwell. She had nothing to lose and everything to gain.

Such impulsiveness was characteristic of her. When she had put on her steel cap and mounted her charger to defeat Moray in the Chase-about Raid, Castelnau had said she was adopting an "all or nothing" approach. This is how she reacted again, and she did so consciously, and not in a daze of lethargy or bewilderment. Mary was a gambler, and this was to be her biggest throw of the dice.

On Bothwell's side, ambition was never far away. Now thirty-one, he was the epitome of tough masculinity. He appeared on the surface to be the military gallant with an insolent swagger and a bristling mustache. He could be suave when it suited him, even if he upset people with his freebooting ways and obscenities. Beneath the surface he was no better than any of the other lords. Mary had not yet seen him as he really was. Her wicked sense of humor may have misled her into finding his past misdemeanors to be more like schoolboy pranks than criminal acts of lawlessness. He may have seemed diverting to her as well as unswervingly loyal. But his single biggest attraction was that, apart from a brief interlude of reconciliation for which Mary herself was responsible, he had always been Moray's mortal foe.

When Bothwell had married Huntly's sister, Lady Jean Gordon, he arguably became the most powerful lord after the Earl of Argyll. He had already made a pact with Morton and the Douglases. Of the other leading power blocs, Châtelherault, the leader of the Hamiltons, was in voluntary exile after the Rizzio plot. Atholl was vulnerable to Argyll, who was keen to score points off his nearest neighbor and rival in the western and central Highlands, and the Lennoxes were in turmoil after Darnley's death.

Bothwell's alliance with Huntly brought him a vast military retinue to add to his own border forces. On top of this, his right as Lord Admiral to a share of the profits of all vessels wrecked off the coast of Scotland meant that he was one of the few Scottish nobles who was financially independent. As sheriff of Edinburgh he had the legal profession and many of the judiciary on his side. He must have calculated that he could protect Mary for as long as she asked him to, provided his pact with Morton and the Douglases could be kept alive.

The day after the murder, Mary emerged from her apartments looking pale and drawn to join the wedding feast of Margaret Carwood, her favorite bedchamber woman. Carwood had married John Stuart, one of Mary's distant relatives, at Holyrood that morning. It was Shrove Tuesday, the last day it was possible to marry by Catholic rites until Easter. The bride's dress was given by Mary, who also paid for the banquet. No doubt she was fulfilling a long-standing promise to her loyal gentlewoman by attending her wedding, but it came at some cost to her reputation.

Mary's attendance made it clear that she had dispensed with the strict rules of court protocol. She should have put the court into mourning for forty days immediately after the murder, but waited for five days before ordering her *deuil* attire and large quantities of black taffeta to cover the walls and windows of her apartments. To some degree, the lapse may be attributed to her decision to move into Edinburgh Castle. But she drew attention to it by attending Carwood's wedding.

Darnley was not given a state funeral. He was buried without pomp or ceremony during the night of Friday, February 14, 1567, his corpse laid to rest in the tomb of the kings in the old Abbey-Kirk at Holyrood. It must have been an eerie occasion; it certainly attracted unfavorable comment. An Edinburgh chronicler recorded that the former king of Scotland was interred "quietly in the night without any kind of solemnity or mourning heard among all the persons at court."

The following Sunday, Mary left Edinburgh for Seton, by the shore of the Firth of Forth some eleven miles east of Edinburgh. She stayed with Lord Seton, the half-brother of one of her Maries and a Catholic who had fought alongside Bothwell in the guerrilla war against the Lords of the Congregation. She was said to be following the advice of her doctors, who felt she had been in a dark and stuffy room for long enough and needed a change of scene. Yet there is no independent evidence that she traveled on medical advice, and far from having spent the previous week in somber mourning, her blackout had been in place for just a day.

Mary's disregard for convention can only suggest that even though she played no part in Darnley's death, she must at some level have been happy to see him gone. Even when her bales of taffeta had finally arrived and the blackout was in place, she could not escape the charge that it was more for show than substance.

When first departing for Seton, Mary left her son in Bothwell and Huntly's care. She traveled with her other lords, returning to Edinburgh three days later, where she lived in seclusion for a week. She then set off again for Seton, this time with Bothwell as her escort. They stayed at Lord Wharton's house on February 26, dining at nearby Tranent. There an archery contest was held in which Mary partnered Bothwell against Huntly and Argyll. To Mary's delight, she and Bothwell won, and the losers paid for dinner.

Long before the end of the official forty days of mourning, Mary and Bothwell were seen outdoors together. She had decided to trust him, which was perhaps prudent as a security measure but politically very unwise. He took command of the royal bodyguards, who from then on were constantly within Mary's sight. Bothwell himself was guarded when he walked through the open streets. If he spoke to anyone he did not know, he kept his hand firmly on his dagger.

There was an acute sense of danger and foreboding in the air. Before leaving for Seton, Bothwell swore an oath that if he ever found the authors of the placards accusing him of Darnley's murder, he would wash his hands in their blood. Previously known for its "joyousity" and light-hearted atmosphere, Mary's court was acquiring menacing and militaristic overtones as he recruited more and more soldiers to guard her palaces.

On March 7, Morton was brought secretly by Bothwell to see Mary late at night. He humbly apologized to her for his part in Rizzio's murder and made his peace with her; she relaxed his curfew and allowed him back to court.

Now the cards were stacking up. Bothwell and Morton were in the top positions. Their closest allies were Huntly and Argyll. Maitland was for the moment siding with Bothwell, nudging Moray aside. Argyll was trying to restore a semblance of normality to government. Atholl had receded into the background, deliberately squeezed out by Argyll.

Moray was shunned by his sister. His disgrace sprang from his refusal either to declare in favor of Bothwell and Morton or to denounce them openly to Mary. This time his attempts to hedge his bets had undone him. Mary suspected him to be the chief instigator of the explosion, for once doing him an injustice. While Bothwell and Morton preened themselves, Moray prepared to travel to exile in France. When he finally left the country, Bothwell danced for joy.

Step by step, Bothwell was seizing control of the available military power. On his advice, Mary dismissed the Earl of Mar as governor of Edinburgh Castle, replacing all the officers and gunners with Bothwell's nominees. She appointed James Cockburn, the Laird of Skirling, Bothwell's servant, as the new captain of the castle. She also made him comptroller of her household, thereby fusing military and civil power in the Edinburgh region. As comptroller, Cockburn ousted Sir William Murray, the Laird of Tullibardine, a noted Lennox supporter whose brother was suspected of being the author of the placards denouncing Bothwell.

And yet Bothwell did not always trust his own men. A month later, he had Cockburn removed from the castle, bringing in Sir James Balfour instead. Almost at once, large supplies of food and munitions were requisitioned as if to prepare against a possible siege, and Bothwell levied fresh companies of soldiers for the royal bodyguard.

Mary was not yet wholly in thrall to Bothwell. She still trusted Mar, with whom she had stayed at Alloa the previous year and to whom she had given her son when she suspected a plot by Darnley and the Lennoxes to kidnap him. Now she put Mar in charge of Stirling Castle. And on March 20, Prince James was again left in his safekeeping. It was almost as if Mary knew that she should do what her own mother had done with her as a baby, when the lords posed a threat to her life.

She then wrote to Lennox, inviting him to file his charges against those he suspected of Darnley's murder. She promised that they would be sentenced to the fullest extent of the law if a jury found them guilty.

Lennox put Bothwell at the top of his list, but Mary was unimpressed. She detested Lennox, well aware that his son's ambition for the crown matrimonial had been the cue for Rizzio's death. She was relying on Bothwell to protect her, and as a signal of her confidence gave him some old church vestments of cloth of gold to recycle to make new suits. Later he got some of her mother's Spanish furs for a nightgown, and she gave him Darnley's horses and finest clothes.

This last gift was naive in the extreme. When Bothwell had the clothes altered to fit him, the tailor remarked, "It was but right and according to the custom of the country for the clothes of the deceased to be given to the executioner." If only Mary had realized the degree to which public opinion was turning against her.

Soon Bothwell presumed too much. He started thinking in the crudest terms: that to guarantee his position in the giddy game of noble factionalism, he must physically possess the queen. Slowly but surely he began to pay court to Mary. Although married, he was well aware that marriages could — with the right influence — be broken almost as easily as they were made. The Catholic Church in Scotland was comparatively tolerant of infidelity. Annulments were difficult but far from impossible to obtain. The Protestant Kirk was able to grant divorces, and Bothwell was a Protestant.

His father, Patrick, a Catholic, when vying to marry Mary's mother a quarter of a century before, had readily obtained an annulment. We know this fact had been on Bothwell's mind, because when the lords had discussed a possible divorce for Mary in their plotting at Craigmillar Castle shortly before Prince James's baptism, he had reminded them of his father's experience.

On Palm Sunday, Mary collapsed at a Requiem Mass for Darnley. She found the occasion too distressing and stayed in bed for several days. On Good Friday, she went with two of her Maries to her private chapel, where she prayed and meditated for four hours. Those who saw her said that she was stricken with "melancholy." But when Easter arrived and the dancing and banqueting resumed, she recovered her spirits and her looks. She moved out of Edinburgh Castle back into her old apartments at Holyrood, and soon seemed more her usual self.

The Privy Council met on Good Friday. Bothwell took his seat as usual, even though the day's business was to approve the final arrangements for his trial. The rumor mill was churning. The stallholders in the busy street markets of Edinburgh gossiped that their queen would marry Bothwell. One day when Mary rode out of the castle past the Lawnmarket toward the High Street, a small group of women minding their stalls cried out, "God save Your Grace if you be innocent of the king's death."

Public opinion was the part of her gamble that Mary had not taken sufficiently into account. A few days before, she had sent for the parish minister of Dunfermline to question him about the most scurrilous and sensational placard to appear so far. It was a pornographic picture of a mermaid and a hare. The mermaid, naked apart from her golden crown and identified by the monogram *MR*, sported a large sea anemone in her right hand and a rolled-up net in her left. These were surrogates for the orb and scepter, but also had deeper, more suggestive meanings. The sea anemone, a giant polyp with petal-like tentacles around the mouth, stood for the female genitalia. And the rolled-up net was to enable her to catch unwary sailors as they passed by, distracted by the sight of the anemone.

Whoever devised this placard must have had an impressive classical education. The symbol of the anemone was used in this way by the Roman poet Ovid, an author familiar to Mary. The other idea of the drawing was the Roman *retiarius,* or net man: the net fighter in the Roman arena who took on the sword-wielding gladiator and aimed to ensnare him in his net before killing him with the trident he bore in his right hand.

Below the mermaid was a hare within a circle of seventeen swords. The hare was the heraldic symbol of Bothwell's family, and the letters *JH* inside the circle identified Bothwell by his name, James Hepburn. The swords signified Bothwell's military standing and love of dueling, but were also positioned in the drawing as phallic symbols.

Mary was mortified by these references. Acting on a tip, she summoned the minister and demanded whether he knew the artist. He said no. She asked, "Who then was likeliest to do it?" He said, "There was none could write so well unless a canon who is a papist and lives in adultery and hath sired in the same manner three children." At this, Bothwell, who was listening nearby, roared with laughter, disrupting Mary's questioning. He loved nothing more than a ribald spat, which he now proceeded to enjoy with the minister, reveling in the smutty innuendo.

The Privy Council fixed the date of Bothwell's trial for April 12, just short of two weeks away. It was to take place at the Tolbooth, with Argyll presiding and Huntly assisting. No wonder Bothwell was so relaxed about it. Edinburgh was already filling up with his supporters, and he was brashly confident. Morton had supplied 300 cavalry to reinforce the palace bodyguard, for which Mary restored him to his stronghold of Tantallon Castle. More ominously, Bothwell met a German mercenary captain whom Mary's ambassador in Paris had recommended. He offered to send 3600 crack troops to Mary and Bothwell in exchange for regular payments.

On April 4, Mary went on a third visit to Seton to enjoy the fresh spring air. While she was there, an old man, one of Darnley's former domestic staff, approached her while she was walking in the garden with Bothwell on her arm. He humbly presented himself, then asked her "to give him some release" in his poverty. Mary was typically generous to old or sick servants, and would doubtless have asked her ladies to give the man a few coins from her purse and a proper meal in the kitchen before sending him on his way.

But Bothwell rudely interrupted her. Railing against the man's effrontery, he turned to him and said, "Thou custrel [i.e., knave], go thy ways! I shall so release you that you shall be sorry with yourself, churl!" He then attacked him viciously until blood poured from his mouth. The man limped home and died two hours later. Before expiring, he said, "I have served in France, England and Scotland, but the like was never said unto me."

It was a shocking incident. Out of the blue, Mary had seen for herself the rough side of Bothwell in all its cruelty and brutality. Up to now, only his smooth side had been visible to her. The other part of his character was either carefully suppressed or cloaked by a veil of gentility. The truth is that his dashing looks and French education were deceptive. He was a swordsman with a taste for violence, by birth and training a border lord, an adventurer, a pirate and a buccaneer. His civility was superficial, his fiery temper encouraged by his sudden rise to greatness. He strove for personal preeminence over his rivals among the lords, which he sought to win in any way he could. His loyalty to Mary and her mother had all along been directed to this end.

Mary's judgment had become clouded. She had not even begun to think about marrying Bothwell, and why should she rush into a third marriage when her marriage to Darnley had been such a disaster? And yet her trust in Bothwell was becoming something more than a purely pragmatic decision. He had started to court her between Good Friday and the end of Easter week, which fits with the fact that a fortnight earlier, he had first sounded out his wife, Jean Gordon, about her reaction to an arranged divorce.

Mary would have abandoned Bothwell to his fate if she had wanted to divest herself of him. His trial was less than a week away. Despite seeing his true character unveiled, she chose deliberately not to do so.

Lennox, meanwhile, was lobbying Cecil to intervene in Bothwell's trial. He was quite certain it would be rigged. After some prevarication, Elizabeth wrote to Mary to request an adjournment. Only four days were left before it was due to begin. The letter was sent posthaste to Berwick-upon-Tweed, where Drury gave it to a courier to take immediately to Holyrood. His man arrived at the gates of the palace at six in the morning on the day of the trial.

At first he was told it was too early to enter, because Mary was asleep. He was advised to wait, so he went into Edinburgh to find some breakfast. When he returned, Mary was still not awake, and so he paced up and down until ten o'clock, when Bothwell's men began mustering in the courtyard. Seeing his opportunity, he approached the entrance of the palace, but was denied access. He asked for permission to deliver his letter, but everyone pretended to be deaf.

At that moment, who should appear but Thomas Hepburn, another of Bothwell's relatives. He brought a message from Bothwell, advising the courier to withdraw, "for the queen was so molested and disquieted with the business of that day that he saw no likelihood of any convenient time to serve his turn until after the assize."

Cockburn, Laird of Skirling, then emerged. He demanded to know if Drury's man brought a letter from Elizabeth or Cecil. On hearing it was from Elizabeth, Cockburn said, "Then ye shall be soon discharged." He ordered Hepburn to escort him off the premises.

At that instant, Maitland and Bothwell came out, and everyone mounted their horses. Maitland spotted Drury's man and demanded Elizabeth's letter. When it was handed over, he and Bothwell went back inside to see Mary, disappearing for half an hour.

When they reappeared, Drury's man asked if Mary had read the letter and what reply he should take back. Maitland said that she was still sleeping. This was untrue, because just then she appeared at an open window, flanked by Mary Fleming and one of du Croc's servants. For once caught in a blatant lie, the "Scottish Cecil" said, "No, I have not delivered the letter, and there will be no convenient opportunity to do so until after the assize."

Bothwell clattered out of the courtyard on Darnley's favorite courser, followed by four thousand retainers. Before leaving, he looked up at Mary's window. She saw him, laughed and gave him a friendly toss of the head as a farewell. A "merry and lusty cheer" was raised at his departure. His company rode in a stately procession to the Tolbooth, preceded by a force of two hundred musketeers to clear the streets. When everyone had gone inside, the musketeers kept guard at the door, so that no one might enter but Bothwell's supporters.

The court sat for over eight hours. The indictment was read, charging Bothwell with the explosion and Darnley's murder, but no further evidence was submitted. The court's debates were on procedural issues. Lennox was too afraid to appear. He had traveled as far as Linlithgow accompanied by three thousand retainers, but did not dare to continue when told he could bring only six of them into Edinburgh. His case was conducted in his absence by two professional advocates, who requested an adjournment for forty days. This motion was discussed at great length. Finally, it was denied and the court moved to a verdict. Bothwell was acquitted by the jurors, who then quickly sought to insure themselves against claims by Lennox for willful error. When this was agreed, Argyll and Huntly brought the proceedings to a close.

Bothwell was overjoyed. Immediately he posted a notice on the door of the Tolbooth, declaring himself cleared of the murder and challenging anyone who claimed otherwise to a duel. He then rode in triumph back to Holyrood. He supposed he was untouchable. He was to find out that he was not: his very acquittal would become a major source of grievance. The danger no longer came from Lennox, who took the hint and fled to England. It came from the other conspirators, who were not in Bothwell's fortunate position. Morton, the most fiendish of the lords, had refused to attend the assize. He should have been one of the jurors, but said that although Darnley "had forgotten his part in respect of nature toward him, yet for that he was his kinsman he would rather pay the forfeit which was £100."

Another villain swiftly recalculating the odds was Sir James Balfour. After Bothwell's acquittal, he felt especially vulnerable. He was, after all, the man who had obtained the gunpowder and whose brother had offered Darnley the use of the Old Provost's Lodging. He was now "minded with full determination to have had an assize for him[self] in like manner." When Mary rejected his request, he decided to change sides and support those lords who were already beginning to mutter against Bothwell.

Balfour for the moment kept his own counsel. He had no wish for an outright confrontation with Bothwell. But he posted guards outside his house night and day. He also murdered one of his servants. The man had been at Kirk o'Field; he was about to break ranks and claim the reward and free pardon that Mary had offered to anyone who would inform on his accomplices. For this he was killed and his body buried secretly at night.

On April 16, Bothwell rode up the Canongate with Mary and the lords on his way to Parliament. He carried the scepter, leaving Argyll to bear the crown and the Earl of Crawford the sword of state. Everyone noticed that he refused to allow the bailiffs of Edinburgh their traditional stations as Mary's guard. His influence was pervasive. On her short journey to the Tolbooth, she was escorted by a force of his own musketeers.

And Bothwell oversaw every aspect of the Parliament. He was determined to entrench his position. As a Protestant himself, he appealed first and foremost to his coreligionists, sponsoring an Act Concerning Religion that brought the Protestant Kirk formally under Mary's protection after seven years of uncertainty. The act did more than maintain the religious status quo: it declared Protestantism to be the queen's official religion, even if privately she worshiped as a Catholic. At this aspect of Bothwell's ascendancy, not even Knox and his adherents in the General Assembly of the Kirk could complain.

Next Bothwell helped his co-conspirators to their bounty. Morton, Argyll and Huntly had all of their ancestral lands confirmed to them, and Argyll even received some of Darnley's former property. Bothwell also helped himself. His grant of Dunbar Castle, received as his reward for assisting Mary to escape after the Rizzio plot, was ratified with all the privileges belonging to it. He then secured a confirmation and enlargement of his hereditary rights as Lord Admiral.

Lastly, a retaliatory Act Against the Makers and Setters Up of Placards and Bills was passed at his insistence, clearly devised to block further attempts by the supporters of the Lennoxes to blacken his name.

By the 19th, Parliament's work was done. That night, Bothwell invited his fellow lords to a supper at Ainslie's Tavern in Edinburgh, where he produced the draft of a bond he wanted them to sign. They were asked to confirm his innocence of Darnley's murder, to declare their willingness to defend him from calumny, and finally to promise that if Mary should just "happen" to choose "James Hepburn, Earl of Bothwell" as her future husband, they would support him.

The Ainslie's Tavern Bond was couched as a petition to Mary:

> . . . weighing and considering the time present, and how our sovereign the Queen's Majesty is now destitute of a husband, in the which solitary state the commonwealth of this realm may not permit Her Highness to continue and endure, but at some time Her Highness in appearance may be inclined to yield unto a marriage . . . [it] may move Her Majesty so far to humble herself, as preferring one of her native-born subjects unto all foreign princes, to take to husband the said Earl . . .

Bothwell wanted everyone to sign, but had missed an important trick. He had helped his co-conspirators to their land grants less to thank them for their role in murdering Darnley than to bribe them to support his future marriage to Mary. Instead of waiting until the grants had been ratified by Parliament, Bothwell should have got the lords to sign his bond first. Not everyone was willing to sign afterward. Cecil, for once, got everything wrong. His "copy" of the bond said that all the lords had signed. Morton and Huntly did sign, but Argyll, Maitland and Atholl refused — in fact, Maitland and Atholl did not turn up at the rendezvous. As to Moray, he had already left the country and was on his way to France.

In any case, it is likely that nothing was signed on the night of the 19th. Next day, Bothwell sent his men to each of the lords individually to demand their signatures.*

* Several of the signatories said later that they had agreed to the bond only after Bothwell told them Mary wanted them to sign. No such claims were made at the time, but the charge may well be true, and Mary herself leveled it against Bothwell. See chapter 22.

When the Ainslie's Tavern Bond was signed, Kirkcaldy of Grange warned his English allies that Mary was infatuated with Bothwell. She had said "that she cares not to lose France, England and her own country for him, and shall go with him to the world's end in a white petticoat ere she leave him."

The day of the Ainslie's Tavern Bond saw an ugly scene at Holyrood. While Mary was watching, the soldiers in the great hall began to mutiny for lack of pay. Bothwell intervened, confronting their captain and seizing him by the throat. When the other soldiers came to the man's rescue, Bothwell was forced to let go. He swore profanely, but promised to pay the men shortly. Mary always hated trouble, and had never experienced anything of this sort. She stepped forward, calling for her embroidered purse, which contained 400 French crowns. She proceeded to walk down the line from man to man, giving them 2 crowns each.

Whereas Mary sought to soothe conflict, Bothwell was overweening and puffed up with pride, behaving as if he were king already. He was never in love with Mary. His efforts to woo her were minimal. He dominated their relationship to the point of brutality, yet she accepted and even seemed to welcome her subordinate role. Mary could be strong and masterful, but it now looked as if she wanted to surrender all her worldly cares to Bothwell, who took the opportunity to usurp her power and authority at every turn.

Why this should have happened remains a mystery. Mary's correspondence dried up in these crucial weeks. Later she gave her reasons. She said that her country, "being divided in factions as it is, cannot be contained in order unless our authority be assisted and set forth by the fortification of a man." She needed Bothwell to deal with the sheer "insolence" of lords, who would otherwise constantly be in rebellion against her.

But in April 1567, she said nothing at all. The only letter she wrote was to the papal nuncio to Scotland, the Bishop of Mondovi, still loitering in Paris. "I beg you," she said, "to speak well of me to His Holiness, and not to let anyone persuade him to the contrary concerning the devotion I have to die in the Catholic faith and for the good of his Church."

Mary's tone was faltering and evasive, reflecting her guilt and unease over her relationship with Bothwell. This was not least because the Act Concerning Religion was read in Catholic circles as a signal of her secret conversion to Protestantism. It was as though Mary were playing a game of chess, but not thinking more than one move ahead.

And then the pact between Bothwell and Morton suddenly disintegrated. Morton resented Bothwell's good fortune, and Bothwell suspected Morton of plotting against his intended marriage. It was a classic falling-out among thieves. Although Morton was willing to sign the Ainslie's Tavern Bond, he did so only on terms that severely restricted Bothwell's powers and prevented him from ever being styled king. Argyll was also deeply offended by Bothwell's presumption, telling his friends that he would soon leave Holyrood for good.

Bothwell had to act precipitately, before the rest of his supporters melted away. With Morton no longer to be relied on, and with Argyll plainly about to withdraw, the gamble was all now on his side. He had to make a move if his bid for power was to succeed, and that meant staging a coup.

On Monday, April 21, Mary rode to Stirling to fetch her son. To her dismay, the Earl of Mar refused to deliver him. A moderate politician with his finger on the pulse of the lords, he knew they would rebel if Bothwell got his hands on the heir to the throne. When Mary entered the castle, Mar would allow her to be accompanied by only two female attendants. To deny the monarch and her entourage access to her own fortress was considered treasonable by Mary, who threatened to punish Mar severely.

Two days later, she kissed her ten-month-old son goodbye. It was the last time she ever saw him. She rode to her birthplace at Linlithgow, where she spent the night. On the 24th, she rose early. Her thoughts were back in France, as it was the ninth anniversary of her wedding to the dauphin. Her plan was to return to Holyrood, but as she crossed the bridge over the River Almond, a few miles outside Edinburgh, Bothwell intercepted her. He took her forcibly to Dunbar, where she was "ravished."

Bothwell was no better than Darnley. He believed that to assure his position, he must own Mary sexually, and if she would not yet marry him, he must conquer her.

His conduct was outrageous. To seize the person of an anointed queen was considered to be sacrilegious as well as treasonable, even in a country so lax in its interpretation of law and order as Scotland. A scandalized Sir James Melville claimed such an act could only have been collusive. One of Bothwell's men was said to have admitted that it was done "with the queen's own consent." Drury, with his resolutely English point of view, took the same line: "The manner of the Earl Bothwell's meeting now last with the queen which though it appeared to be forcible, yet it is known to be otherwise." Kirkcaldy of Grange was most explicit: "She was minded to cause Bothwell to ravish her, to the end that she may the sooner end [his] marriage, which she promised before she caused [the] murder [of] her husband."

But all three hated Bothwell and feared his malign influence on Mary. Kirkcaldy was already an official spokesman for the lords: his comment shows the direction their propaganda was taking. The price she would have to pay if she went ahead and married Bothwell was guilt by association in Darnley's murder.

Some details of the "abduction" can be checked. When Bothwell unexpectedly jumped out and grabbed Mary's horse by the bridle, she was startled and appalled. She instantly ordered her servants to ride to Edinburgh and summon a rescue party. She was definitely abducted against her will. The evidence later produced by the lords to "prove" otherwise would be doctored, which even Cecil could hardly have failed to notice when it was put in front of him. (The annotation of Cecil's secretary on this "evidence" is discussed in chapter 26.)

But was she also raped? The difference is between a woman who was becoming a fool for love and one who was already a political pawn.

Mary was a woman of spirit: high-minded and fully conscious of her "grandeur" as a queen. It is entirely out of character that she would have married Bothwell if he had raped her. It is sometimes claimed that he was the first man who satisfied her sexually. That is quite possible, given the dauphin's ill-health and puny physique and Darnley's sheer selfishness. And yet even if it is true, it is a world apart from saying that Mary could ever have forgiven Bothwell for forcing her into bed against her will.

Despite the heavy pressure exerted on Mary by Bothwell's use of the Ainslie's Tavern Bond, she had not yet decided to marry him, because if she had already given her consent, there would have been no need for him to abduct her. It follows that the sequel at Dunbar was more critical than the kidnapping at Almond Bridge.

Mary stayed with Bothwell in his castle for twelve days and not just a single night. As he was not there the whole time, it cannot seriously be maintained that she was prevented from leaving if she had really wanted to escape. No one else at Dunbar was going to hold their queen a hostage if she had given them an outright command.

Mary and Bothwell were lodged in the state apartments, but they occupied separate rooms. As a chronicler sardonically remarked, there was "no great distance between the queen's chamber and Bothwell's." But while the sexual innuendo is unambiguous, the rooms *were* indeed apart, so that Mary could have locked or barricaded the door to her own room if she had wanted Bothwell kept at bay. She could have shouted for help to her servants, but did not.*

The most likely sequence of events is that Mary was genuinely ambushed by Bothwell on the road at Almond Bridge. She was taken to Dunbar against her will. When they arrived, she was frightened and angry, but he protested his love to her and pleaded with her to marry him. If he had not already done so, he would also have shown her the Ainslie's Tavern Bond, which appeared to be a petition from the lords indicating the unanimity of their support. We cannot know what she answered, but within one or two days, he had won her over. (We will see further evidence of Mary's state of mind at Dunbar in chapters 22 and 26.)

It probably took two days. On April 26, Bothwell galloped at high speed to Edinburgh. There he arranged for his wife, Lady Jean Gordon, to lodge her suit for a divorce in the Protestant court. She filed her petition that very same day. Since her case was set out in graphic detail, it suggests that Bothwell had prepared the documents in advance. He was said to have enjoyed himself in the precincts of Haddington Abbey in broad daylight with the "bonny little black-haired" Bessie Crawford, his wife's maid. He cheerfully confessed to his adultery.

On May 3, the judges issued a decree of divorce. Its effect was immediate, since Huntly had grudgingly been bought off and had already agreed to the terms of his sister's financial settlement with Bothwell.

In parallel, Mary asked the Archbishop of St. Andrews to grant Bothwell an annulment in the Catholic court. Her request was made on April 27, and the decree followed on May 7, issued on the grounds that his marriage to Lady Jean had been invalid from the start for lack of a canonical dispensation.

If Huntly had so far condoned his sister's divorce, he had his limits. When Bothwell returned to Dunbar, he discovered that Huntly and Mary had quarreled. Despite signing the Ainslie's Tavern Bond, Huntly wavered in his support for Bothwell. By the end of May, his reluctance to

---

* Her remaining servants, as well as Maitland, Huntly and Sir James Melville, were also taken to Dunbar.

take up his erstwhile brother-in-law's cause in arms against the rebel lords would be unconcealed.

The lords had first gathered at Stirling three days after Mary's abduction. Morton, Argyll, Atholl and Mar were the instigators, and after four days a new bond was signed. They called themselves the Confederate Lords, denying that they were in revolt but admitting they were in an association to free their queen from "captivity." They pledged to strive by all possible means to set Mary "at liberty," to "preserve the prince and the commonwealth" and to kill Bothwell, whom they now called "that barbarous tyrant" and "cruel murderer."

Bothwell had simply become too powerful. His increasing monopoly of military power was especially feared. He had, claimed the Confederate Lords, "the strengths, munitions and men of war at his commandment." While this was something of an exaggeration, it is true that he had seriously threatened the independent positions of these lords, and so stepped into Darnley's shoes in a way quite different from the one he had imagined.

On May 6, Mary and Bothwell processed in triumph to Edinburgh. On their arrival at the gate called the West Port, the castle guns fired a salute and Bothwell dismounted. He then led Mary's horse by the bridle as they slowly advanced up the hill to the castle. But the crowds were sullen. Mary and Bothwell were visibly displeased, and yet it is hard to see how they could have expected anything else. Mary's popularity did not extend to marrying Bothwell, and his position as her protector was secured at a heavy price.

The same day, Knox's assistant at St. Giles Kirk, John Craig, was asked to proclaim the banns of marriage between Mary and Bothwell. This he bravely refused to do. He demanded a royal writ, which arrived next day, in which Mary volunteered the information that she had been neither raped nor held as a prisoner by Bothwell.

She ordered Craig to make the proclamation. If, however, she had not been raped, as she maintained, she must have willingly consented to sleep with Bothwell at Dunbar, which meant she had committed adultery with a married man. She could not have it both ways.

Craig read the banns the following day, but only after calling heaven and earth to witness that he abhorred and detested the proposed marriage.

On May 9, Bothwell privately summoned Craig, demanding an explanation of his remarks. But far from apologizing, the fearless minister admonished Bothwell. "I laid to his charge," he said shortly afterward, "the law of adultery, the ordinance of the Kirk, the law of ravishing, the suspicion of collusion between him and his wife, the sudden divorcement, and proclaiming within the space of four days, and last, the suspicion of the king's death, which her marriage would confirm."

On Sunday the 11th, Craig repeated his rebuke from the pulpit. Bothwell fell into a frenzy. He summoned the minister again, this time before the Privy Council, accusing him of exceeding his authority. But Craig answered, "The bounds of my commission, which was the word of God, good laws and natural reason, was able to prove whatsoever I spake."

Here was a second Knox! But Bothwell, unlike Mary, refused to bandy words. He silenced the minister and ordered him to leave, threatening to hang him summarily with a cord.

Next day, Mary pardoned Bothwell for abducting her at Almond Bridge, then raised him high enough in the peerage for her to marry him. Heralds in their coat armor led the procession into the Abbey-Kirk at Holyrood. They were followed by the Earl of Rothes, who carried the sword of state, the Earl of Crawford carrying the scepter, and Huntly, who bore the crown. Mary, resplendent in her royal robes and seated on a gilded throne beneath her cloth of state, placed the ducal coronet on Bothwell's head with her own hands. He was clad in a scarlet robe lined and edged with ermine, and was attended by Cockburn, Laird of Skirling, carrying a blue banner with Bothwell's arms emblazoned on it. He was created Duke of Orkney and Lord of Shetland by a seemingly rapturous Mary. After the ceremony, four of his retainers were knighted.

But Bothwell's meteoric rise was likely to be brief. The Confederate Lords were occupying Stirling, where they established a rival court in the name of Prince James. They staged a masque in which Bothwell was tried for Darnley's murder, convicted and hanged. It might have gone unreported, but the poor boy playing Bothwell was hanged so realistically, he was almost suffocated and had to be frantically revived. Bothwell was beside himself with rage when he heard of the masque. He swore a foul oath and threatened to be revenged on the rebel lords.

On Wednesday the 14th, the marriage contract was signed. It justified the wedding on the grounds that Mary was a young widow, "apt and able to procreate and bring forth more children" to maintain the dynasty, who had been petitioned and advised by the "most part of her nobility" to marry. It quoted from the Ainslie's Tavern Bond, justifying Mary's decision to "so far humble herself" as to marry one of her subjects. It noted that Bothwell had been recommended to her by the nobles, and claimed that Mary had "graciously accorded" to their petition.

No longer did Mary seek advice on her marriage from France, Spain or England. Her focus had narrowed to her own realm, and she seemed happy to be known as Bothwell's wife and to accept that the Guise family looked down on her for it.

The next morning, Mary and Bothwell were married in the great hall at Holyrood by Protestant rites. Du Croc, the French ambassador, organized an official boycott, which even some of those attending Bothwell's creation as Duke of Orkney joined in. The ceremony was thinly attended. There were few witnesses beyond the four Maries and Mary's and Bothwell's own servants.

Nevertheless this was no tawdry occasion. The element of spectacle was provided by Mary herself. She wore the *deuil*, as for her marriage to Darnley, yet these were no ordinary widow's weeds. Her dressmakers and embroiderers must have been working furiously night and day for a week.

She had married Darnley in white, but this time she wore a magnificent flowing gown of black patterned velvet in the Italian style, richly embroidered with gold strapwork and gold and silver thread. Her dress was so eye-catching, it hardly seemed to be a mourning dress at all.

She was in the prime of life. She was taller than Bothwell, and they must have looked a slightly odd couple, she with her exceptional height and thin waist, he with his stocky build, mustache and ruddy complexion. It was said that he had never looked more handsome, and yet his language at supper on the eve of the wedding was so profane that Sir James Melville walked away in disgust.

The service was conducted by Adam Bothwell, Bishop of Orkney, one of Bothwell's relatives. First he preached a sermon on a text from Genesis: "Therefore shall a man leave his father and his mother, and shall cleave unto his wife: and they shall be one flesh." Next he declared the bridegroom's repentance for his former sins and wicked life, and affirmed his resolve to make amends and conform to the discipline of the Kirk. Finally, the bride and groom were "handfasted." This may well have been done without an exchange of rings, as the Protestants objected to the use of the wedding ring, which they considered to be a "popish" superstition.

When Mary returned to her apartments, she changed into another new gown, this time of shimmering yellow silk. She did not get the chance to show it off. Unlike at her previous marriages, there was to be no "balling, dancing and banqueting," no cries of "Largesse" from heralds as they showered money on the guests, because there was no wedding banquet or masque, the most obvious sign that the ceremony had been hastily arranged.

Only three months and five days had passed since Darnley was murdered. Only fifteen months had passed since Bothwell and Lady Jean Gordon had been married with the "advice and express counsel" of Mary, who had signed their marriage contract, paid for the reception and presented the bride with her wedding dress.

While Bothwell and his servants were moving into Darnley's old apartments at Holyrood, the Confederate Lords were mustering their forces at Stirling. If what had happened in the past few months was not sensational enough, the most breathtaking and dramatic events of Mary's life were about to unfold.

21

Dénouement in Scotland

ARY SHAPED her own destiny by marrying Bothwell. Watching the downward spiral from afar, Cecil wrote that Scotland was "in a quagmire; nobody seemeth to stand still; the most honest desire to go away; the worst tremble with the shaking of their conscience." But the pattern was clear. Morton had allied with Darnley in a Faustian pact to murder Rizzio. When Darnley betrayed him, Morton allied with Bothwell to take his revenge. Then Bothwell became too powerful. When he threatened the interests of the other lords, Morton broke with him, leading a revolt that was all the more deadly in that Bothwell had enough inside information to condemn all of his accomplices in Darnley's murder, while he himself had been acquitted of the crime. There could be only one survivor of this, Bothwell's final feud with Morton.

On Mary and Bothwell's wedding night, a new placard was nailed to the gates of Holyrood. Quoting Ovid, it declaimed:

> As the common people say,
> Only harlots marry in May.

Mary was stung by the insult, although a moment's thought would have reassured her that, like the drawing of the mermaid and the hare, it could only have been the work of someone versed in Latin poetry, making it less hurtful to her than if it had come from the ordinary people.

But Mary could not think straight. Whatever Bothwell had told her during their twelve days at Dunbar, she was distraught to realize soon after their marriage that he did not really love her. His protestations had been insincere. Her biographers argue that her quarrels with Bothwell had already started before their wedding. Mary must therefore have known what she was taking on when she married him.

This is incorrect. Records show only a single row before the wedding: what Drury called a "great unkindness" lasting half a day when Bothwell returned to Dunbar after getting his wife to file for a divorce. Drury's reports are a key source for Mary and Bothwell's "jars," but the handwritten folios of these important documents were muddled when they were bound into volumes in the nineteenth century. Parts of the same report were filed as different documents under different dates. A paragraph that begins on one folio breaks off in midsentence and may continue on a folio several hundred pages later. When extracts from the bound volumes were edited for publication, crucial passages were misdated. A close examination of all the handwritten reports shows that apart from their one big row at Dunbar, Mary and Bothwell's quarrels all took place *after* their wedding.

Bouts of jealousy and mutual resentment were the cause. Mary wept because Bothwell "would not allow her to look at or be looked on by anybody, for he knew very well that she loved her pleasure and passed her time like any other devoted to the world." Bothwell was "the most jealous man that lives." On May 20, Drury wrote: "There hath been already some jars between the queen and the duke and more looked for. He is jealous and suspicious and thinks to be obeyed." The strain was so great that Mary's beauty was affected. "The opinion of many," continued Drury, "is that the queen is the most changed woman of face that in so little time without extremity of sickness they have seen."

History was repeating itself. Bothwell's attitude toward Mary as a woman was fundamentally incompatible with her view of being a queen. As with her rows with Darnley, one of the bitterest was over a coveted animal. When she gave a particularly desirable horse to the younger brother of Bothwell's old enemy the Earl of Arran, Bothwell ranted and raged, demanding the horse for himself. Mary, in a flood of tears, countered that Bothwell's ex-wife still lived at Crichton Castle, the closest of his several homes to Edinburgh, where he was said to write to her and visit her as if they had never been divorced.

Mary was in turmoil. It was another of the occasions in her life when she said repeatedly, "I wish I were dead." The difference is, this time she was not physically ill. As Drury had carefully noted, she was in great distress, but "without extremity of sickness."

Two independent witnesses heard her threaten to kill herself. One was du Croc, whom Mary summoned late on the evening of her wedding day. "I perceived," he informed Catherine de Medici, "a strange formality between her and her husband, which she begged me to excuse, saying that if I saw her sad, it was because she did not wish to be happy, as she said she never could be, wishing only for death." Two days later, "being all alone in a closet with the Earl of Bothwell, she called aloud for someone to give her a knife with which to kill herself. Those who were in the room adjoining the closet heard her."

One of those in the next room was Sir James Melville. He overheard Mary make her threat. "Or else," she said, "I shall drown myself." According to Sir James, Bothwell brutalized Mary after her marriage. He "mishandled" her in every way. "He was so beastly and suspicious" that he never allowed her a single day's peace, causing her "to shed abundance of salt tears."

There is plenty of evidence of Bothwell's violent temper, but it is also possible that he told Mary the truth — or more likely an expurgated version of the truth — about Darnley's murder. In that event, she would have felt betrayed by the very man whom she had trusted to protect her from the noble factions, whose role as queen's protector she had sealed in marriage. Even if he denied or concealed his own part in the plot, as he certainly would have done, she could never have forgiven him for failing to confess what he knew before the murder, which had shipwrecked all her hopes of asserting her claim to the English succession.

But Mary behaved differently in public. After the wedding, she still took Bothwell's arm. She dressed well and rode out with him. On May 23, she organized a belated celebration of their marriage. A water pageant was staged beside the shores of the Firth of Forth, followed by martial sports in which Bothwell "ran at the ring" and a group of soldiers enacted a mock skirmish. The court at Holyrood seemed as lively and convivial as ever, but a closer inspection showed that only a few nobles were present, and the majority of people who chatted or lounged in the courtyard or great hall were soldiers or Bothwell's servants.

When others were around, Bothwell took care to treat Mary with courtesy and respect. He "openly useth great reverence to the queen," said Drury, removing his hat in her presence, "which she seems she would have otherwise, and will sometimes take his cap and put it on."

And yet in private his temper was unbearable. Mary still clung to him, partly still in love, partly not knowing what else to do, but she was a changed and unhappy woman.

It was a heartbreaking transformation, and it affected Mary's own manners. She became coarser and harsher in her speech. When first told about the Confederate Lords, she even sounded like Bothwell. "For Argyll," she said, "I know well enough how to stop his mouth, and for Atholl, he is but feeble. I will deal well enough with him. And for Morton, his boots are but now pulled off him and not made clean. I will return him again." By this she meant that she would send him back into exile for breaking his promise, barely two months old, to be her loyal servant. Only one of these lords was given the benefit of the doubt. "And for the Earl of Mar," she said, "he hath assured me to be mine and faithfully ever." Despite her severe irritation with him since her last visit to Stirling, his offenses, she believed, were out of character, and perhaps she could even believe them to be in her son's best interest now that she knew Bothwell better.

Of the other lords, Maitland and Huntly were still at Holyrood, but only just. Bothwell's arrogance was too much for either man to stomach. Maitland, ever vacillating, was already making overtures to the Confederate Lords at Stirling. Huntly too asked Mary for permission to leave the court, but she refused. In a fit of pique she exclaimed, "Your desire is but to do as your father before you!" Nothing could have been more offensive to the leader of the Gordon clan after the destruction of his family at the battle of Corrichie and his loyalty to Mary since his recall to help in defeating Moray.

The greatest fear, even among Mary's diehard supporters, was that Bothwell was plotting to be named governor and protector of Scotland, and if possible king. According to Drury, he meant to proclaim himself king as soon as he got his hands on the heir to the throne. It was said that this would be his next big political gamble.

Although Bothwell still had no official title, he was already behaving as if he were king. On May 19, he issued a decree against counterfeit brass money, because an influx of fake foreign coins was driving sound money out of circulation and so preventing him paying his troops. This was a sensible move, but others were more partisan. On the 22nd, he reintroduced an old regulation for the compulsory attendance of privy councilors at Privy Council meetings which sought to force lords boycotting Mary's court to return there.

Next day, he made a blatant bid to win new friends. An act of the Privy Council annulled all religious dispensations obtained from Mary that permitted her nobles or servants to worship in private as Catholics. Here was the Protestant Bothwell attempting to ally with the Kirk. The Catholic clergy had petitioned Mary to restore the Mass, but Bothwell blocked their suit. He sent for Craig, whom he had recently threatened to hang, only to reassure him that he would do everything to assist the Protestants against the Catholics and even attend sermons himself.

What mattered most, however, was military power. The lords were gathering their forces; a civil war was in the offing. On May 20, Mary was seen with Bothwell weeping bitterly. The pressure was increasing; Drury heard from his spies that she was suffering from renewed bouts of vomiting. Two days later, she decided to confront the challenge. She issued a proclamation mustering troops to restore law and order on the borders. But, said Drury, everyone knew they were to form the nucleus of a royal army.

Bothwell, meanwhile, sought to establish his foreign credentials. He wrote to Elizabeth, Cecil and Throckmorton, all on June 5. His letter to Cecil was couched in an offensively regal tone. "Seeing God has called me to this place," he began, "I heartily desire you to persevere in all good offices." It might have been Darnley speaking! Bothwell was less presumptuous in his letter to Elizabeth, but still overconfident. He was aware, he said, of the evil reports she had received of him, but protested they were wholly undeserved. He wanted to restore the amity with England. Men of greater birth might have been preferred to the high estate he now occupied, but none could be more eager for Elizabeth's friendship.

It was a clumsy exercise in diplomacy. Bothwell wrote too much as an equal to Elizabeth, something Mary herself rarely did. According to Drury, Bothwell's aim was to persuade the English to be no more than "lookers on" while he delivered a knockout blow to the lords.

It was to be a vain hope. After a furious exchange with Bothwell, Maitland left Holyrood with his wife, Mary Fleming, and joined the rebels. He avoided saying goodbye to Mary, who was in tears at the defection of the chief of the four Maries, her former playmate and cousin. Her marriage to Bothwell was forcing her to choose between her husband and one of her lifelong companions. And yet for some reason she lacked the will to put Bothwell in his place. She had married him for better or worse, and she would stick by her decision.

The Confederate Lords had by now increased to around thirty. No one doubted that it would be a struggle to the death. The atmosphere was becoming like a performance of *Macbeth*. Rumors and prophecies swept the country from the Highlands to the Lowlands. "There is a witch in the north," said Drury, "that affirms the queen shall have yet to come two husbands more." Bothwell would live but a year, and Mary would be burned as a witch. These prophecies were reported to Mary, who dismissed them out of hand, "and as yet it is said that she fears the same."

A serious problem for Mary was lack of money. Her income as dowager queen of France meant that she had never before had to stint on personal expenses, raising taxes only once in her reign to pay for the banquets and entertainments at the baptism of Prince James. But then, she had never before attempted to muster an entire army independently of the nobles, even during the Chase-about Raid. Bothwell wanted to recruit five hundred professional infantry and two hundred cavalry, which would cost 5000 crowns.

Mary did the unthinkable. She trimmed her household budget to raise money. She stripped her cupboards bare and sent large quantities of gold and silver plate to the mint. Even the font of solid gold that Elizabeth had presented for the baptism was to be melted down and turned into coins worth £3 each. Ironically, it was so big and heavy it refused to melt completely, despite several attempts to do so.

On June 6, Bothwell took Mary from Holyrood to Borthwick Castle to escape from Morton and the Confederate Lords, who were planning an attack. They left in such a hurry, Mary only had time to pack what for her were bare essentials: a silver basin to wash in, a silver kettle for heating water, a small cabinet with a lock and key for her papers and a large supply of pins to hold back her hair.

At first Bothwell meant to withdraw to the Hermitage, but decided to stay at Borthwick, where they would be within easy reach of Edinburgh. The castle was well fortified. It consisted of a single, bleak square tower, three stories high. The walls were thirteen feet thick at the base, narrowing to six feet at the top. The roof was paved, with crenelated battlements from which a lookout could see for up to two miles. Nestled in a hollow, the castle was surrounded on three sides by steeply rising ground and water. It was accessible only by a drawbridge and so impregnable to anyone without artillery, but difficult to provision in a siege.

On the evening of the 10th, when Bothwell was about to go to bed, the

Confederate Lords attempted a raid. Unbeknown to them, he escaped through a postern gate, then galloped away to Haddington to summon reinforcements. The lords called for him to come out, crying "Traitor, murderer, butcher!" When they realized he had foiled them, they camped beside the castle. It was the height of summer, and remained light until almost eleven o'clock. Mary went up to the roof and leaned over the battlements, joining in a shouting match with her enemies below in which insults were traded and she comfortably held her own.

As Drury described it, the lords assailed her with "divers undutiful and unseemly speeches used against their queen and sovereign, too evil and unseemly to be told, which poor princess, she did with her speech defend, wanting other means for her revenge." It is a sign of the almost sacral respect accorded to anointed queens that, having written these words, Drury was so embarrassed by them he attempted to obliterate them with a series of slash marks.

The Confederate Lords retired to nearby Dalkeith, then set out next day for Edinburgh with two thousand men. When they arrived, the gates were barred against them, but a small advance party climbed over the town wall, broke open the Cowgate Port and occupied the town. The provost surrendered and the lords took control, issuing a resounding call to arms:

> That the Earl of Bothwell having put violent hands on the queen's person and shut her up in the castle of Dunbar, having proceeded to a dishonest marriage with Her Majesty after obtaining a divorce from his former wife, having already murdered the late king, and now attempting by his gathering together of forces to murder the young prince also; therefore they command all the lieges to be ready on three hours' warning to pass forward with them to deliver the queen's person [from captivity] and take revenge on the Earl Bothwell.

On June 12, these lords convoked a "secret council" at the Tolbooth, where they declared Bothwell "to be the principal author and murderer of the King's Grace of good memory and ravishing of the Queen's Majesty." It was, in effect, a quasi-judicial verdict reversing Bothwell's acquittal of Darnley's murder and paving the way for his "impeachment" or "trial by combat" on the battlefield.

Winning control of Edinburgh was the key to the lords' campaign. When Huntly and his allies attempted to retake the town, they were forced to seek refuge in the castle, where they supposed they would be protected by Sir James Balfour, its new captain and Bothwell's nominee. Except that, in a move as brazen as it was fatal to Mary, Balfour changed sides. He decided to support the Confederate Lords in return for a promise of a pardon for his part in Darnley's murder. He had been seething with resentment ever since Bothwell's acquittal. His defection ensured that Huntly was kept in the castle — perhaps all too conveniently, as he had burned his bridges with Bothwell — until it was too late for him to assist Mary and Bothwell at the final showdown.

The Confederate Lords were riding high. Argyll, whose military power was greater than that of the other lords, was expected to rendezvous with them shortly. Meanwhile, Morton allowed his men to sack the Abbey-Kirk and the royal mint, where they took what remained of the gold and silver plate that was waiting to be coined, including the font, which was still substantially in one piece.

In these straitened circumstances, Mary recovered her resolve and her wits. She snapped out of her daze and once more became her old daring self. On the night of the 11th, she fled from Borthwick Castle in disguise. She had often enjoyed dressing up and pretending to be a man. This time it was for real. As Drury reported, she put on men's clothes and rode "booted and spurred . . . that same night from Borthwick to Dunbar, whereof no man knew save My Lord Duke [Bothwell] and some of his servants, who met Her Majesty a mile from Borthwick and conveyed her to Dunbar."

Almost as soon as they arrived at his castle, Bothwell rode off again, this time to Melrose, to rendezvous with his border retainers. They were the mainstay of his infantry, but did not all show up. Drury had bribed the Elliots of Liddesdale, the brigands who had ambushed and wounded Bothwell the previous year, to intercept and harry them before they joined Mary's forces. The move was a serious setback. More devastating still was Balfour's treachery, since it was on his advice that Mary, apparently in Bothwell's absence, would deem it safe to leave Dunbar and return to the capital.

On June 13, the Confederate Lords beat the drum in the streets of Edinburgh to levy troops in the name of the lords and Prince James, offering to pay the fabulous rate of 20 shillings sterling a month. When Argyll finally arrived with his Highlanders, an army of three thousand men was in place.

Next day, Mary retaliated. She issued a proclamation commanding her loyal subjects to muster early the next morning at Musselburgh, the eastern gateway to the capital. To encourage her troops, she declared that within twenty-four hours she would be in Edinburgh or Leith. She then rode at the head of some 600 men to Haddington, where Bothwell joined her with 2000 more. She had a force of only 260 men when she set out, but gathered her adherents on the way. She also had the advantage of artillery, bringing three or four brass cannons from the munitions store at Dunbar.

Mary was burning with defiance. No longer was she insulated by the trappings of monarchy. Her hair was in disarray and she wore borrowed clothes. According to Drury, they were "after the attire and fashion of the women of Edinburgh." She wore a "red petticoat with sleeves tied with points, a partlet, a velvet hat and a muffler." The captain of Inchkeith, a Frenchman in Bothwell's service who kept a diary of these events, said that she wore "a red skirt which scarcely reached halfway down her legs."

Mary and Bothwell rode on to Seton, leaving their troops to rest overnight in the fields beside Prestonpans. When the lords learned of their approach, they mobilized their forces and marched through the night toward Musselburgh. Morton and Atholl were in command. Their banner, carried between two spears at the head of their column, was a picture of Darnley lying dead under a tree with a young child kneeling beside him and crying, "Judge and revenge my cause O Lord."

By five the next morning, Mary and Bothwell were on the road. The date was Sunday, June 15, exactly a month after their wedding. Their banners bore the saltire and the lion rampant, emblems of Saint Andrew, the patron saint of Scotland, and of the crown. The rival armies came in sight of each other at Carberry Hill, an elevated ridge close to Inveresk, almost two miles southeast of Musselburgh. Mary positioned her forces a mile or so to the northwest of the village of Elphinstone, inside earthworks thrown up by Protector Somerset during the final campaign of the Rough Wooings. The lords had just crossed the old bridge at Musselburgh when they spotted Mary's army on the higher ground. They marched a few miles up the east side of the River Esk until they reached the vicinity of Cousland, a village nearly two miles southeast of Carberry Hill, where they too established themselves.

It was a very long, hot day. The armies were more or less evenly matched; neither side was willing to risk a charge. After three hours of stalemate, du Croc, who had followed the Confederate Lords' army at a distance, stepped forward and offered to mediate between the two sides. Speaking through an interpreter, he implored the lords to avoid bloodshed, arguing that it was one thing to attack Bothwell but quite another to engage an anointed queen in battle.

The lords heard him out, but replied that there were only two ways to avoid a battle. One was for Mary to leave Bothwell forever. The other was if he would come out into the open ground and fight in single combat.

Du Croc hesitated. Such terms amounted to an ultimatum. He attempted to conciliate the lords, who grew impatient. They protested that they would rather be buried alive in a pit than fail to avenge Darnley's death. A rhetorical game was being played. Morton was particularly artful, saying "that they had not taken arms against the queen, but against the murderers of the late king; and if Her Majesty would either give [Bothwell] up to be punished or remove him from her company, she should find in them a continuation of all dutiful obedience."

On Maitland's urging, the ambassador rode across the open ground and up the hill. He was led straight to Mary, who was sitting on a stone talking to Mary Seton, still wearing her borrowed red dress. He kissed her hands and told her what was proposed. As she listened, her eyes lit with fire. She knew how these lords played their game. "It looks very ill of them," she said, "to go against their own signed bond, after they themselves married me to him, having already acquitted him of the deed of which they would now accuse him."

Mary called on the lords to submit. If they surrendered, she would be merciful. "If they ask me for a pardon, I shall be pleased to give it to them and receive them with open arms."

At this moment Bothwell appeared. Du Croc greeted him, but refused to take his hand. Bothwell spoke loudly so everyone nearby could hear him, insisting on knowing what the lords had said. Du Croc answered just as loudly that they had assured him of their loyalty to Mary. He then lowered his voice and added that as to Bothwell himself, they were his mortal enemies.

Bothwell shouted back, "Is not the bond they gave to me well known to everyone?" Then he said, "I have never meant to offend any of them, but rather to please them all, and they only speak of me as they do out of envy of my high estate." He then spoke the words that must surely be his epitaph: "Fortune is free to those who may profit from it, and there is not a single one of them who would not gladly be in my place."

It seemed to be an impasse. Then Bothwell asked du Croc to return to the lords. He was to tell them that, although he was the queen's husband, he would accept their challenge. He would fight any of them in single combat as long as they were of sufficient rank. "My cause is so just," he said, "I am quite sure that God is on my side."

At this Mary interposed. Fighting back tears, she insisted the quarrel was hers. Du Croc, who was fast getting out of his depth, also rejected single combat. Although he loathed Bothwell and would happily have seen him dead, his instructions were to advance French influence in Scotland, not to risk the overthrow of the monarchy. "In that case," said Bothwell, "there is no need for further talk." Then, making a joke, he gave the example of the unhappy envoy who had striven in vain to arbitrate between Scipio and Hannibal before the battle of Zama. He could do nothing and so stood aside. Before long he was rewarded with "the biggest show he was ever likely to see."

Du Croc returned to the lords, informing them that Mary wished them to sue for a pardon. They too became angry. The Earl of Glencairn retorted, "We are not come here to ask pardon for any offense we have done, but rather to give pardon to those that have offended." His answer confirmed the degree to which a compromise was impossible; du Croc turned on his heels and rode back to Edinburgh.

The standoff continued through the hottest part of the day, giving an advantage to the lords. Whereas the royal army had limited space to maneuver at the top of the hill, the lords could adjust their position below to gain some shade or avoid the glare of the sun in their eyes. They had brought supplies of drinks, probably "small ale," or low-alcohol beer. They had also struck camp within easy reach of a stream.

Up on Carberry Hill, the temperature soared. When Bothwell's scouts went in search of water from a nearby well, they were captured. About midday, some casks of wine arrived from Seton for the royal forces. The soldiers fell back to drink. But the alcohol only increased their dehydration. When ordered to return to their ranks, many disobeyed and slipped away. As they did so, the mood changed. Morale collapsed as numbers fell. Those who stayed began to complain that the quarrel was too personal. If the lords wanted Bothwell to fight in single combat, there was no reason why he should not do so. Why should they risk their lives for him unnecessarily?

Around two o'clock, Kirkcaldy of Grange arrived under a white flag to speak to Mary. Bothwell ordered a soldier to shoot him, but Mary gave a cry and ordered that "he should not do her that shame." Although a friend of Knox and Buchanan, Kirkcaldy was far from an outright republican. In his heart, he was loyal to Mary. He begged her to leave Bothwell, assuring her that if she did so, the lords would disperse. They would honor and serve their queen, but first she had to divest herself of the man who had murdered her second husband.

On the likely assumption that Bothwell had by now given Mary his own expurgated version of the events at Kirk o'Field, she finally knew that it was always going to be his word as a "man of honor" against those of his former accomplices, and that word, she must have realized now, was worth little. When Bothwell stepped forward again, reiterating his offer to fight in single combat, Kirkcaldy nodded with approval. It was what the Confederate Lords had suggested. He would return promptly with their answer.

After the briefest discussion, the lords nominated Kirkcaldy as their champion. Bothwell, however, rejected him. He said that a mere laird was not his equal in honor or degree and "could not be his peer." He made the same answer to Sir William Murray, the Laird of Tullibardine, the ousted comptroller of Mary's household, whose brother was the suspected author of the placards denouncing Bothwell as Darnley's murderer.

Bothwell demanded that he fight with Morton, his partner in Darnley's murder and his ally for two months afterward until they fell out. Bothwell wanted Morton dead. He challenged him "to come forth and fight with him hand to hand between the two armies."

Faced with a fight to the death, Morton blinked. He was fifteen years older than Bothwell and a lot less fit. He had not borne a sword in the heat of battle for twenty years, whereas Bothwell was a skilled swordsman, a man still physically and mentally in his prime.

Lord Lindsay, a relative of Darnley, volunteered to be Morton's surrogate. He was some years younger than Morton and had fought with distinction at the battle of Corrichie. He advanced, and in full sight of both armies knelt on the ground, praying that God would preserve the innocent and punish the cruel and evil assassin who had shed his kinsman's blood.

Morton gave Lindsay his own double-handed sword, a weapon prized by his family as the one wielded by their ancestor the great Earl of Angus, nicknamed Archibald "Bell the Cat." It was a trenchant gesture, since this same ancestor of the Douglas clan had been ousted from his birthright by Bothwell's great-grandfather. This really was going to be a fight to the death.

Drury claimed that Bothwell drew back at the sight of Lindsay, and so proved himself a coward. This was shameless propaganda. Whatever his faults, Bothwell was a brave soldier. Moreover, he believed it was his best chance to vindicate himself from the charge of sole complicity in Darnley's murder, since by the laws of trial by combat, the winner was acquitted of the crime by the "voice," or judgment, of God.

Just when the combatants prepared to advance, Mary intervened. She had first agreed to, but in a split-second decision forbade, the combat with Lindsay. She stopped the fight because, whatever happened, she knew nobody would be satisfied. If Bothwell lost, she would be the lords' prisoner and would have lost her protector. If he won, she would still lose, because she had come to know Morton for what he really was. He would never accept the result. Either the rebel lords would order their forces to charge or else they would send other champions to repeat the challenge until Bothwell collapsed from exhaustion or died of his wounds.

What Mary now intended is disputed. There are two conflicting reports. According to Sir James Melville, who hated Bothwell and wanted him dead, she summoned Kirkcaldy. She offered to surrender. She would separate herself from Bothwell and put herself under the lords' protection if they would promise to guarantee her safety.

According to Bothwell's man, the captain of Inchkeith, her reaction was the exact opposite. Although forbidding Bothwell to fight, she cried out to him to order their army to charge. She "desired nothing more than that her forces should do battle."

It is impossible to judge between these versions. Neither is corroborated, and neither recorded Bothwell's response to Mary's order forbidding the man-to-man combat.

But whatever she intended to happen, her army did not charge and she surrendered. Her forces had melted away, reduced to some four hundred of Bothwell's men. Even the captain of Inchkeith conceded that she and Bothwell were glancing around anxiously, looking for Huntly and his retainers, who they hoped might arrive at the last minute to save the day.

Using Kirkcaldy as her intermediary, Mary reached an agreement with the lords. Bothwell would be allowed to escape. He would be free to go where he pleased, fleeing with a handful of supporters without immediate fear of pursuit. She would then give herself up and return to Edinburgh in exchange for honorable treatment.

Mary was torn by conflicting emotions. She now resented Bothwell for his treachery over Darnley's murder. She had been too trusting. It was in her nature, as she had shown since her teenage years. Far from rising above noble infighting by marrying him, she had become more embroiled in factionalism than ever. He had been dishonest and manipulative. Stirred by ambition and a desire to trump his old rivals, he had betrayed the queen to whom he had pledged loyalty and his heart.

She was also calculating the odds. She knew that if Bothwell escaped, he would be free to fight another day. Mary, at her most ingenious in a crisis, was keeping her options open. For all his faults, Bothwell was her protector. She had been born a queen and was not a woman to surrender her throne lightly, least of all to a junta of factious nobles.

When Mary parted from Bothwell, she was weeping. She was pregnant by him, the result of her sojourn at Dunbar. The fact may have been widely known. It was reported to Cecil on the same day as the battle, and so must have been talked about in Mary's circle for a week or so. Her reluctance to allow Bothwell to engage in single combat might have been at least partly because he was the father of her new baby.

Now it was Bothwell's turn to be agitated. He began to waver and protest, unsure of the conditions Mary had agreed to with her captors. Would he be allowed to escape only to be taken and killed when he was out of her sight? He appealed to her to tell him the truth. Would he have safe-conduct? She answered "Yes" and held out her hand.

Bothwell clasped it in farewell and, turning his horse's head, rode at a gallop toward Dunbar, closely watched but still unopposed by the lords. He was flanked by a dozen or so followers. Only when they had ridden almost two miles and were slipping from view did Mary turn to the waiting Kirkcaldy.

The report of Mary and Bothwell's parting is from du Croc. His account was compiled the next day from his conversations with eyewitnesses and is likely more accurate than anything put out for public consumption by the lords, who claimed that Mary kissed Bothwell passionately in full sight of both armies, and each promised to be true to the other.

Mary never saw Bothwell again. As she turned to Kirkcaldy, she said in a matter-of-fact way, "Laird of Grange, I give myself up to you on the terms you explained to me on the lords' behalf." He knelt and kissed her hand. She mounted her horse and rode down the hill at a walking pace. A few steps behind her, Mary Seton followed dutifully on her pony.

As Mary approached her captors, she held her head high. A colored drawing illustrating the scene was prepared for Drury and sent to Cecil. It shows her in her borrowed clothes, determined to keep her dignity as the soldiers stood motionless in awe, watching a spectacle they would one day tell to their children and grandchildren.

Mary's reception was deferential at first, but once the spell was broken, the Earl of Atholl's men and those of the Laird of Tullibardine cried out, "Burn the whore! Burn the murderess!" Such insults stunned Mary. She turned to the lords in bewilderment, but they looked away.

Mary's honor and reputation had been dragged through the mud. No longer would she keep up a pretense. She hated the lords for the humiliation they had inflicted on her; she would do anything she could to destroy them. As she regained her confidence on the road back to Edinburgh, she "talked of nothing but hanging and crucifying them all." In high dudgeon, she summoned Lindsay, Bothwell's challenger, to present himself. When he arrived, she offered him her hand. As he took it, she snarled, "By the hand that is now in yours, I will have your head for this and therefore [I] assure you."

Mary's threats cost her dearly. She had betrayed her intentions at a moment when she and the Confederate Lords were locked in a struggle for survival. She reached Edinburgh between eight and nine in the evening. The lords had sent a message ahead, and onlookers packed the streets. It was a world apart from her triumphal entry into the town six years earlier, when she was cheered by civic dignitaries and welcomed by "Moors" in yellow taffeta. Then she had ridden in state. Now she was "all disfigured with dust and tears."

The lords' propaganda was in full flood. As Mary passed by, the mob shouted insults at her. As a final indignity, she was not taken to Holyrood, but billeted in the provost's house opposite the Market Cross. "That same night," said Sir James Melville, drawing on information supplied by the lords, she wrote to Bothwell, "calling him her dear heart, whom she should never forget nor abandon for absence." She assured him that she had sent him away for his own safety. She urged him to be comforted, and warned him to be on his guard. She gave the letter to a messenger and paid him to carry it to Dunbar. He took the money, but gave the letter to the lords.

The story is quite improbable. Even Melville doubted it, adding a marginal note: "Some suspect this letter to be invented."

Drury was fed the same story. Then he was told: "The queen hath made a vow she would not eat no flesh [i.e., fast] till she saw Bothwell again." That may be true. It was entirely in character that Mary should refuse to eat when she knew she was cruelly deceived. The lords had promised her honorable treatment only to keep her under arrest. She would rather die than subject herself to such indignity.

Next day, said Drury's spy, Mary appeared at a window and called to the people for help. "She came to the said window sundry times in so miserable a state, her hairs hanging about her ears, and her breast, yea the most part of all her body, from the waist up, bare and discovered, that no man could look upon her but she moved him to pity and compassion." Seeing Maitland go past, Mary begged him to come and speak to her, but he pulled his hat over his ears and pretended not to hear.

Maitland told du Croc a different story. Mary had leaned out the window and demanded to know why she had been separated from Bothwell, "with whom she had hoped to live and die with the full approval of the world."

In a dark intrigue lasting three hours, Maitland put out feelers to see how du Croc would react to Mary's imprisonment or exile abroad while the lords ruled in the name of Prince James. A secret pact was reached. Du Croc agreed that France would not interfere, but only if England did not become involved. If Elizabeth intervened, then Charles IX would be forced to declare in Mary's favor.

The truth was, as long as Mary was still queen, she had the support of a majority of her subjects. An otherwise reliable chronicler, one not in the pay of the Confederate Lords, gave by far the most credible account of this episode. His version was that when Mary leaned out the window and appealed for help, "the people of the town convened to her in great number, and perceiving her so afflicted in mind, had pity and compassion of her estate. The lords perceiving that, came unto her with dissimulate countenance, with reverence and fair speeches, and said that their intentions were no ways to thrall her."

This fits the independently known facts. To disperse a large crowd, the lords had to make an announcement. Mary would be allowed to return to Holyroodhouse, where she would be free "to do as she list." When Mary heard this, she "was so pacified that the people willingly departed."

That evening Mary was escorted to Holyrood by Morton and Atholl, one on either side of her, preceded by a force of musketeers carrying the banner of the murdered Darnley. At a prearranged signal, a carefully selected group of onlookers shouted out, "Burn her, burn her, she is not worthy to live, kill her, drown her."

But the lords were treacherous. Mary was kept for just a few hours at Holyrood, and only then because it was unsafe to move her elsewhere in daylight. As night fell, she was told to make herself ready. With only a mantle thrown over her nightdress, the fearful Queen of Scots was taken by ferry across the Firth of Forth to Fife, and on to the island castle of Lochleven, near Kinross. Shortly after daybreak on Tuesday, June 17, she climbed into a boat and was rowed to the island in the middle of the loch. The castle loomed eerily out of the morning mist: a square tower four stories high with round projecting turrets at the corners, encircled by a loch that then was about fifteen miles in circumference and at least half a mile from the mainland at its narrowest point.

When Mary disembarked at the foot of the stairs leading to the fortress's drawbridge, she was met by Sir William Douglas, the Laird of Lochleven, who took her to a sparsely furnished room on the ground floor. Later she was moved to a more secure place, the round turret at the southeast corner of the castle.

This was not a household in which Mary was likely to thrive. Sir William Douglas was one of Morton's clan. If that were not enough, his widowed mother, who still lived in the castle, had been born Lady Margaret Erskine. She was one of James V's former mistresses, none other than the mother of Mary's half-brother Moray. The laird, himself Moray's half-brother as well as Morton's dependent, had two legitimate brothers and seven sisters, one of them married to Lindsay, Bothwell's challenger whom Mary had threatened on the road back from Carberry Hill. Lady Erskine had more than once claimed to have been officially if clandestinely married to James V. She had said that her son, Moray, was not only legitimate but rightfully King of Scots in place of Mary.

The inference was inescapable. Moray intended to return to Scotland from his self-imposed exile in France to rule in one capacity or another. The Confederate Lords had justified their revolt as a moral crusade to avenge Darnley's murder and secure Mary's release from the "captivity" and "thralldom" of Bothwell. This was pure humbug. What they did after her surrender at Carberry Hill was to imprison her themselves. By way of an excuse, they tried to implicate her in Darnley's murder. The stakes soared when they asserted in their warrant for her arrest that she had "appeared to fortify and maintain" Bothwell in his crimes. She was, they claimed, a woman of "inordinate passion."

For Mary, it was the beginning of the end of her tumultuous reign in Scotland. She had been queen for all but the first six days of her life, and a reigning queen for six years. She was now a prisoner, guarded night and day by her enemies. Apart from a few short but intoxicating weeks in the following year, the rest of her life would be spent in captivity.

22

Mary's Story

WHILE MARY was imprisoned in Lochleven, a black legend was spun. "Moral turpitude" was the charge leveled against her. She was accused of adultery and murder and said to be unfit to rule. The Confederate Lords claimed that she and Bothwell had enjoyed passionate sex for months before Darnley was killed. They said the queen and her lover had jointly planned the explosion at Kirk o'Field, and when Darnley was dead she had organized a fake kidnapping and rape to cloak her brazen designs to marry him.

This story is not Mary's. History is written by the winners, and after her incarceration, she was to be a spectacular loser. The villainy and cunning of the lords are shown by their willingness to accuse her of the crimes they had themselves committed, rewriting history to make theirs the official story.

When the lords wove their damning fiction, Mary's version of history was forgotten. What may come as a surprise is that her story — told in her own words — can be retrieved. She has rarely been allowed to tell it. Few biographers have given it more than a brief mention, and the documents recording it have not been quoted at full length since 1845. Perhaps because her story *is* so different from the supposedly official version, it was not thought worthy of examination.

Shortly after marrying Bothwell, Mary chose two ambassadors to put her case to her royal neighbors. She sent William Chisholm, Bishop of Dunblane, to Paris and Robert Melville to London, to justify and explain her recent actions.

She herself gave them their instructions, dictated to her secretaries and handed to the ambassadors as briefing notes. This was the usual procedure, and the documents reflect Mary's own defense of her position. She had spoken in Scots, which she almost always did on these occasions. Sometimes breathless or containing pendent clauses as was typical of her style under pressure, the documents set out her version of events. Although divergent in focus, their overall line is consistent.

In Melville's instructions, Mary stressed the political advantages of her marriage. She played on the factions and plots that had frustrated her efforts to rule, arguing that only a native-born Scot could discipline the nobles successfully. She began by making excuses to Elizabeth for the unseemly haste of her marriage:

> After that you have presented our most hearty commendations to our said dearest sister, you shall expound and declare unto her the true occasions which has moved us to take the Duke of Orkney [Bothwell] to husband, and therewithal make our excuse for that we so suddenly proceeded to the consummation of our marriage, not making our dearest sister advertisement, nor asking her advice and counsel therein.

Mary was initially convincing in her claims. Her statements about the unruly factionalism of the Scottish lords are far from exaggerated:

> For the first, you shall ground you upon the condition and state of us and our realm, declaring how we were destitute of a husband, our realm not thoroughly purged of the factions and conspiracies that of long time have continued therein. These, occurring so frequently, had already in a manner so wearied and broken us that by ourself we were not able of any long continuance to sustain the pains and labor in our own person that were requisite for repressing of the insolence and sedition of our rebellious subjects. They are, as is known, a people as factious amongst themselves and as factious toward the ruler as any other nation in Europe. For their satisfaction, which could not suffer us long to continue in the state of widowhood, it behoved us, moved by their prayers and requests [i.e., the Ainslie's Tavern Bond], to yield unto one marriage or another.

Where Mary began to delude herself was in arguing that her nobles had supported her marriage to Bothwell and petitioned her to marry him as soon as possible without looking any further for a husband:

Seeing no advantage to follow by protracting of time, but as on the one part they were very well content, yea and earnestly urged us that we should without delay proceed to our marriage, even so on the other side, we perceived by their meaning how unwilling they were that we should choose any foreign husband, but rather [we should] so far humble ourself to be content with some native-born subject of our own for that place, someone acquainted with their traditions and the laws and customs of our realm. For indeed we ourself have had proof and experience of their revolts, when as in the case of our foreign marriage [i.e., Darnley, who was born an Englishman] they have supposed that they would be severely handled by foreigners.

Mary may well have been correct that the lords would have rejected the idea of a foreign marriage, but she inflated the significance of the Ainslie's Tavern Bond to claim she had all along been advised to marry Bothwell. Instead of recognizing that it was she who had chosen him, she tried to cast herself as a political pawn in the hands of the noble factions. She also glossed over Bothwell's true character:

When, therefore, in the eyes and opinion of our people, one of our own subjects was judged most fitting both for us and them, our whole nobility being lately assembled at our Parliament were best content that the Duke of Orkney, then Earl of Bothwell, should be promoted to that place, if so were our pleasure. To that effect they subscribed a letter [the Ainslie's Tavern Bond] with all their hands before or ever we agreed to take him to our husband or that he opened his mind to us in that behalf, whereby we were moved to make our choice of him, as one whose wisdom, gallantry and other good qualities might be well compared or rather preferred to any other nobleman in our realm, and his ancestry honorable and ancient. But indeed his faithful and upright service ever since he came to man's estate spent and bestowed for us and in our cause, for setting forth our authority whosoever gainsaid it, was no small motive in our consent in making of our choice. This was the rather because none, or very few of all the nobles, are able in that point to compare with him, seeing at some time or other, the most part of them had left or abandoned us, he alone excepted.

Mary protested too much. She pretended that she had no idea of Elizabeth's likely hostility to Bothwell. She did not even seem to remember her dueling with her "sister queen" two years before over her choice of Darnley as a husband:

Concerning all these matters, you shall pray and desire her heartily to excuse us, for as we never meant to join in marriage with any that we believed she was not content with, so at this present time, we trust she will not only continue her accustomed favor and mutual intelligence with us, but also, for our respect, will extend her friendship to our husband with whom we are inseparably joined, and to bear him and us no less goodwill than if all had proceeded to this hour with the knowledge and advice of our dearest sister, whom you shall assure to find him ready to do her all the honor and service that she can require of him.

Mary then turned to the most dangerous ground. She had to deal with the fact that Elizabeth had written to her immediately after Darnley's death, charging her to punish his murderers without delay, sparing no one, not even "him whom you have nearest to you," meaning Bothwell, "if he was involved":

> In case the queen our good sister shall think strangely of our marriage with the Duke of Orkney by reason he was suspected and accused of the odious violence committed on the person of the king our late husband, and because she had written to ourself somewhat in that behalf before, it is true that she wrote to us and we sent her an answer, the copy whereof we have enclosed herewith. It will instruct you sufficiently in what you shall answer to this objection in case you should be asked about it.

Mary wished Melville to answer such charges fully. To that end, she asked him to refer to a letter she had belatedly sent to Elizabeth addressing this very point. A copy of the letter was included in his diplomatic brief, and Mary drew his attention to the key passage:*

> I lament more highly the tragedy of my husband's death more than any other of my subjects can do; and if they had allowed, and if I had been permitted to use my authority untroubled by my subjects, I had punished the committers thereof . . . I had never knowledge, art nor part thereof, nor none of my subjects did declare unto me that they who are now holden culpable and principal executors thereof were the principal authors and committers of the same: which if they had done, assuredly I would not have proceeded as I did so far. I suppose I did nothing in that matter but by the advice of the nobility of the realm.
>
> If they allege that my marriage with the Earl of Bothwell will be taken as a presumption against me, I never condescended thereto until the time the greatest part of the nobility had cleansed [acquitted] him by an assize and the same ratified in Parliament, and they had given their plain consent unto him for my marriage, and solicited and persuaded me thereto, as their

---

* This letter from Mary to Elizabeth is missing from the archives. Fortunately its contents can be recovered, as Mary quoted them in 1568. A verbatim extract from that later document is used here.

handwriting [the Ainslie's Tavern Bond], which was shown to me, will testify.

Mary had been forced into a corner. To say that she had never been informed that Bothwell was one of those accused of Darnley's murder would be palpably untrue. Not only had Lennox named Bothwell as the chief suspect in a letter to Mary, the placards affixed to the walls and doors of the Tolbooth had also accused him. The whole force of her statement depends on her use of the words "principal authors and committers." Her argument was that Bothwell may well have been an accessory to the murder, but she had not been made aware of his role at the time as Morton's chief ally in devising the plot. We already know that she had held Moray to be the arch-conspirator, a view reinforced by his decision to seek voluntary exile abroad.

Lastly, Mary tried to deal with the inconvenient fact that Bothwell had been a married man when she had consented to marry him:

> It may be that our good sister shall allege our present marriage not to be lawful insofar as the duke our husband was coupled to a wife before, who yet lives. You shall answer that by the laws received within our realm and often times practiced as is generally known, his former marriage was dissolved and the process of divorce orderly led for resolute causes of consanguinity and others before our marriage with him. And so we might lawfully consummate the same, for it is no new thing neither in Scotland nor England.

The Bishop of Dunblane's instructions were angled differently, offering rationales more likely to sway the French. Mary still defended her marriage on political grounds, but focused on Bothwell's character and role as queen's protector:

> First, you shall excuse us to the king, the Queen Mother, our uncle, and others our friends, in that the consummation of our marriage is brought to their ears by other means, before that by any message from ourself they have been made participant of our intention therein; which excuse must be chiefly grounded upon the true story and report of the Duke of Orkney, his behavior and proceedings toward us before and until this time that we have been made content to take him to our husband . . .

Mary settled down for the long haul. She went right back to the start of Bothwell's career and his role as her mother's defender:

Beginning from his very youth and first entrance to this realm immediately after the death of his father, who was one of the first earls of the realm and one of the foremost in reputation by reason of his nobility and ancestry, and of the great offices which are his by inheritance. At which time the queen our mother [Mary of Guise] being then regent of our realm, he dedicated his whole service to her in our name with such devotion and earnestness, that albeit soon thereafter, the most part of the nobility, almost all of the burghs, and so consequently in a manner the whole substance of the realm, made a revolt from her authority under color of religion; yet swerved he never from our obedience.

Mary went into almost superfluous detail as to how Bothwell had stolen the first installment of gold sent covertly by Cecil to aid the Lords of the Congregation in their revolt. Then she turned to his service since her return to Scotland six years before, glossing over his many misdemeanors and quarrels with Moray and Arran, but stressing his loyalty and resourcefulness:

After our return to Scotland, he gave his whole study to the setting forth of our authority and to employ his person to suppress the insolence of the rebellious subjects inhabiting the counties lying west of the borders of England, and within a short time brought them to a perfect quietness . . .

But as envy ever follows virtue, and this country is of itself somewhat subject to factions, others [i.e., chiefly Moray and Arran] began to dislike his proceedings. They went about so far by bad reports and by misconstruing his doings to put him out of our good grace, that at length — upon colors invented by his evil-willers, for satisfying of them that might not abide his advancement and avoiding of further contention, which might have brought the whole realm in trouble — we were compelled to put him in prison.

Out of the which escaping . . . he passed out of the realm toward France, and there remained until about two years ago, when the same persons who were before the instruments of his trouble began to forget their duty toward ourself, putting themselves in arms [Moray and the rebels of the Chase-about Raid], displaying open banners against our person.

Mary then turned to Bothwell's service during the Chase-about Raid:

At which time by our commandment he was called home and immediately restored to his former charge of lieutenant-general. Our authority prospered so well in his hands that suddenly our whole rebels were constrained to flee the realm and remain in England, until some of them [Châtelherault and his family] upon submission and humble suit were reconciled to us.

Step by step, Mary drew closer to recent events. She said she would skip over the Rizzio plot on the grounds that her uncle the Cardinal of Lorraine was already fully aware of it. But she could not fail to mention Bothwell's role in her daring escape at midnight from Holyrood:

> Yet it is worthy of remembrance with what dexterity he escaped from the hands of those who at that time detained our person captive, and how suddenly by his providence not only were we delivered out of the prison [Mary had been confined to her apartments with guards at her door], but also that whole company of conspirators were dissolved and we recovered our former obedience. Indeed we must confess that service done at that time to have been so acceptable to us that we could never to this hour forget it. He has ever since then prosecuted with the like diligence in all that might content us, so that we could not wish for more fidelity nor good behavior than we have always found in him.

Mary then attempted to explain her conduct after Darnley's assassination. Her remarks made it clear that she knew she had married in haste and Bothwell's behavior was unworthy. Neither his gallantry and bravado nor his unswerving loyalty justified his dubious methods of courtship:

> Until of late, since the decease of the king our husband, when as his ambitions began to be higher, so find we his proceedings somewhat strange. Albeit now since we are so far proceeded with him we must interpret all things to the best, yet have we been highly offended. First, with his presumption, who thought we could not sufficiently reward him unless we should give ourself to him [in marriage] for the recompense of his service. Next, for his practices and secret means, and at length the plain attempting by force to have us in his puissance, for fear to be disappointed of his purpose.

Mary's retrospective characterization of Bothwell was forthright:

> His deportment in this behalf may serve for an example, how cunningly men can cover their designs when they have any great enterprise in head until they have brought their purpose to pass. We thought his continuance in the awaiting upon us and readiness to fulfill all our commandments, had proceeded only upon the acknowledging of his duty, being our born subject, without further hidden respect. This moved us to make him the better visage, thinking that the same was but an ordinary countenance to such noblemen as we found affectionate to our service, and never supposing that it should encourage him or give him boldness to look for any extraordinary favor at our hands.

Mary's side of the story was that she had known nothing of the Ainslie's Tavern Bond before Bothwell had secretly procured it. She claimed he had extracted the lords' signatures deceitfully, pretending she had already sanctioned the bond and thus agreed to marry him:

> And in the meantime, he went about by practicing with the lords secretly to make them his friends, and to procure their consent to the furtherance of his intents. And [he] so far proceeded by these means with them, before that ever the same came to our knowledge, that our whole estates being here assembled in Parliament, he obtained a writing subscribed with all their hands, wherein they not only granted their consent to our marriage with him, but also obliged themselves to set him forward thereto with their lives and goods, and to be enemies to all who would disturb or impede the same. This bond he obtained by giving them to understand that we were content therewith.

As soon as Bothwell had the bond in his possession, he began to hint at a possible marriage, but Mary rejected his suit:

> And the same [bond] being once obtained, he began afar off to reveal his intentions to us and to assay if he might by humble suit purchase our goodwill. But finding our answer nothing corresponding to his desire and casting before his eyes all doubts that usually men use to resolve with themselves in similar enterprises . . . he resolved with himself to follow forth his good fortune.

Mary denied that her abduction at Almond Bridge had been collusive. She had known nothing about it until she was kidnapped:

> He suffered not the matter long to sleep, but within four days thereafter, finding opportunity by reason we were passed secretly toward Stirling to visit the prince our dearest son, in our returning he awaited us by the way, accompanied with a great force, and led us with all diligence to Dunbar.
>
> In what part we took that manner of dealing, but specially how strange we find it of him of whom we doubted less than of any subject we had, is easy to be imagined.

Mary claimed to have been surprised and shocked by her abduction. Once she had reached Dunbar, she had censured Bothwell for his unseemly behavior, but he had paid court to her and won her over:

> Being there, we reproached him on account of the honor he had to be so esteemed of us, the favor we had always shown him, his ingratitude, with all other remonstrances which might serve to rid us out of his hands.

Albeit we found his doings rude, yet were his answer and words but gentle: that he would honor and serve us and no wise offend us. He asked pardon of the boldness he had used . . . and there began to make us a discourse of his whole life, how unfortunate he had been to find men his enemies whom he had never offended; how their malice never ceased to assault him at all occasions, albeit unjustly; what calumnies had they spread upon him touching the odious violence perpetrated on the person of the king our late husband.

Bothwell had then produced his trump card, the Ainslie's Tavern Bond:

And when he saw us like to reject all his suit and offers, in the end he showed us how far he was proceeded with our whole nobility, and what they had promised him in their own handwritings . . .

In the end, when we saw no hope to be rid of him, never man in Scotland once making an effort to procure our deliverance, for that it might appear by their handwritings and silence at that time, that he had won them all, we were compelled to mitigate our displeasure, and began to think upon that he propounded. And then [we] were content to lay before our eyes the service he had done in times past, the offer of his continuance hereafter, how unwilling our people are to receive a foreigner unacquainted with their laws and customs, that they would not allow us long to remain unmarried . . .

Mary's account of her twelve-day stay at Dunbar is deeply disingenuous. She had been a fool for love. She knew she had done wrong in sleeping with a married man, and her excuse that no one had made the effort to rescue her can be countered by the fact that she made no attempt whatever to escape, even when Bothwell was away in Edinburgh, encouraging his wife to file her divorce petition.

Against this, her insistence that she had married Bothwell to seal his role as queen's protector is persuasive. She could make a strong case for deploying him as an instrument against the noble factions, even if his treatment of her as a woman had bruised her dignity and sense of honor. Her explanation is that she felt she had no other option, wearied and broken as she had become by the relentless infighting of the lords and the threat of bloodshed:

[Finally, we realized] that this realm, being divided in factions as it is, cannot be contained in order unless our authority be assisted and set forth by the fortification of a man, who must take pain upon his person in the execution of justice and suppressing of their insolence that would rebel, the travail whereof we may no longer sustain in our own person, being already wearied and almost broken with the frequent uproars and rebellions raised against us since we returned to Scotland . . .

After he had by this means and many others brought us on the way to his intent, he partly extorted and partly obtained our promise to take him to our husband. And yet not content therewith, fearing ever some alterations, he would not be satisfied with all the just reasons we could allege to have the consummation of the marriage delayed as had been most reasonable . . .

But as by his act of bravado in the beginning he had won the first point, so ceased he never until by persuasions and importunate suit, accompanied none the less with force, he had finally driven us to end the work begun at such time and in such form as he thought might best serve his turn. We cannot dissemble that he has used us otherwise than we would have wished or yet have deserved at his hand . . .

Mary always refused to acknowledge the gusto with which she had raced into her third marriage. When it had quickly turned sour, she would be willing to confess that Bothwell had treated her brutally. And yet she had made her choice and would stick by it. Even while enumerating his faults, she still wrote of him in the way she wished him to be:

Now since it is passed and cannot be brought back again, we will make the best of it, and it must be thought, as it is in effect, that he is our husband whom we will both love and honor, so that all that profess themselves to be our friends must profess the like friendship toward him who is inseparably joined with us. And albeit he has in some points or ceremonies behaved imprudently, we are content to impute this to his affection toward us.

A single sentence in the bishop's instructions would encapsulate Mary's understanding of her plight: "We cannot dissemble that he has used us otherwise than we would have wished or yet have deserved at his hand." This realization must, on deeper reflection, have torn her apart in her prison at Lochleven, where a month or so after her arrival she would miscarry twins, her issue by Bothwell.

Mary's diplomatic efforts were doomed to failure. When the bishop reached Paris, he was coolly received. The same was true of Melville's mission to London. In fact, by the time the ambassador presented himself to Elizabeth, Bothwell had already fled from Carberry Hill, making the visit irrelevant.

But if Mary's story fell into oblivion, Elizabeth was utterly scandalized that a fellow ruler had been imprisoned in an island fortress by her rebellious lords. She could not yet bring herself to write to Mary in her own hand, but she dictated a letter in which she threw her weight behind her fellow monarch, cousin and close kinswoman. On the same day, she sent another to the lords, expressing her grief and anger at what they had done.

The difference in the approaches of Elizabeth and her chief minister over how to deal with Mary had never been more striking. Elizabeth sent Sir Nicholas Throckmorton to Scotland as her crisis manager. He, like his former protégé, Leicester, and quite unlike Cecil, had become one of Mary's lesser champions. He had even become a supporter of her claim to the English succession under the right conditions, and had been one of those backing Elizabeth's offer to negotiate a new "treaty of perpetual amity" that would replace the offending clauses of the treaty of Edinburgh after Mary's illness at Jedburgh.

Throckmorton was not welcome in Scotland. From the rebel lords' standpoint, he had to be neutralized. For this they turned to Cecil, who wrote a memo of instructions for Throckmorton that greatly elaborated and substantially contradicted Elizabeth's own. Whereas Elizabeth wanted Mary restored to the throne, Cecil laid down the only terms on which she might be freed. She was to be stripped of her authority, which would be vested in a council of nobles. She might be styled queen, but only nominally. In all other respects, Cecil planned to restore the quasi-republican "States of Scotland" that had governed after the deposition of Mary's mother during the lords' first revolt.

At the end of the memo, Cecil jotted down these words: *"Athalia interempta per Joas[h] regem"* — "Athalia was killed so that Joash could be king." It is one of the most revealing comments he ever made. A quotation from 2 Chronicles in the Old Testament, it is the very same text that Knox used to justify the use of armed resistance against "idolatrous" female rulers. Athalia was the perfect exemplar. She (like Mary) ruled in person as queen of Israel for six years, but because of her moral turpitude, the high priest joined with the nobles to depose and kill her along with her idol-worshiping acolytes. The nobles made a covenant with God, installing the young Prince Joash, then seven years old (for whom read Mary's own son, Prince James, then one year old), in her place. After Athalia was murdered, the nobles ruled in the name of King Joash until he reached the age of majority, just as successive regents and their allies would attempt to rule in Scotland.

When Cecil made that jotting, he had seen the hand of God in history. He read the biblical text (as Knox had done) as a prophecy applying to Mary. His note proves that regicide was already in his sights. For Cecil, the spider weaving his web in London, it was but the shortest of steps from a round tower at Lochleven to a scaffold draped in black at Fotheringhay. His mantra had always been Elizabeth's "safety," and he had regarded Mary as the instigator and intended beneficiary of an international Catholic conspiracy ever since her ushers had cried "Make way for the queen of England" as she walked to chapel with the dauphin.

Throckmorton was caught squarely between the conflicting policies of Elizabeth and Cecil. As he confided to the Earl of Bedford, "I never was in so busy and dangerous a legation in my life." The lords endlessly told him that Mary "will not consent by any persuasion to abandon the Lord Bothwell for her husband, but avoweth constantly that she will live and die with him." But he was never allowed to see her and hear this for himself, nor was she allowed to see him. What the lords said to Throckmorton was what they (and Cecil) wanted Elizabeth to hear. They several times promised that if Mary would divorce Bothwell, she might be restored to her throne. But this was less a serious proposal than a delaying tactic to appease Elizabeth, who was becoming more and more impatient and threatening to go to war to free Mary from her prison. All along, Morton and his staunchest ally, Lord Lindsay, Bothwell's challenger, were set on deposing Mary or forcing her to abdicate.

On July 24, 1567, exactly three months after her abduction by Bothwell, she was lying on her bed, weak and despairing after her miscarriage, when she heard footsteps on the stairs of her tower. Through the door burst Lindsay with a delegation from the lords. Three documents were put before her. One declared that she was so depleted in body, mind and spirit by the responsibilities of government that she could no longer continue, and so abdicated in favor of her son, whose coronation she authorized. By the second document she was to appoint Moray, now hastening home from France, to be regent until Prince James reached the age of majority. By the third she nominated Morton and others to serve as interim regents until Moray took up the reins of power.

At first Mary refused to sign. Seeing her hesitate, Lindsay ordered her to rise from her bed and get ready to depart, swearing that she would be marooned for life on an island in the middle of the sea or else thrown into the loch. Finally, he swore a great oath and threatened to cut her throat.

Mary had no choice; she cannot have known of Elizabeth's threat to go to war to defend her. She put her signature on each of the papers, but managed to blurt out, "When God shall set me at liberty again, I shall not abide these, for it is done against my will."

Five days later, the lords crowned Prince James in the parish church at Stirling. It was the worst-attended coronation in Scottish history. Morton swore the one-year-old child's coronation oath for him, and Knox preached the sermon. Then the lords staged a spectacle. A thousand bonfires were lit in the towns and villages, and in Edinburgh the castle guns fired a salute. But the people were sullen. "It appeared," Throckmorton noted wryly, "they rejoiced more at the inauguration of the new prince than they did sorrow at the deprivation of their queen." But their reaction to the coronation had been so muted, it was clear that they were longing for some stability, and outside Edinburgh there was almost no support for the new regime.

When Elizabeth heard of Mary's forced abdication, she sent at once for Cecil. When he arrived, he was harangued (as he informed Throckmorton) in a "great offensive speech" on the grounds that he had failed to do anything for Mary.

Elizabeth was almost speechless with rage. Cecil answered, he said, "as warily as I could." But to no avail. Elizabeth was so incensed, she once more threatened to declare war on the Scots. She refused all Cecil's protests and counterarguments. It was one of their classic rows, and it turned on the nature and power of monarchy. Mary was an anointed queen, accountable to God alone. Elizabeth wanted it demonstrated that no such coercion would be tolerated. She particularly had her own English subjects in mind. Now it was she and not Mary who was brooding over the potential for a domino effect.

Cecil artfully replied that a declaration of war might precipitate what Elizabeth most feared — Mary's assassination in the dead of night. He was muddying the waters to help his Scottish allies, because he knew that with the passage of time Elizabeth's anger would subside. He had the nerve to joke later that it usually took three to six weeks.

Mary was kept in stricter confinement after her abdication than before. Moray had returned to Edinburgh on August 11, and when he visited his sister a few days later, she reproached him for her treatment, weeping bitterly. Far from showing any sympathy or affection for her, he was as cold and calculating as ever, scolding her and giving her a lecture on good government. They talked until one in the morning, when she was too tired to go on.

Next day, they continued where they left off. Moray was gentler now, but out of cunning rather than kindness. His purpose was clear: he wanted his sister to promise not to try and escape or seek aid from England or France. He used every technique of psychological intimidation to induce a sense of gratitude in her, alternately threatening and comforting her, and promising to mitigate the worst plans of Lindsay and his friends if she cooperated, even though they were really acting on his behalf.

Moray got what he wanted. Mary "took him in her arms and kissed him, and showed herself very well satisfied, requiring him in any ways not to refuse the regency of the realm, but to accept it at her desire." Such was the devious way in which her brother won her acquiescence in what she had previously agreed only under duress. On August 22, Moray was proclaimed regent. He had led the lords who deposed Mary's mother eight years before, and now he had done it again. But the fight was far from over.

When Mary recovered from the trauma of her miscarriage and abdication and started to take stock, she knew she had been cheated. She was determined to recover her honor and her throne. She bided her time, resting and eating properly. She passed the days in sewing and embroidery, which she loved. She played cards. She even danced to the fiddle and the bagpipes. She had managed to obtain a small domestic staff: five or six ladies, four or five gentlewomen, a doctor, a cook and two chamber servants, one of whom was French. Mary Seton, the most faithful of the four Maries, was still continually by her side.

Mary put on weight through lack of exercise, but her wits were as sharp as ever. She began to plan her escape. She sent a ring to Mary Fleming, now the wife of Maitland, himself wavering in his support for the lords. She contacted Mary Livingston, another of the four Maries, who had married John Sempill, the younger son of the Catholic Lord Sempill. Although Moray had won over Lord Sempill to his cause by bribery, his son was loyal to Mary and was at the center of an unsuccessful plan to rescue her by assaulting the castle by night.

Mary then turned to George Douglas, nicknamed "Pretty Geordie," the dashing and handsome younger brother of the Laird of Lochleven. Nine years younger than Mary, he fell in love with her and offered to serve her. He had witnessed her forced abdication and was determined to help her, sending her secret messages through her gentlewomen, for which he was expelled from the castle by Moray and the laird and forbidden to return.

But George's loyalty and Mary's courage and ingenuity overcame such obstacles. On March 25, 1568, nine months after Mary had first been imprisoned at Lochleven, her first serious attempt to escape, in the disguise of a laundress, failed when she was recognized by one of the boatmen halfway across the loch. He tried to snatch at her muffler to see her face better, and when she threw up her hands, he noticed that they were "very fair and white" — the hands of a queen and not a washerwoman. He turned the boat back to the castle, but promised Mary he would not tell the laird about her scheme.

A plot was then hatched in which young Willie Douglas, a page in the castle known as "little Willie" or "orphan Willie" — he was possibly an illegitimate son of Sir William Douglas, the Laird of Lochleven — agreed to row Mary across the loch. Willie was barely sixteen, but was more than willing to help rescue the charismatic queen. He joined forces with George Douglas, who waited patiently at Kinross on the mainland with Lord Seton and a small force.

On the afternoon of Sunday, May 2, while the sun was high in the sky and the Douglas family were enjoying their May Day weekend, Willie Douglas sabotaged all the boats moored beside the castle jetty except one. Then, shortly after seven, when the laird and his family were at supper, he stealthily made off with the key to the main gate of the castle, which the laird had left on the table. He signaled to Mary's turret and she came down to the courtyard. She had exchanged clothes with Mary Seton, who stayed behind to impersonate her in case the alarm was raised.

After they passed through the main gate, Willie locked it behind him and threw the key into the mouth of a nearby cannon. He helped Mary into the boat and rowed her across the loch. They were met as they landed by George Douglas, who had stolen the Laird of Lochleven's best horses, conveniently stabled on the mainland.

Mary rode off at a gallop through Fife to North Queensferry, where she crossed the Firth of Forth. When she arrived on the southern shore, she was escorted to Niddrie, one of Lord Seton's fortresses situated between Edinburgh and Linlithgow, where she rested for the night.

After almost eleven months of captivity, Mary was free again. She was exultant, enjoying every moment of the ride and the late-spring evening. At Niddrie, she scribbled several hasty letters to her friends. She also sent a messenger to Dunbar with orders to fortify the castle in her name.

Early the next day, Mary rode west to Hamilton, where her supporters rallied to her and she established a temporary court. Moray was in Glasgow when he heard the news of her escape. The lords were at first incredulous; then, seeing the danger, they immediately ordered their forces to muster at Glasgow.

Mary was reluctant to engage in a battle. The memory of Carberry Hill was too painful. But she was still popular. The lords' propaganda had been effective only as long as she was in their clutches. Within a week, six thousand men rallied to her cause. Huntly, Seton and the Hamiltons, the family of Châtelherault, were beside her. They urged her to fight, and she agreed. "By battle let us try it," she declared. Already several of the lords were defecting from Moray and joining her camp. They included Argyll, who had been shocked by his queen's deposition but had not dared to oppose it alone.

When Mary rode toward Dumbarton at the head of her army, her forces outnumbered Moray's by a third. When the two armies clashed at Langside, just outside Glasgow, on the morning of Thursday, May 13, the result appeared to be a foregone conclusion. It was not. Moray threw all he could into his attacks. He knew everything turned on this day. Mary now hated her half-brother even more than Morton for his role in her downfall. She was determined to be avenged. She would never forgive him for the way he had played on her fears and emotions at Lochleven. She had been thinking it over repeatedly as she feigned resignation at her captivity. It was plain to her that all along he had wanted her throne for himself. If the victory was hers, Moray would be put on trial for treason.

The result was a crushing defeat for Mary. The battle got off to a disastrous start. Argyll had long suffered from a recurrent illness called "the stone," an omnibus term in the sixteenth century for a variety of internal ailments. Now he collapsed from a sudden fit, throwing Mary's command structure into disarray.

On Moray's side, by contrast, Kirkcaldy of Grange organized a brilliant ambush of her vanguard as it passed along a narrow lane. The fighting lasted only three quarters of an hour. Staring defeat in the face from her vantage point on a nearby hill, Mary fled to Dumfries, riding sixty miles at a stretch. After that, she took cover during the day and emerged only at night; such was her terror of falling again into her brother's hands.

Mary covered the last thirty miles to Dundrennan at night. She hid in the abbey, from where she wrote an urgent appeal for aid to Elizabeth, enclosing the diamond ring that her "sister queen" had sent her in 1563 as a token of love and friendship to be redeemed. What she could not know was that as she put the finishing touches to her letter, Elizabeth was gloating over the most precious of Mary's pearls, plundered from the royal cabinet at Holyrood and sold to her by Moray.

Mary did not wait for Elizabeth's reply. She was too afraid. On the 16th, she embarked on a fishing boat to cross Solway Firth, landing in England at about seven in the evening, close to Workington, about thirty miles from Carlisle.

At dawn, she wrote a second letter to Elizabeth. She asked to see her cousin and to have her aid and support in recovering her throne and defeating her rebels. She expected to gather fresh troops and return shortly to Scotland. But her decision to cross the border was a catastrophic mistake. It precipitated a crisis in England, where it was feared that the northern and overwhelmingly Catholic counties would rise to support her, leading to civil war in both countries. Elizabeth was still acutely sympathetic to Mary; what Moray and the rebel lords had done was unconscionable. They had imprisoned and deposed an anointed queen, a crime against God that was no less heinous than Darnley's assassination, perhaps more so from a monarch's point of view.

But Cecil got there first. He saw instantly the danger posed by Mary's unexpected arrival and had her placed under strict guard in the castle at Carlisle. Her movements were to be closely watched. He was determined that the lords' charges of adultery and complicity in Darnley's murder should be investigated, because his goal, unlike Elizabeth's, was to keep her off the throne.

In another of his unremitting memos, Cecil set out his case in detail. A trial of Mary's crimes, he argued with almost hairsplitting logic, would lead to one of two outcomes. If she was acquitted, conditions must be placed on her return to Scotland. They were none other than his old favorites. Mary would be required to forge a permanent (and Protestant) alliance in which she acknowledged Scotland's status as a satellite state by ratifying the treaty of Edinburgh in its original form.

If she was found guilty, she might — if her culpability was minor — be allowed to go into exile, but only if Moray was permitted to continue as regent in Scotland and rule in the name of Prince James. If her guilt was greater, then the punishment must fit the crime. Mary must "live in some convenient place without possessing of her kingdom, where she may not move any new troubles." By this Cecil meant a prison in England.

For Cecil, Mary's flight to England was an almost providential finale to the reign of the woman he always regarded as his sinister antagonist. For Mary, now confined to her "lodgings" at Carlisle, it was heads he wins, tails she loses. She wrote again to Elizabeth. "Do not," she pleaded, "[be] as the serpent that stoppeth his hearing, for I am no enchantress but your sister and natural cousin." But where Cecil was concerned, it was Mary who was the serpent and the enchantress.

Within a fortnight of her arrival, Mary knew that her whole future lay in Cecil's hands. She wrote to him on May 29 in the vain hope of throwing herself on his mercy:

> Mester Ceciles. The renown that you enjoy of being a lover of equity, and the sincere and faithful service you give to the queen, madame my good sister, and consequently to those who are of her blood and high dignity, invite me to write to you above all others, in my just quarrel, at this time of trouble, in the hope of obtaining the assistance of your good counsel.

She had briefed her messenger to appeal to Cecil's sense of honor and fair play, and her letter ended with the words "Recommending myself to you and your wife, I pray God to keep you in his holy care. Your very good friend, Marie."

When Cecil read this letter, all he could do was laugh. He was already in touch with Moray, from whom he now demanded the "manner of the proofs" and other "evidence" against Mary. Her trial was fast approaching and the spider's web all but complete.

23

Bothwell's Story

BOTHWELL ALSO GAVE his side of the story in the months following his flight. He dictated a full if often unreliable account of everything he alleged had happened to him, starting with Mary's return to Scotland and ending with his flight from the field at Carberry Hill. He then vividly described his escape into exile and his imprisonment in Denmark, a tale of such daring and bravado it excelled his theft of the gold coins sent covertly by Cecil to aid the rebel Lords of the Congregation in 1559.

In many respects, Bothwell's story is an anticlimax: its chief value is to confirm our impression of his self-serving duplicity. But his account is still worthy of our attention. He had always been a man with a rough side and a smooth side, and to allow him the opportunity to tell his own story is at least to refrain from condemning him unheard.

After leaving Mary at Carberry Hill, Bothwell had galloped with a handful of his followers to Dunbar. There he was left alone for a fortnight. He was lulled into a false sense of security; the vengeful and implacable Morton was in no hurry to begin the chase. But the Confederate Lords meant to hunt him down. On July 17, 1567, they outlawed him as a rebel. A bounty of 1000 crowns was offered to anyone who would bring him back a prisoner to Edinburgh.

By then he had already sailed north to Banff, in Aberdeenshire, in the hope of raising troops. This was Gordon country, where Huntly, Bothwell's erstwhile brother-in-law, dominated the local retinues. But

Huntly refused to assist him, saying that "he heartily wished both his sister and the queen rid of so wicked a husband." Bothwell withdrew to Spynie Castle, just north of Elgin, the home of his great-uncle Patrick Hepburn, Bishop of Moray, who had supervised his early education.

When Spynie became too hot to hold him, he fled farther north to Kirkwall, the capital of the Orkney Islands, where he planned to levy a fleet. Since Bothwell was Duke of Orkney and Lord of Shetland, Kirkwall Castle belonged to him, but its keeper, Gilbert Balfour, another of the siblings of his former ally Sir James Balfour, denied him entry and trained the castle guns on his ships.

Bothwell stayed for only two days. He sailed northward again to the Shetland Islands, where Olaf Sinclair, his mother's kinsman, provided him with money and supplies. Bothwell now planned to escape to France, hoping to rally Guise support for Mary.

The lords began the pursuit. On August 19, Kirkcaldy of Grange and Sir William Murray, the Laird of Tullibardine, set sail from Dundee. Their warships were the fastest in Scotland, fitted with cannons and carrying no fewer than four hundred musketeers. Their orders were to seize Bothwell, if they could find him, and execute him on the spot. The last thing Morton wanted was for his deadly enemy to be brought back alive to testify against him in a trial for Darnley's murder. Until this point, Bothwell could not have denounced his co-conspirators without admitting his own guilt, but now he had nothing to lose.

Six days later, the pursuers found their prey. They sailed into Bressay Sound, close to Lerwick, the chief port of the Shetland Islands, where Bothwell's ships lay at anchor.

When Kirkcaldy came into view, Bothwell and many of his men were ashore. Kirkcaldy saw his chance and raced forward, but Bothwell leaped aboard his own ship and cut the anchor cable. With reckless pluck, he sailed over some sunken rocks, grazing the hull, but tempting Kirkcaldy to follow him so that his vessel was holed and sunk.

Bothwell escaped to Unst, the northernmost of the Shetlands, where his squadron regrouped. But Kirkcaldy kept coming. He had three warships left. The rivals met in a battle lasting three hours. All seemed to be over when Bothwell's mainmast was shot away by a cannonball, but no sooner had Kirkcaldy sent a boarding party to capture him than a violent gale blew up.

Bothwell, an expert seaman, escaped with three of his ships. Kirkcaldy chased him for 60 miles, but Bothwell sailed southeast before the storm, putting an increasing distance between himself and his pursuers and covering the 250 miles of the North Sea between the Shetlands and Norway in record time.

Kirkcaldy had to return home empty-handed. Bothwell had gotten safely away. He hove to at Karmøy Island, twenty miles northwest of Stavanger. No sooner had he put down his anchor than he was arrested and his vessels brought north to Bergen. He was first detained on a charge of piracy. The rulers of Europe were watching every stage of the crisis in Scotland. The governor of Bergen Castle, Eric Rosencrantz, believing he might have Bothwell in his clutches, revealed nothing and entertained him lavishly while awaiting instructions from above.

By an amazing coincidence, who should be living in Bergen with her mother but Anna Throndsen, the beautiful Norwegian girl with whom he had dallied seven years earlier (see chapter 14). Anna confirmed Bothwell's identity and promptly sued him for breach of his promise to marry her. Seeing greater troubles ahead, he settled out of court. He promised her an annuity, to be paid in Scotland, and gave her the smaller of his two remaining ships. This was enough to halt the legal proceedings so that he could concentrate on his next hurdle.

Soon Rosencrantz received his orders from the king of Denmark and Norway, Frederick II. He was to arrest Bothwell, whom Frederick wished to exploit as a lever to recover the Norse-speaking Orkney and Shetland Islands from Scotland. The islands had belonged to Denmark until 1469, when they had been pledged as the dowry of Margaret, daughter of Christian I of Denmark, who was to marry Mary's great-grandfather James III. The Scottish Parliament had legally annexed them in 1472.

At first Bothwell denied possessing any jewels or valuables, or any letters or papers. Then, when he realized he would not be allowed back to his flagship, he admitted to hiding certain papers in the ballast. When his letter case was opened, it was found to contain Mary's parchment creating him Duke of Orkney and Lord of Shetland, as well as proclamations and other documents denouncing him as a murderer and traitor. There was said to be a letter from Mary, written after her return to Edinburgh in the hands of the lords, lamenting the treatment she had received. Since the document was never transcribed or filed away, it is impossible to tell if it had ever existed.

On September 23, Bothwell was examined by the magistrates of Bergen, after which he was put on board one of Frederick's ships and taken south to Denmark. He was sent to the castle of Copenhagen, to be held as a state prisoner. The Confederate Lords were pressing for his extradition, and Frederick was caught in the middle.

Bothwell insisted that he was on his way to France to seek help for Mary, to which end he appealed to Charles IX. This time the letter did survive. It is in the Bibliothèque Nationale in Paris, written in fluent French and in Bothwell's immaculate italic script. He asked for urgent aid for Mary and himself, but Moray had beaten him to it. He had already denounced Bothwell as a pirate, a murderer and a traitor. He had also reassured Charles of the entente that existed between the Confederate Lords and France, which he said could prosper only if Charles pressed Denmark to repatriate Bothwell for trial.

Charles decided to stay neutral, and so did Frederick, who treated Bothwell generously. He allowed him to live in comfort, to wear velvet clothes, to read books, and occasionally to go hunting or shooting under guard, although he was otherwise confined to his rooms.

It was then that Bothwell decided to write his memoirs. He believed that Frederick was unduly influenced by the Confederate Lords' propaganda. He was out to justify himself, shaping his story to create the most favorable impression. The main action began with his recall from France ten days before Mary's marriage to Darnley. Bothwell saw himself as the chivalrous knight of medieval romance, hurrying home to rescue his lady from her wicked barons:

> But the seditious lords did all they could to oppose her, because they wished above everything else that the queen should have no children, and because they could not bear that anyone should exercise authority in the realm beside themselves. They could clearly foresee that their own influence would be severely reduced by such a marriage.

On Bothwell's return, Mary gave him an opportunity to prove himself:

> The queen put me in command of an army composed of her loyal subjects and my own particular friends, with whom I did my utmost to drive the Earl of Moray out of Scotland into England. After I had achieved this, Parliament was summoned to enquire and determine as to what goods and estates were to be forfeited to the crown.

Bothwell had fully appreciated that a principal motive of the Rizzio plotters had been to preempt the forfeitures in Parliament of the rebels of the Chase-about Raid:

In order to avert these sentences of forfeiture, those of the Earl of Moray's
allies who were still at the queen's court stirred up fresh troubles by organizing the murder of Signor David, an Italian, which was done at supper-
time in the queen's cabinet at Edinburgh Castle [Bothwell's mistake: it was
Holyroodhouse], when none of her guards was present or even her usual
attendants. And if (to avoid danger) some others including myself had
not escaped through a back window, we should have been no better dealt
with . . . at the very least, we should have been forced to connive at so villainous a deed.

Bothwell did not hesitate to emphasize his loyalty and service to Mary,
but in doing so made a Freudian slip:

When we [Bothwell and Huntly] had escaped from the queen's lodging
and gotten to safety, we mustered some of our best friends and Her Majesty's loyal subjects, in order to rescue her and the king her husband from
the captivity in which they were detained. We did this partly by guile and
partly by force.

Next day, Their Majesties set out together for Edinburgh with a goodly
force, pursuing Moray and his allies so actively that they were forced to flee
into exile. The queen, moreover, was so indignant at such an assassination,
she held them in great hatred, as did also the loyal nobility and the rest of
her subjects. But the king she hated the most of all.

Bothwell had forgotten that it was not Moray but *Morton* whom Mary
had harried into exile after the Rizzio plot. Moray had returned to
Holyrood in triumph on the day after Rizzio's murder and did not leave
Scotland again until after Darnley's assassination. Bothwell's account of
the noble factions is strangely simplistic. Maitland and Argyll are left
out, most likely because they were swinging back to Mary's side after her
forced abdication. And while Morton's thirst for revenge after Darnley's
treacherous betrayal of his co-conspirators in the Rizzio plot is touched
on, it is subordinated to Moray's grander project to usurp the throne.
This was to become the key theme of Bothwell's story: that Moray, the
archvillain, had schemed from the beginning to depose his sister. In his
eagerness to embellish his account, Bothwell muddled up Morton with
Moray, forgetting that Moray had been welcomed home with open arms
by his sister after the Rizzio plot.

In his overriding bid to demonize Moray, Bothwell decided not to
mention Morton, his then ally and leading accomplice, even in his description of Darnley's murder. Instead, he gave a fanciful account of his
own allegedly altruistic efforts to buttress the monarchy in Scotland by

nurturing a spirit of reconciliation among the feuding lords. He even tried to pretend he had considered leaving the hurly-burly of politics:

> After I had negotiated for these exiled lords the favor they sought, and in particular their permission to return to court,* I thought about retiring to lead a quiet life after the imprisonments and exile I had suffered. I was tempted to remove myself from this scene of blood feud and revenge.

Only after the failure of Bothwell's efforts to bring about a lasting reconciliation between the lords did the murder plot begin. His description of the explosion and its aftermath is trite and unconvincing:

> A little while afterward, the king, who was suffering an attack of syphilis,† lodged at a place called Kirk o'Field while he convalesced to avoid danger to the health of the queen and her child. This was done by the common consent of the queen and of the members of her council, who wished to ensure the health of all three.
>
> The traitors saw this as their opportunity. They placed a large quantity of gunpowder under the king's bed, then lit the fuse, blowing him up and killing him. This was done at a house of Sir James Balfour . . .
>
> On the night of the explosion, several of her councilors were lodged as usual at her palace of Holyroodhouse. I was also lodged there, in the quarter where the guard of fifty men is usually stationed. And while I was still in bed with my first wife, the Earl of Huntly's sister, her brother came in the morning to inform me of the king's death, at which I was sorely grieved and many others with me . . .

Bothwell claimed that he had striven tirelessly to uncover the truth, never imagining for a moment that he should himself be suspected. His name, he sought to argue, had been mentioned only when his enemies and rivals had conspired with each other to prevent him from loyally serving Mary by solving the abominable crime.

All this is simply incredible, as is Bothwell's account of his trial:

> When the charges against me had been read, and my accusers (especially their principal, the Earl of Lennox, who had been summoned but did not attend),‡ were convinced that there was no just cause of complaint against

* An oblique reference to Morton, whose curfew Mary lifted at Bothwell's request in March 1567.

† The secretary first wrote "petite vérole," which means smallpox. However, Bothwell, reading through the text, crossed out "vérole" and substituted "roniole." The word "rognole" is still French slang for syphilis.

‡ Not surprisingly, since Bothwell had 4000 retainers and Lennox was allowed only six.

me touching either my person, property or honor, I was, according to the laws and customs of the realm, by the direction of my judges and with the consent of my accusers then present, declared innocent and found not guilty of all that of which I had been accused . . .

Bothwell then claimed — as had Mary herself — that he had married the queen at the behest of the very same lords who were now demanding his extradition. He invoked the Ainslie's Tavern Bond to vindicate himself, arguing that the marriage was forced on him by the nobility. But his account is far more disingenuous than Mary's. Hers, at least in the more candid version given to the Bishop of Dunblane, was tinged by regret at Bothwell's deception, whereas he brazened it out with a barefaced lie. He invented a tale in which the signatories of his bond came willingly and eagerly to his house to urge their petition on him:

> After I had won my case as I have stated, twenty-eight members of the Parliament came to me at my own house of their own free will and without being asked . . .
>
> Each of them thanked me particularly for the friendly manner in which I had behaved toward them, adding that the queen was now a widow, that she had only one child, a young prince; that they would not consent that she should marry a foreigner; and that I appeared to them to be the man most suitable to be her husband.

The reality could not have been more different. Bothwell had been forced to entertain his reluctant "supporters" and had struggled the next day to extort their signatures. Argyll, Maitland and Atholl each refused to sign. Nor could Moray have consented, as Bothwell falsely insinuated, since he had already left the country, on his way to voluntary exile.

Bothwell even managed to describe his marriage to Mary without saying anything about his abduction or seduction of her. And his version of the final showdown with the Confederate Lords is unashamedly self-justifying:

> The queen and I marched out of Dunbar with as many men, her loyal subjects, as we could muster in so short a time. We came to within a short distance of Edinburgh, whereupon the rebel forces sallied out and positioned themselves opposite to us at a distance of roughly a cannon shot.
>
> After a short delay, they sent a mediator to us, who presented us with a written statement of the causes that had brought them there. These were firstly to liberate the queen from the thralldom in which I was said to hold her, and secondly to revenge the murder of the late king of which I and mine were accused, as I have already described.

I replied to the first point that I was not in any way holding the queen in thralldom, but on the contrary that I loved and honored her as she deserved, appealing to her to confirm the truth of what I said.

To the second point, I answered that I continued to deny having participated in, or consented to, the late king's murder. I added that although I had already been plainly and sufficiently acquitted, I was still ready, if anyone of sufficient honor and noble birth was still inclined to accuse me of such a deed, immediately to defend my honor and my life in single combat in the presence of both armies, as promised in the challenge which I had formerly caused to be published in Edinburgh, and according to the ancient laws of war.

It was subsequently answered that Lord Lindsay, one of the lords, was prepared to meet me on the field . . . I so persuaded the queen and all of them by the many reasons I advanced, that they eventually agreed to allow the single combat to begin.

Bothwell could never bear to admit that despite Lindsay's acceptance of his challenge, Mary in a split-second decision had changed her mind and forbidden the man-to-man combat. Instead, he claimed he had waited patiently until late evening for his cowardly adversary to arrive. "He did not, however, turn up."

Finally, Bothwell stated that while he was still ready to fight to the death for his queen and his honor, he had no choice but to accept her desire to avoid unnecessary bloodshed:

As night approached, I prepared to give battle to the enemy, putting my battle line in order, while they did the same on the other side.

The queen, seeing me and her loyal subjects on the one side, and the rebel lords on the other, ready to begin a battle . . . , was anxious to avoid bloodshed at all costs and so surrendered to them. She crossed the field and went to them, escorted by Kirkcaldy of Grange in order to discuss things and see if matters could be resolved peaceably. And, believing that she might go over to them in safety, without fear of treachery, and that no one would presume to lay hands on her, she asked me not to advance further with my troops.

His efforts at self-exoneration complete, Bothwell ended his dictation on or shortly before January 5, 1568. After correcting his secretary's work, he had it copied out in duplicate. One of these copies was sent to Frederick II, who cannot have found it convincing: his reaction was to ship Bothwell across the Sound to the greater security of Malmö Castle, then in Denmark but now belonging to Sweden.

He was lodged in the north wing of the castle, where he occupied a large, oblong vaulted chamber on the ground floor beneath the royal apartments. The main room had two large south-facing windows opposite the inner courtyard. The bedroom faced the Sound, but there was no window for this most experienced of escapees to climb through.

As long as Bothwell was a piece on Frederick's diplomatic chessboard, his requests for luxuries would be met. There was little that was inhumane about his treatment, except he was not free. Of course, Bothwell was not a man who easily bore such a restriction. However often he was allowed to go out shooting, he felt he was locked up in a cage.

He still kept his wits about him. When the Confederate Lords commissioned Captain John Clark, a Scottish mercenary serving in the wars of the Danish king, to negotiate his extradition, Bothwell turned the tables. He told Frederick that when Clark had returned to Scotland in 1567 to enlist more troops, he had misappropriated the king's money, using it to pay reinforcements to line up on the side of the lords at Carberry Hill. The accusation was true, and Clark would later be convicted by a court-martial in Denmark.

The Confederate Lords then sent Thomas Buchanan, the nephew of the brilliant classical scholar and poet who had written Mary's masques, to Copenhagen. He attempted to reinvigorate the lords' demand for Bothwell's extradition by disclosing how Mary had sent Bothwell "certain writings" from England, including the evidence that had helped him to seal Clark's fate. The report is vague and uncorroborated, as Buchanan steadfastly refused to name his sources. Although said to be "men of great estimation" and "worthy of trust," their identities are impossible to pin down. Moreover, since all they ever appear to have said is that Mary had urged Bothwell "to be of good comfort, with sundry other purposes," it is distinctly possible that his "sources" were fabricated.

Bothwell was said to have sent spies to England, who were to try and speak to Mary in her captivity. Their instructions were to procure the documents Bothwell needed in his efforts to trade his freedom for the Orkney and Shetland Islands. If this was true, it described the actions of a desperate man. The Confederate Lords would never have allowed the islands to be handed back to Denmark, even in exchange for their most wanted enemy.

One of these alleged spies was Bothwell's Danish-born page, a man reputed to be easily mistaken for a Scot, as he could speak Lowland Scots like a native. He was said to have carried documents to Mary, whom the reporter described as "that woman." But once again, the details are nebulous and confused.

Intriguingly, the younger Buchanan knew without asking that it was his job to send a copy of his final report to Cecil. There was no solid evidence to justify his claims. And yet the mere rumor of a secret correspondence between Mary and Bothwell was enough to create alarm. In the eyes of the Confederate Lords, any form of communication between them would be "prejudicial and hurtful to both our countries and to the discontentment of the Queen's Majesty of England."

Clark was imprisoned at Dragsholm Castle, a solitary fortress on the edge of Kalundborg Fjord, on the northwest coast of Zealand some sixty miles west of Copenhagen. In June 1573, by which time five years had elapsed since Mary's flight and imprisonment in England, Bothwell was himself sent there. Legend says that he died there insane after five years of solitary confinement, chained to a wall. But Bothwell was always larger than life. He was an opportunist, a gambler who knew that fortune favored the brave and who profited from life as he found it. He and Clark, both Scots and professional military men, buried the hatchet and drowned their sorrows in wine. It is unlikely that Clark ever admitted that his more nefarious activities had included acting as one of Cecil's spies, writing him reports and accepting English pay. If he had, Bothwell — depending on his mood — would have roared with laughter or killed Clark on the spot.

By July 1575, the unbridled revelry at Dragsholm had taken its toll. Clark died of excessive drinking, and even Bothwell's ox-like constitution had started to collapse. As their names were always linked in the bulletins emanating from Denmark, it was reported that Bothwell too was dead. It took another four months for proof to arrive in Paris and London that he "is but great swollen and not yet dead." He was the victim of liver or kidney failure, but not quite ready to die.

The merest whisper of Bothwell's death was enough to animate Mary's supporters on the Continent. They began to claim that he had made a deathbed confession in which he had exonerated her of all responsibility for Darnley's murder and taken the blame himself.

When Mary heard this, she clutched desperately at the chance to prove her innocence. She wrote to her ambassador in Paris, "I have been told of the death of the Earl of Bothwell, and that before he died he made a full confession of his sins, and among the rest, that he acknowl-edged himself guilty of the murder of the late king my husband . . . I pray you, therefore, investigate the truth of it by all available means."

She managed to scrape together 500 French crowns to pay a courier to travel to Denmark and secure a transcript. The man pocketed the money and did nothing. A full-length copy of the "confession" was, meanwhile, sent by Frederick II to Elizabeth, who deposited it in the royal library from which it later disappeared. Mary was furious that Elizabeth refused to share with her or publish a document that she be-lieved could restore her honor and reputation. She was still complaining about this two years later. Catherine de Medici instructed her ambassa-dor to Denmark to obtain a copy, but if it reached her, she too kept its contents to herself.

Only short abstracts now survive. According to what seems to be the superior fragment, Bothwell swore that Mary "never knew nor con-sented to the death of the king." The murder was the work of Bothwell and his friends; it was engineered "by his appointment, divers lords con-senting and subscribing thereunto, which yet were not there present at the deed doing." Their names were Moray, Morton, Lord Robert Stuart and John Hamilton, Archbishop of St. Andrews, "with divers others, whom he said he could not remember at that present."

Bothwell confessed that "all the friendship which he had of the queen, he got always by witchcraft and the inventions belonging thereunto." He had drugged her *eau sucrée* to seduce her, and then "found means to put away his own wife to obtain the queen." A slightly fuller variant of this passage says that he had "bewitched the queen to fall in love with him, and so invented means to get rid of his own wife." Then, "after the mar-riage was consummated, he sought all means how to destroy the infant prince and the whole nobility that would not fall in with him."

Bothwell's "confession" is a blatant forgery. The version that appears to have once been in Cecil's archive and is now in the British Library used these words: he "forgave all the world and was sorrowful for his of-fenses, and did receive the sacrament that all the things he spoke were true, and so he died."

Leaving aside the fact that Bothwell was a Protestant who did not be-lieve in receiving the last rites and had refused to attend Catholic ser-vices even when Mary took him by the hand, he did *not* die. He was alive for almost three more years. One of the copies of the confession says that Bothwell lay "sick unto death in the castle of Malmö." That is also incorrect — he had left Malmö for Dragsholm two years before. Whoever faked these documents did not even check the basic historical facts.

If that were not proof enough, the roster of Darnley's murderers does not add up. Morton is named correctly, while Moray had foreknowledge of the assassination and "looked through his fingers." But Sir James Balfour and the Douglases are not mentioned, whereas Lord Robert Stuart, another of James V's illegitimate children and Mary's half-brother, and John Hamilton, Archbishop of St. Andrews, went down in history only because they attended Mary and Bothwell's wedding and because the archbishop granted Bothwell a decree annulling his marriage to Huntly's sister in the Catholic court. Neither was involved in Darnley's murder, but as both were known to be Bothwell's supporters, their names were put on the list.

One of the great ironies is that Margaret Douglas, the Countess of Lennox and Darnley's mother, was so impressed by Bothwell's confession, she decided that Mary was the victim of a foul injustice. In the same month as the abstracts of the confession were making the rounds, the countess wrote a letter of reconciliation to Mary from her house in Hackney, adding a postscript in her own hand. "I can," she said, "but wish and pray God for Your Majesty's long and happy estate, till time I may do Your Majesty better service . . . I beseech Your Highness pardon these rude lines and accept the good heart of the writer, who loves and honors Your Majesty unfeignedly."

The countess sent Mary a token of her affection. It was a small but delicate and costly piece of *point tresse* embroidery, which Mary would regard as one of her prized possessions. To achieve the intricate, glittering effect of *point tresse*, the pattern had to be painstakingly worked with soft silvery hair that was mixed with the finest silk or flax thread. The countess had used strands of her own hair, so it was a very special gift. Nothing could have been a more powerful signal of the reconciliation between Darnley's mother and her daughter-in-law.

When the countess died, Mary said, "As soon as she came to learn of my innocence, she refrained from any further criticism of me, even going so far as to repudiate anything that was avowed against me in her name." Since Bothwell's confession is so obviously a forgery, the true significance of the countess's reconciliation with Mary is to show how fragile and poorly documented the original proof against her must have been.

The confession was to have a similar effect on Mary's impressionable young son. James never saw his mother again after she kissed him good-bye the day before she was abducted by Bothwell. He remained with the Earl and Countess of Mar, who created a household for him at Stirling Castle and supervised his upbringing. He played regularly with Mar's own children, who taught him to ride and hunt. These would be his favorite sports, but he also loved archery and was presented with innumerable sets of bows and arrows.

James was given the best possible education. He was taught to speak Latin before he could speak Scots, becoming fluent in Greek, Latin and French. He was also brought up as a Protestant. One of his tutors was George Buchanan, who by now ranked among the chief of Mary's vilifiers and whose scholastic regime included regular beatings as well as republican tirades against divine-right monarchy.

James was repeatedly flogged to encourage him to study, which worked to the extent that he could quote classical authors by heart for the rest of his life. But what he must have remembered most vividly were the strictures he received from Buchanan against his mother, who he was assured had been a tyrant, an adulteress and the woman who had murdered his father.

James was ten and a half years old when he suddenly noticed the Laird of Tullibardine avidly reading something and pointing out passages to his friends. James insisted on seeing for himself what was causing so much interest. It was a copy of Bothwell's "confession." At first the boy said nothing, but his mood changed to one of "bon visage." Finally, he said, "Have I not reason to be glad after all the terrible accusations and calumnies against the queen my mother?" Then he added triumphantly to the laird: "I've now seen such a clear proof of her innocence!"

On April 14, 1578, Bothwell died at Dragsholm. As was customary for state prisoners, his body was carried to the promontory that juts into the fjord a mile or so from the castle and buried at the parish church of Faarevejle. It was wrapped in a linen shroud and placed in a wide oak coffin. His head was wrapped in a white linen cloth lined with green silk and laid on a white satin pillow.

We know this because the coffin was opened in 1858, when a mummified body was discovered in near-perfect condition. Aged forty-three at the time of his death, but said to look much older after a decade in prison, Bothwell had stood just over five and a half feet tall. His dark red hair could still be seen, liberally flecked with gray. An English antiquarian who came to view the body pronounced it as belonging to "an ugly

Scotsman." Others reserved judgment, not just over the deceased's appearance, but also over his identity.

Whether Bothwell's or not, the head had posthumous adventures of the sort he would have appreciated. It was first displayed as a trophy on a writing desk at Dragsholm, then used by children as a football. By 1935, it had been reunited with the corpse and reburied in the crypt of the church. After the Second World War, the remains were once more exhumed, this time as a tourist attraction. Until 1975 or thereabouts, the body — by then no more than a bare skeleton — was exhibited in a glass case at Faarevejle until protests were received. Dragsholm Castle is now a hotel, where Bothwell's ghost is said to walk at night. Such was the ironic end of a man who altered the course of history and aspired to be buried among kings.

24

The Lords' Story

WHEN BOTHWELL had been acquitted of Darnley's murder in a rigged trial and he set about canvassing support for his marriage to Mary, a dramatic turnabout had taken place in the manner of the cover-up of the events at Kirk o'Field. Up to then, the lords had closed ranks to suppress any evidence that might help to solve the murder mystery, in particular that of the women living in the nearby cottages. But when Bothwell had abducted and seduced Mary, breaking his pact with Morton, the lords turned their full attention to accusing him, and uncovering as much evidence as possible to prove his involvement, before he became too powerful.

Then, when Mary had been imprisoned at Lochleven and forced to abdicate, she too had to be implicated in the murder. The lords had deposed an anointed queen. It had to be justified, which was done by claiming that she was no longer fit to rule. She was not a proper queen. She had disqualified herself because she was guilty of "moral turpitude." Knox's stereotype that a Catholic woman ruler was by definition motivated by unbridled sexual lust provided the template. Mary's "furious love" for Bothwell had "proved" that she was unable to control her passions. Carnal lust had led her first to commit adultery with a married man and then to conspire with him to murder her husband so that she could be free to marry her lover.

Everything seemed to connect in a steamy story of sex and violence, adultery and murder, designed to titillate as well as to shock. The author chosen by the Confederate Lords to tell their tale was none other than George Buchanan. His credentials were perfect for the role. He was a Calvinist and a republican, the friend and "master" of Thomas Randolph, Elizabeth's first ambassador to Scotland, in his student days in Paris. Later at Holyrood, he had read Livy with Mary in the afternoons. When commissioned to stage the entertainments on the theme of reconciliation for the baptism of Prince James at Stirling Castle, he had reached the pinnacle of his courtly career. Now Buchanan wrote not a masque but an anti-masque. Fact was mingled with fiction to create an artful piece of character assassination.

Mary had rewarded Buchanan generously. How could he betray her so shamelessly? The explanation is that he was first and foremost a Lennox client. His loyalty to Darnley took priority over his allegiance to Mary. He was born in Gaelic-speaking Lennox territory near Glasgow, the son of an impoverished lairdly family. He owed his education to local philanthropy, for which he felt forever indebted.

Buchanan had also kept up his links to his old Huguenot friends in Paris. He was close to Moray, to whom he had dedicated a book on educational reform. Moray in return gave him the lucrative post of principal of St. Leonard's College, the richest, though never the most academically dynamic, of the three colleges of the University of St. Andrews. When Darnley was murdered, Buchanan was thunderstruck. His king was slain; his earliest fealties called him to action. He threw in his lot with the Confederate Lords, especially Moray, playing a key role as moderator of the General Assembly of the Kirk after the battle of Carberry Hill, when he joined Knox and Craig in rallying the Protestants to justify Mary's imprisonment. Then, when she was deposed and Moray became regent, Buchanan found that he had attached himself to the most powerful junta in Scottish politics.

His close alignment to Moray enabled Buchanan to come into his own. For some time he had been striving to surpass Knox's ideas and replace them with a more sophisticated theory that fused the classical concepts of ancient Greece and Rome with Protestantism. His ideal was republican Rome, with its tradition of *civitas libera,* the free state. He was repelled by the tyranny represented by the idea of empire and abhorred ideals of divine-right monarchy, dismissing the kings, popes and emperors of the Middle Ages as charlatans and tyrants. He showed a particular distaste for the colonial idea of empire in the New World, which he had first encountered while teaching languages at the University of Coimbra in Portugal.

His central premise was that rulers were chosen by the people to perform a set of defined functions. If they failed to carry out their obligations, they broke the terms of the contract laid down in their coronation oath. If this happened, the people had the right to depose them and appoint someone better qualified to fulfill the duties of the royal office. Buchanan skillfully reworked the theory of monarchy in a quasi-republican idiom to argue that rulers were accountable to those who elected them. According to his model, the ruler, far from being above the law by royal prerogative, was subject to it at all times: to flout the law was not merely to oppose the will or welfare of the people, but to declare oneself a tyrant and an enemy of God.

If Darnley had been crowned king and not been assassinated, Buchanan would have kept his radical ideas to himself. With a scion of the Lennoxes on the throne, his view of monarchy would have been different. But when his liege lord was savagely murdered, he spoke out. He heartily approved of Mary's forced abdication, which he regarded as one of the best practical illustrations of his theory of royal accountability in eight hundred years of Scottish history.

When Mary had lost the battle of Langside and fled to England, causing Cecil to ask the Confederate Lords to send him the "manner of the proofs" and other "evidence" against her, it was Buchanan whom Moray chose to compile the lords' story. His report would take the form of a dossier: a compendium of the allegations against Mary citing all the relevant facts. Moray would then forward it to Cecil, who would use it to appease Elizabeth, who loathed rebels and was angry and incredulous at Mary's deposition, for which she judged there had never been sufficient grounds.

The dossier was to be the case for the prosecution. By collating all the evidence, Cecil meant to help Moray to destroy Mary. He knew it would be an uphill struggle. Elizabeth was not alone in disputing Cecil's conspiratorial view of Mary or refusing to presume her guilt. The more traditional of the English landed nobles, led by the Duke of Norfolk, the Earl of Arundel and Lord Lumley, agreed with her. Although willing to conform outwardly to the Protestant religious settlement, they had deep-rooted Catholic sympathies. Norfolk, the grandson of Henry VIII's commander who had routed the Scots at the battle of Solway Moss, was barely a Protestant: the merest scratch would reveal a Catholic under-neath. His allies, Arundel and Lumley, were undeniably Catholics. They vociferously supported Elizabeth's refusal to put religion ahead of he-reditary right when considering the succession to the throne. They were convinced that an attack on Mary's authority in Scotland would be the prelude to a similar attack on Elizabeth's power in England.

Norfolk and his allies viewed the Confederate Lords as the natural en-emies of monarchy and royal government. They were deeply suspicious of Cecil and his inner circle. The future of divine-right monarchy in the British Isles hung in the balance if Mary's forced abdication was upheld. Mary was not even English, so how could she be accountable to the queen of England or to English judges?

Cecil evaded such awkward questions by arguing that Scotland was a satellite state of England, and therefore Elizabeth had jurisdiction because she was Mary's feudal superior. It was a tenuous claim. But it was at least a way of justifying a quasi-judicial investigation into her "crimes," which would certainly be necessary if Elizabeth was to be satis-fied that the lords were not mere rebels but men of honor genuinely seeking to right a terrible wrong.

For his part, Moray was crafty and circumspect, nudging Cecil into an irrevocable commitment to attack Mary. The Confederate Lords, he said at first, had all the evidence they needed to implicate Mary in Darnley's murder. As soon as Cecil's attention had been grabbed, Moray back-tracked, asking if the judges to be deputed to hear the case in England could first be invited to examine the evidence *in camera* and give an in-formal prejudgment. Then the Scottish lords would know for certain if their case would stick, and if not, exactly what further material would need to be collected to make it do so.

On June 22, 1568, a month after Mary's flight across the Solway into England, Moray sent Cecil a copy of Buchanan's dossier, which was writ-ten in Latin. No version of this Latin text can be found, but the dossier was shortly afterward translated into Lowland Scots for the Earl of Lennox, whose manuscript has been carefully preserved. It is entitled "An information . . . whereby it evidently appears that Mary, now Dowa-ger Queen of Scots, not only was privy of the horrible and unworthy murder perpetrated on the person of the King of good memory, but also was the very instrument, chief organ and cause of that unnatural cruelty."

As a blend of fact and fiction, Buchanan's story is a masterpiece. He

claimed that Mary's crimes were so far premeditated, she had first imagined them after the Rizzio plot, six months earlier than the Confederate
Lords themselves had previously alleged. Her indiscretions had supposedly begun during her holiday at Alloa, just a few weeks after her son
was born. Then, when she had returned to Edinburgh, instead of abandoning her liaison, she had intensified it. She had continued to gratify
her infatuation for Bothwell while openly shunning her husband.

Buchanan had fixed on the earliest possible date for the beginning of
Mary's alleged affair that avoided any imputation that Prince James was
not Darnley's legitimate heir. He claimed that Mary had flaunted herself
while on holiday, leaping straight into Bothwell's arms:

> What her usage was in Alloa needs not to be rehearsed, but it may be well
> so said that it exceeded measure and all womanly behavior ... But even as
> she returned to Edinburgh [on September 6, 1566, after also visiting the
> Water of Megget and Glenartney] ... what her behavior was, it needs not
> to be kept secret being in the mouths of so many: the Earl of Bothwell
> abused her body at his pleasure, having passage in at the back door ... This
> she has more than once confessed herself ... using only the threadbare ex-
> cuse that the Lady Reres gave him access ...

The innuendo was deliberate. When Buchanan referred to Bothwell's
"having passage in at the back door," he was echoing the main charge of
Ane Ballat (a ballad), issued by the Confederate Lords, accusing the illicit lovers of the "beastly buggery Sodom has not seen." And yet he had
failed to check his facts. A closer investigation of those who had accompanied Mary to Alloa would have revealed that Moray had been there
throughout. He had witnessed everything that happened there, but at
the time had said nothing whatever about Mary's supposed misbehavior
(see chapter 17).

Buchanan was never a man to allow the facts to get in the way of a
good story. Mary, he brazenly continued, had relied on a pimp. She had
used the services of Lady Reres, a niece of Cardinal Beaton. Reres was
middle-aged, "corpulent" and overweight: the stereotype of a common
bawd, who had led Bothwell secretly to Mary's boudoir for their sexual
encounters, one day meeting with a comic accident when she had been
forced to climb a high garden wall, but had fallen off, tumbling into a
lubberly heap on the ground.

It was an incredible yarn, because far from helping Bothwell, Reres
was one of his discarded mistresses. Drury, Bedford's deputy at Berwick-

upon-Tweed, had warned Cecil in one of his handwritten reports that Reres was so jealous when Mary had taken Bothwell as her protector that she was banished from court, her place taken by Bothwell's sister.

Buchanan, meanwhile, was just hitting his stride. Poor Darnley, according to the dossier, had been denied his conjugal rights. All his threats, sulks, plots and paranoia were to be forgotten altogether. Instead, Mary's "shamed" and "cuckolded" husband — who was, of course, the hero, victim and martyr of the Confederate Lords' story — was depicted in this sanitized version of history as a saintly and statesmanlike figure who matched the image propagated by the Lennoxes on their placards and notices nailed to the doors of the churches and public buildings of Edinburgh.

One of the important incidents in Buchanan's narrative was the episode at Jedburgh in early October 1566, where Mary had ridden to preside at her Justice Ayre, or traveling court, and where Bothwell, who had preceded her by a couple of days, had been seriously wounded in a sword fight with his old enemies the Elliots of Liddesdale, a border bandit clan.

According to the dossier, when Mary heard of Bothwell's clash with the Elliots, she galloped furiously to Jedburgh and next day to the Hermitage. She arrived just in time to meet her supposed lover for a tryst before returning to Jedburgh the same day. Her desperate ride was "in the company of such a convoy as no private man of honest reputation would have entered among."

Buchanan simply refused to report the facts accurately. Mary had actually waited a week, until the formalities of the Justice Ayre were completed, before riding to visit Bothwell, with Moray at her side for every step of the journey (see chapter 17). She saw Bothwell in the presence of her Privy Council and stayed for only two hours.

Yet it was Buchanan's version of events that was to come down in history. In the eighteenth century, his story was so popular it became the springboard for one of the most notorious Marian forgeries: the supposedly new or Crawford chronicle, which was concocted by a Presbyterian minister, who inserted his own voyeuristic material into an older source to provide graphic sexual detail.

Back at Jedburgh, as Buchanan falsely alleged, Mary had moved Bothwell "from his accustomed lodging" and placed him "in the queen's house in the chamber directly under her own." There, in spite of his severe injuries and the fact that she had herself been close to death when her gastric ulcer burst, they carried on their sordid affair as if nothing

untoward had happened, until the news reached a shocked and distraught Darnley. Filled with righteous indignation and knowing himself
to be the wronged party, Mary's lawful husband "delayed not but with all
speed came to Jedburgh."

But there is no independent evidence that Bothwell had ever stayed in
Mary's house. And Darnley came to Jedburgh later and reluctantly,
quarreling with his wife as soon as he arrived. Buchanan was out to create the strongest presumption of a motive for murder. He attempted to
show how Mary had plotted to rid herself of Darnley:

> About the fifth day of November, removing from Jedburgh to Kelso . . . she
> said that unless she were quit of the king by any means or other, she could
> never have one good day in her life, and rather than that she should fail
> therein, she would rather be the instrument of her own death.
>
> Returning to Craigmillar beside Edinburgh, where she rested until the
> end of November, she renewed the same purpose which she spoke of before
> at Kelso in an audience with my Lord of Moray, the Earl of Huntly, the Earl
> of Argyll and Maitland, propounding that the way to be quit of the king
> and make it look best was to begin an action of divorce against him . . .
> whereunto it was answered how that could not goodly be done without
> hazard, since by the doing thereof the prince her son should be declared
> [a] bastard . . . , which answer, when she had thought over it, she left that
> conceit and opinion of the divorce and ever from that day forth imagined
> and devised how to cut him away as by the sequel of this discourse more
> plainly shall appear . . .

Buchanan was distorting the known facts to create an interpretation
of almost complete fantasy. The truth was that the lords had suggested a
divorce to Mary, not the other way around. She had said that she wished
herself dead at Craigmillar Castle, but not at Kelso. And she had never
asked to be quit of Darnley "by any means or other." When Maitland's
plan to offer her a divorce from Darnley had been put to her at Craigmillar, she had replied: "I will that ye do nothing whereto any spot may
be laid to my honor or conscience, and therefore I pray you rather let the
matter be in the state as it is" (see chapter 18).

None of this was of any concern to Buchanan. He was determined to
establish a motive for Darnley's murder, and according to his account,
Mary had slept with Bothwell after her son's baptism for eight successive
nights, "using that filthiness almost without cloak or respect of shame or
honesty." While in bed with her lover, she had plotted a "device" to kill
her husband, choosing a deserted house at Kirk o'Field as the place

where the murder would be committed. She had then ridden to Glasgow to fetch her doomed spouse, all the while conveniently laying down a paper trail for posterity in her letters to Bothwell, who had stayed behind in Edinburgh to bait the trap:

> Her mind, as well appears by her letters,* [was] to bring him to Edinburgh to his fatal end and final destruction, which she would never attempt not having her son in her own hands. She left [Prince James] at Holyroodhouse. She then left [for Glasgow] accompanied by the Hamiltons and such others as bore her husband no favor.

Buchanan was enjoying himself, discarding the true chronology of events. In reality, Mary had been escorted by Huntly and Bothwell as far as Callander House on the first stage of her journey to Glasgow to fetch her husband. It was only on the second stage that the Hamiltons — the family of Châtelherault, enemies of the Lennoxes — had been her bodyguards. And no sooner had Bothwell returned to Edinburgh than he departed again for Liddesdale, on the border, which means he could not have been in Edinburgh receiving Mary's final instructions to prepare the scene of the crime at Kirk o'Field at the time Buchanan claimed he was (see chapter 25 for more details of this chronology).

Perhaps aware of a possible discrepancy, Buchanan again alleged that Mary had "devised" with Bothwell to lure Darnley to his death *before* leaving Edinburgh for Glasgow:

> In the meantime, the Earl of Bothwell, according to the device appointed between them, prepared for the king the lodging where he ended his life. In what place it stood, enough [people] know and enough thought even that it was a ruin unsuitable to have lodged a prince into, standing in a solitary place at the uttermost part of the town, separate from all company . . .
>
> He was lodged at Kirk o'Field [for the good air], howbeit in Scotland at the beginning of February a sick man will be better content with a draft-proof and warm chamber as [with] any air in the fields.
>
> Lay she not in the house [in the room] under his on the Thursday and Friday before he was murdered to give the people the impression that she was beginning to entertain him?

Once Darnley had been ensnared at Kirk o'Field, Mary kept him sweet. Then, supposedly on a signal from her trusted servant Nicholas

---

* This refers in particular to the "long Glasgow letter" said to have been written by Mary to Bothwell. For a full discussion of this significant and contested document, see chapter 25.

Hubert, whom she nicknamed "French Paris," she returned abruptly to Holyrood on the pretext of attending the masque in honor of Bastian Pages and his bride:

> As soon as she saw [French Paris],* she knew that the powder was put in the lower [part of the] house under the king's bed, for Paris had the keys both of the front and back doors of that house, and the king's servants had all the remaining keys of the lodging. And so with feigned laughter she said, "I have given offense to Bastian by not attending the masque in honor of his marriage tonight, for which purpose I will return to Holyrood."

She then departed, allegedly spending the rest of the evening with Bothwell. (This sounds very unlikely, since Mary attended the wedding masque.) He later "passed to his chamber and there changed his hose and doublet and put his side cloak about him and passed up to the accomplishment of that most horrible murder."

Mary, insisted Buchanan, was unable to sleep because she was so animated at the prospect of Darnley's murder. She scarcely moved when she heard the fatal "crack": "for she needed not, understanding the purpose as she did."

Hearing the shock of the explosion, Bothwell quickly rose from his bed (apparently he could be in two places at once!), and, accompanied by Huntly, Argyll and others, went to her to explain "how the king's lodging was raised and blown in the air and himself dead, [at] which news her passions were not so great nor her face so heavy as one in her state ought to have been — not even if he had not been her husband but an ordinary man — for the unworthiness and strange example of the deed."

Mary supposedly received the news in silence. She then "took her rest with no sorrowful countenance for anything that had happened." Belatedly she ordered her household into mourning, but could not keep up the pretense for very long: "Of the forty days *deuil* she could not tarry at Holyroodhouse above ten or twelve days, and that with great difficulty being in most great hard case how to counterfeit *deuil* and nothing less in her mind."

In the relaxing surroundings of Seton, Mary disported with Bothwell as if the explosion had never happened. Using her customary guile, she placed her lover in a set of rooms adjacent to the kitchen so that he could creep along a servants' passage used for the delivery of food to her apartments to cavort with her.

* But in reality, Mary met Hubert only as she was leaving the house.

But Bothwell was already married, and rumors of his affair were seeping out. To avert a scandal, it was time for him to divorce his wife and marry the queen. Since a subterfuge would be required to justify such outrageous conduct, Mary had allegedly planned a collusive abduction in which she would pretend to be kidnapped and ravished. And as with her initial conspiracy to murder Darnley, she would herself create the evidence needed by her enemies. One of her own letters written "out of Linlithgow," according to Buchanan's dossier, would prove her intentions:

It could not be without scandal that the queen should go openly to bed with the Earl of Bothwell, who had a married wife of his own. Howbeit of before and then, they spared no time to fulfill their ungodly appetite, yet somewhat to cover her honesty she pretended to be ravished. This was brought to pass shortly after she returned from Stirling to Edinburgh, and whether it proceeded of herself or not, her letter written to the Earl of Bothwell out of Linlithgow can declare. (For a discussion of this alleged letter, see chapter 26.)

As soon as the lovers had reached Dunbar, Bothwell began the proceedings to divorce his wife. After his wedding to Mary, many of the lords went into internal exile. Mary was by then a changed woman: even those lords who stood by her were badly treated.

Being conveyed by him to Dunbar, without delay they caused a divorce [to] be moved in double form against his lawful wife . . . [When they returned to Edinburgh] she declared she was at liberty, and so within eight days passed to the consummation of that ungodly marriage that all the world counts naughty and a mocking of God.

The time was not long between the same pretended marriage which was made on 15 May 1567, and the 15th day of June thereafter, when after the said earl's fleeing, she came to the lords [who had] assembled for revenge of the murder, and yet in that month's space what confusion and corruption was there to behold — it was marvelous! All noblemen for the most part had withdrawn themselves, and such as stayed behind, how affectionately that ever they showed themselves to Her Majesty, yet were they in no better grace than the others that utterly abandoned the court . . .

Wrong facts apart, the Confederate Lords' story would be clear and consistent. It set out the charges plainly. There were three counts directed at Mary:

1. adultery with Bothwell both before and after Darnley's murder;
2. conspiracy to murder Darnley in January 1567;
3. collusive abduction to enable Mary and Bothwell to justify their marriage.

The sensational evidence in the form of eight incriminating letters from Mary to Bothwell, which Moray would produce as the proof of the allegations in Buchanan's dossier, are almost evenly split among these topics. Two letters would be about Darnley's murder, three would be love letters, and three would relate to the abduction. Of the love letters, two would also refer to the murder, and of the murder letters, one would also be a love letter.

Moray first told Cecil he had these letters on the same day as he sent his secretary to London with the dossier. "We have," he said, "such letters [of Mary's] . . . that sufficiently in our opinion proves [*sic*] her consenting to the murder of the king her lawful husband." He would make "copies" of them, translated into Scots, and it was these that he wished the English judges who were to consider Mary's guilt to examine *in camera* and comment on before the original documents in French were finally laid on the table.

The case against Mary had taken a dramatic new twist. How and when did these seemingly devastating documents come into the possession of Moray and the Confederate Lords? What did they actually say? And could they be genuine letters by Mary? To these, the most searching and momentous questions about the life of Mary Queen of Scots, we must now turn, for it is upon them that her honor and reputation, both then and now, depend.

25

Casket Letters I

THE MOST HOTLY debated question about Mary is whether she was involved in Darnley's assassination. Only if she was already an adulteress and Bothwell was her lover is the case against her convincing. The subject is bound up with the controversy surrounding the eight letters produced by Moray to justify the charges in Buchanan's dossier. The sole evidence that she was a party to the murder plot comes from them. There is no other proof. Her guilt or innocence depends on whether the letters are true or false.

The first hint that the Confederate Lords had unearthed incriminating material had come in the summer of 1567. A month after Mary was taken to Lochleven Castle, the Spanish ambassador in England, Guzman de Silva, reported to Philip II that Moray had visited him on his way back to Scotland from exile in France and told him that his sister had known all along about the murder. This had supposedly been proved by a letter she had written to Bothwell from Glasgow.

At this stage, it seems the lords had one letter, not eight. After Mary's imprisonment, Morton's men had looted the apartments she and Bothwell had occupied at Holyrood, which is when they had stolen the pearls later sold by Moray to Elizabeth. No doubt they would have also found some documents, but what these might have been is unknown.

De Silva said that du Croc, the French ambassador to Scotland, had visited London on his way home to France. He was reported as saying that the Confederate Lords "positively assert" she had been an accomplice in Darnley's murder, which was "proved by letters under her own hand." The same claim had reached Throckmorton, then Elizabeth's crisis manager in Scotland. He put in one of his reports that the lords "mean to charge her with the murder of her husband, whereof (they say) they have as apparent proof against her as may be, as well by the testimony of her own handwriting, which they have recovered, as also by sufficient witnesses."

Five months later, when the Confederate Lords summoned Parliament to give legal sanction to their revolt, they passed an act solemnly declaring that "the cause and occasion of their taking the said queen's person upon the said 15th day of June" was "by divers [of] her privy letters" said to be wholly in her own handwriting.

These lords confidently assured Parliament that Mary's letters had been found *before* they had forced her to surrender on the field at Carberry Hill. They justified their taking up arms against their queen by claiming they had already obtained the damning evidence that she was a party to Darnley's murder.

The lords' declaration did not go unchallenged. Mary's supporters vigorously protested: "And if it be alleged that Her Majesty's writing should prove her culpable, it may be answered that there is in no place mention made in it by which she may be convicted, albeit it were in her own handwriting which it is not."

Far more harmful to their cause, the Confederate Lords contradicted their own statement to Parliament in their sworn testimony to Elizabeth and Cecil. When called upon to explain how such incriminating documents had been found, Morton deposed under oath that it was not until June 19, four days *after* Mary had surrendered, that an informer had told him how three of Bothwell's servants had come to Edinburgh and gone inside the castle. One of them was captured and jailed in the Tolbooth.

Next day, this servant confessed under torture that he had taken a silver box or casket from Bothwell's rooms at the castle the day before. The box was then fetched and brought to Morton. It was locked but missing its key, so Morton kept it all night without opening it. Then, on June 21, in the presence of ten witnesses, the lock was broken open and the contents examined. Twenty-two documents were found: the eight letters, two marriage contracts said to prove that Mary had consented to marry Bothwell before he was divorced, and twelve allegedly adulterous sonnets said to have been written by Mary to Bothwell.

One of the biggest challenges in discussing these momentous documents is that the originals, which were all in French, have disappeared. They are now known only from word-for-word transcripts (in French) or Scots or English translations made at the time the originals were examined in England, or else from later printed copies.

The handwritten transcripts and translations provide the most reliable information about them, as the variants later published to vilify Mary were of a glaringly propagandist intent. Even the standard scholarly editions of the Casket Letters contain a worrying number of mistranscriptions. And astonishingly, not all the manuscripts have been edited. Two were missed completely, perhaps because they had not yet been catalogued at the time these various editions were made. They will be discussed here for the first time.

The fate of the original Casket Letters is a mystery. After they were officially scrutinized in England, they were returned to Moray, who took them back to Scotland. In 1571, Morton obtained them, and in 1581 they descended to his heir, only to vanish from sight a few years later. There is no proof that they were deliberately destroyed, but that is the most likely reason for their loss. By then James VI was approaching eighteen and wished to protect his family's reputation.

Nevertheless, their impact when they were first submitted to Elizabeth and Cecil can, if anything, be judged more accurately from the transcripts and translations than from any other source. This is because several of these versions have Cecil's notes or qualifying comments on them. A careful reexamination of these intriguing documents enables us to glimpse his thoughts as he eagerly pored over them. He knew every one of them inside out, instantly recognizing that they were dynamite.

If the Casket Letters were genuine, an anointed queen could justifiably be deposed from her throne, Elizabeth's "safety" would be guaranteed and the threat of an international Guise conspiracy ended forever. Mary would be utterly discredited. It might even be possible to try her for murder and execute her.

If they were forgeries, there was no other proof of her culpability. Elizabeth would be likely to release her. She would refuse to recognize

Moray as regent in Scotland. She might even restore Mary to the throne. The lords would then be forced to flee into exile, destroying all of Cecil's hard work and precipitating a fresh crisis, since the lords would choose England as their refuge, and Mary would angrily demand their extradition.

Why were there so many documents? The reason is that some key elements of the lords' story as it had evolved under Buchanan's surveillance were not covered by the letters. They shed no light, for instance, on Mary's alleged adultery in the six months between her visit to Alloa and purportedly luring Darnley to his doom. The sonnets were meant to plug that gap. Likewise, none of the letters proved she had agreed to marry Bothwell before he was divorced. The marriage contracts were supposed to take care of that oversight.

The sonnets were said to be Mary's own reflections on her adultery. They were intended to prove that her consuming passion for Bothwell gave her a powerful motive for murder. Very few literary experts believe them to be genuine. They are clumsy and would pass only with the greatest difficulty as the work of a native French speaker. As imitations of the genre of courtly love poetry, in which Mary had been trained by Ronsard, they fail every test. The most incriminating poem, "Entre ses mains et en son plein pouvoir," if not said in advance to be inspired by sexual passion, could easily be read as a religious poem:

> In his hands and in his full power,
> I put my son, my honor and my life,
> My country, my subjects, my soul all subdued
> To him, and none other will I have
> For my goal, which without deceit
> I will follow in spite of all envy
> That may ensue. For I have no other desire,
> But to make him perceive my faithfulness:
> For storm or fair weather that may come,
> Never will it change dwelling or place.
> Shortly I shall give of my truth such proof,
> That he shall know my constancy without pretense,
> Not by my tears or feigned obedience,
> As others have done, but by fresh ordeals.

Even if this were a genuine work, the evidence is inconclusive. The addressee is not mentioned, and the sonnet is just as likely to be a pious exercise giving thanks to God after Mary's ordeal in childbirth as it is to be an outpouring of her infatuation for Bothwell. A genuine poem or poems may well have been found by Morton's men when they ransacked Mary's apartments. If so, they adapted what they discovered to suit their own purposes.

The contracts are even less problematic. Royal and aristocratic marriages were always preceded by a written contract, a form of prenuptial agreement that settled the property rights of both parties. Mary and Bothwell signed such a contract on May 14, 1567, the eve of their marriage. But for some reason this was not one of the documents found in the casket. According to the Confederate Lords, their "discoveries" meant the couple must have signed earlier contracts.

One was said to be in Huntly's handwriting, dated April 5, 1567, "at Seton." It is a blatant forgery, because although Mary and Bothwell were indeed at Seton on that day, the wording includes extracts from the Ainslie's Tavern Bond, and yet that document was not ready for another two weeks. The contract even refers to the gathering of the lords at Ainslie's Tavern, because it says that Mary *"among the rest"* had chosen Bothwell as the most suitable man to be her husband. As the italicized words could only refer to the discussions at the tavern, though Mary was not present, it follows that the contract could not have been written earlier than April 19, when the supper had been held.

The other contract, undated but supposedly signed "Marie R," cannot be conclusively pinned down. Yet even if it is genuine, it is innocuous. It is less a contract than a written promise by Mary to marry Bothwell:

We Mary, by the grace of God, Queen of Scotland, Dowager of France, etc., promise faithfully and in good faith, and without constraint, to James Hepburn, Earl of Bothwell, never to have any other spouse or husband but him . . . And since God has taken my late husband, Henry Stuart called Darnley, and that hence I am free . . . and since he [Bothwell] is also free, I shall be ready to perform the ceremonies requisite for marriage . . .

This wording, which is unambiguous, could only have been framed after Darnley's death and Bothwell's divorce. Whatever it might suggest about Mary's feelings, if genuine, it tells us nothing about her complicity in murder or adultery. Moray knew that he was grasping at straws when he submitted it. He acknowledged it was written "without date and though some words therein seem to the contrary, yet it is upon credible grounds *supposed* to have been made and written by her before the death of her husband."

Cecil decided that the sonnets and the marriage contracts were irrelevant. Everything turned on the Casket Letters, especially the first two (letters 1 and 2).* These were potentially the most devastating. In fact, letter 2, usually called "the long Glasgow letter" to distinguish it from "the short Glasgow letter" (letter 1), would, if genuine, be enough by itself to prove that Mary was Bothwell's illicit lover and co-conspirator in Darnley's murder. The only surprise is why, if indeed it was true, the lords needed to cast their net wider.

Both Glasgow letters were said to have been written while Mary was visiting Darnley on his sickbed. Letter 1 is the complaint of a woman who begins by chiding her correspondent for his forgetfulness. He had promised to send her news, but had forgotten to do so. He had left her to go on a journey, and the woman accused him of deliberately delaying his return.

She would accordingly "bring the man" on Monday to Craigmillar, "where he shall be upon Wednesday." "And I go to Edinburgh to be let blood if I have no word to the contrary." Whoever this man was, he was "the merriest that ever you saw, and doth remember unto me all that he can to make me believe that he loveth me."

The woman was suffering from a pain in her side. "I have it sore today. If Paris [Nicholas Hubert] doth bring back unto me that for which I have sent, it should much amend me." She ended with a request for news. "I pray you, send me word from you at large, and what I shall do if you be not returned . . . For if you be not wise, I see assuredly all the whole burden fallen upon my shoulders." The letter was sent "from Glasgow, this Saturday morning."

One of the least satisfactory features of the Casket Letters is that only this one was dated. In addition, not a single letter was directed to a named addressee, and all ended abruptly without any indication whatever of who might have written them.

The short Glasgow letter would be highly incriminating if read in a certain way. The words "from Glasgow, this Saturday morning" meant it

* I have followed the numbers given to the Casket Letters by Henderson (1890), which are those most commonly used.

must, to be genuine, have been written on January 25, 1567, the only Saturday on which Mary had been in Glasgow on her visit to Darnley.

The reader is expected to assume this. We are also asked to guess that "the man" who is to be brought to Craigmillar is Darnley, but this does not quite add up. Even today, "the man" is a common colloquial expression in Scotland for a male child, and the suspicion that Prince James was really meant here is reinforced by the writer's description. To depict the sulky and syphilitic Darnley as "the merriest that ever you saw, and doth remember unto me all that he can to make me believe that he loveth me" beggars belief. On the other hand, a doting Mary could well have been expected to say such things about her growing baby.

But if Prince James was "the man," and the author his mother, the letter could not have been written at Glasgow. The child was never there; he had stayed with Mary at Stirling Castle until she returned to Edinburgh after the baptism celebrations, and afterward he was with his nurse at Holyrood, where Mary had left him when she had set out for Glasgow to visit Darnley. Later he was returned to the safekeeping of the Earl and Countess of Mar.

Then again, the writer was returning to Edinburgh "to be let blood." The letter does not say "to *let* blood." According to the medical understanding of the time, a person's health was governed by correctly balancing the four cardinal "humors," or bodily fluids: blood, phlegm, choler (yellow bile) and melancholy (black bile). Bloodletting was a routine part of a physician's practice, in the same way that he might prescribe rhubarb pills for choler or licorice tablets for digestive disorders. And we know that Mary's doctors practiced it when they felt her state of health required it.

The writer of the letter had sent Paris to fetch something. Paris was, of course, the valet Nicholas Hubert, the man Mary had seen to be unusually dirty on the night before Darnley's murder. The writer had a pain in her side, and the likely interpretation is that Paris had been sent somewhere for medicine. Paris was Mary's servant. He had formerly been Bothwell's servant for twenty years, but Mary's household accounts show that he had changed employers a year or so before.

This could raise suspicions, because Paris was clearly still in Bothwell's pay, as his role in the explosion at Kirk o'Field suggests. In that case, he might have been the missing link between Mary and Bothwell. Could he have been running errands between them while Mary was at Glasgow?

The Confederate Lords always maintained that Paris had been the trusted messenger who assisted the conspiracy. They said he had taken letters to and fro between Mary in Glasgow and Bothwell in Edinburgh, but this is highly problematic. By Saturday, January 25, Bothwell had already left Edinburgh. He had set out for Liddesdale the previous day after escorting Mary to Callander House on the first stage of her journey to Glasgow.

The date of Bothwell's departure is incontrovertible. Moray's own "journal" describing Mary's and Bothwell's movements between January 21 and 30, 1567, admits that Bothwell had left Edinburgh on the evening of the 24th. Moray was brazen enough to submit his diary to Cecil, despite the conflict of this evidence with letter 1.

And Bothwell's movements can be checked. They are corroborated by a handwritten report sent to Cecil four days later by Lord Scrope, the English official based at Carlisle, on the western side of the border. In his report, Scrope vividly described Bothwell's excursion.

This was because Bothwell had been involved in another big fight. He had left Edinburgh accompanied by eighty men for the "reformation" of thieves and other malefactors on the border, traveling as far as Jedburgh on the first night. From there he had continued into Liddesdale, where his men captured a dozen brigands, one of whom was Martin Elwood, the leader of the border gang comprising the associates of the Elliots of Liddesdale.

Elwood got word to his men and a rescue attempt was made. One of Bothwell's men was killed and five more captured. Bothwell went in hot pursuit of his assailants "until his servants and horses were wearied and spent, and he could not make any recovery and so was forced to return to Jedburgh again."

If the logic of the short Glasgow letter is to be believed, Bothwell was supposed to be keeping his promise to write to Mary about the final arrangements for Darnley's murder while he was intent on a sortie to deal with the Elwoods and the Elliots.

More puzzling still, Mary was said to be accusing him of delaying his return from Liddesdale, where he had scarcely arrived. If his absence was a sore point by Saturday the 25th, she must have expected his return *before* Saturday. But he had not left Edinburgh until Friday evening, and Liddesdale was seventy miles away.

And yet several aspects of letter 1 ring true. The writer had a pain in her side requiring medical treatment and a young son who was "the merriest that ever you saw." It is possible — perhaps likely — that this was a draft or copy of one of Mary's genuine letters, written to an unknown recipient at another time and in a different context from that claimed by Moray and his allies. In that case, all the Confederate Lords had to do was to add the words "from Glasgow, this Saturday morning." At the stroke of a pen, an otherwise unremarkable letter could be turned into something incriminating. The reference to Craigmillar is easily explained, as it was one of Mary's favorite retreats where she had regularly stayed for periods ranging from a few days to several weeks.

Letter 2 is the longer of the two letters said to have been written from Glasgow while Mary attempted to persuade Darnley to return with her to Craigmillar. It is also the longest of the eight Casket Letters. Selected by the lords for its seemingly graphic allusions to the murder plot, it is also interspersed with its author's protestations of longing and desire for her lover.

The letter opens on an elegiac note: "Being gone from the place where I had left my heart, it may be easily judged what my countenance was." The author, a woman, had said her fond goodbyes to her lover, then continued on her journey to Glasgow, where she had been met by a gentleman from the Earl of Lennox and by the Laird of Luss with forty men and others, but no one from the town.

When she had arrived, she found that Darnley had already been asking questions about her. When she saw him on his sickbed, she was surprised at how well informed he seemed to be about events at Holyrood. After her first meeting with him, she had gone to eat her supper.

Later that evening, he had recalled her, making all his usual complaints and blaming her for his illness, which he was convinced was brought on by her disdain. He asked her for a reconciliation. He even pleaded for a pardon. She answered that he had been pardoned many times before, only to return to his dissolute ways.

"I am young," he protested. "May not a man of my age for want of counsel fail twice or thrice, and miss of promise [i.e., break his promises], and at the last repent and rebuke himself by his experience?" Darnley offered to amend his behavior if he could receive just one more pardon. "I ask nothing," he said, "but that we may be at bed and table together as husband and wife. And if you will not, I will never rise from this bed."

The writer asked him why, in that case, he had tried to leave the country in an English ship. He denied it, but finally admitted he had spoken to the sailors. She turned next to the rumors about his plotting. She asked him about Hiegate's reports that he was conspiring with the Lennoxes to put her in prison and rule in the name of her son. "He denied it till I told him the very words, and then he said that . . . it was said that some of the Privy Council had brought me a letter to sign to put him in prison, and to kill him if he did resist."

Darnley at first denied all knowledge of his conspiracies, but next day changed his mind and was willing to confess. He made a clean breast of everything, after which he asked for sex with the writer. "He desired much that I should lodge in his lodging. I have refused it. I have told him that he must be purged and that could not be done here."

The writer refused to sleep with Darnley, offering instead to take him to Craigmillar in a horse litter: "I told him that so I would myself bring him to Craigmillar, that the physicians and I also might cure him without being far from my son."

Darnley flatly refused. He wanted no one to see him with his pock-marked skin. He became angry when further questioned about his plotting: "He hath no desire to be seen, and waxeth angry when I speak to him of Walker, and saith that he will pluck his ears from his head and that he lieth."

The writer asked Darnley why he had been complaining about the lords, threatening to kill those who had lately been reconciled to her. "He denieth it, and saith that he had already prayed them to think no such matter of him. As for myself, he would rather lose his life than do me the least displeasure."

Darnley, it seems, was far less afraid of the writer than he was of some of the lords. This makes sense, because the lords most recently pardoned and who had only just crossed the border back into Scotland — that is, Morton and the Douglases — were his co-conspirators in the Rizzio plot. He had brazenly betrayed them, causing them to seek revenge.

Darnley's train of thought led him to attempt a reconciliation with the writer. He "could not believe that his own flesh (which was myself) would do him any hurt." He appealed to her not to leave him. She almost had pity on him until she compared his heart of wax with hers, which was as hard as a diamond.

The letter writer then veered almost randomly from topic to topic. Darnley's father, Lennox, "hath bled this day at the nose and at the mouth. Guess what token that is?" The woman said she had not seen Lennox herself: "He is in his chamber." And she said that she hated deceiving Darnley. "I do here a work that I hate much, but I had begun it this morning."

There followed a passionate outburst from the woman to her lover: "We are tied to with two false races.* The good year untie us from them. God forgive me, and God knit us together forever for the most faithful couple that ever he did knit together. This is my faith. I will die in it." She longed to be in bed with him again. "I am ill at ease and glad to write unto you when other folks be asleep, seeing that I cannot do as they do according to my desire, that is between your arms, my dear life."

She said she was becoming weary, but could not stop herself scribbling while any paper was left. "Cursed be this poxy fellow," she exclaimed, "that troubleth me this much!" His health was stable, but his body was disgusting. His breath stank whenever she approached his bedside. "I thought," she said, "I should have been killed with his breath, for it is worse than your uncle's breath, and yet I was sat no nearer to him than in a chair by his bolster."

At this point, halfway through the letter, a hiatus occurs. A memo appears from nowhere, a set of jottings or headings that the writer referred to as a "memorial" — a list of topics she had so far wished to include in her letter and were obviously meant to jog her memory as she wrote. The letter then continues as if the memo were not there.

The woman said she had remembered that Lord Livingston, during supper, had openly teased her in the presence of Lady Reres about her lover. This (somewhat implausibly) was when she was leaning on him while warming herself at the fire. Livingston had drunk a toast to her inamorato.

In Cecil's copy of the letter, this paragraph is followed by an unexplained double-line space. When the transcription resumes, the woman informs us that she was making her lover a bracelet. "I have had so little time that it is very ill, but I will make a fairer. And in the meantime take heed that none of those that be here do see it, for all the world will know it."

She had been seen working on the bracelet, and was anxious because anyone who later noticed her lover wearing it would guess their secret. This seems a bit odd, since only a few sentences earlier the woman's relationship with her lover had been a topic of general banter.

---

* The innuendo is that Mary is tied to the Lennoxes by her marriage to Darnley, just as Bothwell is to the Gordons, the family of his wife, Lady Jean.

It is usually said that the hiatus caused by the memo was the result of the writer's ending her work on one evening and starting again on the next. That is certainly a possibility, because by the next paragraph it seems that another day had passed and the writer had resumed her "tedious talk" with Darnley.

And once more, she hated herself for her dissimulation. "You make me dissemble so much," she now reproached her lover, "that I am afraid thereof with horror, and you make me almost to play the part of a traitor. Remember that if it were not for ~~your sake~~ obeying you I had rather be dead."

Darnley would not accompany the woman back to Edinburgh unless she would promise to have sex with him. The matter was argued back and forth, and she finally agreed as the only way to persuade him. She promised to sleep with him as soon as he was cured of his pockmarks.

Darnley was still somewhat suspicious, but since he was too afraid of the reprisals of the lords to take any chances, he decided to travel in the litter with her and they would lead a happy life together. "To be short," said the woman, "he will go anywhere upon my word."

She said that she would follow such instructions as she hoped to receive shortly from her lover. "And send me word what I shall do, and whatsoever happens to me, I will obey you." She then added her own chilling suggestion: "Think also if you will not find some invention more secret by physick, for he is to take physick at Craigmillar and the baths also. And shall not come forth of long time."

She still had deep misgivings about what she was doing. "I shall never be willing to beguile one that putteth his trust in me. Nevertheless, you may do all, and do not esteem me the less therefore, for you are the cause thereof. For my own revenge I would not do it."

She did not visit Darnley on the second evening of her stay, as she was too busy finishing her lover's bracelet. She had not found clasps for it yet. "Send me word whether you will have it and more money, and when I shall return and how far I may speak. Now as far as I perceive, I may do much with you. Guess you whether I shall not be suspected."

Changing the subject again, the writer remarked that Darnley would become enraged whenever he heard any mention of his enemies among the lords. His father, Lennox, meanwhile, would not emerge from his room. Darnley himself had asked to see the woman again early the next morning.

She would send her letter by a messenger, who would also tell her lover the rest of the news. She ended with a grim warning: "Burn this letter, for it is too dangerous, neither is there anything well said in it, for I think upon nothing but upon grief if you be at Edinburgh."

A second unexplained double-line space follows in Cecil's transcript of the letter, after which is written a damning last paragraph:

> Now if to please you my dear life, I spare neither honor, conscience, nor hazard, nor greatness, take it in good part, and not according to the inter-pretation of your false brother-in-law, to whom, I pray you, give no credit against the most faithful lover that ever you had or shall have . . . Excuse my evil writing and read it over twice . . . Pray remember your friend and write unto her and often. Love me always.

A second memo concludes the letter. The final headings are: "Of the Earl of Bothwell" and "Of the Lodging in Edinburgh." If letter 2 really had been written on two consecutive evenings, this second memo might not have raised so many suspicions, but the references to Bothwell and the "lodging" (we may assume this means the Old Provost's Lodging at Kirk o'Field) tacked on at the end are too much to swallow. Of all these jottings, these are the only two that fail to match the contents of the letter.

According to the Confederate Lords, Mary was the author of letter 2, Bothwell was its recipient, and the messenger to whom it was entrusted was Paris. When she had stayed at Callander House on the first leg of her journey to see Darnley, she had allegedly told Paris to accompany her to Glasgow, where he was to wait to take letters back to Bothwell in Edin-burgh.

After she had been two nights at Glasgow, she had supposedly given Paris a packet of letters, some for Bothwell and others for Maitland, which Bothwell was also to be shown. On reaching Edinburgh, Paris was said to have found Bothwell and delivered his letters. Bothwell ordered him back to Glasgow with the message that everything was ready and Mary was to lure Darnley to his fate at Kirk o'Field.

Viewed in conjunction with letter 1, this scenario becomes a farce. Mary had left Edinburgh on January 20 or 21. She did not arrive at Glas-gow until the 22nd. Huntly and Bothwell had returned to Edinburgh after leaving her at Callander House. Bothwell then departed for Jed-burgh on the evening of the 24th and continued the next day into Lid-desdale.

If Mary had done no more at Glasgow than write letter 2 on the consecutive evenings of January 22 and 23, it might just have been possible for Paris to catch Bothwell in Edinburgh on the 24th before he left for Jedburgh. But the rest is beyond belief. Paris was supposed to have returned to Glasgow on the 25th to inform Mary that the "place" was to be Kirk o'Field, and then left Glasgow "this Saturday morning," also the 25th, for Edinburgh to deliver letter 1. But Edinburgh and Glasgow are fifty miles apart, a hard day's ride in each direction. He would have had to leave Glasgow on the 25th before he had even arrived. Moreover, had he achieved this miraculous feat, Bothwell would no longer have been in Edinburgh to receive letter 1, because he had already arrived at Jedburgh.

The lords' allegations put Paris firmly in the spotlight. He had sought refuge in Denmark after Mary was taken to Lochleven, but after her defeat at Langside and flight to England, the lords set out to extradite him, and if that failed, to kidnap him. This was not to produce him as a witness before Elizabeth and Cecil. On the contrary, it was to make sure he *never* appeared to give his testimony. Dead men tell no tales.

Paris would be brought back to Scotland in June 1569. He was kept in prison until August 9 and 10, when he was secretly interrogated at St. Andrews by George Buchanan and Moray's private secretary, John Wood. On the first day of questioning, Paris did not say what his inquisitors wanted to hear. On the second day he was tortured, and confessed to everything.

No sooner had Moray received a transcript of this confession than poor Paris was silenced forever. There was no trial: Mary's loyal valet was summarily hanged. But he, and not Moray, had the last word. On the scaffold, he shouted out the truth to the assembled crowd. He denied everything he had said about carrying the two Glasgow letters from Mary to Bothwell.

Moray sent a copy of Paris's confession to Cecil, who instantly wrote in Elizabeth's name demanding that the execution be deferred. As soon as he had trawled through the confession, he must have realized how flimsy and far-fetched it was. He demanded that Paris be kept alive for further interrogation, presumably in England. But it was too late.

Despite the fabrications, lengthy passages of letter 2 can only have been taken from a genuine document Mary is likely to have sent to an unknown recipient while she was visiting Darnley. Several key extracts ring true:

1. "I thought I should have been killed with his breath, for it is worse than your uncle's breath, and yet I was sat no nearer to him than in a chair by his bolster."
2. "He desired much that I should lodge in his lodging. I have refused it. I have told him that he must be purged and that could not be done here."
3. "I told him that so I would myself bring him to Craigmillar, that the physicians and I also might cure him without being far from my son."
4. "To be short, he will not come but with condition that I shall promise to be with him as heretofore at bed and board, and that I shall forsake him no more, and upon my word he will do whatsoever I will, and will come."

The first extract clearly relates to Darnley's treatment for syphilis by salivation of mercury. After a while, the patient's breath began to stink. But who was the mysterious uncle whose breath was almost as bad as Darnley's?

Bothwell no longer had a living uncle, so this part of the letter could not possibly have been addressed to him.* Some historians have suggested that Mary had written a detailed report on Darnley's condition to Moray, and that a whole run of paragraphs from this lost document were spliced into letter 2. If that is true, the "uncle" would have been the Earl of Mar, whom Mary entrusted with the care of her son. That, however, remains a matter of conjecture.

The other three extracts refer to Darnley's characteristic demands for sex, first refused and then accepted by Mary. They confirm that she knew exactly how to handle her debauched and degenerate husband. Her tactics closely resembled her approach after the Rizzio plot, when she had offered to sleep with Darnley to win his affection, but knew he would be too drunk to take advantage of her offer.

If indeed these passages are true, it is less rather than more likely that Mary was involved in Darnley's murder. Although afterward highlighted by Moray to claim that his sister was a party to the plot, the third extract severely undermines the charges against her. To make the fanciful chronology of the Glasgow letters somehow add up, Buchanan had asserted that Mary had "devised" with Bothwell to lure Darnley to Kirk o'Field before ever leaving Edinburgh for Glasgow.

But the third extract confirms that she sought to bring Darnley not to Kirk o'Field but to Craigmillar, even though the charge that she had plotted to lure him to Kirk o'Field was the crux of the case against her.

* His father, Patrick Earl of Bothwell, had no living brothers, and his uncle on his mother's side, Alexander Earl of Buchan, had died in 1563.

Other sections of letter 2 tell us just what it was that Darnley was thinking.

1. "He hath no desire to be seen, and waxeth angry when I speak to him of Walker, and saith that he will pluck his ears from his head and that he lieth."
2. "I asked him of the inquisition [i.e., investigation] of Hiegate. He denied it till I told him the very words, and then he said that . . . it was said that some of the Council had brought me a letter to sign to put him in prison, and to kill him if he did resist."
3. "I asked him . . . what cause he had to complain of some of the lords and to threaten them. He denieth it, and saith that he had already prayed them to think no such matter of him. As for myself, he would rather lose his life than do me the least displeasure."

Such extracts fit hand in glove with what is known of Darnley's plotting and Mary's fears as expressed to her ambassador to France. They also suggest that Darnley had heard whispers of Morton's rendezvous with Bothwell and Maitland at Whittingham Castle, and that he suspected a plot to imprison or kill him.

But even if this means that whole paragraphs of letter 2 are true, others are either forged or interpolated. In particular the final paragraph, where the writer castigates her correspondent's "false brother-in-law," cannot be genuine.

By "your false brother-in-law," Mary was supposed to have referred to Huntly.* If so, the paragraph could not possibly have been taken from a genuine letter written in January 1567, because Huntly was then still her stalwart supporter. He did not quarrel with her until the last week in April 1567, when Bothwell's divorce was pending.

And there is further evidence of cheating. The second memo ended with the jottings "Of the Earl of Bothwell" and "Of the Lodging in Edinburgh." We have already asked, Why would Mary, if she were already writing to Bothwell, have made such a note? Letter 2 runs to more than three thousand words, yet it contains not one word on Bothwell or the "lodging." It seems that these final jottings were added in a blank space on the last page of the letter to make it appear more incriminating. We have already noted that letter 2 looks to be in two parts. It is becoming a

* No one else could have been meant. Lord John of Coldingham, husband of Jane Hepburn, Bothwell's sister, had died in December 1563, and Jane took as her second husband John Sinclair, Master of Caithness, who was loyal to Mary.

distinct possibility that pages culled from different places were surreptitiously spliced together.

This line of inquiry can be pursued further. All the passages in which Mary is said to have openly admitted her love for Bothwell are curiously placed in the text. The woman had said she was "ill at ease and glad to write unto you when other folks be asleep, seeing that I cannot do as they do according to my desire, that is between your arms, my dear life." She had asked her beloved to "Love me always." Elsewhere in the letter, Lord Livingston had supposedly teased her. And yet only a few lines later, she could warn her lover to beware that "none of those that be here" ever saw him with the bracelet she was making for him.

Each of these "love" passages appears either just before or just after the jottings that Mary is supposed to have made when preparing her thoughts for the letter. We already know there are two sets of jottings, which reinforces the suspicion that perhaps the pages of more than one letter had been combined.

Still more suggestively, just before the curious remarks on the bracelet, and then again immediately before the concluding and (as we now know) interpolated paragraph in which the writer castigated "your false brother-in-law" — which also happens to be the very same section in which the words "Love me always" appear — there are unexplained double-line spaces in Cecil's transcript of the letter. These have never been noticed before.

Although Cecil's copy is an English translation of the document, its only gaps are in these two places. Since neither falls at the top or bottom of a page, it is likely that they represent blanks in the original (and unfortunately lost) French version of the letter. It is a remarkable coincidence that these double-line spaces appear exactly at points where the text seems to have been doctored. A plausible hypothesis is that these spaces also appeared in the original French text, possibly at page breaks or other places where spaces had been left on the paper, which enabled forged material to be inserted later. To prevent similar interpolations, it had been Elizabeth's practice to draw hatch marks in pen across the blank spaces of her sensitive and important letters to prevent forged material from being added.

Returning to the charges leveled by the Confederate Lords, their case in the end rested on these passages:

1. "I do here a work that I hate much, but I had begun it this morning."
2. "We are tied to with two false races. The good year untie us from them. God forgive me, and God knit us together forever for the most faithful couple that ever he did knit together. This is my faith. I will die in it."
3. "His father hath bled this day at the nose and at the mouth. Guess what token that is?"
4. "You make me dissemble so much that I am afraid thereof with horror, and you make me almost to play the part of a traitor. Remember that if it were not for obeying you I had rather be dead."
5. "Think also if you will not find some invention more secret by physick, for he is to take physick at Craigmillar and the baths also. And shall not come forth of long time."
6. "Send me word whether you will have it and more money, and when I shall return and how far I may speak. Now as far as I perceive, I may do much with you. Guess you whether I shall not be suspected."
7. "Burn this letter, for it is too dangerous, neither is there anything well said in it, for I think upon nothing but upon grief if you be at Edinburgh."

If Mary wrote all of these extracts, she was involved in a plot that Darnley had good reason to think would put him in mortal danger. If the fifth extract in particular is authentic, it would seem to point to poison as a possible method of assassination, and Mary would rightly stand condemned of conspiracy to murder.

However, apart from the fifth extract, there is no reason to connect any of these passages to a murder plot. The first extract is compatible with an intention to fetch Darnley and keep him under house arrest. The second extract cannot have been written from Glasgow in January 1567, since the phrase "two false races" refers again to Darnley, who is "tied" to Mary, and to Huntly's sister, Lady Jean Gordon, who is "tied" to Bothwell. But Huntly was not "false" until the end of April, which means this passage was interpolated.

Likewise, the fourth, sixth and seventh extracts do not add up. They would, of course, make sense if culled from the pages of genuine letters from Mary to Bothwell in late April or May 1567. Such pages could have been written after her abduction, when she was anxiously waiting at Dunbar for his return from Edinburgh, where he had gone to encourage his wife to file her divorce petition and while Mary was quarreling with Huntly, or they may have been written after she had returned to Edinburgh with Bothwell to call the banns for their marriage.

The sixth extract suggests that Bothwell needed money, possibly as pay for his soldiers or bribes for his divorce. And whereas the seventh extract suggests that Mary expected him to be in Edinburgh while they were apart, that is easier to credit if the letter was written three months after the Confederate Lords said it was. As has already been shown, Bothwell had left Edinburgh for Liddesdale on the evening of January 24. He was, however, in Edinburgh for several days, filing his divorce papers at the end of April, when Mary was at odds with Huntly and anxious for reassurance.

That leaves the third and fifth extracts. The third is difficult to judge. Lennox, Darnley's father, had suffered a nosebleed. What did it signify? Obviously it was some sort of omen. That was exactly Mary's point, assuming the question was not itself interpolated. But since she did not venture an answer, the innuendo was hardly proof of a conspiracy to murder Darnley.

All along, it was the fifth extract that packed the punches. And yet this was far from being what it seemed to be. Although the extract was supposedly the proof that Mary had wanted Darnley to be poisoned at Craigmillar, the lords' story — as scripted by Buchanan in his dossier and written with this very same letter on his desk (see chapter 24) — had made no mention whatsoever of this particular charge.

The whole force of the lords' accusation, as it had been compiled by Buchanan, was not that Mary and Bothwell planned to poison Darnley at Craigmillar. It was that they had conspired to lure him to Kirk o'Field where he was to be blown up in a gunpowder plot. That was why it was so important to add the crucial words "Of the Earl of Bothwell" and "Of the Lodging in Edinburgh" to the jottings on the last page of the letter. Otherwise these details were not adequately covered.

The trouble with the lords' story is that it was getting far too complicated for any one person to remember. An unnecessarily elaborate tale was becoming a mangled one in which plot was piled on plot, making it difficult to keep all the details under control.

The main objection to the genuineness of the fifth extract is that the suggestion that Darnley was to be poisoned at Craigmillar is wholly inconsistent with Buchanan's principal accusation, which is that Mary and Bothwell had already "devised" with each other to lure Darnley to his death at Kirk o'Field *before* Mary traveled to Glasgow to fetch her errant husband. If the fifth extract really was a genuine passage by Mary, it would have been seized upon and given the highest priority in Buchanan's original list of charges. As the most incriminating section of either of the two Glasgow letters, it would have been far too juicy to pass over.

The fact that it *was* missed can only mean that it did not exist by the time Buchanan's material was sent to Cecil in June 1568. It has to be regarded as a later forged interpolation.

This is all the more likely because, after Buchanan's dossier had been compiled but before the Confederate Lords made their final accusations against Mary, they changed their story. They added the additional charge that an attempt had been made to poison Darnley at Stirling after the baptism of Prince James. They argued that because of this, Darnley had fallen victim to a "grievous sickness."

Just two months before the Casket Letters were shown to Elizabeth and Cecil, Moray drew Cecil's attention to Darnley's illness as the supposed "proof" of the fifth extract. By this sleight of hand, Darnley's syphilis had turned into an attempted poisoning.

But Moray's charge does not add up. If the fifth extract was genuine, then Mary and Bothwell must have intended to poison Darnley after she had lured him to Craigmillar, not before at Stirling. And since Darnley was already sick when he arrived at his father's house in Glasgow, the failed attempt must have been made before he had left Stirling.

It is impossible to explain why so clumsy and improbable a charge was added to the roster of allegations against Mary at such a late stage. But the slipperiness of the lords is established. We already know that letter 2 contained interpolations about Huntly and his sister: "your false brother-in-law" and "two false races." The final jottings — "Of the Earl of Bothwell" and "Of the Lodging in Edinburgh" — are also extremely suspicious.

In the absence of the original handwritten pages of letter 2 in the form in which they were submitted to Elizabeth and Cecil, no final and unassailable conclusions about this extraordinary document can be reached. But all the evidence points in a single direction. Between 1500 and 1800 words are likely to be genuine, even if the passage about Darnley's treatment by salivation of mercury could not possibly have been addressed to Bothwell. And between 1000 and 1200 words of the text we now have are likely interpolations from letters that Mary wrote several months later, or else are outright forgeries.

Probably most of the interpolations were from genuine (if later) letters. This shows the true extent of Moray's cunning, because Elizabeth and Cecil were nobody's fools. They wanted the handwriting in the documents to be authenticated. Cecil never relaxed his guard in dealing with them. When the original letters were finally inspected, he demanded a supporting affidavit, signed by Moray, Morton and Lord Lindsay, the chief architects of Mary's forced abdication, to declare that all were Mary's, all in her "own proper handwriting." Thereafter, comparisons with genuine letters from Mary to Elizabeth from the collection in the royal archives were to be made before a verdict could be reached.

The lords knew that they could not expect to get away with crude forgeries. To establish Mary's guilt, they needed to find pages of genuine letters that, if doctored up here and there, would clinch what they wanted to prove.

This hypothesis would also explain the curious incongruities in the contents of letter 2 as old and new pages were spliced together to make up a composite document. It would also explain why the most suspect passages are relatively brief. If they were filled in using blank spaces on the existing sheets, Moray could be fairly confident that they would pass undetected. Since Mary was known to scribble her letters and also to use the most expensive paper as profligately as only a queen might be expected to do, it could prove to be a risk worth taking.

Mary with a hat and feather, copied from a miniature

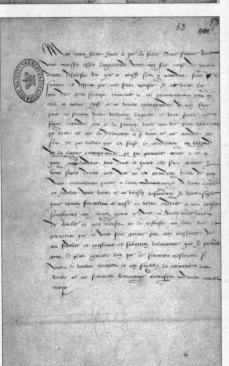

LEFT: The first page of the English translation of the long Glasgow letter (Casket Letter 2)

BELOW LEFT: A transcript in French of Casket Letter 5, in the hand of one of the clerks in Moray's delegation

BELOW: A narrative drawing of Darnley's assassination, sent by Sir William Drury to Cecil, showing the Old Provost's Lodging as a heap of rubble (left center), the bodies in the garden on the opposite side of Thieves Row (top right), Darnley's body being carried away (below left), the cottages built into the garden wall not far from where the bodies were found (lower center), and the future James VI in his cradle, praying, "Judge and revenge my cause, O Lord" (top left)

RIGHT: Placard of the Mermaid and the Hare

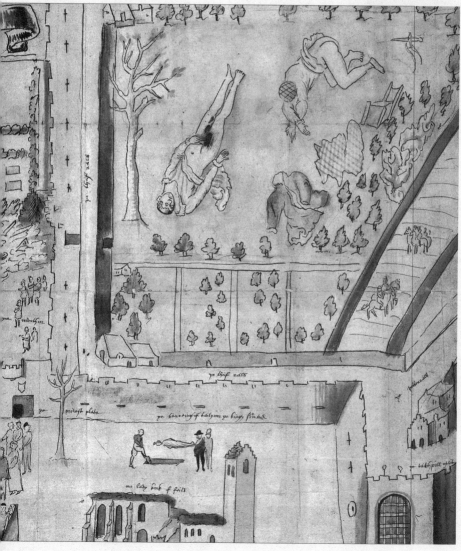

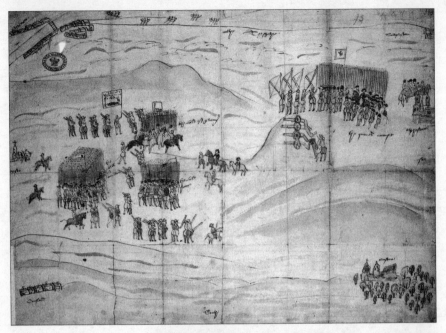

A narrative drawing of Mary's surrender at Carberry Hill, showing Mary on horseback (center), still wearing her borrowed clothes, as she is led across the field toward the army of the rebel lords, followed by Mary Seton on her pony

Mary's smuggled letter to Throckmorton appealing for help "from my prison in the tower of Lochleven"

A version of Nicholas Hilliard's miniature of Mary in captivity, about age thirty-six

Bess of Hardwick, with whom Mary talked and embroidered during the first ten years of her captivity

Two of Mary's
embroideries,
a phoenix and
"a catte," now at
Oxburgh Hall

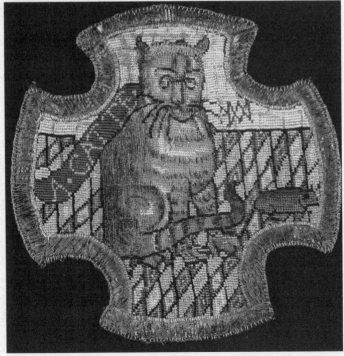

Sir Francis Walsingham,
by John de Critz the Elder

Mary's "alphabet of ciphers,"
seized from Chartley after the
discovery of the Babington plot

A narrative drawing of Mary's trial at Fotheringhay Castle. Mary is about to enter the courtroom (top right). She sits in a high-backed chair (upper right, marked A). Elizabeth did not attend, and her chair of state on the dais is empty. The trial commissioners and clerks are seated, watched by the local gentry. The trial commissioners are identified by numbers. Cecil, directly opposite Mary, is 2, Walsingham, opposite the vacant chair of state, is 28, Sir Ralph Sadler is 29, and Sir Amyas Paulet 31.

Mary's execution at Fotheringhay, Netherlandish school, seventeenth century

26

Casket Letters II

THE TWO Glasgow letters were the principal documents produced by Moray and the Confederate Lords to suggest that Mary had been involved in Darnley's murder. Yet they were not the only evidence. Three of the six remaining Casket Letters were love letters, of which two were also said to have referred to Darnley's murder. The last three letters were said to relate to Mary's abduction at Almond Bridge, which had supposedly been planned collusively. None of the handwritten transcripts or translations of the lost originals makes it possible to verify the real author, the intended recipient or recipients, or the date of composition. And all six of these letters will raise almost as many questions as they answer.

Letters 3 through 5 were said to be the love letters that Mary had written to Bothwell. Letter 3 was written by a woman to her estranged husband. She bitterly complained of his cruelty, absence and neglect, but "for all that, I will in no wise accuse you, neither of your little remembrance, neither of your little care, and least of all of your broken promises or of the coldness of your writing."

The woman was doing her best to keep up appearances and to defer to her husband in the way that he expected. "I am else so far made yours that what pleases you is acceptable to me, and my thoughts are so willingly subdued unto yours that I suppose that all that comes of you proceeds . . . for such [reasons] as be just and reasonable, and such as I desire myself."

She was still a woman of spirit. To make clear her true feelings for her husband, she was sending him a jewel, a ring designed as a locket. But this was no ordinary lover's token. It was not a diamond in the shape of a heart, as such gifts usually were, but a stone in the shape of a sepulcher. Mounted in black enamel and set with small jewels representing tears and bones, the locket contained a ringlet of the woman's hair — she called this "the ornament of the head": "And to testify unto you how lowly I submit myself under your commandments, I have sent you in sign of homage by Paris [Nicholas Hubert] the ornament of the head, which is the chief guide of the other members [i.e., parts of the body]."

The woman's words veered from the ironic to the macabre. She would defer to her husband just as he had commanded, but her deference would be such that she would rather die than lose her sense of dignity and autonomy:

> I send unto you a sepulchre of hard stone colored with black, sewn with tears and bones. The stone I compare to my heart, that is carved in a sure sepulchre or harbor of your commandments, and above all of your name and memory that are therein enclosed, as is my hair in this ring, never to come forth until death grant unto you to a trophy of victory of my bones.

This letter is so different in sentiment and tone from letter 2, with its author's fervent expressions of longing and desire for her lover, it is hardly possible to suppose it was written by the same person. And it is impossible to judge on stylistic grounds whether Mary was the author, because she wrote no other indisputably genuine letters of this kind of emotional nakedness.

The debate has to focus on the letter's contents. In this instance, the woman's spouse was a brute. He had rejected her, and so she had forced herself into a despairing appeal to the better side of the nature she knew he really lacked. Words and phrases like "your broken promises," "the coldness of your letters," "your name and memory" and "the tears for your absence" made it clear that this was a couple whose marriage was in serious trouble.

We have seen that Bothwell *could* behave like a brute. The lords might well have been trying to add some verisimilitude to the quarrels and tears of his relationship with Mary in the month between the date of their marriage and the final showdown with the rebel lords at Carberry Hill.

But that would mean overlooking the problem of the handwriting. It would have been essential, for the letter to look convincing, to discover a genuine letter in which Mary had expressed her feelings in this way. But Mary and Bothwell had scarcely been apart — and certainly not for more than a few days at a time — between their marriage and Mary's surrender. Moreover, their acknowledged moments of separation, as when Mary stood on the flat roof of Borthwick Castle looking down over the battlements and trading insults with the lords, or when she was mustering her forces at Dunbar while Bothwell was heading for Melrose, were so full of feverish activity that it hardly seems likely she would have found time to commission an elaborate jewel from a goldsmith.

If the letter was to pass the handwriting test when exhibited to Elizabeth and Cecil, it would have been advisable to produce a genuine letter that Mary had addressed to Darnley at one of the low points of their relationship, but which the lords would now claim had been written to Bothwell. In this instance, it would most likely have been Mary's draft of the letter, or the document as sent and later ransacked from Darnley's old cabinets in the royal apartments by Morton's men.

Two handwritten transcripts of this letter are extant, one in French, the other in Scots. The latter, until now completely unrecognized, since it was catalogued at the British Library only in 1994, at which time no one seems to have realized it contained unique information, has copies of Cecil's annotations. His comments show that he was baffled by the letter. "A head," "a sepulchre," "a ring with her hair," he carefully noted at the end. What did it all mean?

A further complication is that if the letter *was* genuine but written by Mary to Darnley, the jewel could only have been delivered by Paris, as the writer said it was, after he became her valet rather than while in Bothwell's service. This would limit the date of the letter to one not earlier than the spring of 1566, but that would fit comfortably with the known facts of Mary and Darnley's estrangement and with their rows and lengthy separations as reported by du Croc. The convergence is such that it is highly probable that letter 3 is genuine, but was written to Darnley and not Bothwell, and had nothing to do with a murder plot.

Letter 4 is even less convincing than its predecessor as a supposed love letter to Bothwell. It was also taken by the Confederate Lords to contain chilling references to the murder plot, since the opening lines included the words "I have promised him to bring him tomorrow. If you think it, give order thereunto."

The writer was far from happy with her lover: he had recently betrayed her. He had been unfaithful, ungrateful, and had ordered her not to write to him or contact him. He was sulking, and she was mortified by his latest affair. His new flame did not have "the third part of the faithfulness or voluntary obedience" that she did. She compared herself to Medea, the first wife of Jason, the mythical hero who led the Argonauts in search of the Golden Fleece. It was Jason who, in one version of the myth, had deserted his wife to marry Glauce, only to be forever deprived of happiness when his new love was murdered on her wedding day by the jealous Medea, who also killed her own children.

In theory, this letter could have been written by Mary to Darnley or Bothwell. We know that both men had enjoyed illicit sexual affairs, although Bothwell is not known to have slept with Lady Reres or Bessie Crawford while he was actually married to Mary. But as elaborated by the writer, the comparison has to refer to Darnley. The woman had likened herself to Medea. She was Jason's first wife, not his second, whereas Bothwell was still married to Lady Jean Gordon at the time Mary was said to have written the letter.

Mary was Darnley's first wife, even if he was her second husband. She could very well have imagined herself to be Medea if her sentiments had been addressed to him, and since we know that she had watched the pageant of the ships on the theme of Jason and the Argonauts at her first wedding banquet in Paris nine years before, it is hardly likely that she would have confused Glauce and Medea. She had studied classical literature with the leading experts at the court of Henry II. Either the letter is to Bothwell but is not from Mary, or else it is from Mary but not addressed to Bothwell.

The discrepancy pales into insignificance when compared to the next. The Confederate Lords alleged that letter 4 had been written on February 7, 1567, two days before Darnley's murder. Transcripts in French and English are still among Cecil's papers. The French version ends: "Faites bon guet si l'oiseau sortira de sa cage ou sens [*sic*] son per comme la tourtre demeurera seulle à se lamenter de l'absence pour court quelle soit." This literally translates as: "Take good care lest the bird fly out of its cage or without its mate, as the turtledove shall live alone to mourn its absence, however short it may be." The metaphor may refer back to a poem that Darnley had sent to Mary before they split:

The turtle for her mate.
More dule may not endure,
Than I do for her sake,
Who has mine heart in cure.

("Dule" means grief or mourning; "cure," care or keeping.)

If Mary subsequently sent Letter 4 to Darnley, it could only mean that she was the bird who might fly away, leaving her partner to grieve for her loss. It could well have been sent to him while she was staying at Craigmillar Castle, in the weeks immediately before Prince James's baptism, when Maitland and his allies were lobbying for a divorce between them.

The Confederate Lords, however, insisted that the allusion referred not to a divorce but to the murder plot, even though it is contained within a letter primarily concerned with infidelity. As they advised Cecil, this passage "proved" that Mary had warned Bothwell shortly before the murder to "make good watch that the bird escape not out of the cage."

To clinch it, they doctored the letter. This time, the deception can be pinned down more or less conclusively. Cecil had been given a transcript of the French version of the document by the Scots. This is proved by the handwriting, which belongs to one of the clerks accompanying Moray's delegation. The same hand can be seen in the delegation's official papers and reports, and when compared to that of this transcript, they are identical. But when the clerk prepared the transcript, whether on his own initiative or on Moray's instructions, the word "per" was rendered as "père." "Mate" became "father," meaning Lennox, Darnley's father. (The fact that Lennox was at Glasgow and not Kirk o'Field on February 7 does not seem to have worried the lords.)

The transcript was then given to Cecil's clerk to translate into English. And not surprisingly, he construed it thus: "And watch well if the bird shall fly out of his cage or without his father."

This was too much even for Cecil. He was a brilliant linguist, who had won accolades for languages as a student at St. John's College, Cambridge. He knew the metaphor of the turtledove, which was repeatedly used in literature. When he spotted the sleight of hand, he struck out the word "father" in his clerk's translation and wrote "mate" above it.* Next

---

* His correction is clearly visible. Cecil's *t* can sometimes look like a *k*. The reading is erroneously given as "make" in the modern scholarly editions, which obscures the meaning, and yet no one appears to have gone back to the handwritten transcript to check it.

he corrected "père" to "per" in the French transcript by striking out the last letter. This obliterated the innuendo and should have called everything else into question but for the fact that Cecil wanted so badly for Mary to be incriminated, he either could not see, or else chose to ignore, the wider implications of the deception.

Letter 5 was said to have been written by Mary to Bothwell a few days before the marriage of Margaret Carwood, her favorite bedchamber woman, at Holyrood on Shrove Tuesday 1567, and therefore only a day or two before Darnley's murder. It began with an anguished rebuke: "My heart, alas! Must the folly of a woman whose unthankfulness toward me you do sufficiently know, be occasion of displeasure unto you?"

There is no internal evidence to prove that this referred to Carwood, who, far from being unthankful to Mary, was one of her most loyal and devoted servants. But whoever the woman was, her inconsiderate gossip had shamed and embarrassed the unfaithful lover — still said to be Bothwell — who had bitterly complained and insisted that she be replaced at the earliest opportunity.

The writer — still said to be Mary — said she could do little about it until she knew more precisely what it was that the servant was alleged to have said. In this she had been greatly hindered by the fact that she was not supposed to be in communication with her lover. If he did not write to her by that evening, she would have no choice but to take matters into her own hands. She would confront the servant and look for a replacement. Since the servant was about to be married, this could quickly be arranged.

The writer asked the man not to be so mistrustful. She reminded him of her "faithfulness, constancy and voluntary subjection." Their relationship was more than usually strained. She concluded the letter by saying, "You could do me no greater outrage, nor give me more mortal grief" than to doubt her honesty and her word.

When Cecil questioned the identity of the mysterious female servant, he was informed by the Confederate Lords that "Margaret Carwood was one special in trust with the Scottish queen and most privy to all her most secret affairs." Needless to say, Carwood would not be called to give evidence to Elizabeth and Cecil in support of the lords' interpretation of the letter or to testify to whom it had been addressed.

It would seem that the letter on its own proved nothing. Why then had it been selected for inclusion in the bundle of casket documents sent to England for inspection?

The answer lies in the newly discovered handwritten translation of the letter into English. Until now, the only known handwritten transcript of this document was in French. For some reason, the English translation among Cecil's working papers in the Public Record Office had not been identified in time for inclusion in the scholarly editions of the Casket Letters in the 1890s. Then, when the document was finally catalogued, no one appears to have seen anything new in it, and so it did not seem necessary to remark on it.

As a result, no one knew how Cecil interpreted the words "Je m'en deferay au hazard de la fayre entreprandre ce qui pourroit nuire à ce à quoy nous tandons tous deux." This literally translates as: "I will unburden myself of it at the risk of making her attempt something that could be harmful to what we are both aiming at."

What could Mary — assuming she was the author of the letter — have been talking about? The sentence is so obscure and indirect it is impossible to guess its meaning. All that is clear is that it does not explicitly refer to a murder plot.

The English translation made by someone in Cecil's office read: "I will rid myself of it, *and hazard to cause it to be enterprised and taken in hand*" (italics added). That is a misconstruction, a phony attempt by the English to refer to Darnley's assassination and a misrepresentation of the French. Cecil should have hung his head in shame, because he personally corrected the translation of the letter. At the top, he changed the words "My lord" to "My heart," where the original said "Mon coeur," but left the remainder as it stood.

No one examining this evidence with an open mind could regard letters 3 through 5 as incriminating. They are often said to be forgeries. That is very unlikely. They could easily be genuine letters Mary had written in 1566 to the narcissistic and adulterous Darnley. He had engaged in illicit affairs. He had been forever moping and sulking. He had deserted her and threatened to go and live abroad. His character closely resembles that of the recipient of these idiosyncratic and unusual "love" letters.

In letter 3, the unhappy couple were estranged and living apart, which Mary and Bothwell were not in the month after their marriage. In letter 4, the writer is Medea and not Glauce, the wife and not the lover. Lastly, in both letters 4 and 5, the writer reminds her partner that she has "voluntarily subjected" herself to him — a remark typical of Mary, who always believed that by marrying Darnley she had stooped beneath herself, and told him so when they had quarreled.

That leaves the last three letters, numbers 6, 7 and 8, which relate to the Confederate Lords' charge that Mary was a willing accomplice in her abduction at Almond Bridge. Letter 6 is probably a genuine letter from Mary to Bothwell that the lords had mischievously redated. It was almost certainly written after, not before, the abduction, since it contains references to events that we know to have occurred after Mary had arrived at Dunbar and Bothwell had left for Edinburgh to get his wife to file her divorce petition.

"Alas my lord," it begins, "why is your trust put in a person so unworthy to mistrust that which is wholly yours?" Or, expressed in simpler language, "Why have you gone and trusted someone so unworthy? Their interventions have led you to mistrust the one person who really cares about you." As the writer explained:

> You had promised me that you would resolve all, and that you would send me word every day what I should do. You have done nothing thereof. I advertised you well to take heed of your false brother-in-law. He came to me and without showing me anything from you, told me that you had willed him to write to you . . .

Huntly was this "unworthy" individual, and he was once again described as "your false brother-in-law." The alarm bells ring, because it was only *after* Mary's abduction, on April 24, 1567, that Huntly and Bothwell had fallen out over Bothwell's plan to divorce Huntly's sister and marry Mary.

Mary had quarreled violently with Huntly, but not until after his sister's divorce petition was filed. We also know that Bothwell, when he returned from arranging his divorce, had argued bitterly with Mary. Drury's handwritten reports show that there had been a "great unkindness" between them lasting half a day. It had been their one spectacular fight before their marriage.

The writer said that her lover had "willed" Huntly to offer his advice as to what she should say and "where and when you should come to me." But it had not gone according to plan. Instead, Huntly had given the writer a dressing down. He had railed against her "foolish enterprise," saying that she could never marry a man who not only was married already but had kidnapped her, and that his family, the Gordons, would never allow it.

According to the Confederate Lords, Mary had sent this letter "from Stirling" to warn Bothwell of the risks he was about to take at Almond Bridge. Moray's chronology asserted that she had written it on April 21, 22 or possibly early on the morning of April 23, 1567, before her abduction on the 24th.

But nowhere do the words "from Stirling" appear in the letter. Neither does Moray's claim that the letter preceded the abduction add up. The lords maintained that the writer's purpose was to encourage Bothwell to abduct her, using Huntly as his intermediary to inform her of the time and place where the pretend kidnapping would occur. We are expected to believe that she had so far given Bothwell no idea of when and where the event was to be staged. Bothwell, who had promised to settle everything, could not in the end decide, so he took advice from Huntly, asking *him* to ask the writer what she wished to say about it and "where and when you should come to me."

There are two handwritten transcripts of this letter in Cecil's papers, one in the original French and one in an English translation. The original French reads: Huntly (your "false brother-in-law") "m'a presché que c'estoit une folle enterprise, et qu'avecques mon honneur je ne vous pourries jamais espouser, veu qu'estant marié vous m'amenies et que ses gens ne l'endureroient pas et que les seigneurs se dédiroient." This literally translates as: He "preached to me that it *was* a foolish enterprise, and that with my honor I could never marry you, seeing that being married you *brought* me away, and that his folk would not endure it and that the lords would go back on their word."

Moray maintained that the meaning of the French was: He "preached to me that it *would be* a foolish enterprise . . . seeing that being married you *would bring* me away . . ."

But Mary was a native French speaker. If she had used the past tense, the events she would have been describing were in the past and not the future, as in the interpretation of her accusers. Moreover, the use of the past tense makes far better sense and corresponds to what we know really happened at Dunbar after Bothwell had left for Edinburgh to arrange his divorce, leaving Mary on her own with Huntly.

On the 26th, Bothwell had galloped to Edinburgh. There, he had arranged for his wife to lodge her divorce suit in the Protestant court. Next day, Mary had asked the Archbishop of St. Andrews to grant Bothwell an annulment in the Catholic court. When Bothwell returned to Dunbar, he found that Huntly and Mary had quarreled. Then Mary and Bothwell had their first big row.

If letter 6 was genuine, but was written after rather than before Mary's abduction, it would also add considerably to our understanding of her feelings for Bothwell while he was away in Edinburgh and she was quarreling with Huntly:

> I wish I were dead. For I see everything is going badly. You promised something very different in your prediction, but absence has power over you who have two strings to your bow.* Hasten your answer so that I may not fail and put no trust in your brother[-in-law] for this enterprise. For he speaks and is all against it. God give you good night.

All this makes a great deal of sense. We will never know exactly what it was that Bothwell had promised Mary. But her feelings tally with those she later expressed in telling her story to Robert Melville and the Bishop of Dunblane for their respective diplomatic missions to Elizabeth and Catherine de Medici. There, she knew she had done wrong in sleeping with a married man and was attempting to justify her behavior to herself as much as to others.

But not everything went Moray's way. When Cecil gave the original French version of letter 6 to his clerk to translate, the tenses were rendered accurately. "And thereupon," he wrote, Huntly "hath preached unto me that it *was* a foolish enterprise and that with mine honor I could never marry you, seeing that being married, you *did* carry me away . . ."

When docketing the translation (i.e., labeling it for filing), the clerk put, "Copie from Stirling after the Ravissement . . ." Exactly. Whoever he was, he was an honest man. He had understood the sense of the document he had just translated, which could not have been written before Mary's abduction and so "proved" none of Moray's claims.

And Cecil saw the danger. He was getting seriously worried. This "evidence" was not turning out to be anything near as good as Moray had said it would be. Sometimes the ends have to justify the means. With a stroke of his pen, he crossed out "after" and wrote "afore" above it in his own inimitable scrawl. Now it read, "Copie from Stirling *afore* the

---

* This could mean either that Bothwell could choose between his wife and Mary, or that he could mend his fences with the lords and ditch her.

Ravissement . . ." In a second or two, an innocuous and fully comprehensible document was turned into something that was both incriminating and complete gibberish. This "afore" was perhaps the most important single word that Cecil ever put into any document connected with Mary Queen of Scots.

Letters 7 and 8 are tame in comparison, which may explain why there are no handwritten transcripts of either. They are known only from the versions later printed in Scots by Mary's enemies.

Letter 7 is almost an exact duplicate of the core of letter 6. It is once again supposedly a letter from Mary to Bothwell, written on the eve or a few days before her abduction. "Of the place and the time," it began, "I remit myself to your brother and to you . . . He finds many difficulties."

The letter seems to have been no more than a variant Scots translation of letter 6, included to fill out the case. Perhaps the overriding reason why it was introduced as evidence is that it described Huntly to Bothwell as "your brother" and not "your false brother-in-law," thereby softening the clash with the known fact that Huntly had not quarreled with Bothwell until *after* the abduction, when Mary had consented to marry him.

Letter 8 is the last of the letters said to have been written on the eve of the abduction and the third Mary is supposed to have sent to Bothwell in less than twenty-four hours. It too reiterated the gist of letter 6. But whereas letter 7 had attempted to correct an anachronism in the lords' description of Huntly, this time he was described as "your brother-in-law *that was.*"

The letter begins, "My Lord, since my letter written, your brother-in-law that was, came to me very sad, and has asked me my counsel, what he should do after tomorrow, because there be many folks here, and among others the Earl of Sunderland . . ."

That description of Huntly is less plausible even than "your false brother-in-law." And there was a second careless mistake, because the Earl of Sunderland did not accompany Mary from Stirling to Linlithgow and on to Almond Bridge after her visit to her son, nor was he later at Dunbar. Those attending Mary on her journey were Maitland, Huntly and Sir James Melville. Even Moray's own retrospective account of Mary and Bothwell's proceedings failed to mention Sunderland, who does not figure in the story. It is a mystery why his name ended up in letter 8.

Of letters 6 through 8, the three "abduction" letters, only number 6 is worthy of serious attention. It is likely to be genuine, but far from making it seem that Mary colluded with Bothwell in her abduction, it proves nothing of the sort. "I wish I were dead," she was alleged to have said. "For I see everything is going badly. You promised something very different in your prediction . . ." These are hardly the sentiments one would expect from a willing partner about to engage in a daring escapade.

Beyond that, the past tenses were fatal to Moray's case. He probably knew it and thought something more concrete would be needed if the lords were to win the argument. This led to the manufacturing of letters 7 and 8, which tried to amplify the text of letter 6, but which in the process further exposed the invalid assumptions on which the argument rested.

And yet superficially the Casket Letters were damning if they were genuine and in Mary's handwriting. Two months after first reading Buchanan's dossier and the sample copies of the evidence Moray had supplied to whet his appetite, Cecil decided to act. He persuaded Elizabeth to empower a special tribunal — often ambiguously called a "conference" to avoid the use of the word "court" — to examine the case against Mary.

A legal trial would shortly take place under the auspices of impartial arbitration. Elizabeth was probably sincere when she said that her aim in appointing commissioners was to bring about a "good end to the differences, debates and contentions grown and continued between her dear sister and cousin, Mary Queen of Scots, and her subjects." But the tribunal as Cecil was to engineer it was closer to a special court set up to convict Mary on a charge of conspiracy to murder Darnley. The judges were to be the Duke of Norfolk, the Earl of Sussex and that veteran of Scottish business Sir Ralph Sadler, Henry VIII's former secretary and ambassador to Edinburgh.

Cecil's idea, an ingenious one, was that Elizabeth would appear to act as an unbiased referee over the Casket Letters, whereas in reality she had been persuaded to appoint a committee that was authorized to pronounce them genuine or forgeries, and so determine the extent of Mary's guilt.

Mary was hesitant about accepting the tribunal's legality, but was scarcely in a position to argue. She had everything to gain from a careful scrutiny of documents that she claimed were outright forgeries. She took the Duke of Norfolk's nomination as a positive signal, since his opinion of the Confederate Lords as the congenital foes of monarchy was unconcealed. She therefore accepted the tribunal's terms of reference and named her advocates, even if she did not formally concede Elizabeth's right to detain her in England or to try her.

The tribunal opened on October 4 at York, when Mary's advocates were allowed to speak first and bring their own charges against Moray and the Confederate Lords. They protested against the "great injuries, wrongs and damages" inflicted on Mary and her loyal subjects by the rebel lords, and appealed to Elizabeth for relief as an honorable umpire. If Moray and his allies had any evidence, they said, it should be exhibited in writing so that it could be lawfully examined and challenged.

Then it was Moray's turn. Not surprisingly, he was reluctant to pursue the murder charge or exhibit the evidence he said he had collected. He was watching his back, fearful of Elizabeth's wrath should he accuse his sister of murder and adultery and then fail to prove it. He also wanted guarantees in advance that if he was successful in establishing Mary's guilt, she would never be returned as queen to Scotland.

This was the heart of the matter. Moray feared that even if he *did* prove Mary to be an accomplice in Darnley's death, Elizabeth still would not recognize his appointment as regent, agree to hand over Mary to the lords to be dealt with as they thought fit, or keep her safely locked away in an English prison. Until he was assured of victory one way or the other, he was unwilling to place the originals of the Casket Letters on file before the court as evidence — at least officially.

Moray wanted it both ways. He declined to introduce the letters formally, but happily showed them to the English judges "privately and secretly." The letters were not shown to Mary's representatives, and under the procedure Cecil had laid down for the tribunal, they had no right to demand to see them.

The English judges were unimpressed by Moray's tactics. The allegations were infamous, but were the letters Mary's? The judges had seen nothing to prove it. The most hardheaded of them, the Earl of Sussex, gave Cecil a blunt warning. He wrote on October 22: "If the party adverse to her accuse her of the murder by producing the letters, she will deny them, and accuse the most of them [i.e., the Confederate Lords] of manifest consent to the murder, which could hardly be denied."

Sussex had realized that if Moray and his delegation exhibited the Casket Letters, all Mary's lawyers had to do was deny they were hers. She could turn the tables on the rebel lords, accusing them of complicity in Darnley's murder. Then they would be in the dock, not her. And since a majority of her accusers had been involved, Moray's case would collapse.

Norfolk was even more disturbed. "This cause," he explained to Cecil, "is the doubtfullest and most dangerous that ever I dealt in." Mary's advocates had already hinted that if the Casket Letters came formally to trial, she would demand to attend the tribunal in person. For his part, Norfolk was shocked by the implications of letter 2 (the long Glasgow letter) and letter 6 (the most compelling of the three abduction letters), but it was also perfectly clear, he said, that the lords were not interested in justice, but "seek wholly to serve their own private turns." In their eagerness to vindicate themselves, "they care not what becomes neither of queen nor king!"

Cecil's plan was starting to go wrong. What Sussex and Norfolk had said made it plain that the English judges knew that many of Mary's accusers were themselves accomplices in Darnley's murder, and the last thing Moray wanted was for this to come out.

Cecil hastily intervened. He arranged for the tribunal to be temporarily adjourned. It was moved south to Westminster and five extra judges were appointed, chief among them Cecil and his brother-in-law Sir Nicholas Bacon, shifting the balance of opinion decisively in Moray's favor. The enlarged tribunal resumed its hearings on November 26.

Cecil, meanwhile, talked Elizabeth into giving Moray the assurances he sought. If the Casket Letters were genuine, Elizabeth would hand Mary over to the Confederate Lords, provided they guaranteed her safety, or else keep her in England. She would also recognize Moray as regent. This gave Moray what he wanted, and shortly afterward, he formally charged Mary with complicity in Darnley's murder.

Mary was stunned by her cousin's change of heart. She had been assured that the English judges would not seek to try her, but only to act as umpires in her dispute with her rebel lords. She refused as a queen to be called to account by a jurisdiction she did not recognize, and ordered her advocates to withdraw immediately from the tribunal.

They left on December 6, but it was too late. The next day, Moray laid the originals of the casket documents on the table. They were closely examined in lengthy sessions lasting two full days, and then again on the 14th at Hampton Court.

When correcting the tribunal's minutes describing how the documents had finally been submitted as evidence, Cecil took care to cross every *t* and dot every *i*. "And so," his secretary's draft suggested, "they produced a small coffer of silver and gilt, wherein were certain letters and writings they said [were] of the Queen of Scots to the Earl Bothwell."

It sounds simple enough, but that would not do for Cecil. He rewrote the minute in his own handwriting:

And so they produced a small gilt coffer of not fully one foot long, being garnished in many places with the Roman letter "F" set under a king's crown [the monogram of Francis II), wherein were certain letters and writings which they said and affirmed to have been written with the Queen of Scots' own hand to the Earl Bothwell.

These were not just Mary's letters. They were the letters "affirmed" to be in her "own hand." And it was not just any old casket. The added detail gave credence to the investigation.

The irony is that forever after, the casket containing the famous letters has come down in history exactly as Cecil described it. The Scots' delegation, who had so far thought the object unremarkable and called it just "the box" or "the silver box," would not even be allowed to describe their own trophy.

The tribunal was rigged. Moray's delegation did not submit the originals of the Casket Letters for close examination until Mary's advocates had withdrawn. And under the highly irregular rules that Cecil had engineered, Mary was to be denied access to them, at least until after they had been judged by the tribunal itself.

To her credit, Elizabeth now balked. Deep down, she wanted Mary to be found innocent and restored as queen to Scotland. She disliked and distrusted Moray, and did not believe that the Casket Letters were genuine. The whole point of the tribunal in her eyes had been to uncover the extent to which they had been forged. Now the proceedings were getting out of hand. They were far too one-sided. Her reputation was at stake if she acted unfairly or dishonorably.

She intervened to reassert control over Cecil and her own policy. She suspended the tribunal a second time, then added the Earls of Northumberland, Shrewsbury, Huntingdon, Westmorland and Warwick to the roster of judges, herself supervising the proceedings at Hampton

Court on the 14th, when the letters were examined for the last time. She had the minutes of the tribunal's sessions so far read aloud, then Moray's accusations, and finally the "evidence," notably the Casket Letters.

Cecil took the minutes as usual. "There were produced," he said, "sundry letters written in French, supposed to be written by the Queen of Scots' own hand to the Earl Bothwell":

> Of which letters, the originals* . . . were then also presently produced and perused, and being read, were duly conferred and compared for the manner of writing and fashion of orthography with sundry other letters long since heretofore written and sent by the said Queen of Scots to the Queen's Majesty. And next after these was produced and read a declaration of the Earl Morton, of the manner of the finding of the said letters, as the same was exhibited on his oath the 9 December. In collation whereof no difference was found.

Cecil's minute is too good to be true. It is almost certainly misleading. Read uncritically, it appears that the handwriting test had been passed with flying colors when the original Casket Letters were compared to genuine letters from Mary to Elizabeth that had been filed away over the years in the royal archives.

This is intrinsically unlikely. At least so far as the crucial long Glasgow letter is concerned, the text itself explained that it had been "scribbled" late at night. "Excuse my evil writing," it said, "and read it over twice. Excuse also that I scribbled . . ." Moreover, we have already seen that letters 7 and 8 were forgeries on the evidence of their contents.

It is inconceivable that Mary's scrawling hand could have been compared and collated to the genuine letters she had sent to Elizabeth without some differences being noticed. Anyone who has looked at the dozens of examples of Mary's autograph letters to Elizabeth in the archives would immediately spot that when writing to her "sister queen" she used her very best handwriting. These letters are usually immaculate, with only infrequent signs of haste. If they had been compared to Mary's scribbling hand, it would have been impossible to reach any definitive conclusion one way or the other. At most, a certain resemblance might have been observed.

And what exactly was it that had been "collated"? Had all eight Casket Letters been scrupulously checked against Mary's genuine letters to compare the handwriting letter for letter? Or was it merely that the *list*

---

* This really means the original Casket Letters and not the transcripts or copies.

of casket documents as given in Morton's declaration to the tribunal, explaining the circumstances in which they were found, had been compared to the items laid on the table to see if they corresponded to the description given under oath? Cecil's minute is tantalizingly ambiguous, as every one of Mary's biographers has been obliged to point out.

If, on the other hand, the handwriting test had been cursorily conducted or was based only on a few sample pages, it is conceivable that it could have been passed. We have already seen that apart from letters 7 and 8, Moray is likely to have submitted genuine if artfully chosen pages, some of which were culled from earlier letters to Darnley and others from later letters to Bothwell, and that the most damning interpolations in Letter 2 were most likely fitted into blank spaces on the paper. Since it would have taken many hours to scrutinize every folio of every one of the Casket Letters, and since we know that by the time the tribunal had reached Hampton Court there were too many judges for everyone to sit around a small table and look closely at every single document, it is possible that just enough was done to make it *appear* that the handwriting test had been passed.

When Mary heard of the proceedings at Hampton Court, she demanded to address the tribunal in person, which was refused. But Elizabeth told Moray's delegation that she would allow the judges to continue with their hearings only if her cousin was allowed to depute someone to answer the charges on her behalf, or else speak to a deputation sent to her by Elizabeth.

Mary finally issued a statement. She would decline to answer until she was allowed to appear before Elizabeth in person. Otherwise, she would not recognize the tribunal. She had made too many concessions. "I am not an equal to my rebels," she said, "neither will I submit myself to be weighted in equal balance with them."

Mary stood on her dignity. She refused to be treated so disrespectfully. Moray and his delegation had lied, she protested. They had maliciously charged her with Darnley's murder, "whereof they themselves are authors." The only further thing she would instruct her advocates to say was: "I never wrote anything concerning that matter to any creature. And if any such writings be, they are false and feigned, forged and invented by themselves, only to my dishonor and slander."

If only Mary had known it, Elizabeth largely agreed with her. She had watched the first day of the proceedings at Hampton Court, but saw her-

self as no more than an "observer." She did not think it right for one
queen to sit in judgment of a fellow sovereign. In her own words, the
purpose of these hearings had been "to understand truly and plainly the
state of the cause of the Queen of Scots," but "without prejudicing one
side or the other."

When this had manifestly not occurred and Cecil had written minutes that upheld Mary's guilt, Elizabeth brought the proceedings to an
abrupt end. She refused to allow Mary to come into her presence, but
said that if Mary was unwilling to answer to the charges as they were explained to her by a deputation, she would adjourn the trial indefinitely.
This is exactly what happened. By Christmas 1568, Mary had not been
found innocent, but neither had she been convicted. The matter lay
in abeyance. No conclusion about the truth of the Casket Letters could
be drawn one way or the other. Elizabeth was guided by the principles
of natural justice. Cecil's tribunal had been grossly unfair, and it was
Mary's right to put her side of the story to the judges. The one thing Elizabeth could *not* bring herself to do was meet Mary and listen to her argue her case in person.

Mary found this insulting. It implied that she was Elizabeth's inferior.
She was convinced she had been wronged by her cousin. "Alas, madam,"
she demanded, "when did you ever hear a prince censured for listening
in person to the grievances of those who complain that they have been
falsely accused?"

Mary spiritedly advised Elizabeth to put out of her mind the notion
that she had fled to England to save her life. She went only to recover her
honor and obtain support to be revenged on her rebels. She would not
"answer them as their equal." She would never abase herself in that way.
She had always regarded the English queen as her "nearest kinswoman
and perfect friend." She had supposed that Elizabeth would be honored
to be called "the queen restorer," and had hoped to receive that kindness
from her.

Mary saw now that she was mistaken. And Cecil was behind it all.
"You say," she berated Elizabeth, "that you are counseled by persons of
the highest rank to be guarded in this affair. God forbid that I should be
cause of dishonor to you when it was my intention to seek the contrary."

Following events as best she could from her imprisonment by correspondence, Mary reacted in the only way she knew: the way the Cardinal
of Lorraine would have behaved. She wrote to Philip II, the leader of the
Catholic cause in Europe, to protest her innocence and seek his aid. "I

am deprived of my liberty and closely guarded," she said. In consequence, she wanted all the Catholic princes in the world to know that she was "an obedient, submissive and devoted daughter of the holy Catholic and Roman Church, in the faith of which I will live and die, without ever entertaining any other intention than this."

Her letter was no more than a feeler. But it was a dramatic turnabout for a queen who, apart from a brief interlude before the Rizzio plot, had made religious compromise the cornerstone of her policy. A seismic shift was about to occur, one that discounted her kinship bonds to Elizabeth and linked her cause to that of Philip II and the papacy, and therefore posed a greater threat to Elizabeth's "safety" and the Protestant cause than anything since Mary had returned home to Scotland.

Mary had been bloodied by Cecil, but she was unbowed. A new phase was about to open in her life.

27

Captive Queen

WHEN ELIZABETH rejected Mary's request for a personal audience and Mary refused to discuss the Casket Letters through intermediaries, a stalemate was reached. The tribunal did not reconvene. No decision was ever given. Mary had not won, but neither had she lost, except that Elizabeth yielded to Cecil's insistent demands and grudgingly recognized Moray as regent of Scotland. Otherwise, there would have been a power vacuum in the country.

But Elizabeth still mistrusted Moray. He was fobbed off with a loan of £5000, which from his point of view was woefully inadequate. Even if his sister had been marginalized for the moment, she was alive and well and living as a queen in exile with nothing proved against her. Elizabeth had such a high respect for the ideal of monarchy, she had sworn all the judges at the tribunal to strict secrecy. As far as English public opinion was concerned, the Casket Letters did not exist.

Mary would be held in captivity by Elizabeth for the next eighteen years, a prisoner in the sense that she was under house arrest and forbidden free access to visitors, and yet for much of this time she was allowed many if not all of the courtesies and luxuries due to an exiled ruler and guest. Despite the length of her imprisonment, it was never actually decided that this would be a permanent arrangement. Almost until the end, her privileges included the right to diplomatic representation, the use of ceremonies and protocol appropriate to a royal court (on a strictly reduced scale), as well as the right to exercise and occasionally to ride, but only within a mile or two of her lodgings. In some respects, Mary could behave as she pleased, in others she was severely restricted, her movements watched, her letters intercepted, her person guarded — sometimes rigorously, sometimes in ways that were unbelievably lax — and yet despite her many entitlements, she saw herself as wronged from the very first day to the last.

What rankled was less the nature of her imprisonment than the reason for it. Mary felt that Elizabeth did not have the nerve to do her duty and restore her to her throne. Her deep sense of grievance remained through all the tedium and trauma of confinement. Beyond that, her spirits were often high. Her first custodian, Sir Francis Knollys, said of her, "She is a rare woman; for, as no flattery can abuse her, so no plain speech seems to offend her, if she thinks the speaker an honest man." She was never distant or reserved as long as her "royal estate" was recognized. She talked a lot, but was bold, pleasant and "familiar." "She sheweth a great desire to be avenged of her enemies." She abhorred cowardice, whether in friend or foe. "She sheweth a readiness to expose herself to all perils, in hope of victory." To all other concerns, she was "indifferent."

Mary had arrived in England without money or a change of clothes. She had at first only sixteen attendants, but soon more than a hundred of her old domestic staff congregated around her. Among those returning were Mary Seton, Bastian Pages and his wife, and George Douglas ("Pretty Geordie") and "Little Willie" Douglas, the heroes of her escape from Lochleven. Alone of the four Maries, Seton was happy to share her mistress's dark days for fifteen long years, until her own health gave way, when she retired to the convent of St.-Pierre-des-Dames at Rheims, over which Mary's aunt Renée presided.

Curtailing the size of Mary's household was the most visible way in which the wings of the royal eagle could be clipped. Within a few months, her attendants had been reduced to sixty, a number that was halved within a year. After three years, the number crept back up to forty, then was reduced to thirty when one of the men was denounced as a Catholic priest in disguise. In time the number was halved again, to sixteen, although this excludes the ten or so kitchen and pantry help, laundry staff and stable grooms needed for menial services. Mary's food was cooked in her own kitchens. Her sheets were washed by her own laundresses, and she kept her own horses, even if she was not allowed to ride as often as she wished. In the final and most straitened years of her captivity, when almost every courtesy was denied her, the total number of her staff was still as high as fifty-one.

But Mary was several times forced to say goodbye to servants she very much wanted to keep. When she refused to choose those who were to be retained and those dismissed, it would be done for her without further consultation. "She was exceedingly troubled, weeping and sorrowing," but no notice was taken. "Little Willie" Douglas was one of those let go. With typical generosity, Mary cushioned the blow, sending the laid-off servants to her ambassador in Paris, who was ordered to give them pensions. Sometimes she could find them other positions. When one of her favorite servants, her perfumer, known affectionately as Angel Mary, was dismissed, Mary managed to get her a new job at the French embassy in London.

Within a few weeks of her arrival, Mary began to eat normally and was allowed to venture outdoors from her apartments in the tower of the southeast corner of Carlisle Castle, overlooking the River Eden. She was twice permitted to cheer on her male domestic staff as they played football, ten men on each team, on a nearby village green. Later she went hare-coursing, galloping off so fast that her guards thought she was trying to escape across the border into Scotland.

Despite her initial lack of funds, and until fresh installments of her private income as dowager queen of France were paid, Mary kept up appearances. She was short of cosmetics for the first six months, but took as much trouble as ever over her clothes. When she refused to wear any clothes except her own, a resigned Knollys dispatched a messenger to fetch her wardrobe from Lochleven Castle and also asked Moray to send her coach and a supply of dresses from Edinburgh. Since the royal apartments at Holyrood had been looted by Morton's men, there was little to be had from there beyond a selection of chemises, some perfumed gloves and a clock, but five cartloads and four horseloads of clothes and personal effects arrived from Lochleven, which was a start. Mary was soon commissioning replacement items, and some thirty carts would be needed a year later for her belongings when she moved from place to place.

When Seton reappeared, Mary could begin to look like a queen again. Her hair needed attention, because to avoid recognition in her headlong escape after the battle of Langside, she had cut much of it off. In

Knollys's hearing, she praised the most devoted of her Maries as "the finest 'busker,' that is to say, the finest dresser of a woman's head and hair, that is to be seen in any country." There was nothing she could not do. "Among other pretty devices . . . she did set such a curled hair upon the queen that was said to be a periwig* that showed very delicately." Every day she gave Mary a different style "without any cost, and yet setteth forth a woman gaily well."

Mary spent her twenty-sixth birthday at Lord Scrope's house at Bolton Castle, an isolated spot on the rugged high ground overlooking the picturesque valley of Wensleydale in north Yorkshire. She was taken there from Carlisle in July 1568, a fortnight or so after Cecil began his careful reading of Buchanan's dossier. Her birthday, on December 8, was the day after Moray exhibited the originals of the short and long Glasgow letters at Westminster.

Bolton Castle may have seemed less a place of captivity than one of refuge. Lord Scrope was the senior English official on the western side of the Scottish border. As he was stationed at Carlisle, he was regularly away from home. Mary was entertained by Lady Scrope, the Duke of Norfolk's sister. Her sympathies were with Mary, her religious opinions Catholic. Like her brother, she had a low opinion of Moray, and it must have been reassuring for Mary to have her shoulder to cry on when the Casket Letters were produced.

Moreover, her home was warm and comfortable. Not just a cold and damp fortress, its living quarters had one of the earliest central heating systems of any house in Britain. Mary would not be allowed to stay there long. Within a few months, she would be moved again to a more secure and forbidding location. Cecil not only feared Lady Scrope's influence, he was also greatly exercised about how close the castle was to the border and the sea. Already he was on his guard in case Mary should try to escape.

In late January 1569, she was taken on a long journey south to Tutbury Castle in Staffordshire, a ten-day trip through the steep and muddy pathways of Yorkshire and Derbyshire. Tutbury lay in the Midlands, the heartland of England: far enough away to make a rescue by the northern Catholic gentry difficult, and sufficiently distant from London and the ports to make a dash for freedom unlikely.

---

* This is one of the earliest references in the English language to the use of hair extensions. Later, when Mary's hair became thin and gray, she wore a wig.

Mary was received by George Talbot, Earl of Shrewsbury, who was to be her custodian for the next fifteen years. He was one of Elizabeth's leading noblemen, the highest-ranking peer after the Duke of Norfolk and the ideal candidate, from Elizabeth's viewpoint, to manage an exiled Catholic queen. He was a Protestant, but only just, and knew better than anyone the sorts of contradictions involved in dealing with Mary. He had to win her compliance, but definitely not her affection. He had been present at the concluding sessions of Elizabeth's tribunal, so he knew of Moray's accusations and the Casket Letters. As Mary had never made a secret of her Catholic and Guise connections or her desire to outwit her captors, she was a constant focus of gossip, rumor and innuendo.

Mary came to respect Shrewsbury, whose gruffly expressed opinions of Cecil were closer to those of Sussex and Norfolk than anyone else. He outlived his ordeal, even if after more than a decade in the post he sank into black moods and a sense of paranoia in his dealings with Elizabeth and her courtiers. He felt obliged to denounce Mary periodically as a "foreigner," a "papist" and "my enemy" to maintain his credentials. He also faced the burden of the spiraling costs of entertaining Mary and her suite. It was lucky that he could afford it, because Elizabeth was too tightfisted to bear the full expenses of his charge. Sometimes she would give some money, often none at all, and while Mary could afford to support herself, she refused to contribute any of her French income unless she was granted her liberty. Shrewsbury was caught in the middle, forced to pay the lion's share of what amounted to a gigantic cuckoo in his domestic nest — a mimic royal court.

Shrewsbury had his own domestic ambiguities. Shortly before becoming Mary's custodian, he had married the redoubtable Bess of Hardwick, with whom Mary was to spend a great deal of her time. Bess was a woman of lofty pride, quick jealousy and insatiable ambition for herself and her children by a previous marriage. In her dealings with Shrewsbury, her fourth and last husband, the traditional gender roles were reversed. Her favorite Greek heroine was Penelope, the most independent woman in classical literature, whose picture, flanked by those of Perseverance and Patience, was Bess's theme for her best-loved tapestry. Her ambition, not to mention her unfounded suspicion that her husband was sleeping with Mary, led to the breakdown of her marriage. Whereas in 1568, Shrewsbury called Bess "my own sweetheart," "my dear," "my jewel," ten years later he would castigate her as "my wicked and malicious wife," "my professed enemy" or simply just "wife." He refused to spend a single night under the same roof as her.

Fortunately, the earl had a mistress and seven mansions, while his wife owned houses at Chatsworth and Hardwick, in Derbyshire, in her own right. Tutbury was their least important property, a royal castle belonging to the duchy of Lancaster that was merely leased. It was damp, dilapidated and almost destitute of furniture, and had a particular problem with the drains. Elizabeth sent beds, bedding and a dozen carpets to provide a minimum of creature comforts for Mary on her arrival, but these were inadequate to the task, and within days of crossing the drawbridge into the courtyard, the exiled Queen of Scots was in bed with rheumatism and a fever.

Over the ensuing months, Shrewsbury painstakingly negotiated with Elizabeth and Cecil to transfer Mary from Tutbury to one of his more comfortable residences. She was moved first to Wingfield Manor and then to Bess's house at Chatsworth. Then, in November 1570, she was taken to Sheffield Castle, the earl's principal home, where she was allowed to settle down for more than two years, her longest unbroken stay in any one place.

A mile or so from the castle was Sheffield Lodge, or Manor, a magnificent house on the site of a former hunting lodge which Shrewsbury extended and rebuilt while Mary was in his charge. She first stayed there in April 1573, and over the next eleven years was shuttled between the two Sheffield addresses, making occasional forays to Chatsworth and to a lodge that Shrewsbury built for her at the spa at Buxton — until the earl was replaced as her custodian in 1584.

The close proximity of Sheffield Lodge and the castle was invaluable, since to counterbalance the number of Mary's attendants and cope with the security threat should an escape attempt be made, Shrewsbury had to maintain an even larger staff than she did. This meant that the total size of the joint household was second only to Elizabeth's own court. And in the absence of modern plumbing, long stays at individual houses occupied by large numbers of people were always to be avoided for sanitary and medical reasons. To keep houses free from disease, the drains and latrines had to be cleaned periodically. This involved digging them out and carting the excrement and kitchen waste elsewhere, a noxious chore not easily undertaken while the house was inhabited.

Throughout her long years in Shrewsbury's custody, Mary was treated honorably. Her rooms were luxuriously hung with tapestries and lit at night by candles set in gilt chandeliers, which was just as well, since Mary often refused to go to bed until one o'clock in the morning. Turkish carpets lay on the floor, items so valuable they were normally used only as table covers. Mary's chairs were upholstered with crimson velvet and cloth of gold, while her female attendants sat on low stools, as they had at Holyrood when she was still a reigning queen. At one end of her presence chamber was a high-backed chair under a cloth of state on a dais. A smaller cloth of state was erected over her chair in her bedroom. She slept in a large canopied state bed shrouded by curtains, and her sheets were of the finest linen, changed every day. She would have expected no less.

One of Cecil's emissaries, Nicholas White, who visited her at Tutbury on his way to Ireland, remarked on the layout of her space. She had two main rooms, which she had arranged as a privy chamber and a presence chamber. Mary "came forth of her privy chamber into the presence chamber where I was, and in very courteous manner bade me welcome." She then conducted a royal audience, impressing even this staunchly Protestant visitor with her looks and charm. "She hath withal," he informed Cecil, "an alluring grace, a pretty Scottish accent, and a searching wit, clouded with mildness." The conversation turned to connoisseurship. It was a subject on which Mary felt at home, and she extolled painting as the "most commendable" of the arts.

When White protested that painting, at least in religious art, was "false truth," Mary ended the interview. It was a deliberately provocative remark. "She closed up her talk, and bidding me farewell, retired into her privy chamber." White was left to find his own way out, but before departing, he looked up at her cloth of state. "I noted this sentence embroidered" on it, he said: " 'En ma fin est mon commencement' — 'In my end is my beginning' — which is a riddle I understand not."

Little did he know about Mary's love of emblems and epigraphs. "In my end is my beginning" was the motto of her mother, Mary of Guise, whose emblem, or *impresa*, was the phoenix. This was the mythical bird that set fire to itself and rose anew from the ashes every five hundred years; it stood for unsurpassing beauty or quality, for hope and for ultimate triumph. Up to now, Mary's own *impresa*, chosen while she was in France, had been the marigold turning to face the sun. Her motto was "Its virtue draws me." Later she had used the same motto with an iconographic variant: the lodestone, a naturally magnetic rock used by sailors as a navigational aid.

Now a prisoner in England, Mary changed her motto to her mother's, which was much better suited to her predicament. It was innocuous in the sense that it belonged to her family, but was resonant with prophetic meaning. "In my end is my beginning" was her way of proclaiming that even if she was killed, her dynastic claim would live on in the person of her son, James, who would inherit the English throne.

White's report shows how civilized and luxurious Mary's surroundings were. And yet they were still a prison. She despised the guards who patrolled outside her bedroom window and who escorted her, often carrying pistols, on the rare instances when she rode out in the park or on the moors. Indoors, she was watched night and day by sentries who were summoned by a drumroll at five in the morning. She increasingly feared poison, which was why she insisted on appointing her own kitchen and pantry staff.

But if Mary sometimes lapsed into pessimism, she never forgot she was a queen. She stuck stubbornly to protocol, clinging to the rituals of royal eating whenever she took her meals. Her "diets" were recorded in her household accounts. Like Elizabeth, she enjoyed two "courses" at both dinner and supper, and each course included sixteen separate dishes, individually presented and accompanied by bread, wine, salads and fruit. This meant thirty-two dishes for each meal and sixty-four for each day. Sometimes there were more, sometimes slightly fewer "as the provision serveth" (i.e., depending on what was available). And it was Mary's prerogative to choose whichever of these dishes she wished to eat or taste.

Meat dishes predominated except on "fish days" — usually Fridays and every day of Lent. At dinner, normally served between 11 A.M. and noon, there would be soup, veal, beef, mutton, pork, capon, goose, duck and rabbit for the first course, followed by pheasant, partridge, kid lamb, quail, pigeon, tart and frittered apples or pears for the second. At supper, generally between 5 and 6 P.M. in winter and 7 and 8 P.M. in summer, there would be soup, veal, kid lamb, tripe, tongue, chicken, pheasant, pigeon and rabbit, followed by quail, baked pudding, tart and fruit. On fish days there would be plaice, whiting, haddock, cod, turbot, skate, barbel, ling, eel and pike, followed by carp, sprats, trout, tench, herring and tart for dinner; and salmon, chub, trout, herring, perch, eel, shrimp and crayfish for supper.

Mary dined in state at her own table. Her food was delicately served off silver dishes, her wine poured into crystal glassware. She probably used cutlery made in Sheffield. It was of such high quality, she sent it as gifts to her relatives in France. Before beginning to eat and while taking a break between the two courses, she washed her hands in a silver-gilt bowl. Her principal officers presented her food and wine to her in her presence chamber (which doubled as a dining room) before themselves sitting down to eat at their own table. Mary's gentlewomen were given meals of nine dishes; her secretaries were allowed seven or eight. Most of the lesser servants and their wives and children ate in the kitchens, consuming the leftovers but holding some back for Mary's many dogs.

Inevitably, Mary put on weight. Although her diet was similar to what she had enjoyed in Scotland, there she had exercised daily, usually in the form of long rides, whereas now the combination of limited exercise and heavy meals caused problems. As she became stouter, her shoulders became rounded and she acquired a slight stoop. She suffered from constipation and other digestive disorders, and her old episodic illness returned.

Almost as soon as Mary had reached Tutbury, she was ill for a fortnight. No sooner had she recovered than she was ill again. She complained of severe abdominal pains, which were blamed on "wind" but could easily have been caused by complications arising from her gastric ulcer. Shrewsbury wrote anxiously to Cecil, "Oft times, by reason of great pains through windy matter ascending unto her head and other parts, she is ready to swoon. On Thursday last she received pills devised by her own physician, whereof she was very sick that night, but after the working amended."

A few weeks later, Mary vomited and fainted after taking her pills. Her doctor gave her whisky to bring her around, just as his predecessor had done at the court of St.-Germain ten years before. A petition was sent to Elizabeth that Mary might have two physicians in constant attendance, which was swiftly granted. Almost all her illnesses in captivity seem to have been related to her gastric ulcer, to neuralgia and severe headaches caused by inactivity and frustration, to digestive and bowel imbalances, or to a severe swelling of her leg. This last was perhaps an ulcerated leg or a form of deep-vein thrombosis — again, most likely stemming from inactivity — that is known to have first afflicted her in prison at Lochleven. She also suffered from chronic rheumatism, from pains in her left heel and from recurrent bouts of an unidentified viral illness. One of these attacks occurred at Sheffield in 1575, where she was laid low by what she said was a fever linked to abdominal pains and catarrh.

Mary felt she was aging before her time. She complained of "an incessant provocation to vomit," when she threw up "a very great quantity of raw, tough and slimy phlegm." Her abdominal pains were always on her left side, "under her short ribs." She was unable to sleep, sometimes for ten or twelve days at a stretch. Her doctors prescribed an enema, which caused her to vomit again. She continued to take her pills, without visible signs of improvement. At least some of her illnesses were psychosomatic, brought on by stress at times when it seemed that Elizabeth had forgotten her or refused to extend an olive branch. "No one can cure this malady as well as the queen of England," Mary repeatedly said, not without justification.

She especially yearned for her son. She could glean little news of him beyond the galling fact that he was being tutored by her enemies and brought up as a Protestant. She was not allowed to write to him, and in all the years of her captivity, she had hardly a single letter from him or his guardian. James's first letter to his mother appears to have been written as late as March 1585, when he was eighteen.

Some two or three years before she received it, Mary could contain herself no longer. She blurted out her feelings: "Is this just and right that I, a mother, shall be forbidden not only to give counsel and advice to my oppressed son, but also to understand what distressed state he is in?"

That she longed to see him again is suggested by an item she kept beside her until the day she died. It was a thin gold case with a folding flap, described in her probate inventory as "a book of gold enameled [and] containing the pictures of the late Scottish queen, her husband and her son." Aside from the fact that she wanted her son's portrait miniature close to her, it would be surprising if she had kept it in the same frame as a picture of herself and her dead husband if she had murdered him.

Mary's poor health was exacerbated by the lack of support she felt she had received from France. After two years had gone by, she so far humbled herself as to implore Catherine de Medici to take pity on her. She begged her hardhearted former mother-in-law to listen to "this little word of humble request to have some aid for Scotland." When she wrote to Charles IX recommending a servant for a position in the Garde

Écossaise, she let slip pathetically that "I have not received an answer to any of my letters, which is the reason I did not trouble you about any of my affairs."

In all of the passing years, there was little sign of the acute intermittent porphyria from which some medical experts have claimed Mary suffered. Her illnesses stemmed chiefly from inactivity, stress and depression. To counter the last of these, she wore an amethyst ring, which she claimed had magical properties "contre la melancholie." She wrote to her ambassador in Paris for "mithridate," a substance said to be an antidote to every poison and an instant remedy. She asked him to supply "a piece of fine unicorn's horn, as I am in great want of it." Unicorn's horn was an expensive quack remedy that had a particular cachet among the ruling families of the age. Henry VIII had kept a supply of it with his ointments and salves. It was really made from rhinoceros horn or narwhal tusk, but was believed to be a miracle cure.

Mary was itching to get outdoors, knowing that she would feel much better for it. She first asked to visit the spa at Buxton in 1572. The hot springs there had been famous since Roman times for their curative powers. Robert Dudley, Earl of Leicester, and other courtiers made regular visits, giving the town a cosmopolitan flavor that appealed to Mary's desire to venture into a wider and more sophisticated world. Ironically, she twice met Leicester there, coming face to face at last with the still cherished favorite whom Elizabeth had so bizarrely offered to her as a husband.

And there was always a chance that she would meet Elizabeth herself. In July and August 1575, when the English queen's summer progress took her to the Midlands and so to Staffordshire, the two "sisters" and "cousins" were only a few miles apart. But Elizabeth would not take the risk of meeting Mary, who all along she feared would get the better of her in an argument. When she heard that Mary would be at the spa, she quickly changed her plans.

Between 1573 and 1584, Mary spent many of her summers at Buxton. She enjoyed it so much that a considerate Shrewsbury built a secluded lodge there for her private use. He had not wanted her to go at first; he thought it a security risk. Why did she need the baths when she already bathed regularly with herbs and had started using white wine as a facial toner? Mary's request was sent up to Elizabeth, who handed it to Cecil, who referred it to the Privy Council, and from there back to the queen.

While this buck was being passed, Mary was taken to Chatsworth, a staging post on the road to Buxton, before her stay was finally approved.

Visitors at the spa both drank the water — several pints a day — and bathed in it. When their ablutions were completed, they played games or relaxed in the well-appointed bathhouse (also owned by Shrewsbury), which functioned as a thirty-room hotel. There were chairs around the hot springs, and "chimneys for fire to air your garments in the bath's side, and other necessaries most decent." The men amused themselves at bowling or archery, the women by playing a game of *boules* called "Troule in Madame." Mary would have loved to mingle freely with the other visitors, but as far as possible was kept in isolation. She did sometimes get the opportunity to speak to the other bathers. When the spa was at its busiest, however, she was taken to visit the limestone caves nearby or confined to her lodge and its gardens.

In the first ten years of her captivity, Mary spent much of her time working at her embroidery and talking to Bess of Hardwick. As Shrewsbury reported, Mary saw Bess whenever "she useth to sit working with the needle, in which she much delighteth." Bess, one of four sisters born to a minor gentry family, was twenty years older than Mary, but was eager to be associated with a queen. An eighteenth-century historian, Edmund Lodge, described her as "a woman of masculine understanding and conduct, proud, furious, selfish and unfeeling." She was also an inspired interior designer, who used her links to Mary to obtain scarce stocks of silk and precious fabrics from France. They sat sewing together with their servants and companions, often for days on end.

In 1574, Bess presumptuously married off her daughter, Elizabeth Cavendish, to Darnley's younger brother, Charles, Mary's brother-in-law. Bess had royal pretensions, greatly resented by Elizabeth and by her long-suffering husband, whom she neglected to inform of his stepdaughter's betrothal, and who found himself on the receiving end of Elizabeth's wrath. The marriage caused a sudden chill in Mary and Bess's relationship, as it was undertaken with a dynastic claim in mind.

But until 1577 or thereabouts, they continued to sew together. Embroidery was one of Mary's lifelong pleasures. In her search for designs, she leafed through emblem books, and the superb woodcuts of birds, animals and fish by Conrad Gesner offered a rich choice of subjects that appealed to her artistic taste and love of pets. She embroidered at least three designs from the second edition of his *Animal Illustrations*, published in 1560: the cat, the phoenix and the toucan — which Mary called "a bird of America." She also copied the dolphin from Pierre Belon's *Nature and Diversity of Fish*, published in 1555.

Mary first drew an outline onto her canvas, sketching or marking out the shapes with chalk and then choosing her colors. She then stitched in the silks and colored wools, which might take as little as a week or as much as three months. Finally, she added her *MR* monogram, based on the Greek *mu*, to her work, which was always of high quality. She did not embroider just to pass the time; the work offered her an outlet for her wicked sense of humor. Her themes might seem to be innocently chosen, but often hid a deeper, more subversive meaning for those with eyes to see.

Whereas Elizabeth was flatteringly depicted by a small army of fawning poets and artists as a star or as the sun or moon, Mary cheerfully embroidered panels that showed eclipses. Mary's "A Catte" was a lot more than just a reproduction of Gesner's black and white domestic pet. She — for this cat was unquestionably female — was embroidered in ginger — Elizabeth's red hair was legendary — wearing a miniature gold coronet and closely watching a mouse, details not in Gesner's original design. Mary had several times said that she was the mouse and Elizabeth the cat, who watched and waited before deciding to pounce. The "Phenix" was, in Mary's version, her mother's and now also her own *impresa*, crowned and with the *MR* on either side of its head. Again, her "Delphin" was not just a sea dolphin but a pun on the word "dauphin," and also an emblem of her first husband, Francis II, and so a reminder of Mary's dynastic status.

She also undertook a piece of work on a more sinister theme: a panel of a wall hanging in which a hand descends from heaven with a pruning hook, cutting down a vine, with the motto "Virescit vulnere virtus" ("Virtue grows strong by wounding"). The motto could have referred to Mary's moral outrage at her treatment by Elizabeth, but may well have signaled her determination to survive Elizabeth by whatever means. Most likely the ambiguity was malicious.

Mary's love for animal designs reflected the way, in her solitude, she turned to her pets for solace. Not long after moving to Sheffield Lodge, she asked her ambassador in Paris to find out if her uncle the Cardinal of Lorraine had gone to Lyons. "I feel sure he will send me a couple of pretty little dogs, and you will buy me some also, for besides reading and needlework, I take pleasure only in all the little animals that I can get.

You must send me them in baskets, kept very warm." When they arrived, she confided, "I am very fond of my little dogs, but I fear they will grow rather large."

Mary's love of animals was also a tool of her diplomacy. Castelnau, who had known her since her teenage years, was in 1575 appointed as the French ambassador in London. He wrote to her a few months after his arrival, seeking her help in finding some English hunting dogs. "I at once asked the Earl of Shrewsbury to assist me," she replied. "He has given me three spaniels and two of the others, which he is sure are good ones."

When Castelnau no longer needed the dogs, Mary sent them to her Guise cousins. She recommended them as a present for Henry III, the new king of France, who had succeeded his sickly brother, Charles IX, in 1574 at the age of twenty-two after spending an extraordinary year as the elected king of Poland. Mary's only regret was that she could not try out the dogs herself, as she said somewhat self-pityingly, "because I am a prisoner, and can only testify to their beauty, as I am not at liberty to go out on horseback nor to the chase."

At Sheffield, she decided to start an aviary. She had become used to exotic birds at the court of Henry II, and began by asking her agent in Paris to obtain breeding pairs of turtledoves and Barbary fowl. "I wish," she said, "to see if I can rear them in this country, as your brother told me that, when he was with you, he had raised some . . . I shall take great pleasure in rearing them in cages, which I do all sorts of little birds I can meet with. This will be a pastime for a prisoner."

So often Mary's thoughts were back in France. Despite the surveillance to which she was subjected, she wrote to the officials of the French embassy in London, to the Cardinal of Lorraine, to the Duke and Duchess of Nemours, and less often to her five cousins, the children of her murdered uncle Francis, Duke of Guise. She was unaware at first that her letters were being intercepted and forwarded for vetting or deciphering to the agents of Elizabeth's new principal secretary, Sir Francis Walsingham. Aged forty-four, he had been one of Cecil's protégés since his days as a student at Cambridge.

Walsingham was the most single-minded ideologue at Elizabeth's court, an avowed Calvinist who at every opportunity championed the reformed cause. Cecil had recommended him to Elizabeth as her principal secretary in December 1573. For five years or so, the younger man had worked in his office as an intelligence expert and spymaster. After Mary's letters were painstakingly opened and copies made, they were sent on to their intended recipients, the seals apparently unbroken.

Mary was dismayed by the reactions to her letters. Often they were not even acknowledged. Her aunt Anne, the Duchess of Nemours, formerly the widow of her murdered uncle, was a sympathetic correspondent, but even her letters tailed off. The Guise family had been overshadowed at the French court by the mid-1570s; they were superseded by their rivals, the heirs and successors of Constable Montmorency and his old sparring partner Anthony of Bourbon. The Cardinal of Lorraine's influence had dropped to a particularly low ebb. Both he and Mary's cousin Henry, Duke of Guise, had left Paris for Joinville and Meudon.

Then, in December 1574, the cardinal died at Avignon at the age of fifty, severing Mary's chief link with France. The news reached her in the middle of February, a month after she had written him a long letter urging him to persist in his attempt to recover Catherine de Medici's favor. She was terribly distraught. "Though I cannot, at first, control my feelings or stop the tears that will flow," she said, "yet my long adversity has taught me to hope for consolation for all my afflictions in a better life."

With only one of the sons of her grandfather Claude still living, Mary knew her family's fortunes rested in the hands of a new generation. She was sad and dejected, grieving for her loss. She feared her cousins were not really interested in her personally or in her cause, which seemed to them lost. Catherine, still the power behind the throne, remained an opponent. Equally distant and unhelpful was the family's matriarch, Antoinette of Bourbon, once Mary's guardian angel but now a bitterly disappointed grandmother, who continued to blame her for her ill-advised second and third marriages.

Mary never lost her appetite for the luxuries to which she had grown accustomed in France. She asked her agent in Paris to send her "patterns of dresses, and of cloth of gold and silver, and of silks, the handsomest and the rarest that are worn at court." She ordered headdresses "with a crown of gold and silver, such as were formerly made for me." She also wanted the latest fashions from Italy: "headdresses, veils and ribbons."

In her last letter to her uncle, she had requested "a fine gold mirror to hang from the waist with a chain to attach it to." She wanted a motto to be engraved on the frame combining her and Elizabeth's monograms. She also asked for miniatures of herself, to give to her supporters in England "who ask for my portrait." Shortly after her uncle's death, Mary obtained permission to sit for an artist in Sheffield. She claimed that she wanted to send the miniatures to her friends in France — the right thing to say, if perhaps untrue.

The identification of this likeness has caused much debate. The full-length example at Hardwick Hall in which Mary is shown as a pious Catholic, aged thirty-six, with a cross at her bosom and rosary beads at her girdle, is a posthumous image. Despite its claim to have been painted in 1578, the portrait is not listed in Bess of Hardwick's inventory, which she drew up in 1601, and it was not commissioned until after James VI's accession to the English throne in 1603.

In 1575, Mary sat for an unknown miniaturist, and again in or after 1578 for Nicholas Hilliard, the doyen of the genre, when he returned from France. In both versions she is wearing a soft cambric cap attached to a modish wired veil and a fine lace collar. Her lustrous — but perhaps now darker and artificial — curls are visible at the sides of her cap. Her almond-shaped hazel eyes are still bright and penetrating. Her cheekbones are just as high, her nose slightly aquiline, since the view is partly from the side. Her marble-like complexion is perfect, but her face is filling out, a double chin is starting to appear, and she is no longer in the flush of youth or in her prime. Imprisonment has already taken its toll.

Mary sent expensive gifts to Elizabeth in a vain attempt to win her attention and put her under an obligation to release her. She began with confections of sugar, marzipan and nuts imported from France, which Elizabeth, who had a sweet tooth, enjoyed despite a warning that they might be poisoned. Other presents included a skirt of rich crimson satin, lined with taffeta, that Mary embroidered herself. She had designed an intricately latticed pattern of English flowers surmounted by a thistle. To achieve a stylish and sumptuous effect, she used the most expensive silk and precious metal threads, the best her French suppliers could provide. She asked the French embassy in London to deliver the skirt, which was done despite Shrewsbury's protests.

Elizabeth was said to have admired the gift and "prized it much." She was momentarily softer toward Mary, but it did not last and there was no easing of her restrictions. Mary was not allowed more exercise, the thing she most craved. When consulted on the matter, Shrewsbury advised, "I would be very loath that any liberty or exercise should be granted unto her, or any of hers, out of these gates . . . I do suffer her to walk upon the leads [i.e., the flat parts of a roof] here in open air, in my large dining chamber, and also in this courtyard." That was normally quite enough, in case she was tempted to escape.

Shrewsbury did, however, sometimes bend his own rules, generally when Mary fell sick or burst into tears. One arctic January, when the snow lay deep on the ground at Sheffield, he allowed her to walk in the park, thinking she would refuse his offer and decide not to venture outdoors. Without a moment's thought, she put on her heaviest clothes and went out, even though the snow came well over her shoes and must have soaked her feet.

Such treats were few and far between. Mary's lack of exercise exacted a heavy price. Her legs became so inflamed and her heel so sore she was barely able to walk. In 1582, shortly before her fortieth birthday, she was allowed as a concession to use her coach to ride out to take the air. She was at first exhilarated by this. She was preceded by her secretaries and other principal officers on horseback, guarded by a contingent of Shrewsbury's men armed with loaded pistols. Her route was reconnoitered beforehand by scouts in case anyone tried to meet or attempt to rescue her.

Mary's pleasure was fleeting. Soon she was in pain when she walked even a relatively short distance. Just climbing in and out of the coach could be more than her legs could bear, and in the winter months Mary felt so weak, as she herself said, that she preferred to stay indoors.

Her last years at Sheffield and Chatsworth were tainted by the growing marital discord between Bess and the earl, made worse by Bess's groundless suspicion that her husband was having a secret affair with his prisoner. Gossip was rife at Elizabeth's court, which the queen herself joined in. Castelnau, who always tried to do his best for Mary against the odds, warned her that Elizabeth was telling tales to foreign ambassadors so that they would be spread about. "It is the final poison that your enemies have reserved," he said; "not to poison your body, but your reputation."

Almost all of the information that fueled this gossip came from Bess, and in a fit of pique around 1584, Mary turned the tables. She sent Elizabeth, probably through Cecil, a summary of everything Bess had said that touched her "sister queen," prefacing her remarks with a disclaimer: "I protest that I answered rebuking the said lady [Bess] for believing or speaking so licentiously of you as a thing which I did not at all believe."

The charges were of the raciest sort: that Elizabeth had promised to marry her favorite, Leicester, and was his lover; that she had taken a succession of paramours, including Sir Christopher Hatton (vice chamberlain of Elizabeth's household, later lord chancellor, and a favorite second only to Leicester), and had compromised herself with a French diplomat by visiting him at night, kissing him and enjoying "various unseemly familiarities with him." Not content with this, she had betrayed her own councilors to the French in her pillow talk. Elizabeth, according to Bess, was so vain she had to be flattered by her courtiers "beyond all reason." They would amuse themselves by playing a game in which they tried to outdo each other in offering extravagant compliments to her. It was all they could do to avoid bursting out laughing. Mary knew how to put in the knife. She said that when Elizabeth had been ill for a while, Bess had prophesied her death based on the reports of an astrologer who, "in an old book, predicted a violent death for you and the succession of another queen, which she interpreted as myself."

Grasping at straws in her longing to meet her rival and talk to her face to face, Mary offered to reveal more of Bess's infamy at a personal interview. Elizabeth — if she was ever allowed by Cecil to see this extraordinary document — refused. By the 1580s, the polarization between Catholics and Protestants in Europe was approaching its climax. In the Netherlands, the revolt of the Calvinists against Spain had passed the point of no return. In France, another civil war would erupt in 1584, when the Catholic League allied with Spain to extirpate the Huguenots and block the claim to the throne of their leader, Henry of Navarre. As for Philip II himself, after nearly thirty years in which he had given Elizabeth the benefit of the doubt, he came to believe that the key to the defeat of the Dutch rebels and the mastery of the Atlantic Ocean by Spain was the conquest of England. War between England and Spain was drawing closer.

It was Cecil's worst nightmare. Here was an international Catholic conspiracy with a vengeance. All anyone could think of in the Privy Council, in Parliament and among the "assured" Protestant elite who formed the backbone of his inner caucus was that Mary should be dealt with once and for all. The "preservation" of the Protestant state depended on it. Mary was said to be a bigger threat to her cousin's security than Spain, and although Elizabeth had so far privately supported Mary as her heir apparent, this was too much for Cecil, whose entire career hinged on his almost apocalyptic vision of England's Protestant destiny.

In August 1584, Shrewsbury was recalled to London to attend urgent meetings of the Privy Council. Mary was transferred to the custody of Sir Ralph Sadler and his son-in-law, John Somers. It was meant to be a temporary arrangement; hardly surprising, as this Sadler, now a venerable seventy-seven-year-old, was none other than Henry VIII's ambassador to Scotland who had dandled Mary as a baby on his knee. His instructions were to take her first from Sheffield to Wingfield Manor, and from there to the close confinement of the damp and unhealthy Tutbury Castle, to which she finally returned in January 1585.

On the road to Wingfield, where she was guarded by forty soldiers, Mary asked Somers if he thought she would try to escape. He said that he supposed she would. It was only natural. "No," she angrily retorted, "you are mistaken. I had rather die in this captivity than run away with shame."

Sadler was one of Cecil's most faithful and steadfast supporters, and yet — at least as a private man — he took pity on the captive queen. "I find her much altered," he said, "from that she was when I was first acquainted with her." Her incarceration had ruined her health. "She is not yet able to strain her left foot to the ground, and, to her very great grief, not without tears, finding that being wasted and shrunk of its natural measure and shorter than the other, she feareth it will hardly return to its natural state without the benefit of hot baths."

But Mary's allure was undiminished. When Sadler got his charge to Tutbury, she quickly won him over. Within three months, he had been caught taking her hawking. Called to account by his masters, he confessed: she "earnestly entreated me that she might go abroad with me to see my hawks fly." It was "a pastime indeed which she hath singular delight in, and I, thinking it could not be ill taken, assented to her desire; and so hath she been abroad with me three or four times hawking upon the river here." All the time she had been guarded by forty or fifty men. He had used his discretion and done his best. "But since it is not well taken, I would to God that some other had the charge." Somers confirmed his father-in-law's statement. Mary had been well guarded while out hawking: "if any danger had been offered, or doubt suspected, this queen's body should first have tasted of the gall."

It was not enough. Sadler was too decent and considerate a man to be Mary's keeper. Although a devout Protestant and unswerving in his allegiance to Elizabeth, he was unable to square his political and religious obligations with his feelings as a human being. He accepted an honorable discharge, clearing the way for the appointment of a man better equipped to be a jailer — for this is what the menacing international threat made necessary.

His successor was Sir Amyas Paulet. He was a fluent French speaker, formerly an ambassador to France, where he had colluded with Catherine de Medici to blacken Mary's name. A Calvinist and close ally of Walsingham, he was the sworn enemy of Spain and the Catholic League, a fervent supporter of the Huguenots and the Dutch Protestants. His work in Paris had brought him into contact with several of Mary's agents, about whose malice he never had the slightest doubt. He arrived at Tutbury in April 1585, assuming full responsibility for his prisoner on the 19th, when Sadler returned home to Hertfordshire.

Paulet was not a man likely to allow Mary to ride out hawking. He made no distinction between his private and public duties. "Others," he said, "shall excuse their foolish pity as they may." For his part, as he once bragged, he would rather renounce his claim to a share in the joys of heaven than put any feelings of compassion above his obligations to Elizabeth. He was Mary's sole keeper until November 1586, when his friend Sir Drue Drury — a minor courtier from a prominent legal family, whom Mary described as "most modest and gracious in all things" — was sent to assist him.

Mary found Paulet "one of the most zealous and pitiless men I have ever known; and, in a word, fitter for a gaol of criminals than for the custody of one of my rank and birth." She had quickly gotten the measure of her man.

To begin with, Paulet tore down her cloth of state whenever he had occasion to enter her presence chamber, saying such regalia had no place in her household as there could be only one queen in England. This led to a battle of wills in which he would dismantle the offending canopy and a tearful Mary would have it put back. He opened her letters and packets looking for evidence of a conspiracy, treating anything sent to or from the French embassy with the utmost suspicion. His intrusions extended to random searches of her cabinets, and as a last resort he was willing to break down Mary's bedroom door. All the documents he impounded were instantly forwarded to Walsingham for examination.

Paulet suspected everyone and everything. To avoid being overheard by anyone in the household, he would sometimes confer with messengers sent from London in the open fields. He put Mary's domestic staff into quarantine. They were not allowed to mix with the other servants or to enter or leave the castle unless searched. No one, however menial, was spared the new procedures. Mary's coachmen and laundresses were singled out for special attention. They were to be watched at all times, particularly three women who lodged in a little house in the park adjoining the castle. They had outside contacts as well as access to Mary's rooms. Paulet ordered that they were not just to be searched, but stripped down to their smocks whenever they entered or left the building.

Mary's health continued to deteriorate. She put out a bulletin denying she had dropsy or cancer, but her legs were more swollen than ever. She was allowed to ride out once or twice a month, her coach surrounded by a small infantry detachment bearing loaded muskets and a lighted fuse. At other times, she was carried into the garden in a chair. When she did succeed in walking, two of her secretaries had to support her. Her legs needed poultices and bandages, which had to be applied every day. "Her legs," said Paulet, "are yet weak, and indeed are wrapped in gross manner, as hath appeared to my wife."

Paulet's harsh treatment aroused protests from the French. On Christmas Eve 1585, Mary was moved from spartan Tutbury, notorious for its dark, damp rooms and stinking latrines, to Chartley, a manor house almost palatial in comparison that was chosen because it was surrounded by a large moat.

Mary's mood was changing to one of defiance. She had never made any secret of her aim to be restored as queen in Scotland, if necessary as co-ruler with her son. All her attempts to reach a political accord with Elizabeth foundered over her desire for liberty and to return home. She never gave up on her dynastic claim to the English throne, and talked passionately about it, causing Paulet to protest at her "tediousness" and her "superfluous and idle speeches." When Mary was in full flood, he simply walked away.

Mary's defection to the Spanish interest would finally be her undoing. She believed that France and her Guise relations had deserted her. She had tentatively approached Philip II after the stalemate at Elizabeth's tribunal, and her policy gained a fresh impetus when the Cardinal of Lorraine died. By the 1580s, Mary was beginning to talk openly about linking her cause to Philip's grand European strategy. This was incredibly injudicious, but it put her back in the spotlight. Once again she was about to take center stage in a pan-European drama.

Cecil had bided his time. The spider was poised to catch the fly. Mary felt irrelevant and disposable, and she reacted in the only way she knew.

She reinvented herself as a poor Catholic woman persecuted for her religion alone. It was almost entirely a theatrical pose, but she had nowhere else to go. Meanwhile, Walsingham had recruited a mole in the French embassy. He had pulled off this intelligence coup while Mary was still at Sheffield. The mole was active for little more than eighteen months, but it was long enough for Cecil's spymaster to identify many of her secret agents at home and abroad. It was only a matter of time before Mary's fate was sealed.

28

An Ax or an Act?

MARY HAD FIRST put out feelers to Philip II in November 1568, in the earliest months of her captivity when she was at Bolton Castle. The timing was propitious, because a new Spanish ambassador, Don Guerau de Spes, had arrived in London. He was inexperienced, but a natural conspirator. Within a few months, he had staked out positions that sparked an Anglo-Spanish trade war and threatened a Spanish invasion of England on Mary's behalf.

A fortnight or so after Moray had exhibited the Casket Letters for inspection, de Spes called at the French embassy in London, where he made two proposals. The first was that "he knew of no greater heretic in this world, nor a greater enemy to the Catholic faith, than Master Cecil." He urged France to cooperate with Spain "to make him lose his office and the favor and credit that he enjoyed with his mistress the queen." The second proposal was that France and Spain should place a joint embargo on all English trade until the Catholic faith was restored.

A month later, a Florentine banker, Roberto Ridolfi, who was secretly channeling the funds sent by Pope Pius V to the English Catholics, entered the story. He was a double agent working both for Spain and for Walsingham. In February 1569, he offered to arbitrate between Elizabeth and the Duke of Alba, Philip II's governor-general in the Netherlands, to end the trade war. The following month, he visited de Spes, bringing an exciting and dangerous message from the Duke of Norfolk and his allies, the Earl of Arundel and Lord Lumley. They had said that they were no longer willing to accept exclusion from their rightful places by the lowborn upstarts of Cecil's inner caucus. They wished to overthrow Cecil and force Elizabeth to realign her policy closer to Philip II and Rome. Later Ridolfi told the French ambassador that he was commissioned by the pope to help restore the Catholic faith in England.

Ridolfi contacted Mary through her agent in London, John Lesley, Bishop of Ross, to whom he transferred £3000 from de Spes with Mary's knowledge. The transaction placed her at the center of two converging conspiracies. One was designed to improve her conditions in exile and displace Cecil as chief minister. The other linked her dynastic claim to a plan to restore Catholicism in England.

The Duke of Norfolk had conspired to oust Cecil shortly after the first stage of Elizabeth's tribunal opened at York in October 1568. He had been one of the original English judges and had resented the way Cecil had dictated the proceedings. Cecil had ignored his protests. The tribunal had been shifted to Westminster, and Norfolk was sent off on a futile mission to inspect the northern frontiers until Cecil was ready to reopen the hearings.

But Norfolk was not easily marginalized. He was England's premier duke, whose family stood at the heart of the Catholic party in England. His talent for making advantageous alliances was unrivaled. He had married three heiresses: Lady Mary Fitzalan, daughter of the Earl of Arundel, who died at sixteen; Margaret, daughter of Lord Audley, who died after bearing five children in as many years; and Elizabeth Leyburne, the widow of Lord Dacre. When his third wife died in 1567, Norfolk became the guardian of the heiresses to whom he would marry his three sons. His sister Jane was the wife of the Earl of Westmorland, one of the two predominant landowning nobles in the north of England. Lastly, a second sister, Margaret, married Lord Scrope of Bolton Castle.

Shortly after Elizabeth's tribunal had ended in a stalemate, Norfolk decided that he would try and marry Mary. He had sneaked up to Bolton Castle to take a look at her, using a visit to his sister as an excuse. In the end, he would conspire to put Mary (and himself) on the English throne.

That, however, is not how the plot began. As it was first imagined by that arch-conspirator among the Scottish lords, the insinuating Maitland, the aim was very different. His idea was that if Mary could be married off to an English nobleman, she would be neutralized politically.

She would have been treated honorably, but her power and authority would be vested in her husband. Under English law, she would be a *femme couverte.*

The idea was ingenious, because it appealed to those hereditary nobles who always disliked the way Henry VIII had bypassed the strict order of succession in his will. It also resolved the problem of Mary's dynastic claim in a way that Elizabeth herself had once anticipated: it was analogous to her old scheme to marry her cousin to Robert Dudley, Earl of Leicester.

Mary jumped at the idea. She asked the pope to annul her marriage to Bothwell, which was by Protestant rites and thus canonically invalid. Over the next twelve months, she also wrote Norfolk a series of love letters, despite having seen him only once. "I will live and die with you," she said. "Neither prison the one way, nor liberty the other, nor all such accidents, good or bad, shall persuade me to depart from that faith and obedience I have promised to you."

Norfolk sent Mary a diamond that she was to take with her to Fotheringhay. She described it as something "I have held very dear, having been given to me . . . as a pledge of his troth, and I have always worn it as such." As she told the duke at the time, "You have promised to be mine and I yours . . . As you please command me; for I will for all the world follow your commandment, so you be not in danger for me . . . Your own faithful to death, Queen of Scots. My Norfolk."

We must allow for the conventions of royal marriages in which the custom was to express affection as soon as a betrothal had been arranged in principle. Even so, Mary was once more grasping at straws. If a marriage to Norfolk could enable her to recover her freedom, she would do it without asking any questions. It was a reasonable match, a better dynastic prospect than the old idea of marriage to Leicester. But there was a hopeless catch. No one had dared to tell Elizabeth, without whose consent the plan was worthless and Mary's hopes were raised in vain.

At first, however, even the perfidious Moray signed on. In July 1569, he was in favor of the plan. Then, a month later, he had compelling second thoughts. He saw the snag. Mary would use the marriage to seek her restoration as Queen of Scots, and Norfolk would use it to assert his claim in right of his wife to the throne of England with the aim of restoring Catholicism. With Norfolk at Mary's side, furthermore, Moray would be ousted as regent and forced to seek exile abroad. He would end up changing places again with his sister.

Moray played a dirty trick, sending Elizabeth, still ignorant of the proposed marriage, a certified copy of Norfolk's letter announcing his intention to marry Mary even if Moray objected.

Elizabeth had rarely been so angry. She perhaps disliked the marriage plan less than the fact that it had been devised behind her back. Everyone involved was sent for except Mary, and those like Leicester and Throckmorton, who were agile in a crisis, made their excuses. Badly caught out in this game of musical chairs were the Earls of Northumberland and Westmorland. When these Catholic lords were abandoned by their allies and left to face Elizabeth and possible execution alone, they decided to rebel. Their revolt, the so-called Northern Rising, erupted in November 1569 and was crushed by overwhelming southern forces within six weeks.

The two earls and the Countess of Northumberland took refuge in Liddesdale, but were denounced by Bothwell's old adversary, the Laird of Ormiston. Westmorland and the Countess of Northumberland fled to the Netherlands, where they remained in exile. Northumberland was seized by Moray and executed by Elizabeth after the Scots sold him back to England for £2000. The Countess of Westmorland retreated to her brother's estates in Norfolk, where she lived out her remaining days in obscurity.

Mary was not blamed by Elizabeth for causing this, the most serious rebellion of her reign — at least not yet. But all along there had been two separate conspiracies. The plan to marry Mary to Norfolk had failed and the duke was a prisoner in the Tower. Ridolfi, meanwhile, was still plotting. In May 1569, he drew up a plan to depose Elizabeth, called the Enterprise of England, and took it to the Spanish embassy. Philip II was tentatively listening. These were plans that fit his grand strategy for Europe and the domination of world trade. He wondered if they could possibly succeed.

Ridolfi then visited Walsingham. As long as he was careful and the two sides did not meet and compare notes, he could hope to get away with his double-dealing and claim rewards from both sides. He had information to sell, because his banking services provided the only link among de Spes, Norfolk and Mary's agent in London. He had also supplied the cipher used to encode the communications between de Spes, Mary and the English lords. Whoever had a copy of this cipher could read all their intercepted letters.

Ridolfi was arrested on the eve of the Northern Rising. Yet he was released after six weeks. And he was not detained in the Tower, the usual place for political prisoners, but at Walsingham's house. Ridolfi had been "turned" by Walsingham. It was a stroke of luck for Cecil, because in February 1570 Pope Pius V spurred on by the prospect of reversing the English Reformation, did what his predecessors had always refused to do. He published a decree, *Regnans in Excelsis*, depriving Elizabeth of her "pretended title" to the English throne and releasing her subjects from their allegiance. This was Mary's golden opportunity and what she had sought in her youth, when she had first been married to the Dauphin Francis.

Cecil's reaction was immediate, his logic as inexorable as ever. Protestants were now loyalists and Catholics traitors. By his definition, Catholics agreed with the pope that Elizabeth should be deposed. When Parliament met in April 1571, he introduced an oath to ensure that all Catholic members were excluded. He then drafted a government bill to disqualify any candidate for the succession — principally Mary — who at any time for the rest of Elizabeth's life might claim the throne or usurp its insignia. In debating this bill, he would never let Parliament forget that Mary's heraldic arms in France had been quartered with the royal arms of England and that her ushers had cried "Make way for the queen of England" as she walked to chapel.

During 1571, Cecil and Walsingham closely monitored the Enterprise of England. In an attempt to put the main conspirators off their guard, Norfolk was released from the Tower on parole. Mary's correspondence was more rigorously vetted, and by April, Walsingham — temporarily appointed the English ambassador to Paris, where most of the plotting was centered — had acquired with suspicious ease a copy of the cipher supplied by Ridolfi to the conspirators. There was no need to struggle to crack the code: Walsingham was given the solution on a plate.

On August 4, Philip sent Alba his detailed instructions. The plan was to capture Elizabeth in the final phase of her summer progress to Hertfordshire and Essex. This would unleash a general rising of English Catholics, who would liberate Mary and marry her to Norfolk. A Spanish fleet would then sail with six thousand crack troops from Alba's forces in the Netherlands to secure the country. Mary and Norfolk would then ascend the throne. To meet the costs of the expedition, Philip II was sending 20,000 ducats to replenish Alba's treasury.

The plan leaked when one of Philip's councilors disclosed it to a merchant in the Anglo-Iberian trade who was another of Walsingham's double agents. The man returned at once to England. He arrived on or about September 4 and was sent directly to Cecil. The next day, Cecil issued a warrant to put Norfolk back in the Tower. He also wrote to the Earl of Shrewsbury, Mary's custodian. The letter was endorsed "sent from the court, the 5th of September 1571 at 9 in the night." No Elizabethan document had a higher priority. It was marked "haste, post haste, haste, haste, for life, life, life, life."

Shrewsbury was warned of the plot and of Elizabeth's reaction to it. She knew, Cecil said, that Mary was planning to escape and wished to go to Spain rather than to France or Scotland. She knew Mary had offered her son, James VI, in marriage to one of Philip II's daughters by Elizabeth of Valois, Mary's recently deceased childhood friend and playmate whom she greatly mourned.

At none of this, said Cecil, had Elizabeth taken umbrage. Mary was a captive queen. These were no more than "those devices tending to her liberty." What rankled, and what Elizabeth "understandeth certainly," so it was said, were her cousin's "labors and devices to stir up a new rebellion in this realm and to have the king of Spain to assist it." Of all these facts, concluded Cecil, Elizabeth was "certainly assured and of much more."

He greatly exaggerated. His information about Philip's invasion plan came from Walsingham's double agent and was strictly verbal. Later he claimed to have confessions from Norfolk and the Bishop of Ross, a damaged letter from the pope to Mary and an encrypted letter from her to Norfolk, found under a mat in the duke's house, containing "great discourses in matters of state, more than woman's wit doth commonly reach unto." But there was no proof that Mary had endorsed a plan to depose and kill Elizabeth. The most incriminating letter, intercepted from Mary and written a few months before the Northern Rising, did not say what Cecil needed it to say.

The letter comes to us in the form of a copy of the deciphered original. In it, Mary spoke generally and allusively. She thanked her unnamed supporters for their "care how to enlarge our liberty, to restore us to our

rightful seat, to cease our daily griefs, to suppress our usurping and undeserved foes, to quench the rage of erroneous tyrants, to the furtherance of God's word, to the releasement and comfort of Christians."

> What works could be more acceptable to God than to succor the Catholic
> Church, to defend the rightful title of a prince, to deliver afflicted Chris-
> tians from bondage, and to restore justice to all men, by cutting of the most
> faithless antichrist and usurper of titles, the destroyer of justice, the perse-
> cutor of God and his Church, the disturber of all quiet states, the only
> maintainer of all seditious and mischievous rebels of God and all Catholic
> princes, having a way made by our Holy Father. Wherefore we beseech you
> to proceed in God's name and our Blessed Lady's, with the assistance of the
> whole company [i.e., of all the saints in heaven].

Nowhere did the letter mention Spain or Elizabeth by name. The
"antichrist" referred to could also have been meant literally. Mary had
neither called explicitly for an attempt on her cousin's life nor requested
aid from Philip II or the pope. "Works . . . acceptable to God" were no
more than the workings of divine Providence, and what earthly creature
could resist them? And to "proceed in God's name" was not in itself an
incitement to assassination or invasion.

Mary's meaning is, of course, perfectly clear. But such evidence would
not have stood up in a court of law, supposing that she, who was neither
English nor a subject of Elizabeth, was bound by English law.

Cecil did not have enough evidence to put Mary on trial. To obtain
more damaging admissions, Shrewsbury was ordered to confront her
and "tempt her patience in this sort to provoke her to answer somewhat." But she could not be caught out in this way. She could be impulsive or naive, but was rarely stupid. In dealing with Ridolfi, she had followed the Cardinal of Lorraine's maxim, "Discretion sur tout," to the
letter.

If Cecil could not put her on trial, he could isolate her diplomatically.
He sent his friend and fellow privy councilor Sir Thomas Smith to
Catherine de Medici, playing up the Spanish threat in Europe and urging her to persist in the dissociation of France from Mary's cause. This
was the start of an Anglo-French entente that lasted for a decade. It survived even the setback of the massacres of St. Bartholomew's Day (August 24, 1572), when some three thousand Huguenots were slaughtered
in Paris and ten thousand more killed in provincial France over a period
of three weeks.

The massacres caused panic in England. The Bishop of London advised Cecil "forthwith to cut off the Scottish queen's head." Cecil only wished he could comply. When he had visited Mary at Chatsworth in October 1570 in an attempt to silence her demands to return to Scotland, she had burst into tears and their interviews were inconclusive. How he impressed her and how she impressed him are maddeningly not recorded. But Cecil was impervious to women's tears. He had no more room than Paulet for "foolish pity" and precious little scruple in his dealings with Mary. He had already followed Knox in identifying her as "Athalia" — a biblical precedent for regicide. And he had been advised by Knox before he left for Chatsworth that if he "struck not at the root, the branches that appeared to be broken would bud again with greater force." To give his warning maximum impact, Knox informed his old ally that he had written his letter "with his one foot in the grave."

When Cecil met Mary, he had piously claimed how Elizabeth had "always forborne to publish such matters as she might have done to have touched the queen of Scotland for murder of her husband." If Elizabeth had such reservations, Cecil did not. The judges who had examined the Casket Letters had been sworn to secrecy, but now Cecil broke their silence. He arranged for his friend Thomas Wilson, author of the acclaimed treatise *The Art of Rhetoric,* to prepare a vernacular edition of Buchanan's dossier for the press, telling the lords' side of the story and translating it from Latin into imitation Scots, in order to create the false impression that it was authorized by the lords in Scotland and not by anyone in England.*

A proof copy was ready by late November or early December 1571, to which Wilson craftily appended a damning but anonymous "oration" against Mary and translations of the Casket Letters. Cecil was delighted. He had the book rushed out by his "tame" printer, John Day, under the tortuous but highly informative title *A Detection of the doings of Mary Queen of Scots, touching the murder of her husband, and her conspiracy, adultery, and pretended marriage with the Earl of Bothwell. And a defence of the true Lords, maintainers of the King's grace's action and authority.* The king of the title was James VI.

Cecil had smeared Mary with the old and legally unproven allegations

---

* A Latin version was also anonymously published, without identifying the publisher or place of publication. A genuine Scots translation followed in 1572, printed at St. Andrews, after which the Casket Letters were in the open.

of the Casket Letters. An adulteress who could plot to murder her husband would have no qualms about conspiring to depose and kill Elizabeth. It made her guilt as a co-conspirator in the Enterprise of England seem more credible. It also hit out collaterally at Norfolk shortly before his trial for treason began. The innuendo was that anyone prepared to take Mary as his wife was tarred with the brush of Darnley's murder.

In publishing the *Detection* in this underhand way, Cecil was playing with fire. It is unlikely Day was allowed to sell the book openly in his shop. Elizabeth would have been enraged at Cecil's meddling had she been able to link the work directly to him. It was circulated privately to his inner caucus, at least at first. A final exhortation set the tone:

> Now judge Englishmen if it be good to change Queens. Oh uniting confounding! When rude Scotland has vomited up a poison, must fine England lick it up for a restorative? Oh vile indignity! While your Queen's enemy liveth, her danger continueth. Desperate necessity will dare the uttermost . . .

And this was the point. The *Detection* set the keynote for the Parliament that met in May 1572 to debate Elizabeth's "safety." When Norfolk was convicted of treason and sentenced to death, Elizabeth dithered over his execution, whereas Cecil and his Protestant allies wanted Mary dead too. They had secured this Parliament to make a bid to execute her as well as the duke. Their aim was a bill of attainder by which Mary would be "convicted" of treason by an act of Parliament and without a formal trial, thus avoiding the trouble of providing proof.

In a memo on the eve of the Parliament, Cecil rebuked Elizabeth for her "doubtful dealing with the Queen of Scots." His verdict was chilling. The only "good" Mary was a dead one. Elizabeth's mistake was her desire to give her cousin the benefit of the doubt. The threat she posed, far from diminishing, was increasing. All the Catholics supported her dynastic claim, as did anyone who emphasized hereditary right. Mary enjoyed Spanish and potentially Guise support. Her "party" maintained her title to be the legitimate queen of England. All she needed were the forces and the opportunity to launch a coup d'état.

When Parliament assembled, the members of Cecil's inner caucus called vociferously for Mary's execution. It was an unusually hot spring, the temperature in the House of Commons even hotter. Speaker after speaker rose to denounce her, echoing each other often almost word for word.

Mary, they argued, had disqualified herself. She was no longer queen, but "the late Queen of Scots." She was "a Queen of late time and yet through her own acts now justly no Queen." She was Elizabeth's kinswoman "and yet a very unnatural sister." "She hath sought to dispossess the Queen's Majesty of her crown . . . [She] hath made so small account of the Queen's goodness towards her as she deserveth no favor." She "is but a comet, which doth prognosticate the overthrow of this realm." "She is no Queen of ours, she is none of our anointed. The examples of the Old Testament be not few for the putting of wicked Kings to death."

Cecil's circulation of the *Detection* had done its work. Mary was "this Jezebel," this "Athalia," this "idolatress," this "most wicked and filthy woman." She was "the monstrous and huge dragon and mass of the earth." There was no safety for Elizabeth as long as Mary lived. "She hath been a killer of her husband, an adulteress, a common disturber of the peace of this realm, and for that to be dealt with as an enemy. And therefore my advice is to cut off her head and make no more ado about her."

Precedents from legal and historical treatises were quoted to justify her trial and execution:

Every person offending is to be tried in the place where he committeth the crime . . .
A King passing through another's realm or there resident is but a private person . . .
A King deposed is not to be taken for a King . . .
A King though not deposed may commit treason . . .
Punishments ought to be equal to the offences committed . . .
Death is the penalty for treason.

One of Cecil's protégés, Thomas Norton, summed up Parliament's mood. "The execution of the Scottish Queen is of necessity, it may lawfully be done . . . A general impunity to commit treason was never permitted to any . . . You will say she is a Queen's daughter and therefore to be spared; nay then, spare the Queen's Majesty that is a King's daughter and our Queen." This speaker even quoted from the *Detection:* "desperate necessity dareth the uttermost . . ." Who knew what mischief the serpent might do if she was allowed to live?

But could Mary, the anointed Queen of Scots, commit treason in England? Elizabeth did not yet think so. She refused to hand Parliament an ax for Mary's execution, instead encouraging members to seek an act excluding her from the succession. When the bill was passed, however, she vetoed it. She claimed it was not technically a veto, but in this she played with words. She could not bring herself to proceed against an anointed queen, yielding only to Parliament's pleas to execute the Duke of Norfolk, who went to the block on June 2, a month before Parliament ended.

Cecil was wholly frustrated. As he wrote to Walsingham, then still in France: "All that we have labored for and had with full consent brought to fashion — I mean a law to make the Scottish queen unable and unworthy of succession to the crown — was by Her Majesty neither assented to nor rejected, but deferred." And in a second letter, he made a shrill complaint about how the "highest person" in the realm (meaning Elizabeth) had failed to act, and so brought shame on her councilors.

Cecil was not foolhardy enough to tell Elizabeth this to her face. He did not need to, as she was well aware that she had invoked the royal prerogative to defend her cousin in defiance of Parliament, exonerating herself by promising that the bill of exclusion might be revived at a later date. But the appointed day came and went, and almost four years elapsed before Parliament met again. Mary was left untouched. Cecil had gotten neither an ax nor an act.

This most intrepid of spiders did not let up for a moment. Lady Catherine Grey, the heir apparent under the terms of Henry VIII's will, and her sister were both dead. Cecil turned his full attention to achieving Mary's exclusion from the succession by fair means or foul. He sent his brother-in-law, Henry Killigrew, on a secret mission to Scotland to see whether, if Mary was handed back to them, the Protestant lords would put her on trial and execute her. To fulfill this delicate task, Killigrew was recalled from France, where his instructions had been to drum up support for Mary's execution among the enemies of the Guise family.

Another of Cecil's intermediaries was Robert Beale, Walsingham's brother-in-law. He set about demonizing Mary and lobbying for an integrated policy linked to English support for the Protestant cause abroad and the "extirpation" of Catholicism at home. Where the Catholics were concerned, Beale advised that "their chiefest head must be removed." "I mean," he said, "the Queen of Scots, who as she hath been the principal cause of the ruin of the two realms of France and Scotland, hath prettily played the like part here."

It was a matter of waiting until Mary was trapped. Patience and surveillance were needed. When Walsingham returned to England, the interception of her letters was stepped up. A huge pile of deciphered transcripts accumulated, much of it concerning pensions granted by Mary to Catholic exiles abroad. She had also generously rewarded those who had fled after the debacle of the Northern Rising. "I pray you," she wrote in 1577 to her agent in Paris, "give the pope to understand as soon as you can that all I have remaining from my dowry (as you are able to inform him) is not enough for the maintenance of my household affairs and to furnish the necessity of the banished English and Scots whom I am constrained to relieve."

The same year, she appealed to Philip II to "take care of those that are banished out of England, and especially the Earl of Westmorland." Her cousin the young Duke of Guise was to be a recipient of similar pleas.

Step by step, Walsingham pieced together a jigsaw puzzle linking Mary's agents to Philip II, the pope, the Guise family, the grand master of Malta and the Spanish ambassador in Rome. But nothing out of the ordinary was found for him to spin malice from until 1579.

That year, Esmé Stuart, Sieur d'Aubigny, first cousin of Mary's second husband, Lord Darnley, returned to Scotland from France. His courtly style and handsome looks captivated the impressionable young James, who showered him with gifts and created him Duke of Lennox. Within a month, James abandoned his schoolroom at Stirling Castle and took up his place at Holyrood. D'Aubigny reformed the court and the royal household on the French model, then turned his attention to his enemies among the remaining Confederate Lords. In December 1580, Morton was arrested on a charge of complicity in Darnley's murder and executed the following June. Mary was overjoyed; she believed that events were turning in her favor and that she would soon be restored as Queen of Scots.

She made contact with Castelnau at the French embassy in London. They agreed to work together to achieve the independence of Scotland under the protection of France and with Mary as queen. Their project was to revive the "auld alliance." As Castelnau reminded Catherine de Medici and Henry III — at no small cost to himself, since his attachment to Mary would bring about a premature end to his career — "It behoves Your Majesty to preserve the alliance with Scotland, which has always been the bridle for England."

All Cecil's work since the treaty of Edinburgh in 1560 seemed about to be undone. D'Aubigny had the support of the Guises, who were fast making a comeback in the 1580s in their capacity as Philip II's allies in the French Catholic League. Would their combined influence on James be sufficient to persuade the susceptible teenager to invite his mother back to Scotland as queen?

In November 1581, Elizabeth sought to recover lost ground. She sent Beale to Sheffield to see if Mary could be used as a distraction against her son. Mary was thrilled to be back in the limelight, but did not intend to play such games. She put an alternative scheme to Beale, one by which she would be restored to her throne as co-ruler with her son. She used every weapon in her arsenal to obtain Beale's support for her idea, but he was impervious to her powers of persuasion.

When Mary's tactics failed, she tried a more theatrical approach. She exaggerated her illness, exploiting it to dramatic effect by conducting her interviews from her sickbed. She claimed to be dying. She had been vomiting and was partially immobilized because of lack of exercise, but she was secretly elated at the prospect of recovering her freedom.

Beale was not so easily fooled. When he found Mary and her gentle-women "weeping in the dark," he withdrew, later commenting that "the parties are so wily with whom a man deals." He left Sheffield with Mary's assurances that she would recognize Elizabeth as the lawful queen of England and not have dealings with foreign powers or rebels. Within a few months, however, she was writing to Bernardino de Mendoza, de Spes's successor as Spanish ambassador in London, asking for information about new plots devised by d'Aubigny and the Jesuits. She was determined to keep all her options open.

Her hopes were dashed when a reversal in Scotland led to d'Aubigny's exile. In August 1582, while out hunting, James was lured into Ruthven Castle, where he was imprisoned by the Protestants and d'Aubigny ousted.

Mary was distraught at the news. "When I heard that my son was taken and surprised by rebels as I was myself certain years ago," she said, "I cannot but pour out my grief out of a just fear that he should fall into the same estate as I am in."

But all was not lost. James by now was sixteen and finding his feet. In 1583, he escaped from his captors and declared his minority to be at an end. No longer would a regent rule in Scotland. Almost overnight, Henry III of France started a competition with Elizabeth to dazzle him. Henry sent two ambassadors to Scotland with instructions to renew the "auld alliance." He even recognized James as king, a move that ensured a lavish welcome for his ambassadors. Some progress was made in this diplomacy, and in retaliation Elizabeth talked of granting Mary her freedom under strict conditions. Until French influence on James could be neutralized, the Queen of Scots was still a card worth playing.

And her imprisonment dragged on. Beale was sent back to Sheffield, where Mary again proposed a deal in which she and James would share the throne. They would recognize Elizabeth as queen of England for as long as she lived, while reserving their own dynastic claim to the succession. It was hardly a new idea. But Elizabeth was willing to consider it. She even sent Walsingham, very much against his will, on an embassy to Scotland to float the plan.

This was shortly after he had pulled off his most significant intelligence coup by recruiting a mole at the French embassy. He was Laurent Feron, one of Castelnau's clerks. As Feron had good reason to visit Walsingham in the normal course of his duties, it was easy to slip him secret documents over and above the papers he was really meant to show him. Walsingham was watching the embassy closely, because a young and headstrong Catholic gentleman, Francis Throckmorton (a nephew of Sir Nicholas, a staunch Protestant, who had died in 1571), who had also traveled to Madrid and Paris, was observed making regular visits. He was arrested in November 1583, when he confessed under gruesome torture that the Duke of Guise was preparing to invade England and Scotland with Spanish and papal support, and that Mendoza was the impresario of the plot.

In January 1584, Mendoza was summoned before the Privy Council and expelled from England. He left in a rage, shutting the door of the Spanish embassy in London for the rest of Elizabeth's reign. Mary now relied exclusively on Castelnau to get her encoded letters and instructions delivered to her agents in Paris. This is where Walsingham's mole came in. Whatever she sent or received through the embassy could be copied and sent to Walsingham. Not much of the evidence against Throckmorton could be used against her, because if it were, the mole's cover would be blown and Walsingham could not continue reading her secret correspondence. A trap was nevertheless baited.

Mary was naive about the danger she faced. After so many long years of imprisonment, she had come to occupy an enclosed mental space in which her sense of reality was ebbing away and intrigue became a substitute for activity. She worried about security measures even as she risked her own security. "The best recipe for secret ink," she informed Castelnau (as if he did not already know), "is alum dissolved in a little clear water twenty-four hours before it is required to write with. In order to read it, the paper must be dipped in a basin of water, and then held to the fire; the secret writing then appears white, and may easily be read until the paper dries." Naturally, the mole sent a copy of this letter straight to a much-amused Walsingham, in whose papers it remains.

But Mary was soon undeceived. After a mysterious silence lasting six weeks, she wrote, "Through the discovery of all my agents who have visited your house, many people greatly suspect that one of your servants has been 'turned,' which to speak the truth, I rather think myself."

Fortunately for Walsingham, Castelnau did not smell a mole. He carried on business as usual, enabling the spymaster to keep on reading his letters. They showed that Mary had indeed encouraged Throckmorton and promised to reward him. She had urged Castelnau to do all he could to assist him, which was enough to prove that she was dabbling in conspiracy while she was negotiating with Elizabeth to be restored as co-ruler in Scotland. Elizabeth's fury at this deception ended all further prospects of a political accord with Mary. From this point on, there was no more talk of a rapprochement.

Then, in July 1584, a catastrophe occurred. William of Orange, the leader of the Dutch Protestants in their revolt against Philip II, fell victim to a Catholic assassin's bullet. The shockwaves reverberated across Europe. This, and a series of assassination plots against Elizabeth, created a frenzy in which the Catholics and their allies were portrayed as terrorists by the Protestants.

In October, Cecil and Walsingham drew up the Bond of Association. It was modeled on Scottish and Huguenot examples, extended to include the whole of the Protestant elite and others who wished to subscribe to it. At its core was the notion of Protestant citizenship. Signatories were to comprise "one firm and loyal society . . . by the majesty of Almighty God." They were to take a solemn oath by which they entered into a national covenant to defend Elizabeth's life and the Protestant succession.

The bond was not just an oath of loyalty. It was a license to kill. Signatories swore to "pursue as well by force of arms as by all other means of revenge" anyone who tried to harm Elizabeth. Retribution was to be exacted on the spot. Moreover, no "pretended successor by whom *or for whom* any such detestable act shall be attempted or committed" was to be spared (italics added). This clause was the most arbitrary. It included the heirs and successors of the intended beneficiary. Thus if anyone threatened Elizabeth's life in the interest of the Stuart succession, both Mary and James VI would be executed, whether privy to the attempt or not.

Parliament met in November to discuss the Bond of Association. What followed was a protracted battle of wills. Elizabeth was fifty-one. Her father, Henry VIII, had died at fifty-five, and her sister, Mary Tudor, at forty-two. Quite apart from the threat of assassination and the looming war with Spain, Cecil was keen to settle the vexed question of the succession once and for all. He wanted to create a radical constitutional mechanism that would automatically exclude Mary and enable a Protestant ruler to be selected by Parliament when Elizabeth died. And Elizabeth was determined to prevent him.

Cecil proposed a Great or Grand Council that would come into effect on the queen's death, governing as a council of regency and summoning a Parliament that would choose a Protestant successor, whose authority would be confirmed by a statute. It was a quasi-republican solution to the succession issue, one that guaranteed Mary's exclusion, since Catholics were ineligible to sit in Parliament after the legislation of 1571. But it was a risky proposition, and in Elizabeth's view an almost scandalous subversion of the principles of monarchy and hereditary right.

She wielded her power and instructed Cecil to drop his plans for the succession. The Act for the Queen's Safety, as it was passed in March 1585, dealt instead with two contingencies, each devised with Mary in mind and paving the way for her trial and execution. The first concerned a claimant to the throne (i.e., Mary) who was involved in an invasion, a rebellion or a plot. In such a case, a commission of privy councilors and other lords of Parliament, assisted by judges, would hear the evidence and promulgate its verdict by royal proclamation. Those found guilty were to be excluded from the succession, and all subjects "by virtue of this act and Her Majesty's direction in that behalf" might exact their revenge by killing them.

The act then considered Elizabeth's assassination. In such a case, the commissioners would investigate and proclaim their sentence as before, whereupon the intended beneficiary (i.e., Mary) was to be proscribed, and she and her accomplices would be hunted down and killed according to the Bond of Association.

But if the Act for the Queen's Safety confirmed the bond as it related to Mary, its most draconian sanctions were moderated. Whereas the bond made no provision for a public trial, the act insisted that offenders be tried by a commission. And whereas the bond had referred to heirs and successors (i.e., James VI), Elizabeth exempted James unless he was "assenting or privy" to a plot. This was not mere altruism. She was about to send an expeditionary force under Leicester's command to aid the Dutch in their revolt. That would mean outright war with Philip II, leading up to the battle with the Spanish Armada. To protect her northern frontier, Elizabeth opened negotiations with James. And she dangled before him the prospect of what so far she had always withheld. When, after the tribunal at which Moray had exhibited the Casket Letters, she had recognized a regent of Scotland, she had not recognized James VI as king. Now she opened the door to this and also did not rule out the further possibility of succession to the English throne, tempting the nineteen-year-old with the prospect of this glittering prize.

James was male, Protestant and available. His flirtation with d'Aubigny and his Jesuit friends had been little more than a teenage crush. He was deeply resentful of the constraints to which he had been subjected by the lords, especially those imposed by Buchanan, his hated tutor. He longed for a throne more powerful than that of Scotland, and with so spectacular a reward as England within his reach, he decided it was not in his interest to think of sharing his dynastic claim with a mother he could not even remember.

In the same month that Elizabeth made her decisive interventions over the Act for the Queen's Safety, James informed his mother that he would always honor her with the title of Queen Mother. But that was as far as he would ever go. There could be no question of joint sovereignty or her return to Scotland as queen.

To Mary it was the cruelest of blows. She could not believe what she was reading. She fell, in turn, into paroxysms of vomiting, distress and rage. "I pray you to note," she fulminated as she fought back her tears, "I am your true and only queen. Do not insult me further with this title of Queen Mother . . . There is neither king nor queen in Scotland except me."

Mary wrote at once to Castelnau demanding that, in his dealings with

James, he should not call him king. She threatened to disinherit her son and to curse him if he ignored her and made a separate treaty with England. It touched her most sensitive nerve, destroying everything she had fought for since returning to Scotland from France. "I think," she said, "no punishment, divine or human, can equal such enormous ingratitude, if he is guilty of it, as to choose rather to possess by force and tyrannically that which justly belongs to me, and to which he cannot have any right but through me."

And yet, a year later, the treaty with England was signed. James, perhaps without ever fully realizing it, made his mother's execution at Fotheringhay inevitable. With his signature, he made her irrelevant and disposable. To Mary, who had suffered so many setbacks when her enemies and rivals thought she was in their way, it was the ultimate rejection. She became desperate. The prophecy of the *Detection* would become self-fulfilling: "Desperate necessity will dare the uttermost," it had said.

From now on, Mary was prepared to listen to any plot that might offer a chance of escape, however implausible it might seem and however obscure its advocates. And Walsingham was waiting for her. No longer did he need his mole at the French embassy, because Castelnau was effectively blackmailed. To spare him public exposure as an accessory to the Throckmorton plot, he was required to show all letters in his possession to or from Mary to Walsingham. Then, when Paulet succeeded Sir Ralph Sadler as Mary's keeper and she was returned to the greater security of Tutbury, all her contacts with the outside world were sealed. There was no further need for pretense. Her letters were sent directly to Walsingham to be forwarded. Mary was indignant at this blatant intervention, but to no avail. No longer were any of her letters remotely confidential, because Cecil's spymaster was reading them openly.

By the time a genuine — if totally unrealistic — conspiracy to assassinate Elizabeth was spawned under the leadership of a gullible young man named Babington, Walsingham's network was complete. His trap was baited and sprung. As he wrote to Leicester, "If the matter be well handled, it will break the neck of all dangerous practices during Her Majesty's reign." This time no mistake would be made. They had entered the endgame. The evidence would be obtained by fair means or foul, and Elizabeth compelled to proceed under the Act for the Queen's Safety. Cecil would be victorious, and Mary sent to the block.

29

Nemesis

MARY FELT she had been torn in pieces when her son, James VI, rejected her to clear the way for his own glittering dynastic prospects. Sir Amyas Paulet, her jailer at Tutbury and Chartley, coolly predicted her reaction. As he cautioned Walsingham, it was "when she was at the lowest [that] her heart was at the greatest, and being prepared for extremity, she would provoke her enemies to do the worst." Despite her sorely inflamed legs and physical weakness, this was still the same Mary who had ridden in her steel cap at the head of her army during the Chase-about Raid and dealt so ingeniously and courageously with Darnley after the Rizzio plot.

She guessed instantly the reason for Paulet's severing of her links to the outside world. Walsingham realized that his success in checkmating Castelnau had been all too complete. If he wanted to entrap Mary, then far from sealing off her correspondence, he would have to find her a new postman whom she trusted and believed to be safe. Castelnau could no longer fulfill this role, since he was discredited in both England and France for his collusion in the Throckmorton plot and his aid to Mary over and above his instructions. He was replaced as the French ambassador in September 1585 by Guillaume de l'Aubépine, Baron de Châteauneuf.

Walsingham's genius as a spymaster lay in his ability to penetrate the networks of his Catholic opponents and turn them to his advantage. He recruited a defecting Catholic refugee, Gilbert Gifford, to establish a monitored channel of communication between Mary and the French embassy. Gifford was a friend of one of Mary's agents in Paris, who vouched for his credentials. The operation was in place by the end of January 1586, and Walsingham was soon accumulating a fresh pile of intelligence papers, larger and more informative than before. Mary trusted Châteauneuf, whom she believed to be a zealous Catholic of the sort who might help gain her freedom and take seriously the papal decree of 1570 that had declared Elizabeth to be excommunicated and deposed. She wrote candidly to him, blithely unaware that Gifford was working for Walsingham and forwarding her letters to his office for inspection before they were delivered to the embassy.

Mary urged Châteauneuf to look out for "spies" and "moles" among his secretaries. She had also learned her lesson about using alum as a secret ink. It was too easily discovered, she said, "and therefore do not make use of it except in an emergency." If there was no alternative, she suggested hiding secret messages "in such new books" as she was sent, "writing always on the fourth, eighth, twelfth and sixteenth leaf, and so continuing from four to four . . . and cause green ribbons to be attached to all the books that you've had written in this way."

Mary asked that letters meant for her should be packed, tightly wrapped, in the soles or heels of the fashionable new shoes she still wore and of which she received regular consignments, or else placed between the wooden panels of the trunks and boxes that were used to transport her silks or other goods from London and Paris. At the height of Walsingham's plan of entrapment, Gifford found himself using a small watertight box that he inserted through the bunghole of a beer cask, where it floated on top of the beer. Paulet was sent by Walsingham to intercept the casks, which were delivered weekly by the brewer. Then, at the crucial moment, Walsingham's chief decipherer rode to Chartley and slipped incognito into Paulet's household to read and copy the intercepted documents, after which they were carefully sealed again and returned to their box in the barrel.

Walsingham's early breakthrough was in March 1586. Mary sent Châteauneuf the key to a new cipher, perhaps her own creation, that she insisted he use for their correspondence, as the old ciphers supplied by Ridolfi and Castelnau were compromised. Once again Walsingham's chief decipherer did not have to crack the code, which was inadvertently supplied to him, this time by Mary herself. It was just the first of a series of misjudgments that led to her downfall. As Châteauneuf laconically explained in his final report on the debacle to Henry III, "The Queen of Scots and her principal servants placed great confidence in the said Gifford . . . and thence came the ruin of the said queen." Instead of looking for spies at the French embassy, Mary should have been less trusting when Gifford appeared out of nowhere to offer his services as a postman.

Shortly afterward, a madcap plot took shape. Anthony Babington was a rich young Catholic gentleman who had time on his hands and had served the Earl of Shrewsbury at Sheffield as a page. Now twenty-five, he was married with a young daughter. He became embroiled on the fringes of Catholic conspiracy when he visited France in 1580 to further his education. On his return to England, he ran errands for a number of Catholic priests and missionaries as a favor. Above all, he forwarded five packets of confidential letters to Mary before she was transferred from Shrewsbury's custody to Sadler's.

Babington's role as a conspirator might have stopped there, but his connections to Mary's agents in Paris drew him steadily into an underground cabal after Mendoza was expelled as the Spanish ambassador in London. Mendoza went to live in Paris, and it was from there that he hatched the idea of a coup d'état. Mendoza's plan aimed to combine a revolt of the English Catholics, a Spanish invasion, Elizabeth's assassination, and the final liberation and triumph of Mary.

In May and June 1586, Babington first conspired with John Ballard, a soldier-priest and fanatic, Gifford and others. At a meeting at his rooms in London on June 7, it was agreed that Elizabeth should be seized and Mary freed with foreign aid. After some thirteen conspirators had been recruited, many of them more committed to the plot than he was, Babington reluctantly proposed Elizabeth's assassination by a group of "six gentlemen," although the names of the six were never settled.

But would Mary approve of the plot? Babington wrote to ask her on July 6. Thanks to Gifford's role as her postman, the letter was intercepted and Walsingham's men were alerted. Walsingham knew this was the opportunity he had been waiting for to entrap Mary. The plot was not in itself a "projection"* to frame her; it really existed, but rather than nipping it in the bud, Cecil's spymaster allowed it to gather momentum

---

* That is, a plot using agents provocateurs to foment a conspiracy that was then conveniently "detected."

so that he could collect the written evidence to put her on trial for her life. Walsingham was involved from start to finish, interviewing Babington no fewer than three times after the plot had developed, but before arresting him, to see if he would be willing to defect and implicate Mary. The plot was highly disorganized — almost entirely a product of fantasy. Babington himself was wracked with doubt, and Walsingham even had to send Gifford to buoy him up when he suddenly panicked and wanted to back out.

Walsingham sent Thomas Phelippes, his clerk and chief decipherer, to the vicinity of Chartley on July 7. He waited until the 10th, when he received Babington's letter. After Phelippes had successfully decoded it, the letter was returned to the secret box hidden inside the beer cask and smuggled back into Chartley. In readiness for Mary's reply, the decipherer sat and waited.

Mary ruminated for a week before sending it. Her anguish and despair led to recklessness. She weighed her options, which by now were very limited, and decided to take a gamble. Foremost in her mind was the fear that she was likely to be quietly murdered. "She could see plainly," wrote Paulet to Walsingham, "that her destruction was sought, and that her life would be taken from her, and then it would be said that she had died of sickness."

Her mind made up, Mary was excited. Her new animation was observed by Phelippes, who noted her change of mood with grim satisfaction. Her health improved, and she was allowed out in her coach to enjoy the summer sunshine. As she was driven out of the gates, Mary passed the decipherer, who acknowledged her. "I had," he cynically informed Walsingham, "a smiling countenance, but I thought of the verse:

> When someone gives you a greeting,
> Take care that it isn't an enemy."

Mary first worked out her ideas for the fatal reply in her head. She then sketched some headings on paper and sat at the table in her study with her two secretaries to discuss them. The senior of the pair was Claude Nau, the brother of the surgeon who had so expertly saved her life after her gastric ulcer burst at Jedburgh. He took notes, then drafted a letter to Babington in French for her approval. Afterward, he translated it into English. The "authentic" final version of the reply was in English, not French, on this occasion. The reason was that the code

Babington used was based on English. Nau's original French draft has not survived,* and even if it had, it would not have been the document Walsingham wanted most, because he needed the text as it was actually sent to Babington. Still, this was in cipher, and not written in Mary's own hand.

Mary's letter was finished late in the evening of July 17. It was posted early the next day, when it was quickly retrieved from the beer cask and brought to Walsingham's agent to decode. His deciphered text in English — the momentous evidence that would be produced against Mary at her trial — was sent to Walsingham on the 19th. To indicate its urgency and as a token of the decipherer's black humor, a gallows was drawn on the outside.

As soon as Walsingham read Mary's letter, he knew it to be far more incriminating than her earlier appeal to her foreign supporters before the Northern Rising. There she had spoken generally and allusively. Here she was almost explicit:

> The affairs being thus prepared and forces in readiness both without and within the realm, then shall it be time to set the six gentlemen to work taking order, upon the accomplishing of their design, I may be suddenly transported out of this place, and that all your forces in the same time be on the field to meet me in tarrying for the arrival of the foreign aid, which then must be hastened with all diligence.

Mary's meaning is quite clear. She had consented to Elizabeth's assassination and a foreign invasion. She had not strictly specified what the "work" of the six gentlemen was to be, but the letter from Babington to which she was replying included the graphic passage "For the dispatch of the usurper, from the obedience of whom we are by the excommunication of her made free, there be six noble gentlemen, all my private friends, who for the zeal they bear to the Catholic cause and Your Majesty's service will undertake that tragical execution."

When the two letters are read together, Mary's complicity in the plot is undeniable. She protested at her trial that the evidence against her was purely circumstantial. She demanded to be judged only by her own words and writing, saying that in her own words there would be found no consent or incitement to assassination. She refused to accept that the two letters should be taken together.

* The surviving French versions of the letter are all enemy translations, used either to discredit Mary in France or for the interrogation of her French servants.

This became the crux of her defense, and not the later allegation of forgery. There has been much confusion over a postscript that Walsingham's chief decipherer added to the "authentic" final version of her letter before he returned it to its box in the beer cask for onward delivery to Babington. A whole conspiracy theory has been built on this brief postscript, one in which the decipherer is accused of "doctoring" the main body of the letter to incriminate Mary in the murder plot. But there is no evidence to support the claim that the main text of the letter was altered, and the postscript — a blatant and audacious forgery of which Phelippes cheekily left his draft in the archives — was not used against Mary. It was a clumsy tactical move, an attempt to entice Babington to disclose the "names and qualities" of the "six gentlemen" and their accomplices so that Cecil and Walsingham could arrest them all. It did not work, because Babington's suspicions were aroused and he fled. Ten days later, he was caught in a barn with his hair cropped short and his face grimed to make him look like a farm hand.

Walsingham took his time to put the conspirators on trial. His main quarry was Mary, who was brought to Sir Walter Aston's house at Tixall, about three miles from Chartley. A subterfuge was devised to separate her from her secretaries and seize all her papers and ciphers before she learned that Babington and his accomplices were to be executed. She was in unusually high spirits. When Paulet invited her to ride out and join a deer hunt in Sir Walter's neighboring park, she jumped at the chance. Her legs were so much better, she was able to mount her horse for the very last time in her life.

On August 11, she set out with Paulet and his company for the hunt, attended by her secretaries, her loyal valet Bastian Pages and her physician, Dominique Bourgoing, to whom we owe a vivid account of the final seven months of her life.

After riding a short distance, Paulet and some of his men dropped back. On the horizon, a troop of horsemen had appeared. Mary's heart must have leapt. Her vision was almost apocalyptic, because it was just such a group that she had imagined Babington would be dispatching to free her. The captain of the troop conferred earnestly with Paulet, then rode forward. He dismounted and told Mary that a plot to kill Elizabeth had been uncovered and his instructions were to arrest her secretaries and conduct her securely under guard to Tixall.

Mary tried to lie her way out of trouble. She angrily expostulated that Elizabeth had been misled. "I have always shown myself her good sister and friend," she insisted. She ordered her servants to draw their swords to defend her, but they were heavily outnumbered by men with loaded pistols. After all were disarmed, Mary dismounted and sat on the ground. She refused to move, saying she preferred to die where she was. Paulet threatened to send for her coach and forcibly remove her. Mary demanded to know where she was to be taken. She sat firmly glued to the spot until Bourgoing came to her side to comfort and assist her. At length she agreed to move, but first she knelt against a tree and prayed loudly. Paulet and his men were obliged to wait until she had finished.

Mary was kept at Tixall for a fortnight. While she was there, her rooms at Chartley were searched and her papers packed into three large coffers. Her secretaries were escorted first to a nearby village and then to London for interrogation. Her papers followed. Everything that was found was removed: letters, drafts, notebooks, minutes, memos, and the keys and tables of sixty or so ciphers. All were brought to Walsingham, who presided over a committee of privy councilors assigned to scour the entire archive looking for evidence that could be used to convict her.

On August 25, Mary was brought back to Chartley. As she passed through the gates at Tixall, a small crowd of onlookers was gathered. Seizing her opportunity, she cried out to the beggars, "I have nothing for you. I am a beggar as well as you. All is taken from me." And to the rest she said, weeping, "I am not witting or privy to anything intended against the queen."

But Mary's humiliation had scarcely begun. When she reached Chartley, she was incensed to discover her ransacked drawers and cabinets. Fighting back the tears, she screamed, "Some of you will be sorry for it!" Then she declared that there were two things that could not be taken from her, her royal blood and her religion, "which both I will keep until my death."

On September 5, Paulet was told to confiscate Mary's money and isolate her as much as possible from her servants. Even Walsingham was alarmed at these measures, in case they should "cast her into some sickness." If Mary died unexpectedly, she would become a martyr to the Catholic cause, the center of a glare of publicity. He wished to avoid this at all costs. But Elizabeth was adamant. For the first time since she had quarreled with her cousin over the ratification of the treaty of Edinburgh, she had hardened her heart. At last she took Cecil's dire warnings about her safety seriously. She feared she would fall victim to poison or an assassin's bullet, and had already gone into hiding at Windsor Castle, then a fortress rather than a pleasure palace and one she hardly ever visited except when she felt her life was in danger. She knew that sooner or later Mary would have to be put on trial. The Privy Council was lobbying for Parliament to be summoned, and it was obvious what Cecil's inner caucus would be demanding after their strident calls for Mary's execution in 1572.

Unlike Cecil and Walsingham, Elizabeth preferred Mary to die of natural causes. Her idea was that by seizing her money and keeping her as much as possible in solitary confinement, she would become so seriously demoralized, her existing illnesses and afflictions would fatally worsen.

Elizabeth could afford to turn against Mary, who was shunted aside after James VI signed a separate treaty with England. Yet she was loath to put her cousin and "sister queen" officially on trial. She was not squeamish, but did not want the responsibility of executing an anointed queen, with all the implications that would have for undermining the ideal of monarchy. If Mary did not die naturally, Elizabeth's preference was barely masked. She wanted her hunted down and killed under the terms of the Bond of Association.

Elizabeth had a clear grasp of the issues. She knew that regicide authorized by a statute made in Parliament would alter the future of the monarchy in the British Isles. It would tend to make the ruler accountable to Parliament, diminishing forever the "divinity that hedges a king." This was of slender concern to Cecil, whose aim for nearly twenty years had been Mary's execution and a guaranteed Protestant succession.

Mary, by this time, was ill in bed. When Paulet came to impound her hoard of cash, she railed against him and his political masters. In a harrowing scene, she refused to hand over the key of her closet, but Paulet ordered his servants to break down the door. Mary then yielded and asked her gentlewomen to hand over the key. As Paulet's men took away the money, Mary rose pathetically from her bed and "without slipper or shoe followed them, dragging herself as well as she could to her cabinet." She pleaded with them for some time, but Paulet ignored her.

The Privy Council, meanwhile, met daily in closed session at Windsor. Elizabeth called for retribution against Mary, but objected to everything that was proposed. Cecil wrote to Walsingham, who was ill at his house in London: "We are still in long arguments, but no conclusions do last, being as variable as the weather . . . and so things are far from execution for the bringing of the Scottish queen to some apt place where her cause and herself might be heard." Elizabeth had rejected the Tower as being too close to London. Hertford Castle had been agreed on for a day, then Elizabeth changed her mind.

Fotheringhay Castle in Northamptonshire was at last chosen. Since Henry VIII's death, it had been used mainly as a prison but was still in a reasonable state of repair. It was accessible from London, but not too close. And it was not far from Staffordshire, reducing the risk of Mary's escaping en route.

Mary left Chartley for Fotheringhay on September 21. She was prematurely aged at forty-three. Her rheumatism and swollen legs were getting worse, and the pace was slow. She took four days to reach her final destination, a physically broken woman, but one who still insisted on as much "grandeur" as could be preserved. Even in her presently reduced state, lacking money to pay for necessities as well as luxuries, some twenty mules and carts were required for her baggage train.

Cecil's preparations for the trial were meticulous. Every detail was carefully planned: measurements of the rooms were taken, furniture was requisitioned and a seating plan made. Sufficient food and fuel were brought in, and the sleeping and dining arrangements settled. According to the Act for the Queen's Safety, Mary was to be tried by a commission of no fewer than twenty-four nobles and privy councilors, who would be advised and assisted by common-law judges. In the event, more than forty commissioners were appointed, but seven or eight did not turn up. To Elizabeth's great displeasure, a number evaded the summons on the excuse of illness. They too wanted no part in a regicide.

The place chosen for the courtroom was the second-floor presence chamber of the old state apartments, a space sixty-nine feet long and twenty-one feet wide, which was divided into two unequal halves by a rail at waist height. The larger area in front of the rail was to be the courtroom. At the upper end sat a chair on a dais beneath a cloth of state emblazoned with the royal arms of England. This symbolized Elizabeth's throne and was left empty throughout the proceedings. Benches were arranged on the other three sides of the space for the nobles and privy councilors, with more benches and a table in the center for the judges, lawyers and notaries. A high-backed chair with a red velvet cushion was positioned for Mary at the side, behind the senior judges and in front of the nobles seated to the right of the throne if viewed from the center of the space. The other, smaller area behind the rail was used as standing room for the knights and gentlemen of the county, who were allowed to watch the events.

Most of the commissioners gathered at Fotheringhay on Tuesday, October 11, leaving Cecil to arrive early the next morning. He took charge from the outset, sending a small delegation to Mary in her privy chamber. They handed her a letter from Elizabeth, informing her that she was to be put on trial. After reading the letter, she flatly refused to appear before the commission. "I am an absolute queen," she said, "and will do nothing which may prejudice either mine own royal majesty, or other princes in my place and rank, or my son." She was calm and composed during the interview, speaking slowly but confidently. "My mind is not yet dejected," she said, "neither will I sink under my calamity."

The delegation withdrew, leaving Cecil to reconsider his tactics. He spent most of Thursday, the next day, trying to induce her to change her mind. He led a larger committee that made several trips to and from her privy chamber. In the morning, she repeated her objections. "I am a queen," she said, "and not a subject . . . If I appeared, I should betray the dignity and majesty of kings, and it would be tantamount to a confession that I am bound to submit to the laws of England, even in matters touching religion. I am willing to answer all questions, provided I am interrogated before a free Parliament, and not before these commissioners, who doubtless have been carefully chosen, and who have probably already condemned me unheard."

Mary warned the commissioners to think about what they were doing. "Look to your consciences," she said, "and remember that the theater of the whole world is wider than the kingdom of England."

On hearing this, Cecil (whom Bourgoing in his narrative describes as "homme plus véhement") brusquely interrupted her. He reminded her of Elizabeth's kindnesses to her, then told her that he had legal advice that the commissioners could proceed to judgment in her absence. "Will you therefore," he said, "answer us or not? If you refuse, the commissioners will continue to act according to their authority."

Mary hit back instantly. "I am a queen," she said. To which Cecil retorted, "The queen, my mistress, knows no other queen in her realm but herself." Cecil carried on for several minutes in this vein, but Mary ignored him.

The committee returned in the afternoon, when Sir Christopher Hatton tried a more conciliatory approach. He told her that royal majesty in the case of such a crime as she was charged with would not exempt her from answering, adding artfully, "If you be innocent, you wrong your reputation in avoiding a trial."

Mary replied that she did not refuse to answer. She would not appear before the commissioners, but would plead before a full Parliament, as long as her protest against the legitimacy of the proceedings was admitted and her rights were acknowledged as Elizabeth's nearest kinswoman and the heir apparent to the English throne.

Cecil decided this was enough. The commissioners, he said, "will proceed tomorrow in the cause, even if you are absent and continue in your contumacy."

"Search your consciences," Mary said. "Look to your honor! May God reward you and yours for your judgment against me."

But when she slept on it, she was torn between two conflicting positions. She dreaded appearing as a defendant in a public trial, yet she realized that the commissioners would convict her in her absence of conspiring to murder Elizabeth.

Early on Friday morning, she demanded to see a new and expanded committee. Among them was Walsingham, whom she met for the first time. After some give and take on both sides, Cecil asked her whether she would appear if her formal protest was received and put in writing by the commissioners. Mary reluctantly agreed. She was, she said, so anxious to purge herself of the accusations against her that she was willing to accept his terms.

Soon after nine o'clock, the commissioners took their places in the courtroom, and Mary arrived wearing a gown of black velvet with her distinctive white cambric cap on her head to which was attached a long white gauze veil. As she entered slowly and purposefully through the door, the commissioners removed their hats as a mark of respect. She graciously acknowledged them and took her seat, her eyes darting around eagerly to see who was there and which of them might be on her side.

When the court was brought to order, Mary registered her protest, after which the prosecuting counsel opened the case against her. He delivered what the official transcript of the trial describes as "an historical discourse" of the Babington plot. He argued point by point that she knew of the plot, approved of it, assented to it, promised her assistance and "showed the ways and means."

One of Mary's objections to the court's legitimacy was that its procedure was that of a treason trial. She was not allowed a lawyer, she was not able to call witnesses, and she was not allowed to use notes or examine documents in the course of conducting her own defense. Despite these unbending restrictions, which swung the balance heavily against her, she was ready to enter her plea.

She offered a robust denial. The court reporter was favorably impressed, remarking on her "stout courage." "I knew not Babington," said Mary. "I never received any letters from him, nor wrote any to him. I never plotted the destruction of the queen. If you want to prove it, then produce my letters signed with my own hand."

"But," replied the prosecutor, "we have evidence of letters between you and Babington."

"If so," countered Mary, "why do you not produce them? I have the right to demand to see the originals and the copies side by side. It is quite possible that my ciphers have been tampered with by my enemies. I cannot reply to this accusation without full knowledge. Until then, I must content myself with affirming solemnly that I am not guilty of the crimes imputed to me."

What Mary could not yet have known was the full extent of Babington's confession and the evidence collected by Walsingham. Babington had pleaded guilty at his own trial, but that was kept from her. All she knew was that her reply to him had been sent in code by one of her secretaries and was not written in her own hand. She did not know that the letter had been intercepted en route. She assumed (correctly) that Babington had burned the document after reading it, as she had ordered him to do. As the "final" and most incriminating text was the one he had destroyed, she may have thought the case against her was weak, especially regarding Elizabeth's assassination, as she well knew that she had not specified in her own letter what the "work" of the six gentlemen was to be.

Mary was sadly deluded. She had seriously underestimated Cecil's spymaster. He had the deciphered transcript in English of her fatal reply to Babington. He had a copy of this English text as authenticated by Babington himself under interrogation (happily without the need for torture), and without the forged postscript, so the evidence could be produced in court. Most important, he had a craftily "reciphered" copy of Mary's original letter, to replicate the missing document she had actually sent. That was Walsingham's *pièce de résistance.* A facsimile of the lost original had been reconstructed by Phelippes to stand in for the evidence Babington had burned. The facsimile was so brilliantly done, it looked exactly like the "final" version of the original letter and could easily be taken for it. When it was shown to Mary's secretaries during the last stages of their interrogation, they broke down and confessed everything. Their confessions were at once taken down. Thereafter, their statements were the crucial "corroboration" that the reconstituted cipher was the true text of Mary's original letter — the one she had actually sent to Babington! And of course the contents of the facsimile exactly matched the English transcript on which the chief decipherer had earlier drawn a gallows. By this sleight of hand was Walsingham able to persuade the commissioners that the case against Mary was invincible.

Mary must have been stunned when the prosecuting counsel went on to produce what now seemed to be watertight evidence against her. Piece by piece, the evidence was read out. As these damning documents were placed before the court, she could no longer contain her emotions. She burst into tears of despair. But despite her distress, she kept her wits about her. She turned to Walsingham, taxing him for the uncanny perfection of these proofs. "It was an easy matter," she said, "to counterfeit the ciphers and characters of others." She knew there had been trickery. She was not sure just what it was or how it had been done, but all a forger had to do, she said, was to consult her lately purloined "alphabet of ciphers" to discover the codes, assuming he did not already know them.

It was Walsingham's turn to be stung; he rose to defend himself. "I call God to record," he said, "that as a private person I have done nothing unbeseeming an honest man, nor as I bear the place of a public person have I done anything unworthy of my place." It was a Delphic explanation of the arcane rules of politics, worthy of a Machiavelli. Only Cecil and his spymaster could know where the boundaries had been drawn in this instance between private and public duties. Did the ends justify the means? Walsingham hinted at the correct answer when he admitted that he had always been "very careful for the safety of the queen and the realm."

Mary accepted Walsingham's reply with extraordinary good grace. Perhaps its subtler meaning had escaped her under the pressure of this courtroom drama. She asked him not to be angry with her, and to give no more credit to those who slandered her than she did to those who accused him. She then wept again. Drying her eyes, she cried out, "I would never make shipwreck of my soul by conspiring the destruction of my dearest sister."

After this electrifying exchange, the court adjourned at about one o'clock for dinner. When everyone returned, the rest of the evidence was read out, notably the confessions of Mary's secretaries. She was deeply shocked. She saw how damaging their statements were, and tried to counter them by suggesting that passages might have been added to her letters after she had approved the final drafts. She drew herself up to her full height and with all the remaining dignity she could muster said, "The majesty and safety of all princes falleth to the ground if they depend upon the writings and testimony of their secretaries." She paused for a moment, then locked horns with the prosecution. "I am not," she said, "to be convicted except by mine own word or writing." She pointed out that her secretaries had not been called as witnesses and so could not be cross-examined. And she observed in a gently mocking tone that her initial memo of rough headings for the letters she had discussed with her secretaries on the crucial day had disappeared. She sensed a weakness in the prosecution's case, because she knew these notes should have been retained in the archives seized by Walsingham's men from her rooms at Chartley. (Walsingham had, in fact, searched in vain for the notes and also for Nau's initial draft in French of the reply to Babington.)

Although forced to defend herself without being allowed to subject any of the documents exhibited against her to legal or forensic scrutiny, Mary stood up remarkably well to her ordeal. The afternoon debates continued until late in the evening, and resumed on Saturday morning. The case turned on whether she had consented to Elizabeth's assassination: the crux was the court's insistence that her reply to Babington be read in conjunction with his letter proposing that "the usurper" be "dispatched" in a "tragical execution." This way, the "work" of the six gentlemen was shown to be the plot to kill Elizabeth.

Mary contended that the circumstances of her guilt might be proved, but never the fact. She does seem genuinely to have convinced herself of it, because even after the trial she made a clear distinction in her letters between "intending to assassinate" and "leaving to God and the Catholics under God's Providence." She had not, she said, at any point specified the work that the six gentlemen were to undertake.

But she was grasping at straws, since in the coded letter to Babington she had appealed for foreign (i.e., Spanish) aid to assist her on the field of battle after her liberation. Her rationale was that she was an independent queen wrongfully held in captivity. From her point of view, even an act of war was legitimate if it allowed her to recover her freedom. If she was not an independent queen, she was guilty. If she was, and Elizabeth's death was no more than a providential incident in her legitimate struggle to regain her rights, she was innocent. This was how she saw it, but the commissioners could hardly be expected to agree with her.

No one who attended Mary's trial could ever have forgotten it. Perhaps the most memorable clash was on the last day with Cecil himself. Becoming more and more impatient with what he saw as her futile semantics over her right to make a bid for freedom, Elizabeth's chief minister finally put in the knife. With scant respect for the majesty and dignity of a queen, he told Mary that all her failed efforts to liberate herself were the result of her own actions and those of the Scots, not those of Elizabeth. Hearing this, Mary turned to him. "Ah, I see you are my adversary."

"Yea," he replied. "I am adversary to Queen Elizabeth's adversaries." In that brief but fiery exchange was captured all the ardor of a contest that had lasted for almost thirty years.

From then on, the trial, in Mary's opinion, was over. While she continued to participate, her hopes had faded. "I will hear the proofs," she said, "in another place and defend myself," by which she clearly meant she awaited judgment in heaven. Bourgoing, whose account is slightly different, says she exclaimed, "My lords and gentlemen, my cause is in the hands of God."

She kept her seat until the proceedings drew to a close, but when she was asked if she had anything more to say, she ended as she had begun. "I again demand to be heard," she said, "in a full Parliament, or else to speak personally to the queen, who would, I think, show more regard of another queen." She then stood up (as the official reporter noted) "with great confidence of countenance," spoke disparagingly to Cecil, Walsingham and Hatton about the conduct of her two secretaries, then swept out of the room.

After she had departed, Cecil adjourned the commission for ten days. He had received a letter from Elizabeth ordering him to delay sentence if Mary was found guilty. Elizabeth wanted nothing to be done in haste. The commissioners were told to reconvene on October 25 in the Star Chamber at Westminster. There the evidence was reviewed in full, and Mary's secretaries had to swear under oath that their written statements were true. They were then examined *viva voce* by the commissioners to check their stories. Mary was found guilty in her absence.

Now Cecil called for the verdict to be publicly proclaimed according to the Act for the Queen's Safety so that Mary's execution warrant could be issued. But Elizabeth stayed his hand. Cecil and Walsingham drew up a memo, yet nothing happened. When Parliament reassembled on the 29th, a stormy session was guaranteed. The debates turned immediately to Mary's sentence, and she was denounced in a series of prepared speeches. All the old charges of adultery with Bothwell and Darnley's murder were resurrected alongside the new. Mary was demonized in a frenzy of invective couched in the language of biblical fundamentalism.

Parliament, steered by Cecil and Walsingham, openly petitioned Elizabeth to execute Mary. Behind the scenes, however, a battle royal was in progress. Elizabeth was amenable to a petition, but insisted on the Bond of Association as the basis of action against her cousin. She preferred her to be quietly smothered by a private citizen, someone who had subscribed to the bond, whereas Cecil wanted Elizabeth to sign a warrant to justify a public execution as a means to validate regicide. At stake was the future of divine-right monarchy in the British Isles. If Mary was to be executed by a private citizen who had signed the bond, he would act in a private capacity, whereas an official execution sanctioned by Elizabeth under the terms of the Act for the Queen's Safety would justify regicide as a legal precedent and permanently cede to Parliament some measure of the ruler's prerogative.

Elizabeth asked Cecil to make sure that the parliamentary petition referred to the Bond of Association. Cecil gave a dishonest answer. He said such a change of wording could not be made for lack of time. In reality, the wording that Elizabeth expected had been included in the first draft of the petition, but Cecil had personally deleted it.

The result was deadlock. When Elizabeth gave her answer to the petition for execution, she cloaked her meaning in mist. "If I should say I would not do what you request, it might peradventure be more than I thought; and to say I would do it might perhaps breed peril of that you labor to preserve." She herself called this an "answer answerless."

Cecil, determined to get his way, struck out on his own. On December 4, the guilty sentence against Mary was publicly proclaimed. Elizabeth agreed to this, but for her own reasons. When overruling Cecil the previous year, she had insisted on putting her own wording into the Act for the Queen's Safety. According to this, when the verdict against a conspirator was proclaimed, then the guilty person, "by virtue of this act and Her Majesty's direction in that behalf," was to be hunted down and killed. By the final wording as Elizabeth had approved it, an execution warrant was unnecessary. Those who had signed the Bond of Association were already empowered to take the responsibility once she signaled her desire to them.

And this is how Elizabeth planned Mary's end. The deadlock lasted for six weeks. It was broken by Cecil and Walsingham, who visited Châteauneuf at the French embassy in London. Again France refused to intervene in Mary's defense: Catherine de Medici and her son Henry III by now regarded her as a dangerous embarrassment. She simply had to go.

Châteauneuf connived with Cecil in a shabby little conspiracy. They pretended that another plot to kill Elizabeth had been discovered. It was actually an old plot, known to Châteauneuf for over a year, that had never amounted to anything, but it served its purpose now. Cecil helped to foster a rumor that Spanish troops had landed in Wales, and ordered justices of the peace to instigate the hue and cry.

When Elizabeth was told to double her bodyguards, she momentarily caved in. On February 1, she sent for her secretary and asked him to bring the warrant for Mary's execution. She called for pen and ink, then signed. She even made a joke. Walsingham was ill again at home. "Communicate the matter with him," she said, "because the 'grief' therefore would grow near to kill him outright!"

The idea that Walsingham would die of grief at Mary's death was darkly amusing. Except that Elizabeth did not jest in vain. She never intended the warrant to be used. Instead, she told her secretary to order Walsingham to write a letter in his own name to Paulet, asking him to do away with Mary without a warrant. Paulet was to act on his own initiative, just because he had been told it was a good idea. Elizabeth wanted Mary dead, but without taking any of the responsibility. Paulet had been among the first to sign the Bond of Association, and this letter from Walsingham was to serve as the "direction" referred to by the Act for the Queen's Safety. And yet if Paulet acted on it, he would kill Mary as a private citizen, with all the risks that entailed.

Paulet was shocked. He had once proudly boasted that he would rather forgo the joys of heaven than disappoint Elizabeth in his duty.

When forced to live up to his claim, he ate his words. Robert Beale, who was later responsible for delivering the execution warrant to Fotheringhay on Cecil's orders and without Elizabeth's knowledge, tells the story:

> When I was come to Fotheringhay, I understood from Sir Amyas Paulet and Sir Drue Drury that they *had been dealt with by a letter* if they could have been induced to suffer her [Mary] to have been violently murdered by some that should have been appointed for that purpose. But they disliked that course as dishonorable and dangerous, and so did Robert Beale. And therefore [they] thought it convenient to have it done according to law, in such sort as they might justify their doings by law. One Wingfield (as it was thought) should have been appointed for this deed . . . Her Majesty would fain have had it so, alleging the Association . . . [Italics added.]

When Paulet had protested, "God forbid that I should make so foul a shipwreck of my conscience," Elizabeth had stormed at his "daintiness."

Cecil, meanwhile, pressed ahead on his own. When Elizabeth had signed the warrant even though it was not really to be used, he took over, arranging for it to be quickly sealed. He then convened a secret meeting of ten privy councilors in his private rooms, and within two days they had ordered the warrant to be dispatched to Fotheringhay. The Earls of Shrewsbury and Kent were chosen to direct Mary's execution, their letters of appointment drafted by Cecil on his own initiative and signed on the authority of his fellow councilors. Lastly, the councilors agreed among themselves that they would not tell Elizabeth about the execution "until it were done." A covering letter to the earls, to which Walsingham added his signature from his sickbed, justified this as "for [the queen's] special service tending to the safety of her royal person and universal quietness of her whole realm."

Such a rationale was needed, because Elizabeth had sent for her secretary again, telling him that she had dreamed of Mary's death. She spoke elliptically, but made it clear that what she wanted was for Mary to be assassinated. That was her main aim, but by now she had skillfully contrived things so that she would win whatever happened. If Mary was killed under the Bond of Association, Elizabeth could disclaim responsibility. If Cecil covertly sealed the warrant and sent it to Fotheringhay behind her back, she could claim she had been the victim of a court conspiracy.

Her secretary kept quiet. He knew that the warrant was already on its way to Fotheringhay. Beale had been sent posthaste to find the Earls of

Shrewsbury and Kent at their houses, to show it to them and hand them their letters of appointment. Also riding north was the executioner, disguised as a "servingman," his ax concealed in a trunk. Walsingham had personally selected him and promised to pay his fee and a bonus.

Cecil was utterly implacable; he had acted clandestinely. It was a truly historic moment. He was prepared to take no chances and chose to go ahead regardless. He had waited so long for this day. In this respect, he was far more than just the adversary of the ill-fated Queen of Scots. He really was her nemesis. And now that the time had arrived, not even the queen of England was going to stop him.

30

❦

The Final Hours

MARY WAS IN surprisingly good spirits in the few remaining weeks and months after her trial. Her apartments at Fotheringhay Castle were comparatively spacious and comfortable, and her money had been returned to her, which allowed her to purchase some additional luxuries. Perhaps this, combined with Elizabeth's "answer answerless" to Parliament and the delay in issuing the proclamation against her, raised her hopes and encouraged her to think that no one would in the end dare to put the verdict of the trial commissioners into effect.

An open sore on one of her shoulders and a stiff right arm added to the pain she suffered from her other ailments, but she was otherwise cheerful. Paulet found her "taking pleasure in trifling toys, and in the whole course of her speech free from grief of mind in outward appearance." She kept going over the events of the trial: who had said what in the courtroom, and what she had overheard whispered by those commissioners sitting nearest to her.

Then, on Saturday, February 4, 1587, Robert Beale reached the vicinity of Fotheringhay. His first task was to find and brief the Earls of Shrewsbury and Kent, which took up most of the next three days, as they were visiting their estates and on the move. When this was done and all three had arrived at Fotheringhay Castle, they approached Mary's privy chamber, where they would inform her that she would be executed shortly after eight o'clock the following morning.

Paulet and his assistant, Sir Drue Drury, led the earls upstairs. When they were admitted, the man who so obviously relished his role as Mary's jailer barged in and pulled down her cloth of state for the last time. According to Beale's account, the warrant for her execution was read while she listened in silence.

She sat still for several minutes, then suddenly frowned as she remembered how the deposed King Richard II had been quietly murdered at Pontefract Castle. She asked if this would be done to her, to which Drury, an honorable man who was gentler and kinder than Paulet, answered, "Madam, you need not fear it, for that you are in the charge of a Christian queen."

How little did Mary guess that Elizabeth's firm intention, as Drury knew very well, had been that she should be surreptitiously done to death by "one Wingfield," a hired assassin, and that she owed the relative privilege of a public execution almost entirely to the will of Cecil and Walsingham.

Mary calmly addressed the earls. "I thank you for such welcome news. You will do me a great good in withdrawing me from this world, out of which I am very glad to go." She spoke at some length, recalling her ancestry and dynastic claim, and giving her own account of her attempts to reach a political accord with Elizabeth over the years and her willingness to compromise. All her overtures, she said, had been rejected. There was nothing more that she felt she could have done. "I am of no good and of no use to anyone," she concluded. For all these years, perhaps ever since the death of her first husband, she had been in someone else's way.

But she had discovered a new role. She would die as a martyr for her Catholic faith. She crossed herself in the name of the Father, the Son and the Holy Ghost. "I am quite ready and very happy to die, and to shed my blood for Almighty God, my savior and my creator, and for the Catholic Church, and to maintain its rights in this country."

Mary asked that her own chaplain should be allowed to comfort her, but her request was cruelly denied. She then inquired about her place of burial. Would she be allowed to lie next to her first husband at the royal mausoleum at St.-Denis or beside her beloved mother at the convent of St.-Pierre-des-Dames at Rheims?

Shrewsbury answered that while nothing had been decided, she could hardly expect Elizabeth to allow her to be buried in France. "At least then," said Mary, "my requests in favor of my servants will be granted?"

Since her clashes as a teenager, she had been famous for her generosity to her servants, and she now expressed her earnest wish to reward those of her gentlewomen and domestic staff who had loyally stood by her for so long. As the earls had received no instructions on this point, they offered no objection.

When they had taken their leave and Mary was alone with her gentlewomen, she kept up her defiance. "Weeping," she told them, "is useless." She ate little at supper, but knelt to pray for an hour or so. Then she gathered her strength and set to work. She made her last will. She said that she died in the true Catholic faith, and left instructions for Requiem Masses to be said for her soul in France, which all her servants might attend. She directed that all her debts should be paid, and whatever money was left over used to reward her servants. She named as her principal executor her cousin Henry, Duke of Guise.

Next Mary consulted the inventory of her wardrobes and cabinets and distributed their contents among her gentlewomen and servants. Bourgoing, whose narrative is the most reliable account of Mary's final hours, received two rings, two small silver boxes, two lutes that Mary herself had sometimes played, her music book bound in velvet, and her red valances and bed curtains.

When Mary had distributed her possessions, she went to her writing desk. She had already said goodbye to the Duke of Guise, whom she regarded as the head of her family. "I bid you adieu," she had written a little over two months before and a month after the guilty verdict, "being on the point of being put to death by an unjust judgment, such a one as never any belonging to our house yet suffered, thanks be to God, much less one of my rank." She hoped that her death would bear witness to her Catholic faith and willingness to suffer "for the support and restoration of the Catholic Church in this unfortunate island."

She sent her love to all her relatives. The bitterness she still felt toward her son, James VI, for his rejection of her was barely concealed. "May the blessing of God, and that which I should give to my own children, be upon yours, whom I commend to God not less sincerely than my own unfortunate and deluded son."

It was time for Mary to finish her goodbyes. She had always loved to send and receive letters, and allowing for documents that no longer exist she must have written perhaps two or three thousand over the course of her forty-four years. Now she sat down to write her very last one. It was naturally to be sent to France, the country she regarded as home, even while she was a reigning queen in Scotland. The chosen recipient was her brother-in-law, Henry III, whom she had known since he was a baby in the royal nursery at St.-Germain and Fontainebleau.

It was almost two o'clock in the morning when she began. There were only six hours left before her execution, but despite her usual tendency to scribble, the letter is in her best handwriting. Some slight blotches on the first page may mark the places where her tears fell onto the paper as she wrote.

Today, after dinner, I was advised of my sentence. I am to be executed like a criminal at eight o'clock in the morning. I haven't had enough time to give you a full account of all that has happened, but if you will listen to my physician and my other sorrowful servants, you will know the truth, and how, thanks be to God, I scorn death and faithfully protest that I face it innocent of any crime . . .

The Catholic faith and the defense of my God-given right to the English throne are the two reasons for which I am condemned, and yet they will not allow me to say that it is for the Catholic faith that I die . . .

I beg you as Most Christian Majesty, my brother-in-law and old friend, who have always protested your love for me, to give proof now of your kindness on all these points: both by paying charitably my unfortunate servants their arrears of wages (this is a burden on my conscience that you alone can relieve), and also by having prayers offered to God for a Queen who has herself been called Most Christian, and who dies a Catholic, stripped of all her possessions . . .

Concerning my son, I commend him to you inasmuch as he deserves it, as I cannot answer for him . . .

I venture to send you two precious stones, amulets against illness, trusting that you will enjoy good health and a long and happy life.

Mary at last lay down on her bed and tried to sleep. She could barely doze, but managed to keep still until six o'clock, when the candelabra were lit and she briskly rose and began to prepare herself. Her gentlewomen had been busy throughout the night, making ready her clothes, makeup and wig.

She dressed and stayed seated on a stool until her gentlewomen had finished their work. She then gave orders that all her household should assemble in her presence chamber. Bourgoing read her will aloud, after which she signed it and gave it to him to deliver to the Duke of Guise. She bade everyone farewell, then knelt to pray with her servants.

She had barely begun to mouth the words when a loud knocking was heard at the outer door. It was Thomas Andrews, sheriff of the county of Northamptonshire, with the Earls of Shrewsbury and Kent beside him.

Mary's last hour had come. The earls had arrived to escort her down the stairs to the great hall on the ground floor. She was about to take her final walk, the one for which she will always be remembered.

She picked up her ivory crucifix in one hand and her illuminated Latin prayer book in the other. She almost forgot the prayer book, but Bourgoing reminded her. She kissed the crucifix, then approached the door.

This was the moment. She had been the star of so many glittering spectacles during her life, beginning with her wedding to the dauphin at Notre-Dame when she was not yet sixteen, and she knew that she could keep her nerve. Her most compelling act of theater awaited her. She stepped forward and crossed the threshold. The rest, whatever view is taken of the extent to which she truly ranks as a martyr for the Catholic faith and for the ideal of monarchy, forever settles her place in the pantheon of history as a fully realized tragic heroine.

ELIZABETH I DIED shortly before three o'clock on the morning of March 24, 1603. She had lived to her seventieth year — sixteen years after the death of her cousin queen — and was the first English ruler to survive to that age. She was still unmarried and had steadfastly refused to identify her successor, at least officially. The myth that she named Mary's son, James VI of Scotland, as her successor on her deathbed is unsupported by solid evidence. As she had by then lost the power of speech, the most she might have done was to signal her assent by a gesture. And even that is guesswork.

All the same, James was acknowledged as king-in-waiting. He was proclaimed James I of England and Ireland with beguiling ease. The formalities took no more than a few hours. At ten o'clock, a group of nobles and privy councilors appeared with the heralds at the gates of Whitehall Palace and in the City of London to declare the new king's accession.

The proclamation is a memorable document. It said that James was rightfully king of England because he was "lineally and lawfully descended" from Margaret Tudor, daughter of Henry VII. He was the great-great-grandson of the founder of the Tudor dynasty, the great-grandson of Margaret Tudor, who — the heralds took considerable pains to stress — was Henry VIII's sister. He was therefore king "by law, by lineal succession and undoubted right."

Mary, of course, was not mentioned. But the proclamation is luminously clear that James succeeded by virtue of his hereditary rights. Henry VIII's will was disregarded. This was little short of a recognition of Mary's own claim to be Elizabeth's lawful successor had she lived. She had finally won. Her victory was more conclusive than even she might have dared to hope, because every subsequent British ruler has been descended from her, and all derive their claim to the throne from her and not Elizabeth.

Once James had arrived in London and established himself as king, he came to regret his rejection of his mother, for which he attempted to atone. After her execution, Mary's embalmed body had been kept in a lead coffin at Fotheringhay Castle for six months before it could be buried. A series of highly charged debates took place as to whether she should be buried obscurely in the local parish church or allowed a state funeral at Peterborough Cathedral. In the end, she was given a state funeral, but with a strictly limited number of mourners, the ritual of interment performed in the dead of night. She was placed on the south side of the chancel, not far from the tomb of Catherine of Aragon, Henry VIII's first and most unhappy queen.

Soon after his accession, James commissioned two magnificent monumental tombs, each with a recumbent effigy, to be built in Henry VII's chapel at Westminster Abbey, one for Elizabeth and the other for his mother. Elizabeth's memorial was built in the north aisle of the chapel and Mary's in the south. James did not envisage a pair of precisely matching designs. The tombs were similar in style, but Mary's was larger and cost the astronomical sum of £2000, whereas only £765 was spent on Elizabeth's.

In October 1612, Mary's body was exhumed from Peterborough and reinterred at Westminster. An exercise in mythmaking was under way. When Elizabeth died, the Venetian ambassador reported that many portraits of her were taken down and replaced by those of Mary. She was a queen worthy of honor again, whereas Edward VI and Mary Tudor, Elizabeth's brother and sister, were eclipsed. James built no memorials for them. He left Edward where he lay, and although Elizabeth had originally been laid to rest in her grandfather Henry VII's grave beneath the altar of his chapel, James moved her to the aisle to share her elder sister's grave. He then built her monument over the site as if she were its only occupant. Only a short Latin verse on the side of the tomb indicated that Mary Tudor was also tucked away beneath.

His mother, Mary Stuart, was then carried in a solemn procession from Peterborough and reburied in the south aisle along with the bodies of Margaret Beaufort, Henry VII's mother, and James's paternal grandmother, the Countess of Lennox. This was a shameless piece of dynastic revisionism, intended to put James himself at the hub of British history.

Yet it was very successful. The new dynastic symbols were literally set in stone: to this day they are among Westminster Abbey's biggest tourist attractions. And the "Mary" who occupies the larger and grander monument is Mary Queen of Scots! At a stroke, James honored his two "parents," his natural mother and his political one, and in the process legitimized the Stuarts as the founders of what James loved to call his "empire" of Great Britain.

The new king also encouraged the leading historian of the age, William Camden, to complete his unfinished *Annals* of Elizabeth's reign. Camden was a serious and independent-minded scholar who worked from original documents. He was highly respected, and his account of Mary's reign in Scotland was carefully researched, providing the perfect foil to the vilification of her by his Scottish counterpart, George Buchanan, in his *History of Scotland* and elsewhere.

This was what James was looking for. Camden eulogized Mary to the point where the first English abridged edition of his *Annals* could be published in 1624 as a *History of the Life and Death of Mary Stuart, Queen of Scotland* and not as a *History of Elizabeth* at all. His interpretation of Mary's rule in Scotland flatly contradicted Buchanan's traducement of her.

According to Camden, Mary was "fixed and constant in her religion, of singular piety towards God" and possessed "invincible magnanimity of mind, wisdom above her sex, and admirable beauty." Her political catastrophe could not be discounted. She had to be ranked among those rulers "which have changed their felicity for misery and calamity," but that was not through her own defects of character, but because she was a princess "tossed and disquieted" by fortune. She was a victim of her "ungrateful and ambitious subjects," chiefly her half-brother and leading councilor, James Stuart, Earl of Moray.

Camden swiped at Buchanan, whom he accused by name. He was particularly skeptical of the lords' story as presented in the dossier that Buchanan had supplied to Moray and through him to Cecil after Mary's flight to England in 1568. It was this same dossier that had underscored the accusations of the Casket Letters and the *Detection of the doings of*

Mary Queen of Scots, touching the murder of her husband . . . , which
Cecil had authorized for publication in imitation Scots.

"What Buchanan hath written," said Camden trenchantly, "there is no
man but knoweth by the books themselves printed." This was far more
insulting than it sounds today. Camden was saying he could find nothing
in the archival sources to justify Buchanan's allegations. His remarks
were stinging because his knowledge of Cecil's and Walsingham's papers
— to which he and his collaborators had enjoyed a uniquely privileged
access — was known to be encyclopedic.

A more reckless and tendentious defense of Mary came from the pen
of Adam Blackwood, a Catholic Scot and ultraroyalist exiled in France,
the doyen of her apologists at the time of her execution. He even traveled
to Peterborough to hang an inscription on a pillar next to her grave, eulogizing her as "the ornament of our age" and a martyr to "the majesty of
all kings and princes." She was "a light truly royal . . . by barbarous and
tyrannical cruelty extinct." He published a defense of her martyrdom,
Martyre de la Royne d'Escosse, in Paris in 1587, starting a debate that
fiercely intensified when her tomb at Westminster came to be revered as
the shrine of a canonized saint and was associated with a number of
miracles.

As Mary's champions could scarcely fail to remark, few of the Scottish
lords who had thwarted her during her turbulent reign had lasted long.
Almost all came to a sticky end. Moray's quick intelligence, bluff humor
and infamously "regal manner" were not enough to see him through. He
lived to enjoy his coveted position as regent of Scotland for less than
eighteen months. He was assassinated in January 1570 while riding
through the streets of Linlithgow.

He had partly brought it on himself by his own unscrupulous decep-
tions. He had supported the Duke of Norfolk's plan to marry Mary, then
betrayed it to Elizabeth. This was too much for Norfolk's allies, who
tried to murder Moray on his journey home after Elizabeth's tribunal to
inquire into the Casket Letters. He evaded his assassins, and on arriving
in Edinburgh claimed that he was as devoted as ever to the marriage
project and that his accusations against his sister had been forced on
him by Elizabeth and Cecil.

Having gained time in this way, he reasserted himself as regent and
secured a formal indemnity for all his proceedings against Mary. In
April 1569, he threw the Duke of Châtelherault, the leader of the Hamil-

tons, into prison. The duke had returned to Scotland after two years of voluntary exile after the Rizzio plot, when he had taken up Mary's cause alongside Huntly and Argyll. Moray would be assassinated nine months after signing the warrant for the duke's imprisonment. The fatal bullet, which went right through his body, was fired by James Hamilton of Bothwellhaugh. Hamilton also had a personal grudge, because Moray had deprived his wife of some property near Edinburgh and allowed her to be forced out of her house.

Moray was barely forty when he died. His passing went unmourned by Mary, who seems to have made no reference whatsoever to it. The funeral sermon at St. Giles Kirk was preached by Knox, and Buchanan devised a Latin epitaph praising Moray as a man of virtue and a Scottish patriot. It was inscribed on a brass plate set above the tomb.

Maitland, "the Scottish Cecil," otherwise nicknamed "Mekle Wylie" and "the chameleon," died three years after Moray. He had soon fallen out with his old ally over the skullduggery surrounding the Casket Letters, copies of which he may have leaked to Mary's advocates at Elizabeth's tribunal, which he had attended as a member of the Scots' delegation.

On his return to Scotland, Maitland was accused by Moray of complicity in Darnley's murder, but secured a discharge and his freedom on the evening of the murdered regent's funeral. He rejoined Huntly and Argyll, Mary's chief supporters, but his efforts came to nothing, and in 1573 he was forced to surrender while seeking refuge in Edinburgh Castle. Morton had stormed the castle with massive English aid, and Maitland gave himself up to Drury, the former English border official at Carlisle who had been promoted to command these forces. Maitland's surrender availed him little. He was said to have committed suicide in prison at Leith. According to Sir James Melville, "he took a drink and died as the old Romans were wont to do."

Morton, the most sinister of the leading lords, who had thirsted for revenge on Darnley after the Rizzio plot and offered to do Cecil "such honor and pleasure as lies in my power" as he traveled to rendezvous with Bothwell at Whittingham Castle, survived until 1581. Strongly backed by Elizabeth and Cecil, he became regent of Scotland for a full six years after 1572, but his fiscal and sexual rapacity made him many enemies. He had lived by factionalism and was to die by it. He was ousted when d'Aubigny returned to Scotland, then executed for his role in Darnley's murder.

The Earl of Lennox, Darnley's father, would fare no better. He had first appeared on the scene as the elder Earl of Bothwell's rival for Mary's mother's hand. He would be appointed regent, succeeding Moray, then stabbed in the back fourteen months later by one Captain Calder during a skirmish after a surprise raid on Stirling by Mary's supporters in 1571. Wounded, he rode to the safety of the castle, but bled to death within a few hours. He died in agony. More than most, however, he died content. His last words were "If the bairn's well, all's well." He meant that as long as his grandson, James VI, remained alive, the Lennox claim to the English throne would be vindicated, so that his and his wife's efforts to obtain a crown for their family would not have been in vain.

As to the indomitable John Knox, he at last showed his true colors. Maitland plucked up the courage to tell him, "You are but a drytting [i.e., shitting] prophet." When he was fifty Knox had married a girl of sixteen and never recovered from the scandal. Thomas Randolph, who had been sidelined by Elizabeth as postmaster-general but who returned briefly as ambassador to Scotland during Mary's captivity, quipped gleefully how Knox had gone "quiet," having more on his mind than sermons. The preacher had reemerged at James VI's coronation, to expound on the text "I was crowned young." He suffered a stroke in 1570, which explains his letter to Cecil written "with his one foot in the grave." He died in 1572 at the age of fifty-eight and was buried in the kirkyard of St. Giles. Two years later, his young widow married Andrew Ker of Fawdonside, the man who had pointed a loaded pistol at Mary during the Rizzio plot, then lurked in the alleyway close to Kirk o'Field. The mind boggles at what they must have talked about in bed.

Perhaps surprisingly, Walsingham, the man who led Mary to her destruction in the Babington plot, went unrewarded. Elizabeth had little time for the murky world of spies and intelligence. By 1586, she wanted Mary dead but resented the methods used to entrap her. When Walsingham died in 1590, he was described by Camden as "a man exceeding wise and industrious . . . a diligent searcher out of hidden secrets, who knew excellently well how to win men's minds unto him and to apply them to his own uses." This was cutting rather than flattering. Walsingham had many more talents than this, but it is as Cecil's spymaster that he is remembered.

The man who lived longest and enjoyed himself the most was William Cecil. Not even Mary's execution tempted him to retire. He suffered from gout and bad teeth in his old age, but the profits of office were lucrative, and he had several luxurious houses and amassed a fortune. Elizabeth had raised him to the peerage as Lord Burghley in 1571, and he liked nothing more than to ride around his gardens on a mule, admiring his ornamental trees and plants. Shakespeare lampooned Cecil as old Polonius, the establishment bureaucrat whose idea of politics was haplessly eavesdropping behind the arras. In his last sickness, Elizabeth sat at his bedside and fed him with a spoon. He died in 1598, a few weeks short of his seventy-eighth birthday.

More than anyone else, Cecil was Mary's nemesis. The volatile factionalism of the lords had gradually worn her down, and their refusal to put the interests of Scotland above their private feuds seriously weakened the monarchy. But it was Cecil who had actively encouraged their first revolt of 1559–60. He afterward stood behind Moray, Maitland and Morton, whom he covertly aided and with whom he constantly corresponded, and without his backing they would have made little headway.

Cecil had an apocalyptic, almost messianic vision of England as a Protestant state. When the revolt of the Lords of the Congregation erupted, he knew it gave him a unique opportunity to transform the British Isles into a single Protestant community. He was, of course, English through and through. He treated Scotland as a satellite state of England, just as much as Henry VIII and Protector Somerset had before him. The role of the Scottish nationalist fell most conspicuously, if perhaps ironically, to the swashbuckling adventurer the Earl of Bothwell.

Most of all, Cecil feared the Guise dynastic plan for a Franco-British empire. He considered Mary to be his and his country's most dangerous adversary from the moment her uncle the Cardinal of Lorraine ordered the royal arms of England to be quartered with those of Scotland and France on her badges and escutcheons. Although Mary was barely sixteen, he saw her as the instigator and main beneficiary of an international Catholic conspiracy to depose and kill Elizabeth and destroy the true faith. Cecil believed this so strongly he referred to it in the epitaph he composed for his tomb in his hometown of Stamford in Lincolnshire. His life's achievement, he declared, had been to "safeguard" the queen and the Protestant state.

The collapse of Mary's rule in Scotland was not an accident. All along, Cecil had been following a script of his own creation. It is one of the most remarkable documents in the whole of British history, because it became a template for the actual course of events. He had written it two years before Mary's return from France. It took the form of a memo dated "31 August 1559" and entitled "A memorial of certain points meet for restoring the realm of Scotland to the ancient weal." The wording seems innocuous. The contents were dynamite, because by "the ancient weal" Cecil meant his view of Scotland as a "satellite," as he was convinced it had been under Edward I, who had claimed the feudal overlordship of Scotland in the 1290s.

Scotland, said Cecil, was not to be administered by a governor or regent, in the case of an absentee ruler such as Mary, but by a council of nobles appointed "to govern the whole realm." And if Mary, already now queen of France, "shall be unwilling to this," then quite simply, said Cecil, she should be deposed. "Then is it apparent," he wrote portentously, "that Almighty God is pleased to transfer from her the rule of that kingdom for the weal of it."

Two years before Mary left France, Cecil had already taken his first tentative steps toward her forced abdication. His ideal was for the Protestant lords to call themselves by the name of the "States of Scotland" in order to supplant her. Cecil, like Knox, portrayed Mary from the start in the language of biblical prophecy. She was "Jezebel" and "Athalia," and in his heart he was a supporter of Knox's theory of armed resistance to "tyrannous" (i.e., Catholic) rulers.

Cecil went on to serve Elizabeth for forty years. He was her "loyal subject" and "humble servant," as he fervently protested, even as he went behind her back. And yet his chief priorities were to exclude Mary from the English succession by fair means or foul, while undermining her rule in Scotland by destabilizing her at critical moments — whereas Elizabeth respected Mary's rights as independent Queen of Scots and was repelled by Henry VIII's cavalier disregard for the principles of hereditary succession in his will. Repeatedly, Cecil complained that Elizabeth had been far too generous and understanding to Mary and far too willing to compromise.

The deeper the modern scholar digs into the Elizabethan State Papers, the more Cecil demands scrutiny. The well-entrenched interpretation that sees Elizabeth and Mary as rival queens goes only so far. Of course there was rivalry, and Elizabeth was ruthless in her attempts to dictate the terms and the object of Mary's marriages. But the two queens had much more in common than this reductionist model allows. In particular, they had a clear understanding of the ideological issues. That is, when female monarchs had to deal with male councilors in a dynamic political environment informed by religious sectarianism, more than just business as usual was at stake.

What we glimpse in Elizabeth's relationship with Mary are the contradictions inscribed in a monarchy where the vagaries of dynastic succession competed with loyalties to an ideal of an exclusively Protestant commonwealth. When Elizabeth spoke in her own voice, hereditary rights took priority over religion, but when Cecil did the talking, it was always the other way around. And whereas Elizabeth stood for the ideal of monarchy and was prepared to defend Mary's rights as an anointed queen, Cecil was working toward a definition of Protestant citizenship and toward a framework in which Parliament had the sovereign right to determine the succession in order to defend its citizens' religious beliefs.

Despite the electrifying stage confrontation between Mary and Elizabeth dramatized in Friedrich von Schiller's *Mary Stuart* in 1801, itself the inspiration of Donizetti's opera first performed in 1835, the two "British" queens never met. Elizabeth in the end would not grant Mary the personal interview she had always craved. And as the years passed, the real reason became apparent from her many lame excuses. She feared that the younger, possibly more beautiful Queen of Scots was so magnetic, so brilliant in conversation, that she would overshadow or surpass her.

Mary herself was a mass of contradictions, but some qualities abided. She was glamorous, intelligent, gregarious, vivacious, kind, generous, loyal to her supporters and friends, and devoted to her Guise relations, whether or not they returned her love. She could be ingenious and courageous with a razor-sharp wit, and never more animated and exuberant than when riding her horse at the head of her army wearing her steel cap.

But she had deep emotional needs. She expected love and needed to be loved. And to a large extent she got what she demanded: from her Guise family as a child, from her Maries as an adult, from her domestic servants and, until she married Bothwell, from her people, who were spellbound by her youth, beauty and glamour. Maitland came closest to the mark when he predicted that the ordinary people of Scotland would be captivated by her merest smiles or frowns. But as queen she lacked the love of a partner, an equal, who could have bolstered her in her anxieties and tempered her impulsiveness. And this hunger for a partner, a husband, a king, led her to her most grotesque and uncharacteristic miscalculations.

Although her rank meant that she was never alone, loneliness must often have consumed her, and it was a sign of her emotional isolation during her later years that her pets became everything to her. Her final reckless throw of the dice in 1586, endorsing a madcap plot in which not even the motives of the principals were clear, is a reflection of her desperation.

Beyond this, Mary was a genuine celebrity. She brought out the crowds to her wedding at the cathedral of Notre-Dame in Paris and to her triumphal entries into Edinburgh and Perth. After her return to take up her throne, she brought something different and altogether more vibrant and compelling to the drab routine of Scottish government. When she was led through the streets of Edinburgh for the last time before her journey to Lochleven, the cries of "Burn her, burn her . . . kill her, drown her" came not from the masses but from a group of dissidents handpicked by the Confederate Lords.

For these lords, with their honor code based on tribal loyalties and regional ties, the rules of the game were quite different. Love and loyalty could be bought and sold like a commodity. For Mary, it was to become an unequal contest. The portrait that emerges of her is not of a political pawn or a manipulative siren, but of a shrewd judge of character who could handle people just as masterfully as her English cousin and counterpart. She relished her role as queen and, for a time, managed to hold together a divided and fatally unstable country. Contrary to Knox's well-worn stereotype, she knew how to rule from the head as well as the heart. In fact, she made the transition from France back to Scotland so successfully that within six months Maitland could report to Cecil: "The queen my mistress behaves herself so gently in every behalf as reasonably we can require. If anything be amiss, the fault is rather in ourselves."

Mary was a queen to the last fiber of her body and soul. One of her regal attributes was her desire to defend her honor and keep up appearances. Yet she could be willful as well as astonishingly naive. She was naive in thinking that blood would be thicker than water, that her uncle and her half-brother, the Cardinal of Lorraine and Moray, would not put their own interests before hers time and time again. She was naive in expecting Bothwell to love her simply because she had fallen in love with him. She was naive in fleeing to England after losing the battle of Langside and expecting Elizabeth to help her to recover her lost throne. She was perhaps most naive in expecting a son, who could not remember anything about her, never to betray her.

She had an innate belief in her destiny. However many times she was let down by her uncles or the Scottish lords, she tried to rebuild her bridges, until Darnley's murder made it impossible for her to do so. Her courage has never been in doubt. Even Knox applauded what he called her "manly" ability to stand her ground against Darnley after the Rizzio plot, when she won him over and escaped with him at midnight from Holyrood, riding through the night to Dunbar while heavily pregnant and stopping only to be sick. She made two escape attempts from Lochleven in a rowboat, the second successful, and after the battle of Langside rode for sixty miles at a stretch.

She stuck as best she could with her unhappy marriage to Darnley despite his intolerable behavior. She decided to put him under house arrest at Craigmillar Castle only when she was faced with the prospect of a coup d'état. She kept up appearances with Bothwell after their marriage, even when his true colors emerged and his violent temper raged unrestrained. She allowed nothing to slip during her captivity. Her household followed the strict protocol of a royal court in exile, and she always contrived to look her best, even when in the privacy of her bedroom she must have watched with sadness and dismay as her hair thinned and her waist thickened. She was determined to live up to her image, though her youth and beauty were fading, and spent extraordinary sums and energy to acquire the most sumptuous clothes and jewels to wear in the closed world of her confinement.

Her "solution" to the issue of female monarchy was hardly a radical one. "Not to marry," she told Randolph at St. Andrews shortly before she married Darnley, "you know it cannot be for me." She did what the (male) councilors in all the European dynastic monarchies expected of a woman ruler: she married and settled the succession in her country. Her choice of her first and second husbands is explicable solely on dynastic criteria. The enigma relates to her third husband. Here the truth is more complex. She first saw Bothwell in the role of queen's protector against the incessant infighting of the lords, and then married him to seal the bond. It was a calculated move. In the kaleidoscopic world she had inhabited since her return to Scotland, Bothwell seemed to offer the one chance of stability. "This realm," she said, "being divided in factions as it is, cannot be contained in order unless our authority be assisted and set forth by the fortification of a man." Where she went disastrously wrong was in allowing Bothwell, still a married man, to seduce her at Dunbar. Her worst mistake was to allow herself, a queen, to fall in love.

Mary was the unluckiest ruler in British history. A more glittering and charismatic queen could not be imagined, and yet Scotland was a small and divided country, prey to its larger neighbors. On top of this, the Protestant Reformation had combined with the factionalism of the lords to create a moment when the monarchy was more than usually vulnerable. "Mary Queen of Scots got her head chopped off" is still a familiar children's skipping rhyme in Scotland. But to let the end of her life overshadow the whole is an injustice. The odds were stacked against her from the beginning.

In England and throughout the English-speaking world, Mary is known to almost everyone, even if they do not realize why. One of the best known children's nursery rhymes relates to her:*

> Mary, Mary, quite contrary,
> How does your garden grow?
> With silver bells and cockleshells
> And pretty maids all in a row.

The garden refers to the ornamental garden at the palace of Holyroodhouse. The silver bells are the Sanctus bells used in Mary's private chapel at Mass. The cockleshells refer to the pilgrim badges beloved of all devout Catholics, especially those obtained at the shrine of Saint James in the cathedral of Santiago de Compostela in Spain. And the pretty maids are the four Maries, Mary's playmates and companions for as long as she could remember, who shared so many of her joys and sorrows.

To begin with, Mary's enemies won the argument. While she was alive, Buchanan was Scotland's (and England's) official historian. Thereafter, the debate has raged and will continue to do so for as long as she

---

* Some experts have questioned this on the grounds that no written proof has been found of the rhyme before 1744, but all this means is that it was not printed or anthologized until then.

exerts a fascination on biographers. When Blackwood described her as "by barbarous and tyrannical cruelty extinct," he completely missed the point. If Elizabeth had triumphed in life, Mary would triumph in death. Far from disappearing into oblivion, as Cecil had intended, she rose from the ashes to become one of Britain's most celebrated and beguiling rulers. In choosing the phoenix as her last emblem, she had written her own epitaph: "In my end is my beginning."

CHRONOLOGY

July. Mary's progress to the far north

Sept. Castelnau's mission to Elizabeth and Mary; Lennox returns to Scotland

1565 Feb. 17. Mary meets Darnley at Wemyss

July 19. Bothwell recalled by Mary

July 29. Mary marries Henry, Lord Darnley

Aug.–Sept. Chase-about Raid

Sept. 17. Bothwell lands at Eyemouth

1566 March 9. David Rizzio murdered at Holyrood

June 9. Mary summons the lords to hear her will

June 19. Prince James (later James VI of Scotland and James I of England) born at Edinburgh Castle

Oct. 15/16. Mary rides from Jedburgh to the Hermitage

Oct. 17. Mary falls ill at Jedburgh

Dec. 17. Baptism of James at Stirling

Dec. 24. Mary pardons the Rizzio plotters

1567 Jan. Darnley is treated for syphilis at Glasgow

Jan. 20/21. Mary rides to Glasgow to visit Darnley

Feb. 10. Darnley assassinated at Kirk o'Field

April 12. Bothwell acquitted of Darnley's murder

April 19/20. Ainslie's Tavern Bond

April 24. Mary abducted by Bothwell at Almond Bridge

May 15. Mary marries Bothwell

June 15. Mary and Bothwell confront the lords at Carberry Hill; Mary surrenders and Bothwell flees

June 17. Mary imprisoned at Lochleven Castle

July 24. Mary forced to abdicate

July 29. James VI crowned at Stirling

July/Aug. Bothwell sails to the Orkneys and Shetland, then escapes to Norway and Denmark

Aug. 22. Moray proclaimed regent

1568 Jan. Bothwell taken to Malmö Castle

May 2. Mary escapes from Lochleven

May 13. Mary defeated at the battle of Langside

May 16. Mary crosses the Solway Firth to Workington in Cumberland

May 18. Mary is at Carlisle Castle

July 13. Mary leaves Carlisle for Bolton Castle in Wensleydale

Oct.–Dec. Commissioners to examine Mary's "guilt" meet at York and Westminster, concluding at Hampton Court; Casket Letters produced by Moray

1569 Jan. 10. Mary neither found guilty nor exonerated

Jan. 26. Mary sets out for Tutbury Castle, Staffordshire

April 20. Mary taken to Wingfield Manor, Derbyshire

May 25. Mary moved to Chatsworth

June 1. Mary returns to Wingfield

Sept. 21. Mary returns to Tutbury

Nov. 25. Mary arrives in Coventry

1570 Jan. 2. Mary returns to Tutbury

May 24/25. Mary moved to Chatsworth

Oct. Mary visited by Cecil

Nov. 28. Mary moved to Sheffield Castle

1571   Sept. Ridolfi plot discovered by Cecil

[Nov.–Dec.] The case implicating Mary in Darnley's murder published by Cecil in imitation Scots in *Detection of the doings of Mary Queen of Scots . . .* , including Casket Letters

1573   April 25. Mary taken to Sheffield Lodge (or Manor) in Sheffield Park

June. Bothwell moved to Dragsholm Castle

Aug. 21/22. Mary sets out for Buxton

Sept. 27. Mary moves to Chatsworth

Nov. Mary returns to Sheffield Castle

1574   June. Mary at Buxton again

July 9. Mary returns to Sheffield

1575   June–July. Mary at Buxton

1576   March. Mary moves to Sheffield Lodge

June. Mary at Buxton again

July 30. Mary returns to Sheffield

1577   Jan. Mary back at Sheffield Lodge

Feb. 11. Mary makes her will

May. Mary at Chatsworth

July. Mary back at Sheffield Lodge

Sept. Mary at Chatsworth

Nov. Mary returns to Sheffield Castle

1578   April 14. Bothwell dies at Dragsholm Castle

Aug.–Sept. Mary at Chatsworth

Oct. 5. Mary at Sheffield Lodge

1579   June. Mary at Chatsworth

Sept. Mary at Sheffield

1580   May. Mary at Sheffield Lodge

July 26. Mary goes to Buxton

Aug. 16. Mary returns to Sheffield

1581   May. Mary at Sheffield Lodge

July. Mary at Chatsworth

1582   June. Mary goes to Buxton

July. Mary returns to Sheffield

1583   Throckmorton plot

1584   July. Mary at Buxton

Aug. 8. Mary returns to Sheffield

Aug. Sir Ralph Sadler replaces Shrewsbury as Mary's custodian

Sept. 2. Mary taken to Wingfield

Oct. Bond of Association

1585   Jan. 4. Sir Amyas Paulet first named as Mary's custodian

Jan. 14. Mary arrives at Tutbury Castle

April. Paulet arrives at Tutbury

Dec. 24. Mary is taken to Chartley

1586   Spring–early summer. Babington plot

July 17. Mary writes to Babington

Aug. 11. Mary taken to Tixall, her papers and ciphers seized

Aug. 25. Mary brought to Chartley

Sept. 25. Mary arrives at Fotheringhay, Northamptonshire

Oct. 8. Commission for Mary's trial named

Oct. 11. Commissioners arrive at Fotheringhay

Oct. 12–15. Commission sits at Fotheringhay

Oct. 14–15. Mary appears before the commission

Oct. 15. Commission adjourned to Star Chamber

Oct. 25. Commission sits in Star Chamber, finds Mary guilty

Mid-Nov. Sir Drue Drury appointed to assist Paulet

1587   Feb. 1. Elizabeth signs Mary's death warrant

Feb. 2. Elizabeth expresses reservations about the warrant

Feb. 3. Privy Council meets and decides to dispatch the warrant without telling Elizabeth

Feb. 7. Robert Beale and the Earls of Shrewsbury and Kent arrive at Fotheringhay and tell Mary of her planned execution

Feb. 8. Mary executed

EVENTS IN THE BRITISH ISLES AND FRANCE

1509   Accession of Henry VIII

1513   Death of James IV of Scotland; minority of James V

1542   English defeat the Scots at Solway Moss; death of James V; minority of Mary Queen of Scots

1543   Treaty of Greenwich between England and Scotland

1544   Henry VIII begins Rough Wooing of Scotland, campaign of terror to unite the crowns by marrying Mary to the future Edward VI

1546   Cardinal Beaton assassinated

1547   Death of Henry VIII; accession of Edward VI; death of Francis I; accession of Henry II; English defeat of the Scots at the battle of Pinkie

1553   Death of Edward VI; accession of Mary Tudor

1558   Death of Mary Tudor; accession of Elizabeth I

1559   Elizabethan religious settlement; treaty of Cateau-Cambrésis; death of Henry II; accession of Francis II; revolt of the Protestant Lords of the Congregation against Mary of Guise; Cecil sends covert aid to the lords; Bothwell steals money sent by Cecil

1560   Elizabeth sends expeditionary force to Scotland; treaty of Edinburgh; Scottish official Reformation; Protestant Kirk created; Catholic Mass abolished; death of Francis II; accession of Charles IX

1562   Elizabeth almost dies of smallpox; English intervention in the first War of Religion in France

1569   Northern Rising against Elizabeth I and in favor of Mary Queen of Scots

1570   Papal bull *Regnans in Excelsis* excommunicates Elizabeth and declares her deposed; assassination of the Earl of Moray, regent of Scotland; Earl of Lennox appointed regent

1571   Ridolfi plot to overthrow Elizabeth; Lennox killed and succeeded by the Earl of Mar

1572   Execution of the Duke of Norfolk; death of Mar; Earl of Morton appointed regent of Scotland; death of John Knox; massacre of St. Bartholomew in France

1574   Death of Charles IX; accession of Henry III

1578   Personal rule of James VI begins

1581   Trial and execution of Morton for Darnley's murder

1583   Throckmorton plot to assassinate Elizabeth discovered
1584   William of Orange assassinated
1585   Elizabeth sends aid to the Dutch and so precipitates war with Philip II of
         Spain
1586   Babington plot to kill Elizabeth implicates Mary
1588   Philip II sends the Armada against Elizabeth
1589   Death of Catherine de Medici; assassination of Henry III
1603   Death of Elizabeth

NOTES

Abbreviated citations of printed primary and secondary materials identify the works listed in the Bibliography, where full references are given. For example, Cust (1903) refers to L. Cust, *Notes on the Authentic Portraits of Mary Queen of Scots* (London, 1903); Dawson (1986) refers to "Mary Queen of Scots, Lord Darnley and Anglo-Scottish Relations in 1565," *International History Review* 8 (1986), pp. 1–24. Manuscripts are cited by the call numbers used in the relevant archive, record office or library. In citing manuscripts or rare books, the following abbreviations are used:

AN	Archives Nationales, Paris
BL	British Library, London
BNF	Bibliothèque Nationale de France, Paris
CP	Cecil Papers, Hatfield House (available on microfilm at the BL and Folger Shakespeare Library)
CUL	Cambridge University Library
FF	Ancien Fonds Français
Folger	Folger Shakespeare Library, Washington, D.C.
HEH	Henry E. Huntington Library, San Marino, California
Lambeth	Lambeth Palace Library, London
MS	Manuscript
NAF	Nouvelles Acquisitions Français
NAS	National Archives of Scotland, Edinburgh
NLS	National Library of Scotland, Edinburgh
PRO	Public Record Office, Kew
SP	State Papers

Note on dates: In giving dates, the old style has been retained, but the year is assumed to have begun on January 1 and not on Lady Day, the feast of the Annunciation (i.e., March 25), which was by custom the first day of the calendar year in France, Spain and Italy until 1582, in Scotland until 1600, and in England, Wales and Ireland until 1752.

Note on transcription: The spelling and orthography of primary sources in quotations are always given in modernized form. Modern punctuation and capitalization are provided where there is none in the original manuscript.

PROLOGUE

The most reliable English sources for Mary's execution and its setting are those from Beale's collected papers in BL, Additional (hereafter Add.) MS 48027, fos. 636–41, 642–58^v. These include a copy of the official report of the earls and their assistants (fos. 649^v–50). Robert Wingfield's eyewitness report to Cecil is from Dack (1889), where authorship is discussed and the narrative printed from the Loseley Park MS. Other copies are BL, Lansdowne MS 51, fos. 99–102; Ellis (1824–46), 2nd series, vol. 3.

The fullest English descriptions of Mary's clothes are from BL, Add. MS 48027, fos. 658^{r-v}, and the eyewitness report of Edward Capell (Shrewsbury's servant) at BL, Stowe MS 159, fos. 108–11. Beale's ink and pencil drawing of the execution is now recatalogued at BL, Add. MS 48196 C, and is printed by Cust (1903). Lambeth, Fairhurst MS 4267, fos. 21–32, is a full summary of the context and proceedings. Other material is from PRO, SP 53/21, nos. 9–10, 13, 16, 20; BL, Harleian MS 290; BL, Cotton (hereafter Cott.) MS, Caligula (hereafter Calig.) C.9; BL, Cott. MS, Titus C.7; the appendix to Nicolas (1823); Morris (1874); Collinson (1987a). Some of these documents are summarized in *CSP Scotland* (1898–1969), vol. 9. A version of the execution from a contemporary commonplace book kept by members of a family in Ledbury, Herefordshire, is Folger MS, E.a.1, fos. 21^v–22.

The best contemporary French account is the "Vray Rapport," written by one of Mary's attendants and printed by Teulet (1862), vol. 4, which is essential for Mary's dress. Also useful, but less accurate, as he was not present in the great hall, is the report of Bourgoing, Mary's physician, printed by Chantelauze (1876). A French translation of Andrews's account is printed by Labanoff (1839). The French ambassador's report to Henry III is printed by Strickland (1844), vol. 2.

1. THE FIRST YEAR

The key political documents are from *Sadler State Papers* (1809), vol. 1; *Hamilton Papers* (1890–92), vols. 1–2; *Letters and Papers* (1862–1932), vols. 17–18; *Foreign Correspondence* (1923). Valuable secondary accounts are Hay Fleming (1897), Bonner (1998), Merriman (2000). The best studies of Mary of Guise are by Marshall (1977) and Ritchie (2002).

Useful background works are Cameron (1998), Edington (1994), Wormald (1981 and 1985), Goodare (1999), Guy (1988), Elton (1977). More anecdotal, but still worth consulting, are Mignet (1852); Strickland (1888), vol. 1; Ruble (1891); Stoddart (1908). For the Guise family and their affinities, I have relied on Croze (1866), Romier (1913–14) and Carroll (1998). The notes in *Lettres Inédites de Dianne de Poitiers* (1866) fill in gaps. Standard accounts of France include Knecht (1994), Garrisson (1995), Potter (1995a).

2. THE ROUGH WOOINGS

The outstanding treatment of the Rough Wooings is Merriman (2000). Further detail, notably from French sources, is from Bonner (1998). The documents are from *Letters and Papers* (1862–1932), vols. 18–21; *State Papers* (1830–52), vol. 5; *Foreign Correspondence* (1923); *Hamilton Papers* (1890–92), vol. 2; *Diurnal of Occurrents* (1833); *APS* (1814–75), vol. 2; *PCS, 1st series* (1877–98), vol. 1. Hay Fleming (1897) is brief but to the point; Sanderson (1986) is essential for Beaton's murder, and key documents are from *State Papers*, vol. 5; *Letters and Papers*, vol. 21, pt. 1. Bonner (1996) is definitive on the recovery of St. Andrews Castle. Ruble (1891), Stoddart (1908) and Bryce (1907) are useful for Mary's departure for France. Somerset's links to Cecil, and Cecil's to Knox, under Edward VI are worked out from *Revised CSPD, Edward VI* (1992).

3. ARRIVAL IN FRANCE

Mary's character unfolds when her correspondence begins. Her letters, very few in number before 1553, increase rapidly thereafter. They are cited from the edition by Labanoff (1844); those for 1550 are from vols. 1 and 7. Letters written to Mary of Guise are taken from *Foreign Correspondence* (1925).

The organization and personnel of Mary's household were worked out from the manuscripts in Paris: BNF, MS NAF 9175; BNF, MSS FF 7974, 11207, 25752.

Ruble (1891) and Stoddart (1908) remain useful, but Stoddart is garbled. Romier (1913–14) and Croze (1866) offer the best accounts of the Guises apart from Carroll (1998), who is definitive on their clientage. In the absence of recent works, Diane de Poitiers is followed in *Lettres Inédites de Dianne de Poitiers* (1866). Bryce (1907) discusses Mary's voyage and reception. The Rouen fête is fully discussed by Merriman (2000). Bonner (1999a) offers much on Henry II's policy in Scotland. The foiled plot to murder Mary and her mother's return to Scotland are from *CSPF, Edward VI* (1861), Merriman (2000) and Ritchie (2002).

4. ADOLESCENCE

The Cardinal of Lorraine's letters about Mary to her mother are taken from Labanoff (1844), vol. 1, and *Foreign Correspondence* (1925). The latter also contains the letters of Antoinette of Bourbon, Anne d'Este, Lady Parois and Mary's administrative officials; a key letter from d'Oysel about the decision to end her minority is printed as appendix A. Mary's own letters are from Labanoff (1844), vol. 1, apart from two letters in vol. 7, one to Antoinette and the other to Elizabeth I. The last is wrongly dated May 1566; the correct date is Nov. 1562, as is clear from internal evidence, not least the references to the Earl of Huntly's death at the battle of Corrichie. Details of Mary's officers are from BNF, MS NAF 9175. Further details about payments and the treasurer's accounts are from Ruble (1891), while Stoddart (1908) fills in some gaps. Melville's comment on Mary's generosity is from his *Memoirs* (1827).

Mary's holograph letter to Mary Tudor is PRO, SP 51/1, no. 7. Her health is worked out from *Foreign Correspondence* (1925), Turnbull (1845) and [*British Medical Journal*] (1968). "The sweat" or "quartan ague" is described by Caius (1552) and Thwaites, Taviner and Gant (1997). Her unexplained attacks are from PRO, SP 52/8, no. 76, and SP 52/10, no. 59. As to whether she really had porphyria, we can never know. Fraser (1969) resolves the issue with particular success. Mary's ring is from Scarisbrick (1995).

Material on the resumption of war with Spain, Guise policy and the loss of Calais is derived from Croze (1866), Romier (1913–14), Carroll (1998) and Potter (1983). Mary's betrothal to the dauphin is from *Discours du Triumphe* (1558), described more fully by Ruble.

5. EDUCATION

Mary's education is described by Ruble (1891) and Stoddart (1908), supplemented by Robertson (1863) and Durkan (1987). Her themes are edited by Montaiglon (1855). Further details are from Ambrogini (1520), Selve (1543), Fouquelin (1557), Amyot (1559). General background on the humanist curriculum is from Skinner (1978). Reports on Mary's progress are printed in *Foreign Correspondence* (1925). Barwick (1901) discusses her copy of Ptolemy's *Geography*, about which the Quaritch sale catalogue (1906) at HEH provides invaluable further information. A copy of the same edition as that owned by Mary is HEH 31230.

Mary's letters are edited by Labanoff (1844), vol. 1, but the letter at pp. 5–8 belongs to the early months of 1556, not 1552 as stated (see also Stoddart, appendix B). A further letter is printed by Hay Fleming (1897), appendix A. Mary's links to the Pléiade are summarized by Ruble and Stoddart, supplemented by Plattard (1947), Phillips (1964), Fraser (1969), Garrisson (1995), Davidson (2001).

Ruble discusses fashion and embroidery with lists of Mary's clothes and accessories, but Swain (1986) is essential on embroidery and emblem books. The chalk drawings of Mary at nine and a half (from Musée Condé at Chantilly), at twelve or thirteen (from the library of the Ossolinski National Institute of the Polish Academy of Sciences, Wroclaw), and shortly before her first marriage (from BNF) are discussed (apart from the recently discovered Wroclaw drawing) by Cust (1903). Background is from Stoddart and *Foreign Correspondence*. Art historical information is from Jollet (1997).

Mary's concern for discretion and the use of codes is from her letters, as is her supplying blanks to Mary of Guise. Her handwriting is discussed by Stoddart and Robertson, but the quotations are from Mary's own letters. The emblem of the marigold is from Paradin (1557), with explanation by Swain. Examples of the use of her motto, anagram and *impresa* are taken from Barwick (1901) and Way (1859).

6. A DYNASTIC MARRIAGE

The accounts of Mary's marriage, the processions and state banquet are based on *Discours du Triumphe* (1558) and the civic records of Paris, printed in Teulet (1862), vol. 1. An English translation of the *Discours* by Weber (1969) is seriously mangled. Ruble (1891), Stoddart (1908) and Strickland (1888), vol. 1, are useful, but garbled in different ways. Mary's letter to her mother on her wedding day is printed in the appendix to Hay Fleming (1897). Phillips (1964) provides a full account of the panegyric literature and points up the differences of emphasis in France and Scotland.

The account of the Scottish commissioners and parliamentary proceedings is from Keith (1844–50), vol. 1, and Hay Fleming (1897) and Ritchie (2002). The marriage contract is from BNF, MS FF 4781. Copies of Mary's deeds, dated April 4, concerning the marriage are from BNF, MS FF 6606, also printed in Labanoff (1844), vol. 1.

Merriman (1987) offers a summary of the Scots and English reactions to Guise dynastic policy. The documents are cited from *CSPF, Elizabeth* (1863–1950), vol. 1, *CSP Scotland* (1898–1969), vol. 1, and Mary's letters in Labanoff, vol. 1. The account of Mary's heraldic arms, from Throckmorton to Cecil, is taken from the dispersed portion of Cecil's papers in BL, Cott. MS, Calig. B.10.

Further background on the Guises, French foreign policy, the treaty of Cateau-Cambrésis and the death of Henry II is derived from Croze (1866), Ruble (1891), Romier (1913–14), Knecht (1984), Garrisson (1995) and Carroll (1998).

The documents for the treaty of Upsettlington are from *CSPF, Elizabeth*, vol. 1. Reports on Mary's health in the final months of Henry II's reign are from *CSPF, Elizabeth*, vol. 1, and *CSPS, Series 2* (1892–99), vol. 1.

7. BETRAYED QUEEN

The politics of Francis II's reign and of the Guise ascendancy are badly covered by the literature, but sketches by Croze (1866), Knecht (1984 and 1989), Garrisson (1995) and Potter (1995a) provide an outline. Régnier de la Planche (1836) is full but undiscriminating. The outstanding treatment of the Guises in Normandy is Carroll (1998).

Attempts to characterize Mary by Stoddart (1908) and Strickland (1888), vol. 1, are anecdotal and of limited value. Ruble (1891) is better, since he read the important section of documents (the "series K") from the Estado series in the Archivo General at Simancas, which were removed to Paris in 1810 on the orders of Napoleon. He consulted these documents at AN, where they were kept until returned to Simancas in 1941. Transcripts and (now illegible) photocopies are retained in the AN.

The events of 1558–60 in France, Scotland and England are worked out from the documents printed by Forbes (1741), vol. 1; *CSPF, Elizabeth* (1863–1950), vols. 1–3; *CSP Scotland* (1898–1969), vol. 1; *Sadler State Papers* (1809), vol. 1; Stevenson (1837); [*Négociations*] (1841); *Diurnal of Occurrents* (1833). Francis II's coronation is worked out from Menin (1727 and 1775), with key MSS at Rheims from [*Négociations*] and Throckmorton's dispatches from Forbes, vol. 1. Background for Francis II's reign comes from Régnier de la Planche, vol. 1, and on his coronation from Jackson (1984).

William Cecil's memos are taken from BL, Cott. MSS, Calig. B.9 and B.10; Lansdowne MS 4. French diplomacy over Scotland is from Teulet (1862), vols. 1–2, used in conjunction with *CSPF, Elizabeth*. Mary's letters are from Labanoff, vol. 1. Dispatches of the Spanish or Venetian ambassadors, including reports of Mary's health and Jane Dormer's visit to the French court, are from *CSPS, Series 2* (1862–1954), vol. 1; *CSPV* (1864–1947), vol. 7. Essential for establishing the political contexts, especially the revolt of the Lords of the Congregation, are Keith (1844–50), vol. 1; Read (1955); MacCaffrey (1969); Alford (1998a); Dawson (1989 and 2002). Marshall (1977) and Ritchie (2002) are useful on Mary of Guise.

Throckmorton's later dispatches, notably his account of Mary's *deuil blanc* portrait and his audiences, are from Forbes, vols. 1–2; *CSPF, Elizabeth*, vol. 3. The chalk drawing (BNF) and oil panel portrait (Royal Collection) are discussed by Cust (1903). Later dispatches about the ratification of the treaty of Edinburgh and related matters are from *CSPF, Elizabeth*, vols. 3–4, and BL, Add. MSS 35830–31. The ambassador's playful aside is printed by Stevenson (1837), correctly attributed in *CSPF, Elizabeth*, vol. 3. The Cecil quotation is from BL, Cott. MS, Calig. B.10.

8. RETURN TO SCOTLAND

The death of Francis II, the fall of the Guises and the ascendancy of Catherine de Medici are described by Croze (1866), Romier (1913–14), Garrisson (1995), Carroll (1998). On Mary's conduct, Ruble (1891) is more useful than Stoddart (1908), but both accounts cross the line to the romantic. The notes in Hay Fleming (1897) are a valuable corrective, but Mary's role (as opposed to that of her uncles) has never been properly established and must be pieced together from Chéruel (1858), her own letters in Labanoff (1844), vol. 1, and the diplomatic reports. The most accessible are from Teulet (1862), vol. 2; *CSPF, Elizabeth* (1863–1950), vols. 3–4; *CSPV* (1864–1947), vol. 7; *CSPS, Series 2* (1862–1954), vol. 1. Catherine de Medici's interventions are taken from [*Négociations*] (1841). The meeting between Mary and Darnley, described by Strickland (1888), vol. 1, is fanciful.

The inventory of Mary's jewels is BNF, MS FF 5898, and a sample account for her jointure is from BNF, MS FF 3335. The ordinance establishing it is in Teulet (1844), vol. 2, and its arrears have been studied by Greengrass (1987).

Mary's audiences with Bedford and Throckmorton are described in the latter's words in *CSPF, Elizabeth*, vol. 3, and Throckmorton's later interviews in vol. 4. D'Oysel's role in Mary's household as dowager queen of France is noted by Throckmorton and described by Melville (1827).

The missions of Lesley and Lord James and their contexts are reconstructed from *CSP Scotland* (1898–1965), vol. 1; Rose (1905); and Hay Fleming. Rose prints Lord James's letter to Mary from BL, Add. MS 32091. Events in Scotland, in particular the role and aspirations of Lord James and his allies, were established from PRO, SP 52/6, especially nos. 16–17, 21–22, 28, 35, 42. Mary's letter to Maitland is SP 52/6, no. 45. Morton's letter, memo and draft are from BL, Add. MSS 23108–9. Keith (1844–50), vol. 2, has extensive background, but is not completely reliable. Donaldson (1983) is precise, if impenetrable, on the factions in Scotland.

The accounts of Throckmorton's final meetings with Mary and of d'Oysel's mission are from *CSPF, Elizabeth*, vol. 4, and his correspondence with Cecil and Elizabeth is in BL, Add. MSS 35830–31. Cecil's letter of July 14 is printed (not entirely accurately) in *CSPF, Elizabeth*. I am grateful to Dr. Stephen Alford for the reference to the manuscript and for a transcript.

Mary's last weeks in France and the preparations for her voyage are worked out from *CSPF, Elizabeth*, vol. 4, and the extensive notes in Hay Fleming. The quotation from Castelnau's *Memoirs* is translated from Castelnau (1838), also cited by Mignet (1852), vol. 1.

9. INTO THE LABYRINTH

Mary's arrival at Leith is from *Diurnal* (1833) and Hay Fleming (1897). Knox's account is contradicted by the *Diurnal*, while that of Strickland (1888), vol. 1, is fanciful. The pageants for the triumphal entry are from *Diurnal* (1833); Robertson (1863); Keith (1844–50), vol. 2; supplemented by Randolph's descriptions in BL, Cott. MS, Calig. B.10, and PRO, SP 52/6, no. 63, the latter supplying the full text of the verses. Knox's version of his clash with Mary is from Knox (1949), vol. 2, complemented by Randolph's report (which establishes the correct date) in BL, Cott. MS, Calig. B.10.

Accounts of Mary's first Mass at Holyroodhouse and the subsequent proclamation are from Keith, vols. 2–3, and Hay Fleming. The summary of Knox's political theory is from his own writings, especially those in [Knox] (1994). The best interpretation of Knox's resistance theory is by Mason (1998a and 1998b). Mary's progress is worked out from Hay Fleming and Randolph's reports in BL, Cott. MS, Calig. B.10, and PRO, SP 52/6, no. 66. The vignettes of Buchanan and Randolph rely on Mason (2000), Phillips (1948–49) and Randolph's own dispatches, in particular BL, Cott. MSS, Calig. B.9–10; SP 52/6, nos. 75, 79, 82, 89; NLS, Advocates MSS 1.2.2, 6.1.13, 31.2.19.

Mary's letters are from Labanoff (1844), vol. 1, but more important for these months are documents arising from the debate on the treaty of Edinburgh: SP 52/6, nos. 58, 61–62, 73–74, 84, 86A, 88, 91. Other detail is from *CSP Scotland* (1898–1969), vol. 1.

The diplomacy of Lord James and his allies in advance of Mary's return and their letters to Elizabeth and Cecil are from SP 52/6, nos. 52–53; BL, Cott. MS, Calig. B.10; Keith, vol. 3, appendix to book 2; [Haynes and Murdin] (1740–59), vol. 1; *CSPF, Elizabeth* (1863–1950), vol. 4; [Salisbury MSS] (1883–1976), vol. 1. Extracts from the Privy Council registers are printed by Keith, vol. 2. Mary's letter commissioning Maitland for his visit to London is printed by Stevenson (1837), and the account of his proceedings by Pollen (1904), appendix 1. The diplomatic context, notably the position of the Guises, is established from BL, Add. MSS 35830–31; SP 52/7, nos. 4, 6, 15; Chéruel (1858); supplemented by Pollen. Cecil's letter to Throckmorton is from *CSPF, Elizabeth*, vol. 4. Maitland's revised assessment of Mary is from SP 52/6, no. 81.

10. A MEETING BETWEEN SISTERS

Mary's court at Holyroodhouse has been surprisingly neglected, but the gist can be pieced together from Randolph's reports in PRO, SP 52/6–12, and BL, Cott. MSS, Calig. B.9–10, supplemented by the inventories printed by Robertson (1863). Also essential are Hay Fleming (1897), Durkan (1987) and Lynch (1990). Mary's conversation in the garden at Holyroodhouse is from BL, Cott. MS, Calig. B.10; her letter to the Duke of Guise is edited by Pollen (1904), which also includes Maitland's account of his earlier diplomacy with Elizabeth at appendix 1.

Maitland's request for the interview is SP 52/6, no. 88. The meeting is from SP 52/7, nos. 3–7, and Randolph's reports in SP 52/7, nos. 2, 9–10, 15, 19, 25, 31, 36–37, 39, 41, 43, 45, 47, 51, 53, 56, 58, 62. Additional reports are from BL, Cott. MSS, Calig. B.9–10. Mary's letters are from SP 52/7, nos. 3–4, 6, 42, 88, and Pollen, appendix 9, rather than Labanoff, vol. 1. Elizabeth's correspondence is from SP 52/6, no. 86A (also printed in Pollen, appendix 8); SP 52/7, nos. 7, 81, 84 (also printed in Pollen, appendix 11). Further detail is from *CSP Scotland* (1898–1969), vol. 1.

The exchanges of tokens and verses is from SP 52/7, nos. 7, 56, 58, 62. Mary's verses are from the facsimile of the manuscript printed in *Mary Stuart* (English National Opera: London, 1998); the translation is my own. Maitland's departure for London is from SP 52/7, no. 42, and *Diurnal of Occurrents* (1833). Background on the succession issue and the Wars of Religion is taken from Read (1955), MacCaffrey (1969), Levine (1973), Knecht (1989), Garrisson (1995), Carroll (1998) and Alford (1998a). The dispatches of Cecil and Throckmorton regarding the intervention in Normandy and consequent delay of the interview are from *CSPF, Elizabeth* (1863–1950), vols. 4–5, supplemented by BL, Add. MS 35831.

Cecil's memos and the details of the interview and the decision to postpone it are from BL, Cott. MS, Calig. B.10, and SP 52/7, no. 63. Sir Henry Sidney's report of his audience with Mary is from SP 52/7, no. 65.

Mary's progress to the northeast is largely based on Randolph's reports, supplemented by the full account by Hay Fleming. Randolph's reports are from SP 52/7, nos. 74, 76–77, 79, 82, 85, 86–87. A condensed "news" report is SP 52/7, no. 90.

The reports of the Privy Council's deliberations on Elizabeth's attack of smallpox are from BL, Cott. MS, Calig. B.10; *CSPS, Series 2* (1892–99), vol. 1; Keith (1844–50), vol. 2. Maitland's own intelligence and subsequent diplomacy are from BL, Cott. MS, Calig. B.10; SP 52/8, no. 1; and Keith, vol. 2. Randolph's reports are from SP 52/8, nos. 3, 6, 7,

9, 10, and in particular his assessment of Mary dated Dec. 3, 1562, which is from BL, Cott. MS, Calig. B.9, printed in full by Stevenson (1837). Moray's letter to Cecil is from SP 52/8, no. 11.

11. A SEARCH FOR A HUSBAND

Maitland's first and second instructions are printed in Labanoff (1844), vol. 1, and Keith (1844–50), vol. 2. His departure is recorded in *Diurnal of Occurrents* (1833). Cecil's view of Elizabeth's marriage is from a memo of 1572 in BL, Cott. MS, Calig. C.3, and his parliamentary experiment of 1563 to find a constitutional mechanism to exclude Mary is from PRO, SP 12/28, no. 20, discussed by Guy (1988) and Alford (1998a). Sadler's speech to Parliament is from [*Proceedings in Parliament*] (1981–95), vol. 1, also cited by Neale (1953–57), vol. 1.

The changed situation at Mary's court is described in Randolph's reports, especially PRO, SP 52/8, nos. 4, 6, 14, 16–17, 30, 31, 35–38, 42, 45–46, 52, 59, 67–68, 75; and BL, Cott. MS, Calig. B.10. Mary's letters to the cardinal and the pope are printed by Labanoff, vol. 1, and Turnbull (1845).

The account of Maitland's negotiations with de Quadra is pieced together from Hay Fleming (1897), *CSPS, Series 2* (1892–99), vol. 1, and Maitland's report in BL, Add. MS 32091. His fascinating report from Chenonceaux is from NLS, Advocates MS 6.1.13.

The rival diplomacies of Mary and the cardinal are taken from Labanoff, vol. 1; *CSPS, Series 2*, vol. 1; *CSPF, Elizabeth* (1863–1950), vol. 6; MacCaffrey (1969); Doran (1996). The account of du Croc's mission is from Randolph's reports, SP 52/8, nos. 36–38.

The clash with Knox is from his own account in Knox (1949), vol. 2. The Chastelard incident is from Randolph's reports, SP 52/8, nos. 13–14, supplemented by Hay Fleming and Keith, vol. 2. The adultery of Mary's French chaplain is from SP 52/8, no. 42.

Randolph's conversations with Mary about the Duke of Guise and Elizabeth's letter of condolence are from SP 52/8, nos. 17 and 30. Smith's letter to Cecil is from *CSPF, Elizabeth*, vol. 6. Randolph's new (Aug.) instructions are from SP 52/8, no. 53, and Keith, vol. 2. His account of his audience with Mary is from SP 52/8, no. 59. His further (Nov.) instructions are from SP 52/8, nos. 70–71, the documents where the respective contributions of Elizabeth and Cecil are made plain and the scorings out and alterations become crucial. A fair copy in Cecil's hand, signed by Elizabeth, is in BL, Cott. MS, Calig. B.10. An unsatisfactory text may be found in Keith, vol. 2. Mary's comment on her uncle's betrayal of her diplomacy is from SP 52/9, no. 69.

12. "MY HEART IS MY OWN"

Reports of the gift of a jewel to Mary, her reaction to the English conditions for her marriage, and her illness in late 1563 and early 1564 are taken from Randolph's dispatches in PRO, SP 52/8, nos. 75–76, 79; SP 52/9, nos. 1, 4–5, 11, 13, 15, 18–21, 22–24. The Shrovetide masque is from Robertson (1863) and Keith (1844–50), vol. 2. The scene in which Mary and the lords taunted Randolph is from SP 52/8, no. 79. Mary's words quoted from Randolph's letter to Elizabeth are from SP 52/9, no. 18.

Knox's trial is from his own account in Knox (1949), vol. 2, and the aftermath as reported in SP 51/9, nos. 15, 22. The membership of the working Privy Council is deduced from *CSP Scotland* (1898–1969), vol. 2, and Keith (1844–50), vols. 2–3. Essential background on these advisers is provided by Donaldson (1983) and Goodare (1987).

The reforming ordinance for the Court of Session is from SP 52/9, no. 19. Kirkcaldy of Grange's letter is from SP 52/9, no. 27(1). Further speculation on Darnley's candidacy is from SP 52/8, no. 79; SP 52/9, nos. 15, 26; further evidence is from *CSPS Scotland*, vol. 2, and Knox (1949), vol. 2. Elizabeth's earliest letter requesting Lennox's recall is SP 52/8, no. 43. The most compelling recent accounts of Darnley's release and its consequences are by Dawson (1986) and Adams (1987). Hay Fleming (1897) and MacCaffrey (1969) offer important background and references.

The key documents concerning Mary's proposed marriage to Dudley and the meeting at Berwick are SP 52/9, nos. 22–24, 31, 34, 54, 57, 67, 69, 72, 75, 78–79. Cecil's abstracts of these and other (in some cases now missing) documents are from BL, Cott. MS, Calig. B.10. References to the events leading up to Dudley's suit are from the notes to Hay Fleming. Mary's incredulity at the initial suggestion is from SP 52/9, no. 24, which includes Moray's joke.

Castelnau's missions to Elizabeth and Mary are described in his own words in Castelnau (1838); further detail is from Chéruel (1858). Melville's embassy to Elizabeth is described by himself in Melville (1827). Lennox's reception and restoration are worked out from SP 52/9, nos. 52–53, 62, and the notes in BL, Cott. MS, Calig. B.9. The presentation of Lennox's jewel (possibly the Lennox Jewel) and other gifts is from SP 52/9, no. 62. Further detail on the jewel is from Way (1859). Cecil's letter to Maitland and Moray after the Berwick meeting is SP 52/9, no. 78, to which their reply is no. 79.

13. A MARRIAGE OF CONVENIENCE

Elizabeth's note to Cecil of Sept. 23 is from PRO, SP 52/9, no. 48. The collapse of English policy is described by MacCaffrey (1969) and Dawson (1986), and the evidence of Cecil and Dudley's role in Darnley's release is from SP 52/10, nos. 15–16. Cecil's report of a "device" is from Ellis (1824–46), 2nd series, vol. 2. Dudley's "uncertain [i.e., double] dealing" is criticized by Randolph in NLS, Advocates MS 1.2.2. Throckmorton's assessment of Mary is from *CSPF, Elizabeth* (1863–1950), vol. 3. Dawson makes an excellent case for Throckmorton as Darnley's advocate, citing the postscript to SP 52/10, no. 31A.

Mary's remarks at St. Andrews are from SP 52/10, no. 11. Several different sets of Cecil's abstracts of Scottish papers are found in BL, Cott. MS, Calig. B.10, and the one in his own hand that includes the contents of his reply at Elizabeth's request to Moray is also printed in Stevenson (1837). Raulet's dismissal and Rizzio's employment are from SP 52/9, no. 76, and SP 52/10, nos. 5, 22.

Darnley's arrival, his reception by Mary, her game of billiards, her nursing of him during his sickness, the wedding and his proclamation as king are taken from Randolph's reports, scattered among three different archives: BL, Cott. MSS, Calig. B.9–10; SP 52/10, nos. 16–17, 20, 22–23, 27–29, 31, 31A, 32, 35–37, 39, 39(1), 42–44, 46(1), 54–55, 55(1), 56, 59–61, 65, 73–75, 78, 83, 85; NLS, MS 3657. The NLS MS dispatch to Dudley is printed by Frescoln (1973–74) and is essential for Mary's reaction to Elizabeth's letter of March 5.

Randolph's letter of March 31 to Sir Henry Sidney is from NLS, Advocates MS 1.2.2. Further abstracts of Randolph's dispatches, some in Cecil's hand, are from BL, Cott. MSS, Calig. B.9 and (especially) B.10, which include some documents not otherwise extant. One set from Calig. B.10 is printed by Stevenson (1837). Randolph's report on Mary's marriage from Cott. MS, Calig. B.9, is printed (partially only) in Ellis (1824–46), 1st series, vol. 2.

Melville's account of Darnley's arrival is from Melville (1827). The summary of Darnley's symptoms at Stirling is from SP 52/10, nos. 32, 37(1), and the diagnosis of syphilis is based on the opinion of the medical expert who examined his reputed skull and thigh bone in the museum of the Royal College of Surgeons, printed by Armstrong Davison (1965), appendix A. Reports of Mary's ill health are from SP 52/10, nos. 54, 59, which also illustrate Darnley's bad behavior. Further detail is from the notes to Hay Fleming (1897) and Keith, vol. 2. The intimacy of Darnley and Rizzio is from SP 52/10, nos. 42, 59; Keith, vol. 2; Read (1955).

Elizabeth's letter to Mary recalling Lennox and Darnley is SP 52/10, no. 70, but those to Lennox and Darnley are not extant. They are known from abstracts in BL, Cott. MS, Calig. B.10, and their recipients' reactions as reported in Keith (1844–50), vol. 2. Throckmorton's two sets of instructions are from SP 52/10, nos. 38, 41. His reports are from SP 52/10, nos. 46, 48, 52–53, 53(1).

Mary's Spanish diplomacy is from Chéruel (1858). Her conversations with Castelnau are described by Castelnau (1838) and briefly summarized by Chéruel. Castelnau's activities were noticed by Randolph in SP 52/10, no. 26. Mary's letters are from Labanoff (1844), vol. 1, but unfortunately for these months they are entirely formal, often requests for diplomatic passports. Her retrospective memo on the Darnley marriage, in SP 52/11, no. 80, is unrevealing, although it does confirm that both Protestants and Catholics were behind it, leaving Moray and his allies on a limb (printed in Labanoff, vol. 1, and Turnbull [1845]).

Maitland's love for Mary Fleming is from SP 52/9, nos. 47A, 62; SP 52/10, no. 53(1); Keith, vol. 2. The "Determination" of the Privy Council is from SP 52/10, no. 40. The minutes of the June 4 Privy Council meeting are from SP 52/10, nos. 62–63, and BL, Cott. MS, Calig. B.10. Thomworth's two sets of instructions are from SP 52/10, no. 90, and SP 52/11, no. 1. Mary's answer to Thomworth is from SP 52/11, nos. 9–10, and Keith, vols. 2–3. Mary's offer to Elizabeth on the succession and other issues is from SP 52/11, nos. 11–13.

A useful breakdown of the noble factions at the time of the return of the Lennoxes is SP 52/10A. The report of Darnley looking at a map of Scotland is from BL, Cott. MS, Calig. B.10. The letters of Moray, Argyll and Châtelherault to Cecil and Throckmorton are from SP 52/10, nos. 80, 80A, and their appeal to Elizabeth is documented by no. 81.

By far the best modern interpretations of the period before and shortly after the Darnley marriage, summarizing Mary's political and religious aims and the rival factional alliances, are Dawson (1986) and Goodare (1987). Still invaluable are the notes to Hay Fleming and Keith, vol. 2.

14. ENTER BOTHWELL

Mary's reassurances to the Protestants on her religious policy and her skillful use of propaganda against Moray and his allies are explained by Dawson (1986 and 2002), supported by Goodare (1987). Her letters and a few proclamations are from Labanoff (1844), vols. 1, 7. Other documents are from Keith (1844–50), vols. 2–3; *CSP, Scotland* (1898–1962), vol. 2. The notes to Hay Fleming (1897) are invaluable. The Palm Sunday incident is from PRO, SP 52/10, nos. 37A, 39(1), discussed by Lynch (1981) and

Dawson (1986). The dispositions of the rebel forces are discussed by Hay Fleming and Dawson (2002). Moray's appeal to Cecil through Robert Melville is SP 52/11, no. 41. Cecil's memo on Bothwell's recall is from BL, Cott. MS, Calig. B.10, printed by Stevenson (1837). Bothwell's backstory is worked out from a wide range of primary sources. Schiern (1880) and Gore-Brown (1937) must be used with caution and have been avoided. Background is from *CSP, Scotland*, vols. 1–2; *CSP, Borders* (1894–96), vol. 1; *CSPF, Elizabeth* (1863–1950), vols. 1–7; Keith, vols. 1–2. Detail and quotations are from the original documents in BL, Cott. MSS, Calig. B.9–10; SP 52/7, nos. 32, 32A, 36, 77, 93; SP 52/8, nos. 4, 6, 14, 31, 38, 75, 79; SP 52/9, nos. 5–8, 15, 17, 27(1); SP 52/10, nos. 22, 27, 31, 31A, 39(1), 60; SP 52/11, nos. 45, 60, 63, 84; SP 59/9, fos. 13–14, 15–16, 17–18, 37–38, 73–74.

The recall and restoration of Lord Gordon is from SP 52/11, nos. 2, 60, 63; *PCS, 1st series* (1877–98), vol. 1; *Diurnal of Occurrents* (1833); and the notes to Hay Fleming. The Chase-about Raid is from Randolph's reports in SP 52/11, nos. 20, 22, 24, 28–29, 35–36, 45, 49, 54, 59, 60, 63–65; *PCS, 1st series*, vol. 1; Hay Fleming; Dawson (2002). The minutes of the English Privy Council meetings of Sept. 24 and 29 are from SP 52/11, no. 52. Mary's message for Elizabeth is taken from SP 52/11, no. 30 (English version is no. 31), quoted verbatim with Elizabeth's reaction to it — in a dispatch from Paul de Foix, the resident French ambassador in London, to Catherine de Medici — in Teulet (1862), vol. 2. Châtelherault's submission and the heralds' proclamation summoning Moray and his allies are from Hay Fleming and the report in *Diurnal of Occurrents*.

15. A MARRIAGE IN TROUBLE

Castelnau's "Discourse" and his letters to Charles IX and Catherine de Medici, and to Paul de Foix, are from BNF, MS FF 15971. Many of these documents are printed, with some textual variants, in Teulet (1862), vol. 2. Chéruel (1858) offers a brief discussion. Castelnau (1838) is relevant from the standpoint of hindsight.

Darnley's role, his relations with Rizzio, Yaxley and others, and his conspiratorial activities and pro-Catholic policy are pieced together from Randolph's reports in BL, Cott. MSS, Calig. B.9–10; PRO, SP 52/10, nos. 42, 59; SP 52/11, nos. 44–45, 59–60, 65, 82–85, 84A, 93, 96, 101–3; SP 52/12, nos. 5, 6A, 9, 11, 17, 21. Several dispatches from Cott. MSS, Calig. B.9–10, are printed by Ellis (1824–46), 1st series, vol. 2; Stevenson (1837). A few are printed or summarized in Keith (1844–50), vol. 2, whereas Randolph to Dudley (Feb. 14, 1566) is now found only in Edinburgh as part of NLS, MS 3657. Bedford's dispatches of Feb. 8 and 14, 1566, are printed by Stevenson. Drury's dispatch of Feb. 16, 1566, is from Keith, vol. 2. Yaxley's visit to Spain and Philip II's reaction (including a series of documents printed from the archives at Simancas) are from Mignet (1852), vol. 1, and appendix E. Invaluable are Dawson (1986 and 2002) and Lynch (1981 and 1990).

Mary's claim to be queen of England and switch to a Catholic policy after the arrival of her uncle's agent is worked out from Randolph's dispatches (as above), and in particular the rediscovered reports of Feb. 7 and 10, 1566, from Randolph to Throckmorton from NLS, Advocates MS, 1.2.2, nos. 39–40. Sir James Melville's advice is from Melville (1827). Further background is from Keith, vol. 2; the best modern treatment of Mary's so-called Catholic interlude is Goodare (1987), which also has an invaluable discussion of the origins of the Rizzio plot.

Bedford's report of Mary's attempt to lead Bothwell and Huntly by the hand is from

Stevenson. Her ill health and pregnancy are from SP 52/11, nos. 85, 87, 93, 96; SP 52/12, no. 9; NLS, Advocates MS, 1.2.2, no. 39; BL, Cott. MS, Calig. B.10. The details of her marital breakdown are from the reports of Randolph and Bedford as indicated above.

Maitland's smoking gun and recommendation for Robert Melville are from SP 52/12, nos. 10, 12. The plot is from Randolph's dispatches (as above); those written jointly by Randolph and Bedford are from SP 52/12, nos. 26, 27, 28, 30. The bonds for the plot are from SP 52/12, nos. 28(1–2). Further background is from *CSP Scotland* (1898–1969), vol. 2; *CSPF, Elizabeth* (1863–1950), vols. 7–8; Hay Fleming (1897). The charge against Randolph and his dismissal from Scotland is from SP 52/12, nos. 17, 29.

16. ASSASSINATION ONE

There are six more or less independent accounts of the Rizzio plot and its aftermath: those by Mary, Ruthven, Randolph and Bedford, Sir James Melville, the *Diurnal of Occurrents* and Claude Nau. Some invaluable comments are from the slightly later chronicle known as the *Historie and Life of King James the Sext*. The notes to Hay Fleming (1897) are useful. Brief but important analysis is by Goodare (1987) and Dawson (2002).

The fullest and most valuable accounts are Ruthven's and Mary's. Ruthven's narrative is in several manuscripts, of which the three most important are BL, Cott. MS, Calig. B.9; Add. MS 48043; Lansdowne MS 9. Printed texts include Keith (1844–50), vol. 3, appendix to book II; and [Ruthven] (1891). I have used Keith and [Ruthven] and Add. MS 48043. Mary's fuller account is from her letter of April 2, 1566, to James Beaton, Archbishop of Glasgow, in Labanoff (1844), vol. 1. Her angry and condensed comments to Elizabeth dated March 15 are from the same volume, as is her letter of early May to Anne d'Este.

Randolph and Bedford's extensive report to Cecil of March 27 with a list of the names of the conspirators is from BL, Cott. MS, Calig. B.10, printed in Ellis (1824–46), 1st series, vol. 2. A further report of the same date to Dudley and Cecil is from BL, Cott. MS, Calig. B.9. The chronology of the discharge of Parliament and the aftermath of the plot is from *Diurnal of Occurrents* (1833).

The accounts of the plot by Sir James Melville and Claude Nau, from Melville (1827) and [Nau] (1883), should be treated cautiously. Nau's is retrospective and far from accurate. The *Historie and Life of James the Sext* links Maitland to the plot. I have used the manuscript in NAS, MS GD 1/371/3. A less satisfactory printed version is [James VI] (1825). There are three Italian reports relating to Rizzio in the Medicean Archives in Florence, printed in Labanoff, vol. 7, but their value is slight.

Other reports by Randolph and Bedford are from PRO, SP 52/12, nos. 33, 35, 36, 39, 40, 44, 47. Darnley's declaration protesting his innocence is from BL, Cott. MS, Calig. B.9, printed in Ellis, 1st series, vol. 2. Morton's and Ruthven's letters to Cecil, Dudley and Throckmorton are from SP 52/12, nos. 41, 45; NLS, Advocates MS 22.2.18. Randolph's report on the reconciliation of lords at the end of April is from SP 52/12, no. 51.

17. RECONCILIATION

Castelnau's role is documented by Castelnau (1838) and discussed by Chéruel (1858), with documents and extracts printed by Labanoff (1839) and Keith (1844–50), vol. 2.

Bothwell's political comeback and the tensions among the lords despite Mary's efforts at reconciliation are from PRO, SP 52/12, nos. 49A, 51, 64, 68, 75, 89, 94A, 99; BL, Cott. MS, Calig. B.10, of which extracts are printed by Stevenson (1837).

The account of Mary's deteriorating relationship with Darnley is from SP 52/12, nos. 47, 51, 61, 64, 65, 75, 77; BL, Cott. MS, Calig. B.10, where the story of the dog is found. The report of Bedford's anonymous source is from SP 52/12, nos. 99A, 99A(1), extracts of which are printed by Hay Fleming (1897).

Mary's pregnancy and delivery are from SP 52/12, nos. 71, 74, 75–76, with further detail from Armstrong Davison (1965) and Dawson (2002). Information about her will is from SP 52/12, nos. 68, 77. Her inventory, at NAS, is edited with sample facsimiles and introduction by Robertson (1863). Modern analysis is by Donaldson (1983).

Mary's letter to Cecil rebuking him for his role in the Rokesby affair is SP 52/12, no. 106, printed in Labanoff (1844), vol. 7. Darnley's plotting and letters to European rulers and the pope are from SP 52/12, no. 82; SP 52/13, no. 6; [Nau] (1883); CSP Scotland (1898–1969), vol. 2; Keith, vol. 2; Hay Fleming. The Rokesby affair is pieced together from SP 52/12, nos. 56, 61, 65, 70, 70(1), 71, 72, 75–76, 79, 81–82, 88, 92, 106.

Killigrew's instructions are from SP 52/12, no. 72. His interviews with Mary and visits to Prince James are from SP 52/12, nos. 75–76, 80. His meeting with Moray is from SP 52/12, no. 77.

Mary's visit to Alloa and hunting trips are described by Keith, vol. 2, where documentary extracts are given. Further detail is from SP 52/12, no. 99, and the notes to Hay Fleming. The removal of Prince James to Stirling and Mary's reconciliation with Maitland are from SP 52/12, nos. 99A(1), 102–3, 105; Keith, vol. 2. Darnley's arrival at the gates of Holyrood and the Privy Council proceedings with du Croc in attendance are from Keith, vol. 2.

The account of Mary's planned Justice Ayre and the collapse of her health at Jedburgh and its aftermath is worked out from SP 52/12, nos. 108–9, 112; SP 59/12, fos. 52–145ᵛ, where the reports of Lord Scrope, Sir John Forster and the Earl of Bedford provide reliable information for the period between Aug. 9 and Dec. 11, 1566. Keith, vols. 2–3, is invaluable, especially the letters of du Croc and Lesley in the appendix to book 2 in vol. 3. Keith was unfortunately misled by the forged "Crawford" chronicle. This was an embellishment of the *Historie of King James VI* in [James VI] (1825), of which the earliest and possibly the most authentic manuscript is NAS, MS GD 1/371/3. The discussion in Armstrong Davison has the benefit of medical expertise.

Du Croc's further reports from Jedburgh and Craigmillar, the latter essential for Mary's mental depression, are from Teulet (1862), vol. 2, and Keith, vol. 1. Maitland's insinuations to the Archbishop of Glasgow are from Hay Fleming. The lords' view of the advantages of Mary's rule is established by Lynch (1990).

The terms of the reconciliation between Mary and Elizabeth are pieced together from the following documents: (1) Mary's letter to the English Privy Council of Nov. 18, 1566, from BL, Cott. MS, Calig. B.10, and BL, Add. MS 48043, printed in Ellis (1824–46), 1st series, vol. 2, and Labanoff (1844), vol. 1. (2) Elizabeth's instructions to Bedford dated Nov. 7 and her letter of the 9th from BL, Cott. MS, Calig. B.10, printed in Keith, vol. 2. (3) Mary's letter to Elizabeth of Jan. 3, 1567, from SP 52/13, no. 1, printed in Labanoff (1844), vol. 1. (4) Mary's "heads" of proposals for Bedford from SP 52/13, no. 5, printed in Keith, vol. 2. Melville's instructions of Feb. 8, 1567, are not extant.

18. PLOT AND COUNTERPLOT

Darnley's assassination is treated in literally thousands of books and articles. The most reliable sources are the manuscripts, chiefly PRO, SP 52, 53 and 59 (see below for document nos. or fos.), and BL, Cott. MSS, Calig. B.9–10 and Calig. C.1 and C.6. Classes SP 52 and 53 are based on those portions of Cecil's original working archive now held in the PRO, from which the documents in the Cottonian MSS were extracted in the seventeenth century, ending up in the BL. Both SP 52 and 53 are artificial classes, carved out of Cecil's archive by Victorian archivists.

Class SP 59 comprises the so-called Border Papers, the documents sent by Bedford and Drury from Berwick-upon-Tweed to London, but papers were indiscriminately pulled out to fill in gaps in SP 52 and SP 53. The Border Papers are only sketchily calendared in *CSPF, Elizabeth* (1863–1950), vols. 1–9, and thereafter in *CSP Borders* (1894–96). They are not included in *CSP Scotland* (1898–1969), which has caused massive confusion. When SP 52, 53 and 59 were put into their present arrangement, the original order of the documents was completely destroyed and papers were shamelessly moved around to shape the different classes.

Cecil's further collections in the Cecil Papers (CP) at Hatfield House provide additional material (see also notes to chapters 25 and 26). Other portions of his archive are in the Lansdowne MSS at the BL, but these contain little of significance for Darnley's murder. Other information is taken from Robert Beale's collections on Mary, now BL, Add. MSS 48027, 48043 and 48049, and from NLS, Advocates MS 31.2.19.

Further printed primary sources are the *Diurnal of Occurrents* (1833), essential for dates; [Nau] (1883), which is far from reliable; Melville (1827); *CSPS, Series 2* (1892–99), vol. 1; *CSPV* (1864–1947), vol. 7. Mary's letters are from Labanoff (1844), vol. 1. Other printed collections are by Anderson (1727–28), vols. 1–4; Keith (1844–50), vols. 2–3 and the "Advertisement to the Reader" in vol. 1; Teulet (1859). Modern discussions include Henderson (1890), Peyster (1890), Hay Fleming (1897), Lang (1902), Mahon (1930), Turner (1934), Diggle (1960), Armstrong Davison (1965), Fraser (1969) and Villius (1985).

The plotting at Craigmillar is from Huntly and Argyll's protestation in BL, Cott. MS, Calig. C.1, as critiqued by Moray in the same volume, and printed in Keith, vol. 3, appendix to book 2, no. 16. The aftermath is from the Dumbarton declaration, printed in Strickland (1888), vol. 2, appendix 3, which exonerates Mary.

The baptism of Prince James is from SP 52/12, nos. 121–23, 126, 128–30, 132; SP 59/12, fos. 146–47; *Diurnal of Occurrents;* Melville (1827); Keith, vol. 2 and the "Advertisement" in vol. 1; Lynch (1990). Forster's gibe at Bothwell is from SP 59/12, fo. 146.

The pardon of the Rizzio conspirators is from SP 52/12, no. 133; SP 52/13, no. 3; notes to Hay Fleming, and documentary appendix, pp. 502–4; see also Lynch (1990). Morton's letter to Cecil is from SP 52/13, no. 4. Bedford's to Cecil is from SP 52/13, no. 3. Du Croc's account of Mary's illness at Stirling is from Keith, "Advertisement" in vol. 1. The passage from the enlarged edition of Holinshed's *First and Second Volumes of Chronicles* is from Holinshed (1587), vol. 2, p. 429 (BL copy, LR.400.b.23), where it is inserted into Morton's 1581 confession.

The meeting at Whittingham Castle is from BL, Cott. MS, Calig. C.6, with other versions in BL, Add. MSS 48027 and 48049; printed by Calderwood (1842–49), vol. 3. Drury's reports to Cecil between the baptism and the murder, including the report of

Jan. 23, are from SP 52/12, fos. 146–210. His report on the links between Morton and Bothwell is SP 59/13, fos. 5–7.

Darnley's syphilis is from Armstrong Davison (1965), appendix A. Mary's offer to have sex with Darnley is from her own statements in genuine passages of the longest of the Casket Letters (the long Glasgow letter), cited from the handwritten transcript at SP 53/2, no. 65. For a full discussion of the provenance of the Casket Letters, see chapters 25 and 26.

Mary's movements are from the *Diurnal of Occurrents* and Drury's reports (the dates are close, if not exactly the same). Mary's letter to the Archbishop of Glasgow, her ambassador in Paris, is from Keith, "Advertisement" in vol. 1. Her journey to Glasgow is from the *Diurnal of Occurrents* and the notes to Hay Fleming.

Detail on Kirk o'Field and its layout is from Mahon (1930); Anderson, vols. 1–2; Keith, vol. 2 (which is useful but very inaccurate); Robertson (1863). That it was Darnley's decision to lodge there is proved by Mahon (1930). The location and movement of furniture and tapestries are from the inventories edited by Robertson, also discussed in his introduction. Darnley's letter to Lennox and the background to the murder plot are from Mahon. Drury's reports on the gunpowder transactions and explosion are from SP 52/12, especially fos. 192, 201–2, 207–10; see also Anderson, vols. 1–2.

19. ASSASSINATION TWO

The facts of Bastian's marriage were established by Robertson (1863) and Hay Fleming (1897). Mary's movements and the last hours at Kirk o'Field before the explosion are pieced together from *CSPS, Series 2* (1892–99), vol. 1; *CSPV* (1864–1947), vol. 7; the first deposition of "French Paris"; the deposition of Thomas Nelson; and other confessions, especially that of John Hepburn. These are far from ideal sources. Paris's first deposition is from BL, Cott. MS, Calig. B.9, and BL, Add. MS 48027, printed in Teulet (1859); Nelson's deposition is from NLS, Advocates MS 31.2.19, printed by Anderson (1727–28), vol. 4, and Howell (1816), vol. 1. Others are printed by Anderson, vol. 2, and Howell, vol. 1. Mary's remark to Paris is from [Nau] (1883).

The explosion and Darnley's murder are worked out from *Diurnal of Occurrents* (1833); NAS, MS GD 1/371/3, printed unsatisfactorily in [James VI] (1825); and the reports of Moretta, Clernault and Drury. Moretta's evidence is taken from Labanoff (1844), vol. 4, and *CSPV*, vol. 7, in the first case as reported by the Bishop of Mondovi, in the second by Giovanni Correr, which may explain the inconsistencies. Clernault's report dated Feb. 16, 1567, is from PRO, SP 52/13, no. 13. Drury's extensive and detailed reports, which also cover the placards and the night prowler, are from SP 59/12, fos. 192, 201–2, 207–10, 211, 222–25^v, 235, 243–44, 245–46^v; SP 59/13, fos. 5–7, 19–20, 31–34, 37–38^v, 81–85^v. The references to Andrew Ker of Fawdonside are from SP 59/13, fos. 84, 136. Drury's colored drawing of Kirk o'Field and the events of the murder is PRO, MPF 366. Drury's report on the links between Morton and Bothwell is SP 59/13, fos. 5–7, and Melville's comment on Bothwell is from Melville (1827). For Shakespeare's use of "pack," see *Titus Andronicus*, IV.ii.155; *Comedy of Errors*, V.i.219; *Much Ado*, V.i.308.

Mahon (1930) is invaluable, but his theory that the explosion was a plot by Darnley to kill Mary is fanciful and unsupported by the evidence. His interpretation of Ker of Fawdonside is wrecked by failure to notice the reference (SP 59/13, fo. 136) to Ker as "a great carrier of intelligences and letters" for Bothwell. Ker also took Bothwell's side at

Carberry Hill before fleeing (SP 59/13, fo. 159). The depositions of the women in the cottages are from BL, Add. MS 33531, fos. 37–38, a volume of the papers kept by Alexander Hay, clerk of the Privy Council, that includes the "Book of Articles" and other documents shown to Cecil in 1568. The quotation about the women's "blabbing" is from Buchanan (1571c).

Mary's letter to her ambassador in Paris after the explosion is from Keith (1844–50), "Advertisement to the Reader" in vol. 1, and Labanoff (1844), vol. 2. Both editors date it Feb. 11, but Mary's letter of the 18th shows that it was written on the 10th. Mary's letter of the 18th is from Labanoff, vol. 2, and Stevenson (1837). Her move back to Edinburgh Castle and the proclamation are from Keith, vol. 2; PCS, 1st Series (1877–98), vol. 1; and Hay Fleming.

The Venetian report of Feb. 21 is from CSPV, vol. 7. De Silva collected suspicions about Mary's role in CSPS, Series 2, vol. 1. The letter from her ambassador in Paris warning of the accusations against her is from Stevenson (1837) and Keith, "Advertisement" in vol. 1, dated March 11 by Stevenson and the 9th by Keith. The letter from Catherine de Medici and Charles IX is from SP 59/12, fos. 243–44. The letter from the Cardinal of Lorraine to Moray is from SP 59/13, fo. 84. Mary's letter to the Duke of Nemours is from BNF, MS FF 3637. The Bishop of Mondovi's letter is from Labanoff (1844), vol. 7.

Elizabeth's letter to Mary is from SP 52/13, no. 17, printed by Labanoff, vol. 7. Killigrew's report to Cecil is from SP 52/13, no. 19. Lennox's letter to Cecil (dated March 9) is from SP 52/13, no. 21. Morton's letter (dated March 10) is SP 52/13, no. 22, and letters from Moray and Maitland to Cecil (dated March 13) are from SP 52/13, nos. 25–26.

20. A LOVE MATCH?

The essential facts are established by Tytler (1828–42), vol. 7; Hay Fleming (1897), especially the notes; Keith (1844–50), vol. 2; Donaldson (1983); Wormald (1985 and 1988); Dawson (2002). Beyond this, I have drawn extensively on the State Papers, Scotland (PRO, SP 52), and more importantly the originals of Drury's handwritten reports in the Border Papers (PRO, SP 59). Hay Fleming cited the Border Papers only from the brief printed extracts in CSPF, Elizabeth (1863–1950), vol. 8, where a mass of relevant detail is omitted or garbled.

Mary's letters to Lennox and the Bishop of Mondovi are from Labanoff (1844), vol. 2, where her marriage contract of May 14 is also printed. The key chronicles, which also record the bare facts of Darnley's funeral, are Diurnal of Occurrents (1833) and the Historie and Life of King James the Sext in [James VI] (1825). The Diurnal is invaluable for establishing dates. I have used the original manuscript of the Historie in NAS, MS GD 1/371/3, where the earliest and fullest version of the text is given, confirming the facts of Mary's abduction and the proximity of her and Bothwell's bedrooms.

De Silva's assessment of Mary is from CSPS, Series 2 (1892–99), vol. 1. Killigrew's report of his dinner with the lords and audience with Mary is from SP 52/13, no. 19. Drury's reports of Mary's and Bothwell's movements, including the archery contest, are from SP 59/12, fos. 198, 201–2, 207–10. Hostile public opinion and the placards, in particular the mermaid and the hare, are from SP 59/12, fos. 211, 222–25ᵛ, 235; SP 59/13, fos. 81–85ᵛ; the drawing itself is SP 52/13, no. 60. A rougher sketch is SP 52/13, no. 61. Mary's interview with the minister of Dunfermline is from SP 59/12, fos. 243–44. Bothwell's attack on Darnley's ex-servant is from SP 59/13, fo. 85. The alignments of the lords before they assembled at Stirling, including Moray's exile, are from SP 59/12, fos. 234, 235, 243–44; SP 59/13, fos. 55–56, 62–63ᵛ. Mary's remark that Moray went away for debt is from SP 59/13, fo. 84. Her illness in late March is from SP 59/12, fos. 245–46ᵛ. Her gifts of clothes to Bothwell are from Robertson (1863) and the notes to Hay Fleming.

Bothwell's military deployments are from SP 59/12, fos. 245–46ᵛ; SP 59/13, fos. 19–20, 45–46. Morton's interview with Mary and assurances to Bothwell are from SP 59/12, fos. 222–25ᵛ; SP 59/13, fos. 5–7. Lennox's appeals to Mary, his request for Elizabeth's intervention, and Bothwell's trial are from SP 52/13, nos. 28–30; SP 59/12, fos. 243–44; SP 59/13, fos. 5–7, 8, 19–20, 31–34, 85ᵛ. The English Privy Council meeting is from SP 59/13, fos. 13–14. The report of Drury's officer is extracted from SP 59/13, fos. 31–34, 92–93.

The meeting of Parliament, the Ainslie's Tavern Bond, the soldiers' mutiny and the defection of Bothwell's allies are from SP 59/13, fos. 37–38ᵛ, 41–42ᵛ, 81–85ᵛ; SP 52/13, no. 33. Wormald (1985) settles the bond's correct signatories from the Leven and Melville muniments. NLS, Advocates MS 22.2.18, fixes the most likely date but has additional signatories. Cecil's mistaken version from BL, Cott. MS, Calig. C.1, is printed in Keith, vol. 2. Kirkcaldy's report of Mary's supposed infatuation is from SP 52/13, no. 35. His later reports are from SP 52/13, nos. 37, 40. Drury's reports of the abduction and Bothwell's divorce are from SP 59/13, fos. 45–46, 52–53, 62–63ᵛ, 64–65, 84ᵛ. Summaries of documents on Bothwell's divorce are from Stevenson's introduction to [Nau] (1833). Sir James Melville's comment is from Melville (1827). His brother Robert's report to Cecil, with an account of the lords' assembly at Stirling, is from SP 52/13, no. 42. Drury's report of the assembly is from SP 59/13, fos. 55–56, 64–65, 68–69; Kirkcaldy's report is SP 52/13, no. 43.

The masque of boy actors at Stirling is from SP 59/13, fo. 88. Drury's reports of Craig's calling of the banns, his sermon and Bothwell's reaction are from SP 59/13, fos. 88–89, 90, 91. Bothwell's pardon is from NLS, Advocates MS 31.2.19. His creation as Duke of Orkney and the wedding are from SP 59/13, fos. 91, 94, 98, 99, 103. The account for Mary's clothes is printed in the appendix to Hay Fleming. Du Croc's account is from Teulet (1862), vol. 2.

21. DÉNOUEMENT IN SCOTLAND

The wide disparity of interpretations of the events in the weeks between Mary's third marriage and her imprisonment in Lochleven are shown by Tytler (1828–42), vol. 7; Keith (1844–50), vol. 2; Strickland (1888), vols. 1–2; Hay Fleming (1897); Froude (1912), vol. 2; Fraser (1969); Wormald (1988). I have gone back to the archives to reconstruct an account based on PRO, SP 52; PRO, SP 59 (avoiding the inaccurate and often misdated summaries in *CSPF, Elizabeth* [1863–1950], vol. 8); Teulet (1862), vol. 2; Labanoff (1844), vol. 7; Stevenson (1837). Melville (1827) is essential but not always accurate, as are *Diurnal of Occurrents* (1833) and the *Historie and Life of King James the Sext* in [James VI] (1825), which I have cited from the manuscript NAS, MS GD 1/371/3.

Cecil's and Throckmorton's comments are from Alford (1998a); Elizabeth's remarks on Grange are from Tytler (1828–42), vol. 7. The placard quoting Ovid is from MS GD 1/371/3 and Keith, vol. 2. Drury's reports on Mary's rows with Bothwell and her threat to kill herself are from SP 59/13, fos. 81 (correctly dated), 82 (incorrectly placed and misdated — it belongs after the marriage), 103 (correctly dated after the marriage),

114–15; du Croc's report of May 18 to Catherine de Medici from Teulet, vol. 2 (reprinted in Labanoff, vol. 7); Melville (1827); Keith, vol. 2. Drury's comments on Mary and the horse are from SP 59/13, fos. 82 (misdated), 103.

Mary and Bothwell's keeping up appearances and the triumph are from SP 59/13, fos. 106–7, 108–9, 114–15. Her renewed bouts of sickness and the prophecies are from SP 59/13, fo. 104. Her harsh words about the lords are from SP 59/13, fo. 105. Bothwell's ambition to be king is from SP 59/13, fo. 103. The views and dispositions of the lords are from SP 59/13, fos. 103, 105–6, 110–11, 112–13, 114–15, 134–35. The coining of Mary's plate and the font at the mint are from SP 50/13, fos. 112–13, 114–15. Bothwell's official acts are from SP 59/13, fos. 106–7, 112–13, 116; Keith, vol. 2. His letters to Elizabeth and Cecil are from SP 52/13, nos. 50–51.

The rival musters and events leading up to Carberry Hill are pieced together from SP 59/13, fos. 120–21, 134–35, 136, 140–41, 144–45, 146–47; SP 52/13, no. 65; Keith, vol. 2; Dawson (2002). The report of the captain of Inchkeith is from Teulet, vol. 2. The lords' proclamations and act of "Secret Council" are from SP 52/13, nos. 64–67; Keith, vol. 2. Mary's slanging match with the lords is from SP 59/13, fos. 140–41. The sacking of the mint is from SP 59/13, fos. 144–45.

The description of the battle of Carberry Hill is worked out from SP 59/13, fos. 157, 159, 165; SP 52/13, no. 64; the report of the captain of Inchkeith covering June 7–15; Melville (1827); Keith, vol. 2; and especially du Croc's reports of June 17 to Charles IX and Catherine de Medici, printed in Teulet, vol. 2; Labanoff, vol. 7. The colored drawing of Mary's surrender is PRO, MPF 366. Mary's pregnancy is from SP 59/13, fos. 148–49. Her return to Edinburgh and committal to Lochleven are from SP 59/13, fos. 156–58; SP 52/13, no. 65, 69; du Croc to Catherine de Medici in Teulet, vol. 2; *Diurnal of Occurrents;* [James VI] (1825), cited from manuscript NAS, MS GD 1/371/3; [Nau] (1883); Keith, vol. 2. The warrant for Mary's imprisonment is from the notes to Hay Fleming. Further insights into the lords' mindset are from Stevenson (1837).

22. MARY'S STORY

Mary's instructions to the Bishop of Dunblane are from NAS, MS GD 1/371/3, fos. 277^v–79^v, and BL, Royal MS 18.B.6, fos. 242^v–66, printed in Labanoff (1844), vol. 2, and Keith (1844–50), vol. 2. Her instructions to Melville were printed in Labanoff, vol. 2, and Keith, vol. 2, from what Keith calls a "shattered MS." This usually relates to one of the Cottonian MSS burned in the fire of 1731, which damaged or destroyed a quarter of the collection. Keith first published in 1734, which makes this explanation likely. I have been unable in this case to track down the original MS, but as Keith's transcript bears all the marks of authenticity, and Melville's mission was independently reported by Drury, there is no reason to question the text. The spelling and orthography is modernized, and I have occasionally altered the word order or turned archaic Scots usages into modern English to make the transcripts comprehensible.

Since the copy of Mary's letter to Elizabeth given to Melville as part of her explanation is lost, I have used her extracts from it, given to the Bishop of Ross and her commissioners in England in 1568, from BL, Cott. MS, Titus C.12, printed in Labanoff, vol. 2. Drury's notes establishing the dates of the missions are from PRO, SP 59/13, fos. 114–15, 120–21.

Throckmorton's dispatches to Bedford, Cecil and Elizabeth were mostly printed by Stevenson (1837). Others are from BL, Cott. MS, Calig. C.1, and from Keith (1844–50), vol. 2. Elizabeth's letter to Mary is from SP 52/13, no. 80; her letter to the lords is from

SP 52/13, no. 82. Her instructions to Throckmorton are from SP 52/13, nos. 81, 83. Cecil's memo to Throckmorton is SP 52/14, no. 1. Elizabeth's subsequent outrage is from PRO, SP 52/14, nos. 39A, 53A, 53B. Cecil's jotting is at the foot of SP 52/14, no. 1. He cites "4 Regum," meaning the fourth book of Kings, which today is known as 2 Chronicles. The story of Athalia is from 2 Chronicles 23:11-21 and 2 Kings 11:1-20. Knox's sermon was reported by Throckmorton. The documents of demission and abdication that Mary signed are printed by Anderson (1727-28), vol. 2; Keith (1844-50), vol. 2. They are discussed by Hay Fleming (1897). The account of Lindsay's behavior is from Melville (1827), [Nau] (1883) and Anderson, vols. 3-4. James's coronation is from Throckmorton's reports in Stevenson (1837); *Diurnal of Occurrents* (1833); [James VI] (1825); Keith, vol. 2; Hay Fleming. Elizabeth's dressing down of Cecil is from SP 52/14, no. 53B.

Moray's visit to Mary is from Throckmorton's dispatch of August 20 from BL, Cott. MS, Calig. C.1, printed by Keith, vol. 2. The proclamation of the regency is from *Diurnal of Occurrents* and Keith, vol. 2. Mary's pastimes at Lochleven are from Throckmorton's dispatches and Drury's reports to Cecil in PRO, SP 59/14. The evidence for Mary's escape from Lochleven is confused and contradictory. My account is pieced together from [Nau] (1883); *Diurnal of Occurrents;* a report to Cosimo I from Labanoff, vol. 7; the Venetian ambassador's report in *CSPV* (1864-1947), vol. 7; Tytler (1828-42), vol. 7; Keith (1844-50), vol. 2.

Mary's mustering of her forces and the battle of Langside are from Cecil's notes in BL, Cott. MS, Calig. C.1, supported by the sources printed by Keith, vol. 2, which include Drury's reports. Further detail is from *Diurnal of Occurrents;* [James VI] (1825); Tytler (1828-42), vol. 7; Teulet (1862), vol. 2; Dawson (2002). Mary's flight to England is from Keith, vol. 2; [James VI] (1825); Ellis (1824-46), 1st series, vol. 2. Cecil's memos, in particular that of late May (fos. 97-100ᵛ), are from BL, Cott. MS, Calig. C.1, printed by Anderson (1727-28), vol. 4. Mary's letters are from Labanoff, vol. 2, and Strickland (1844), vol. 1. Cecil's position is treated at length by Alford (1998). The sale of Mary's pearls to Elizabeth is from SP 53/1, no. 46; BNF, MS FF 15971 (fo. 112); [Nau] (1883); Labanoff, vol. 7.

23. BOTHWELL'S STORY

Bothwell's escape is pieced together from Strickland (1844), vol. 1; Stevenson (1837); *CSPF, Elizabeth* (1863-1950), vol. 8. Schiern (1880), Peyster (1890) and Gore-Brown (1937) may be used with caution. Hay Fleming (1897) is accurate but brief. Strickland (1888), vol. 2, uses Danish sources, but her argument is biased toward Mary. Bothwell's letter to Charles IX is from BNF, MS FF 15971 (fo. 168); Moray's instructions to his envoy are from the same MS (fo. 84).

Bothwell's "declaration" is from the Bannatyne Club edition, [Hepburn, J.] (1829). The translation is my own, but uses that by Strickland (1844), vol. 1, as a model. Other translations can be found in *New Monthly Magazine* 13, pp. 521-37, and Drummond (1975). My annotations adopt suggestions by Armstrong Davison (1965). Bothwell's later years at Malmö and Dragsholm are from Schiern, Peyster and Gore-Brown. Captain Clark's activities and reports to Cecil are from *CSPF, Elizabeth* (1863-1950), vols. 8-10. The report of the Confederate Lords' ambassador about Bothwell's communications with Mary is from PRO, SP 52/20, no. 5 (fos. 8-9).

There are two main versions of Bothwell's deathbed "confession." One was once apparently in Cecil's papers, now in BL, Cott. MS, Titus C.7, edited by Strickland (1844), vol. 1; the other was said to be from "a merchant of good faith and reputation," perhaps marketed as the equivalent of a "penny dreadful," edited by Keith (1844–50), vol. 3.

Mary's letter to her ambassador in Paris is from Labanoff (1844), vol. 4, and Keith, vol. 3. Mary's protests about Elizabeth's suppression of her full-length copy of the "confession" are from NLS, Advocates MS 22.2.18, no. 8; Keith, vol. 3.

The Countess of Lennox's letter to Mary is from PRO, SP 53/10, no. 71. Mary's inventory at Chartley in 1586 listing the prized piece of *point tresse* is from Labanoff (1844), vol. 7. Her letter of May 2, 1578, after the countess's death, is from NLS, Advocates MS 22.2.18, no. 11, where no. 7 is the account of James VI's reaction to the "confession." Biographical detail on James is from Croft (2003). Labanoff, vol. 5, and Keith, vol. 3, also printed Mary's letter. The posthumous history of Bothwell's corpse is from Gore-Brown and tourist offices at Malmö, Faarevejle and Dragsholm.

24. THE LORDS' STORY

Buchanan's life and political ideas are from [Buchanan] (1950), Burns (1992 and 1993) and Mason (2000). His dossier against Mary, in the version translated for Lennox into Lowland Scots as "An Information," is from CUL, MS. Dd.3.66, edited with critical annotation by Mahon (1923). Moray's communications to Elizabeth and Cecil contextualizing the dossier and the lords' charges against Mary are taken from PRO, SP 52/15, nos. 39–41. The quotation about the incriminating "letters" is taken from no. 41.

Drury's reports to Cecil on Lady Reres are from PRO, SP 59/13, fos. 84, 104. Modern works consulted were Henderson (1890), Read (1955), Diggle (1960), Armstrong Davison (1965), Donaldson (1969) and Villius (1985).

25 AND 26. CASKET LETTERS I AND II

The casket documents are discussed from the handwritten transcripts as follows: The eight letters are numbered 1–8 after the numbers allocated by Henderson (1890). Transcripts previously unknown, and discussed in this book for the first time, are marked with an asterisk. Letter 1 (English): PRO, SP 53/2, no. 62 (fos. 134–35^v). Letter 2 (English): SP 53/2, no. 65 (fos. 139–42^v). Letter 3 (Scots): *BL, Add. MS 48027, fo. 276^{r-v} (a transcript in Beale's papers from Cecil's copy, with copies of his annotations also at fo. 279^v). Letter 3 (French): SP 53/2, no. 66 (fos. 143–44^v). Letter 4 (French): CP 352/3. Letter 4 (English): CP 352/4. Letter 5 (English): *SP 53/2, no. 64 (fo. 138^{r-v}). Letter 5 (French — transcript supplied by the Scots): SP 53/2, no. 63 (fos. 136–37^v). Letter 6 (French): CP 352/1 (italic hand). Letter 6 (English): CP 352/2. Letters 7 and 8: no handwritten transcripts (texts derived from Scots printed versions in [Buchanan] (1572a), BL C.55.A.26). Handwritten transcripts of the alleged marriage contracts between Mary and Bothwell are from *BL, Add. MS 48027, fos. 277–79^v; BL, Cott. MS, Calig. C.1, fo. 271 (supplied by the Scots, and in the same hand as SP 53/2, no. 63). The sonnets are from CUL, MS Oo.7.47, discussed with a sample facsimile by Davidson (2001). The sonnets have been counted variously by historians as twelve, eleven, or as one long poem. The "twelfth" is only six lines long and is either an unfinished sonnet or a postscript to the others.

The reported discovery of alleged incriminating letters by Mary is from *CSPS, Series 2* (1892–99), vol. 1; Keith (1844–50), vol. 2; Henderson (1890). Moray's affidavit is from BL, Cott. MS, Calig. C.1, fo. 354; Morton's declaration is from BL, Add. MS

32091, fo. 216 (printed by Henderson, appendix A). Cecil's minute with his description of the casket is from SP 53/2, no. 60. The key Cecil memoranda of Dec. 14–15, about the collation of the casket documents and the handwriting test, are from BL, Cott. MS, Calig. C.1, fos. 355–58ᵛ.

The earlier stages of the proceedings at York and Westminster, including the severe criticisms leveled against the Scots delegation and the Casket Letters by the Duke of Norfolk and Earl of Sussex, are from PRO, SP 53/2, nos. 5–10, 14–22, 55–60; BL, Cott. MS, Calig. C.1, fos. 227–358ᵛ; BL, Add. MS 33531, fos. 41–70; CP 4/41, 42–43, 49–51; CP 138/44–48, 70–73; CP 155/123, 125, 128–29, 130–31, 140–44; CP 156/1–3, 7; CP 198/127; printed in *CSP Scotland* (1898–1969), vol. 2; [Haynes and Murdin] (1740–59), vol. 1; [Salisbury MSS.] (1883–1976), vol. 1. The final and revised charges against Mary as presented to Elizabeth and Cecil by Moray in the Book of Articles are taken from BL, Add. MS. 33531, fos. 51–63.

The later proceedings at Westminster and Hampton Court are from CP 155/141–42, 143, 144; CP 156/1–2, 3, 5; BL, Cott. MS, Calig. C.1, fos. 317–58ᵛ; SP 53/2, nos. 55, 60, 74, 78, 81. Elizabeth's letter to Mary is from SP 53/2, no. 79. Knollys's reports to Elizabeth are at CP 155/74; SP 53/2, no. 82; SP 53/3, no. 1. Mary's letters are from Labanoff (1844), vol. 2, and Strickland (1844), vol. 1. Other documents from Cecil's papers are from [Haynes and Murdin] (1740–59), vol. 1. Summaries of CP documents are in [Salisbury MSS.] (1883–1976), vol. 1, and of PRO and Cott. Calig. C.1 documents in *CSP Scotland* (1898–1969), vol. 2.

The relevant entries from Moray's journal of Mary's and Bothwell's movements between January 21 and 30, 1567, are printed in Turner (1934), pp. 166–67. Lord Scrope's original handwritten report of Bothwell's journey to Liddesdale, describing his fight with the Elwoods when the Glasgow letters were alleged to have been received by him in Edinburgh, is from SP 59/12, fos. 175–76ᵛ.

None of the modern works is definitive, but those I found most helpful are Henderson (1890), Diggle (1960), Villius (1985). Armstrong Davison (1965), with his theory of the "other woman," on which Fraser (1969) heavily relies, is too fanciful to take seriously. Information about "French Paris" and the "confessions" extorted from him at St. Andrews on August 9 and 10, 1569, is from BL, Cott. MS, Calig. C.1; BL, Cott. MS, Calig. B.9; BL, Add. MS 48027; Teulet (1859); Teulet (1862), vol. 2; Robertson (1863); Mahon (1930); Donaldson (1983). Cecil's draft of Elizabeth's letter to Moray demanding that Paris's execution be deferred is from SP 52/16, no. 52.

27. CAPTIVE QUEEN

Mary's letters, including those sent to Castelnau, her uncle the Cardinal of Lorraine and her agent in Paris, are from Labanoff (1844), vols. 2–5 and 7. Those appealing to Catherine de Medici and Charles IX are from HEH, MSS HM 21712, 21716. Some are translated by Turnbull (1845), others by Strickland (1844), vol. 1, with a documentary appendix to vol. 2. HEH, MS HM 21712 was apparently unknown to Labanoff. Mary's letter of Nov. 8, 1582, about her son is from a number printed by [Camden] (1624), p. 134f.

Mary's domestic and household arrangements, including her attendants, clothes, diet and accounts, guards and surveillance, exercise, and the use of her coach after 1582, are put together from Lodge (1791), vol. 2; Robertson (1863); [Mary, Queen of Scots] (1867); Morris (1874); Leader (1880); Lang (1905); Collinson (1987a). Mary's framed set of family miniatures is discussed by Way (1859) and listed in the inventories in Labanoff (1844), vol. 7. Way mistakes a single object for two separate ones.

Leader's impressive study is by far the most detailed and invaluable for the years of Shrewsbury's custody, printing in full the key documents from PRO, SP 53/3–13. Further information about the severity of Paulet's regime is from Beale's papers in BL, Add. MS 48027.

Knollys's letters to Cecil are from BL, Cott. MS, Calig. B.9, printed by Strickland (1844), vol. 2. The account of Mary Seton's "busking" is from fo. 345. Nicholas White's description to Cecil of his interview with Mary is from CP 155/100–102, printed in [Haynes and Murdin] (1740–59), vol. 1.

Mary's health is from SP 53/3, nos. 62, 80, 83, 88, 105; SP 53/4, nos. 4, 58, 59, 63; Morris (1874); Leader (1880); Armstrong Davison (1965). Her visits to Buxton are from Leader. Her needlework and use of animal templates are from Swain (1986) and Jones and Stallybrass (2000). Detail on her pets is from her letters; that on Bess of Hardwick, including the so-called scandal letter, from Lodge (1791), vol. 2; Strickland (1888), vol. 2; Chamberlin (1925); Girouard (1996). The Hardwick canvas and portrait miniatures of Mary are from Cust (1903) and Strong (1983).

28. AN AX OR AN ACT?

Mary's letters are from Labanoff (1844), vols. 3–7; some were translated by Turnbull (1845), others by Strickland (1844), vols. 1–2. Key extracts concerning Guise, Spanish and papal intrigue were collected and collated by Beale, whose summaries in BL, Add. MS 48049, offer an invaluable index to the passages found to be incriminating. For Walsingham's role, see Read (1925b), vol. 2; Read (1960), vol. 2; Bossy (2001). On Mary's European diplomacy, Chéruel (1858) is invaluable, as is Castelnau (1838). Events in Scotland and the diplomacy leading to Elizabeth's recognition of James as king of Scots are worked out from Chéruel (1858); Read (1925b), vol. 2; Basing (1994); Croft (2003).

My account of the Ridolfi plot relies heavily on the outstanding account by Parker (2002). Further detail is from Edwards (1968), Alford (1998a), Lockie (1954), Beckett (2002). Mignet (1852), vol. 2, appendix L, documents Philip's intentions and the advice he received.

Norfolk's warning to Cecil is from PRO, SP 53/2, no. 19. His letters to Moray, of which the second was leaked, are from BL, Cott. MS, Calig. C.1, fos. 429, 434; NLS, Advocates MS 31.2.19, fos. 235, 245. Background on the marriage plot is from [Haynes and Murdin] (1740–59), vol. 1. Further extracts from Mary's letters to Norfolk are from [Nau] (1883). The gift of the diamond is from Mary's letter of Nov. 23, 1586, to Mendoza, printed in Labanoff (1844), vol. 6. Cecil's letter to Shrewsbury of Sept. 5, 1571, is from Lambeth, MS 3197, fos. 33–36. Mary's intercepted letter to her foreign supporters is from BL, Add. 48049, fos. 266–67, printed in Basing (1994).

Cecil's visit to Chatsworth is from BL, Add. 48049, fo. 157; Read (1960). Knox's advice is from PRO, SP 52/17, no. 3. Wilson's role in the *Detection* is from Mahon (1923) and the annotations on the documents in SP 53; what I have called a proof copy is perhaps a pilot first edition issued to the Privy Council alone. The imitation Scots edition in its approved form is BL, G.1724(1), HEH 59850. According to its title page, the book was "translated out of the Latin which was written by G[eorge] B[uchanan]." But Buchanan complained about the "over-officiousness of my friends" who "precipitated the publication of what was yet unfit to see the light." The editors had "altered many things and corrupted others according to their several humors." Cecil was finally forced to concede that the book was "written" by Buchanan "not as of himself, nor in his name, but according to the instructions given him . . . by the lords of the Privy Council in Scotland."

Cecil's memo on the eve of the 1572 Parliament is from BL, Cott. MS, Calig. C.3, fos. 457-60. Speeches are from [*Proceedings in Parliament*] (1981-95), vol. 1; Neale (1953-57), vol. 1; Collinson (1987a). The use of Killigrew and Beale by Cecil and Walsingham was established by Taviner (2000). My account of Scottish politics relies on Chéruel (1858), with selected documents in [Haynes and Murdin] (1740-59), vol. 2. Beale's missions to Sheffield are explained by Basing (1994), who prints material from BL, Add. MS 48049.

Walsingham's recruitment of Feron is from Bossy (2001). Those of Mary's letters ending up with Walsingham via the mole to which I refer are SP 53/13, no. 1, and BL, Harleian MS 1582, fos. 311-13. The Bond of Association is from PRO, SP 12/174/1-11, 14-18; BL, Add. MS 48027, fos. 248-51ᵛ. The bond and the Act for the Queen's Safety are printed by Howell (1816), vol. 1, discussed by Neale (1953-57), vol. 2; Cressy (1982). Cecil's drafts of the act and his plans for a quasi-republican regency council to exclude Mary from the succession are from CP 205/128; CP 210/17; PRO, SP 12/176/22, 28-30; HEH, Ellesmere MS 1192, discussed by Collinson (1987b and 1995) and Guy (1995). Debates are from [*Proceedings in Parliament*], vol. 2.

29. NEMESIS

Mary's letters are taken from Labanoff (1844), vols. 6-7; Turnbull (1845); Strickland (1844), vol. 2. Paulet's letters to Walsingham and Cecil are from Morris (1874). The account of the Babington plot relies on the outstanding work of Pollen (1922), where most of the documents are edited. Bossy (2001) provides an invaluable brief overview, especially where the French embassy is concerned. Châteauneuf's report to Henry III is from Turnbull (1845). Some background is taken from Read (1925b), vol. 3, and Read (1955), vol. 2. The key documents in the Babington plot are from PRO, SP 53/18, nos. 32-34, 38, 48, 51-56, 61; SP 53/19, nos. 9-12. The so-called gallows letter is SP 53/18, no. 53; the copy of the forged postscript is SP 53/18, no. 55, discussed by Read (1909).

Mary's removal to Tixall, the confiscation of her papers and money, and the arrest of her secretaries are mainly from Morris (1874) and Chantelauze (1876). The independent account of d'Esneval, the French ambassador to Scotland, is printed in Morris. Scott (1905) offers a modern summary of Bourgoing's narrative, which must be used with caution.

Mary's return to Chartley is from Morris (1874) and Strickland (1844), vol. 2. The preparations for her trial are worked out from Ellis (1824-36), 1st series, vol. 3; Morris (1874); Read (1955), vol. 2.

The proceedings at Fotheringhay and in the Star Chamber are taken from BL, Add. MS, 48027, fos. 492-510, 540-54, 557ᵛ-68; BL, Cott. MS, Calig. C.9, fos. 477-95; BL, Harleian MS 290; Howell (1816), vol. 1; [Salisbury MSS.] (1883-1976), vol. 3; Chantelauze (1876). Cecil's preparatory drawing for the trial is from BL, Cott. MS, Calig. C.9, fo. 635, printed by Dack (1889). Beale's drawing of the actual events (which is not wholly accurate) is from BL, Add. MS 48196 C, art. 7 (formerly Add. MS 48027, fo. 569).

The Parliament of 1586 is from [*Proceedings in Parliament*] (1981–95), vol. 3, and Neale (1953–57), vol. 2. The battle for the wording of the petition is meticulously reconstructed by Heisch (1992); the key documents are HEH, Ellesmere MS 1191; BL, Add. 48027, fos. 651–53; BL, Cott. MS, Calig. C.9, fos. 664–65.

The account of the struggle over whether to assassinate Mary using the Bond of Association or to publicly execute her comes from Beale's papers in BL, Add. 48027, fos. 636–41, 642–58^v; Morris (1874); Nicolas (1823). Beale tells his story at BL, Add. 48027, fos. 639^v–40, partly printed in Read (1925a). In reconstructing these events I have greatly benefited from Taviner (2000), a masterly account of the sending of the warrant.

The extant versions of the execution warrant and the covering letters to Shrewsbury and Kent are from BL, Add. MS 48027, fos. 643, 644, 645–64; Lambeth, Fairhurst MS. 4267, fos. 19–20; Sotheby's sale, London, Dec. 16, 1996, lots 40, 42; Ellis (1824–46), 2nd series, vol. 3. The key documents for the dispatch of the warrant are from BL, Add. MS 48027, fos. 636–41; BL, Harleian MS 290; CP 164/9, printed in [Salisbury MSS.] (1883–1976), vol. 3; Nicolas (1823).

30. THE FINAL HOURS

Mary's health and her postmortems on her trial are from Morris (1874). There are two sources for the events on the night before the execution: one is Beale's account in BL, Add. MS 48027, fos. 639^v–41, 646^v–49; the other is Bourgoing's as printed in Chantelauze (1876). There are discrepancies between the two, which I have resolved as seems most likely. Scott (1905) adds useful detail, but gives too much credit to Blackwood and Jebb. Mary's will and letter to Henry, Duke of Guise, are from Labanoff (1844), vol. 6, and Strickland (1844), vol. 2. Her last letter to Henry III is taken from NLS, Advocates MS 54.1.1.

The best candidate for the Latin book of hours or prayer book carried by Mary to the scaffold is HEH, MS 1200. This is an illuminated book of hours bound in crimson velvet with fine brass clasps. It was not made for Mary, but was sent as a gift to her by Pope Pius V while she was in captivity in England. A note at the front in James II's handwriting (visible only under ultraviolet light) says, "This book belonged to Queen Mary of Scotland and she carried it at her death upon the scaffold." The book was among the items purchased from the collections at the Scots College in Paris by Charles Mostyn shortly before the French Revolution, and was sold to Sir Gregory Osborne Page-Turner in or about 1822.

EPILOGUE

The proclamation for James I's accession is from Larkin and Hughes (1973). Mary's funeral is taken from the documents printed in Dack (1889) and Cust (1903). The monumental tombs at Westminster Abbey are described by Woodward (1997) and Walker (1998). Cecil's memo of August 31, 1559, is taken from BL, Lansdowne MS 4, fos. 26–27, printed as appendix 1 by Alford (1998a). Elizabeth's views of monarchy and hereditary right are from BL, Add. MS 32091, fos. 168^v–69. The quotations are from Dack (1889) and Camden (1630), the latter illuminatingly discussed by Collinson (1998a and 2003).

BIBLIOGRAPHY

The printed primary and secondary works cited in the Bibliography are intended to provide a checklist of full references to the works cited in the Notes, and also to include books and articles that were extensively consulted in researching this book.

PRIMARY WORKS

Ambrogini, A. (1520). *Illustrium Virorum Epistolae,* Paris

Amyot, J. (1559). *Les Vies des Hommes Illustres,* Paris

Anderson, J. (1727–28). *Collections Relating to Mary Queen of Scots,* 4 vols., Edinburgh

[Anonymous] (1571? 1572?). *A treatise of treasons against Q. Elizabeth, divided into two partes,* [Louvain]

APC (1890–1964). *Acts of the Privy Council of England,* new series, ed. J. R. Dasent et al., 46 vols., London

APS (1814–75). *Acts of the Parliament of Scotland,* ed. Thomas Thomson and Cosmo Innes, 12 vols., Edinburgh

[Aylmer, John] (1559). *An harborowe for faithfull and trewe subiectes, agaynst the late blowne blaste, concerninge the government of wemen,* London

[Buchanan, G.] (1571a). *Ane admonitioun direct to the trew Lordis mantenaris of the kingis grace authoritie,* Stirling

[Buchanan, G.] (1571b). *De Maria Scotorum Regina totaque eius contra Regem coniuatione, foedo cum Bothuelio adulterio, nefaria in maritum crudelitate et rabie, horrendo insuper et deterrimo eiusdem parricidio: plean et tragica plane historia,* [London]

[Buchanan, G.] (1571c). *Ane Detectioun of the duinges of Marie Quene of Scottes, touchand the murder of hir husband, and her conspiracie, adulterie, and pretensed mariage with the Erle of Bothwell,* London

[Buchanan, G.] (1572a). *Ane Detectioun of the Doingis of Marie Quene of Scottis tuiching the Murther of hir husband, and hir Conspiracie, Adulterie, and pretensit Mariage with the Erle Bothwell,* St. Andrews

[Buchanan, G.] (1572?b). *The copie of a letter written by one in London to his friend concernyng the credit of the late published detection of the doynges of the ladie Marie of Scotland*, [London]

[Buchanan, G.] (1958). *The Tyrannous Reign of Mary Stewart: George Buchanan's Account*, ed. W. A. Gatherer, Edinburgh

Caius, J. (1552). *A Boke or Counseill against the Disease commonly called the Sweate or Sweating Sicknesse*, London

Calderwood, D. (1842–49). *The True History of the Church of Scotland from the Beginning of the Reformation unto the End of the Reign of King James VI*, 8 vols., Wodrow Society, no. 7, Edinburgh

[Camden, William] (1624). *The Historie of the Life and Death of Mary Stuart, Queen of Scotland*, trans. William Udall, London

Camden, William (1630). *The Historie of the Most Renowned and Victorious Princess Elizabeth*, trans. R. Norton, London

Castelnau, Michel de (1838). *Mémoires de Michel de Castelnau*, ed. J.-F. Michaud and J.-J.-F. Poujoulat, in *Nouvelle Collection des Mémoires pour servir à l'Histoire de France*, 1st series, vol. 9, Paris

[Cecil, W.] (1571). *Salutem in Christo*, London

Chantelauze, M. R. (1876). *Marie Stuart, Son Procès et son Exécution d'après le Journal Inédit de Bourgoing*, Paris

Complete Peerage (1987). *The Complete Peerage of England, Scotland, Ireland, Great Britain and the United Kingdom by G. E. C[okayne]*, 6 vols., Gloucester

CPR (1939–73). *Calendar of Patent Rolls, 1558–1575*, 6 vols., London

CSP Borders (1894–96). *Calendar of Letters and Papers Relating to the Affairs of the Borders of England and Scotland*, 2 vols., London

CSP Scotland (1898–1969). *Calendar of State Papers Relating to Scotland and Mary, Queen of Scots, 1547–1603*, 13 vols., London

CSPD (1856–72). *Calendar of State Papers, Domestic: Edward VI, Mary, Elizabeth I, and James I*, 12 vols., London

CSPF, Edward VI (1861). *Calendar of State Papers, Foreign: Edward VI*, London

CSPF, Elizabeth (1863–1950). *Calendar of State Papers, Foreign: Elizabeth*, 23 vols., London

CSPF, Mary (1861). *Calendar of State Papers, Foreign: Mary*, London

CSPS, Series 1 (1862–1954). *Calendar of Letters, Despatches, and State Papers Relating to the Negotiations Between England and Spain*, 13 vols., London

CSPS, Series 2 (1892–99). *Letters and State Papers Relating to English Affairs Preserved Principally in the Archives of Simancas*, 4 vols., London

CSPV (1864–1947). *Calendar of State Papers and Manuscripts Relating to English Affairs, Existing in the Archives and Collections of Venice and in Other Libraries of Northern Italy*, 38 vols., London

Dawson, J. (1997). *Campbell Letters, 1559–1583*, Edinburgh

Destray, P. (1924). *Un Diplomate Français du XVIᵉ siècle, Philibert du Croc*, Nevers

Discours du Triumphe (1558). *Discours du Triumphe faict au Mariage de François de Valois et Princesse Marie d'Estreuart*, Paris

Diurnal of Occurrents (1833). *A Diurnal of Remarkable Occurrents that have passed within the Country of Scotland since the Death of King James IV till the Year 1575*, Bannatyne Club, Edinburgh

[Elizabeth I] (2000). *Collected Works*, ed. L. S. Marcus, J. Mueller, and M. B. Rose, Chicago and London

Ellis, H. (1824–46). *Original Letters, Illustrative of British History*, 3rd series, 11 vols. London

[Feria's Dispatch, 1558] (1984). "The Count of Feria's Dispatch to Philip II of 14 November 1558," ed. M. J. Rodríguez-Salgado and S. L. Adams, Camden Society, 4th series, vol. 29, London, pp. 302–44

Forbes, P. (1741). *A Full View of the Public Transactions in the Reign of Queen Elizabeth or a Particular Account of all the Memorable Affairs of that Queen*, 2 vols., London

Foreign Correspondence (1923). *Foreign Correspondence with Marie de Lorraine, Queen of Scotland. From the Originals in the Balcarres Papers, 1537–48*, ed. M. Wood, Scottish History Society, 3rd series, vol. 4, Edinburgh

Foreign Correspondence (1925). *Foreign Correspondence with Marie de Lorraine, Queen of Scotland. From the Originals in the Balcarres Papers, 1548–57*, ed. M. Wood, Scottish History Society, 3rd series, vol. 7, Edinburgh

Fouquelin, A. (1557). *La Rhétorique Françoise*, Paris

Frescoln, K. P. (1973–74). "A Letter from Thomas Randolph to the Earl of Leicester," *Huntington Library Quarterly* 37, pp. 83–88

Grafton, Richard (1548). *An epitome of the title that the kynges maiestie of Englande, hath to the sovereigntie of Scotlande, continued upon the auncient writers of both nacions, from the beginnyng*, London

Grose, F. (1791). *The Antiquities of Scotland*, 2 vols., London

Hamilton Papers (1890–92). *The Hamilton Papers. Letters and Papers Illustrating the Political Relations of England and Scotland in the Sixteenth Century*, 2 vols., London

[Haynes and Murdin] (1740–59). *A Collection of State Papers . . . left by William Cecil, Lord Burghley*, ed S. Haynes and W. Murdin, 2 vols., London

HC 1509–1558 (1982). *The House of Commons, 1509–1558*, ed. S. T. Bindoff, 3 vols., London

HC 1558–1603 (1981). *The House of Commons, 1558–1603*, ed. P. Hasler, 3 vols., London

[Henry IV] (1614). *L'Ordre des Ceremonies du Sacre et Couronnement du Très Chrestien Roy de France et de Navarre, Henry IIII*, Tours

[Hepburn, J.] (1829). *Les Affaires du Conte de Boduel, L'An MDLXVIII*, ed. H. Cockburn and T. Maitland, Bannatyne Club, vol. 29, Edinburgh

Holinshed, R. (1587). *The First and Second [and Third] Volumes of Chronicles, first collected by Raphael Holinshed, William Harrison and others, now newly augmented and continued to 1586 by John Hooker alias Vowell and others*, 3 vols., London

Howell, T. B. (1816). *A Complete Collection of State Trials and Proceedings for High Treason and Other Crimes and Misdemeanors from the Earliest Period to the Year 1783*, 21 vols., London

[James VI] (1825). *The Historie and Life of King James the Sext*, Bannatyne Club, vol. 13, Edinburgh

Keith, R. (1844–50). *History of the Affairs of Church and State in Scotland from the Beginning of the Reformation to the Year 1568*, ed. J. P. Lawson, 3 vols., Spottiswoode Society, Edinburgh

Knox, John (1949). *John Knox's History of the Reformation in Scotland*, ed. W. C. Dickinson, 2 vols., London and Edinburgh

[Knox, John] (1994). *On Rebellion*, ed. R. A. Mason, Cambridge

Labanoff, A. (1839). *Lettres Inédites de Marie Stuart 1558-1587*, Paris

Labanoff, A. (1844). *Lettres, Instructions et Mémoires de Marie Stuart, Reine d'Écosse; publiés sur les Originaux et les Manuscrits du State Paper Office de Londres et des Principales Archives et Bibliothèques de l'Europe*, 7 vols., London

Larkin, J. F., and P. L. Hughes (1973). *Stuart Royal Proclamations, vol. I, 1603-1625*, Oxford

Leicester's Commonwealth (1985). *Leicester's Commonwealth. The Copy of a Letter Written by a Master of Arts of Cambridge (1584) and Related Documents*, ed. D. C. Peck, Athens, Ohio

[Lesley, John] (1569). *A defence of the honor of the right high, right mighty, and noble princesse, Marie queene of Scotlande*, Rheims

Letters and Papers (1862-1932). *Letters and Papers, Foreign and Domestic, of the Reign of Henry VIII*, ed. J. S. Brewer, J. Gairdner and R. H. Brodie, 21 vols. in 32 parts, and addenda, London

[*Lettres Inédites de Dianne de Poitiers*] (1866). *Lettres Inédites de Dianne de Poitiers*, ed. G. Guiffrey, Paris

Lodge, Edmund (1791). *Illustrations of British History, Biography, and Manners, in the Reigns of Henry VIII, Edward VI, Mary, Elizabeth, and James I, Exhibited in a Series of Original Papers*, 3 vols., London

[Mary, Queen of Scots] (1867). *Accounts and Papers Relating to Mary, Queen of Scots*, ed. A. J. Crosby and J. Bruce, Camden Society, 1st series, vol. 93, London

[Mary, Queen of Scots] (1960). "The Execution of Mary, Queen of Scots," ed. G. R. Batho, *Scottish Historical Review* 39, pp. 35-42

Médicis, Catherine de (1880-1963). *Lettres de Catherine de Médicis*, ed. Hector de la Ferrière, 11 vols., Paris

Melville, J. (1827). *Memoirs of His Own Life, by Sir James Melville of Halhill*, ed. T. Thomson, Bannatyne Club, vol. 18, Edinburgh

Montaiglon, A. de (1855). *Latin Themes of Mary Stuart*, Warton Club, Edinburgh

Morris, J. (1874). *The Letter-Books of Sir Amias Poulet, Keeper of Mary Queen of Scots*, London

[Nau, C.] (1883). *The History of Mary Stewart from the Murder of Riccio until her Flight into England*, ed. J. Stevenson, Edinburgh

[*Négociations*] (1841). *Négociations, Lettres et Pièces Diverses relatives au Règne de François II*, Paris

Nicolas, N. H. (1823). *Life of William Davison, Secretary of State and Privy Counsellor to Queen Elizabeth*, London

Paradin, C. (1557). *Devises Heroiques*, Lyon

Park, W. (1955). "Letter of Thomas Randolph to the Earl of Leicester, 14 February 1566," *Scottish Historical Review* 34, pp. 135-39

PCS, 1st Series (1877-98). *The Register of the Privy Council of Scotland, 1545-1625*, ed. J. H. Burton and D. Masson, 1st series, 14 vols., Edinburgh

Pollen, J. H. (1901). *Papal Negotiations with Mary Queen of Scots During her Reign in Scotland*, Scottish History Society, 1st series, vol. 37, Edinburgh

Pollen, J. H. (1904). *A Letter from Mary Queen of Scots to the Duke of Guise, January 1562*, Scottish History Society, 1st series, vol. 43, Edinburgh

Pollen, J. H. (1922). *Mary Queen of Scots and the Babington Plot*, Scottish History Society, 3rd series, vol. 3, Edinburgh

[Ponet, John] (1556). *A short Treatise of Politike Pouuer, and of the true Obedience which subiectes owe to kynges and other civile Gouernours, with an Exhortacion to all true naturall Englishe men*, Strasbourg

[*Proceedings in Parliament*] (1981–95). *Proceedings in the Parliaments of Elizabeth I*, ed. T. E. Hartley, 3 vols., Leicester

[*Progresses of Elizabeth I*] (1788–1805). *The Progresses, and Public Processions, of Queen Elizabeth*, ed. J. Nichols, 3 vols., London

Ptolemy (1490). *Liber Geographiae, cum tabulis*, Rome

Read, C. (1909). *The Bardon Papers: Documents Relating to the Imprisonment of Mary Queen of Scots*, Camden Society, 3rd series, vol. 17, London

Revised CSPD, Edward VI (1992). *Calendar of State Papers, Domestic Series, of the Reign of Edward VI, 1547–1553*, rev. ed., ed. C. S. Knighton, London

Revised CSPD, Mary (1998). *Calendar of State Papers, Domestic Series, of the Reign of Mary I, 1553–1558*, rev. ed., ed. C. S. Knighton, London

Robertson, J. (1863). *Inventaires de la Royne d'Escosse, Douairiere de France, 1556–1569*, Bannatyne Club, vol. 111, Edinburgh

[Royal Scottish Geographical Society] (1919). *The Early Views and Maps of Edinburgh, 1544–1852*, Edinburgh

[Ruthven, P.] (1891). *The Murder of Riccio, being Lord Ruthven's Own Account of the Transaction*, Holyrood Series, Edinburgh

Sadler State Papers (1809). *The State Papers and Letters of Sir Ralph Sadler*, ed. Arthur Clifford, 2 vols., Edinburgh

[Salisbury MSS.] (1883–1976). *Historical Manuscripts Commission, Calendar of the Manuscripts of the Most Honorable the Marquis of Salisbury*, 24 vols., London

[Scotland] (1718). *Theatrum Scotiae: Containing the Prospects of His Majesties Castles and Palaces*, London

Scottish Correspondence (1927). *The Scottish Correspondence of Mary of Lorraine. Including some Three Hundred Letters from 20 February 1542–3 to 15 May 1560*, ed. A. I. Cameron, Scottish History Society, 3rd series, vol. 10, Edinburgh

Selve, G. de (1543). *Les Vyes de Huict Excellens Personnaiges Grecz et Romains*, Paris

State Papers (1830–52). *State Papers During the Reign of Henry VIII*, 11 vols., Record Commission, London

STC2 (1976–91). *A Short-Title Catalogue of Books Printed in England, Scotland and Ireland, and of English Books Printed Abroad*, ed. W. A. Jackson, F. S. Ferguson and K. F. Pantzer, 2nd ed., 3 vols., London

Stevenson, J. (1837). *Selections from Unpublished Manuscripts in the College of Arms and the British Museum Illustrating the Reign of Mary, Queen of Scots, 1543–1568*, Maitland Club, vol. 41, Glasgow

Strickland, A. (1844). *Letters of Mary, Queen of Scots and Documents Connected with her Personal History*, 2nd ed., 2 vols., London

Tennyson, A. (1894). *The Works of Alfred Lord Tennyson*, New York and London

Teulet, A. (1859). *Lettres de Marie Stuart*, Paris

Teulet, A. (1862). *Relations Politiques de la France et d'Espagne avec l'Écosse au XVIe siècle*, 5 vols., Paris

Turnbull, W. (1845). *Letters of Mary Stuart, Queen of Scotland*, London

Weber, B. C. (1969). *The Marriage of Mary Queen of Scots to Francis the Dauphin of France 1558*, Greenock

Wood, M. A. (1846). *Letters of Royal and Illustrious Ladies of Great Britain*, 3 vols., London

SECONDARY WORKS

Adams, S. L. (1973). "The Protestant Cause: Religious Alliance with the West European Calvinist Communities as a Political Issue in England, 1585–1630," unpublished Oxford D.Phil. dissertation

Adams, S. L. (1987). "The Release of Lord Darnley and the Failure of the Amity," *Innes Review* 38, pp. 123–53

Adams, S. L. (1991). "Favorites and Factions at the Elizabethan Court," in *Princes, Patronage and the Nobility: The Court at the Beginning of the Modern Age, c. 1450–1650,* ed. R. G. Asch and A. M. Birke, Oxford, pp. 265–87

Alford, S. (1996). "William Cecil and the British Succession Crisis of the 1560s," unpublished St. Andrews Ph.D. dissertation

Alford, S. (1997). "Reassessing William Cecil in the 1560s," in *The Tudor Monarchy,* ed. John Guy, London, pp. 233–52

Alford, S. (1998a). *The Early Elizabethan Polity: William Cecil and the British Succession Crisis, 1558–1569,* Cambridge

Alford, S. (1998b). "Knox, Cecil and the British Dimension of the Scottish Reformation," in *John Knox and the British Reformations,* ed. R. A. Mason, Aldershot, pp. 201–19

Armstrong Davison, M. H. (1965). *The Casket Letters: A Solution to the Mystery of Mary Queen of Scots and the Murder of Lord Darnley,* London

Axton, M. (1977). *The Queen's Two Bodies: Drama and the Elizabethan Succession Question,* London

Barwick, G. F. (1901). *A Book Bound for Mary Queen of Scots,* Bibliographical Society, London

Basing, P. (1994). "Robert Beale and the Queen of Scots," *British Library Journal* 20, pp. 65–82

Bassnett, S. (1988). *Elizabeth I: A Feminist Perspective,* New York

Beckett, M. (2002). "The Political Works of John Lesley, Bishop of Ross," unpublished St. Andrews Ph.D. dissertation

Bernard, G. W. (2000). "Amy Robsart," in *Power and Politics in Tudor England,* ed. G. W. Bernard, Aldershot, pp. 161–74

[Bibliothèque Nationale] (1931). *Collection de Manuscrits, Livres, Estampes et Objects d'Art Relatifs à Marie Stuart, Reine de France et d'Écosse,* Paris

Bonner, E. (1996). "The Recovery of St. Andrews Castle in 1547: French Naval Policy and Diplomacy in the British Isles," *English Historical Review* 111, pp. 578–98

Bonner, E. (1998). "The French Reactions to the Rough Wooings of Mary, Queen of Scots," *Journal of the Sydney Society for Scottish History* 6, pp. 1–161

Bonner, E. (1999a). "The *Politique* of Henri II: *De Facto* French Rule in Scotland, 1550–1554," *Journal of the Sydney Society for Scottish History* 7, pp. 1–107

Bonner, E. (1999b). "Scotland's 'Auld Alliance' with France, 1295–1560," *History* 84, pp. 5–30

Bossy, J. A. (1959). "English Catholics and the French Marriage, 1577–81," *Recusant History* 5, pp. 2–16

Bossy, J. A. (1975). *The English Catholic Community, 1570–1850,* London

Bossy, J. A. (2001). *Under the Molehill: An Elizabethan Spy Story,* New Haven and London

Bowler, G. (1984). "An 'Axe or an Act': The Parliament of 1572 and Resistance Theory in Early Elizabethan England," *Canadian Journal of History* 19, pp. 349–59

[*British Medical Journal*] (1968). *Porphyria: A Royal Malady*, British Medical Association, London

Bryce, W. M. (1907). "Mary Stuart's Voyage to France in 1548," *English Historical Review* 22, pp. 43–50

Burns, J. H. (1992). *Lordship, Kingship and Empire: The Idea of Monarchy, 1400–1525*, Oxford

Burns, J. H. (1993). "George Buchanan and the Anti-Monarchomachs," in *Political Discourse in Early Modern Britain*, ed. N. Phillipson and Q. Skinner, Cambridge, pp. 3–22

Burns, J. H. (1996). *The True Law of Kingship: Concepts of Monarchy in Early-Modern Scotland*, Oxford

Cameron, J. (1998). *James V: The Personal Rule, 1528–1542*, East Linton

Carroll, S. (1998). *Noble Power During the French Wars of Religion: The Guise Affinity and the Catholic Cause in Normandy*, Cambridge

Chamberlin, F. (1925). *The Private Character of Queen Elizabeth*, New York

Chéruel A. (1858). *Marie Stuart et Catherine de Médicis*, Paris

Collinson, P. (1987a). *The English Captivity of Mary Queen of Scots*, Sheffield

Collinson, P. (1987b). "The Monarchical Republic of Queen Elizabeth I," *Bulletin of the John Rylands University Library of Manchester* 69, pp. 394–424

Collinson, P. (1995). "The Elizabethan Exclusion Crisis," *Proceedings of the British Academy* 84, pp. 51–92

Collinson, P. (1998a). "One of Us?: William Camden and the Making of History," *Transactions of the Royal Historical Society* 8 (6th series), pp. 139–63

Collinson, P. (1998b). "John Knox, the Church of England and the Women of England," in *John Knox and the British Reformations*, ed. R. A. Mason, Aldershot, pp. 74–96

Collinson, P. (2003). "William Camden and the Anti-Myth of Elizabeth: Setting the Mould?," in *The Myth of Elizabeth*, ed. S. Doran and T. S. Freeman, London, pp. 79–98

Cowan, I. B. (1972). *The Enigma of Mary Stuart*, London

Cowan, I. B. (1987a). *Mary Queen of Scots*, Edinburgh

Cowan, I. B. (1987b). "The Roman Connection: Prospects for Counter-Reformation during the Personal Reign of Mary, Queen of Scots, *Innes Review* 38, pp. 105–22

Crane, M. T. (1988). "'Video et Taceo': Elizabeth I and the Rhetoric of Counsel," *Studies in English Literature* 28, pp. 1–15

Cressy, D. (1982). "Binding the Nation: The Bonds of Association, 1584 and 1696," in *Tudor Rule and Revolution*, ed. D. J. Guth and J. W. McKenna, Cambridge, pp. 217–34

Croft, P. (2003). *King James*, London

Croze, Joseph de (1866). *Les Guises, Les Valois et Philippe II*, 2 vols., Paris

Cust, L. (1903). *Notes on the Authentic Portraits of Mary Queen of Scots*, London

Dack, C. (1889). *The Trial, Execution and Death of Mary Queen of Scots*, Northampton

Davidson, P. (2001). "The Casket Sonnets: New Evidence Concerning Mary Queen of Scots," *History Scotland* 1:1, pp. 28–34

Dawson, J. (1986). "Mary Queen of Scots, Lord Darnley and Anglo-Scottish Relations in 1565," *International History Review* 8, pp. 1–24

Dawson, J. (1989). "William Cecil and the British Dimension of Early Elizabethan Foreign Policy," *History* 74, pp. 196–216

Dawson, J. (1991). "The Two John Knoxes: England, Scotland and the 1558 Tracts," *Journal of Ecclesiastical History* 42, pp. 556–76

Dawson, J. (1995). "Anglo-Scottish Protestant Culture and Integration in Sixteenth-Century Britain," in *Conquest and Union: Fashioning a British State, 1485–1725*, ed. S. G. Ellis and S. Barber, London, pp. 87–114

Dawson, J. (2002). *The Politics of Religion in the Age of Mary, Queen of Scots: The Earl of Argyll and the Struggle for Britain and Ireland*, Cambridge

De Peyster, J. Watts (1882). *A Vindication of James Hepburn, Earl of Bothwell*, Philadelphia

De Peyster, J. Watts (1890). *Mary Stuart, Bothwell and the Casket Letters*, New York

Diggle, H. F. (1960). *The Casket Letters of Mary Stuart: A Study in Fraud and Forgery*, Harrogate

Donaldson, G. (1969). *The First Trial of Mary Queen of Scots*, London

Donaldson, G. (1971). *Scotland: James V to James VII*, Edinburgh

Donaldson, G. (1974). *Mary Queen of Scots*, London

Donaldson, G. (1983). *All the Queen's Men: Power and Politics in Mary Stewart's Scotland*, London

Doran, S. (1996). *Monarchy and Matrimony: The Courtships of Elizabeth I*, London

Drummond, H. (1975). *The Queen's Man: James Hepburn, Earl of Bothwell and Duke of Orkney, 1536–1578*, London

Duncan, T. (1905). "The Queen's Maries," *Scottish Historical Review* 2, pp. 363–71

Durkan, J. (1987). "The Library of Mary, Queen of Scots," *Innes Review* 38, pp. 71–104

Edington, C. (1994). *Court and Culture in Renaissance Scotland: Sir David Lindsay of the Mount*, Amherst, Mass.

Edwards, F. (1968). *The Marvelous Chance: Thomas Howard, Fourth Duke of Norfolk, and the Ridolfi Plot, 1570–1572*, London

Elton, G. R. (1977). *Reform and Reformation: England, 1509–1558*, London

Forbes-Leith, W. (1885). *Narratives of Scottish Catholics under Mary Stuart and James VI*, Edinburgh

Fraser, A. (1969). *Mary, Queen of Scots*, London

Froude, J. A. (1912). *The Reign of Elizabeth*, 5 vols., London

Garrisson, J. (1995). *A History of Sixteenth-Century France, 1483–1598*, London

Geddie, J. (1911). *Romantic Edinburgh*, 2nd ed., London

Girouard, M. (1996). *Hardwick Hall*, London

Goodare, J. (1987). "Queen Mary's Catholic Interlude," *Innes Review* 38, pp. 154–70

Goodare, J. (1999). *State and Society in Early Modern Scotland*, Oxford

Gore-Brown, R. (1937). *Lord Bothwell*, London

Greengrass, M. (1987). "Mary, Dowager Queen of France," *Innes Review* 38, pp. 171–94

Guy, John (1988). *Tudor England*, Oxford

Guy, John (1995). *The Reign of Elizabeth I: Court and Culture in the Last Decade*, Cambridge

Harrison, J. (1919). *The History of the Monastery of the Holy-Rood and the Palace of Holyrood House*, Edinburgh and London

Hay Fleming, D. (1897). *Mary Queen of Scots from Her Birth to Her Flight into England*, London

Heisch, A. (1992). "Arguments for an Execution: Queen Elizabeth's 'White Paper' and Lord Burghley's 'Blue Pencil,'" *Albion* 24, pp. 591–604

Henderson, T. F. (1890). *The Casket Letters and Mary Queen of Scots*, 2nd ed., Edinburgh

Holmes, P. J. (1987). "Mary Stewart in England," *Innes Review* 38, pp. 195–218

Houlbrooke, R. A. (1994). "Henry VIII's Wills: A Comment," *Historical Journal* 37, pp. 891–99

Ives, E. W. (1992). "Henry VIII's Will: A Forensic Conundrum," *Historical Journal* 35, pp. 779–804

Ives, E. W. (1994). "Henry VIII's Will: the Protectorate Provisions of 1546–47," *Historical Journal* 37, pp. 901–14

Jackson, R. A. (1984). *Vive le Roi!: A History of the French Coronation from Charles V to Charles X*, Chapel Hill and London

Jollet, E. (1997). *Jean et François Clouet*, Paris

Jones, A. R., and P. Stallybrass (2000). *Renaissance Clothing and the Materials of Memory*, Cambridge

Jordan, C. (1987). "Women's Rule in Sixteenth-Century British Political Thought," *Renaissance Quarterly* 40, pp. 421–51

Kantorowicz, E. H. (1957). *The King's Two Bodies: A Study in Medieval Political Theology*, Princeton, N.J.

Knecht, R. J. (1984). *French Renaissance Monarchy: Francis I and Henry II*, London

Knecht, R. J. (1989). *The French Wars of Religion, 1559–1598*, London

Knecht, R. J. (1994). *Renaissance Warrior and Patron: The Reign of Francis I*, Cambridge

Lang, A. (1902). *The Mystery of Mary Stuart*, London

Lang, A. (1905). "The Household of Mary, Queen of Scots in 1573," *Scottish Historical Review* 2, pp. 345–55

Leader, J. D. (1880). *Mary Queen of Scots in Captivity: A Narrative of Events*, Sheffield

Levine, M. (1973). *Tudor Dynastic Problems, 1460–1571*, London

Lockie, D. McN. (1954). "The Political Career of the Bishop of Ross, 1568–80," *University of Birmingham Historical Journal* 4, pp. 98–145

Lynch, M. (1981). *Edinburgh and the Reformation*, Edinburgh

Lynch, M. (1988). *Mary Stewart, Queen in Three Kingdoms*, Oxford

Lynch, M. (1990). "Queen Mary's Triumph: The Baptismal Celebrations at Stirling in December 1566," *Scottish Historical Review* 69, pp. 1–21

MacCaffrey, W. T. (1963). "Elizabethan Politics: The First Decade, 1558–68," *Past & Present*, no. 24, pp. 25–42

MacCaffrey, W. T. (1969). *The Shaping of the Elizabethan Regime: Elizabethan Politics, 1558–72*, London

MacCaffrey, W. T. (1981). *Queen Elizabeth and the Making of Policy, 1572–1588*, Princeton, N.J.

MacCaffrey, W. T. (1993). *Elizabeth I*, London

MacCaffrey, W. T. (1997). "The Newhaven Expedition, 1562–1563," *Historical Journal* 40, pp. 1–21

MacDonald, A. A. (1991). "Mary Stewart's Entry to Edinburgh: An Ambiguous Triumph," *Innes Review* 42, pp. 101–2

MacKay, J. (1999). *In My End Is My Beginning: A Life of Mary, Queen of Scots*, Edinburgh

Mackie, J. D. (1921). "Queen Mary's Jewels," *Scottish Historical Review* 18, pp. 83–98

Mahon, R. H. (1923). *The Indictment of Mary Queen of Scots*, Cambridge

Mahon, R. H. (1930). *The Tragedy of Kirk o'Field*, Cambridge

Marshall, R. K. (1977). *Mary of Guise*, London

Mason, R. A. (1994). "The Scottish Reformation and the Origins of Anglo-British Imperialism," in *Scots and Britons: Scottish Political Thought and the Union of 1603,* ed. R. A. Mason, Cambridge, pp. 161-86

Mason, R. A. (1998a). *Kingship and the Commonweal: Political Thought in Renaissance and Reformation Scotland,* East Linton

Mason, R. A. (1998b). "Knox, Resistance and the Royal Supremacy," in *John Knox and the British Reformations,* ed. R. A. Mason, Aldershot, pp. 154-75

Mason, R. A. (2000). "George Buchanan and Mary Queen of Scots," *Records of the Scottish Church History Society* 30, pp. 1-27

McCullough, P. E. (1998a). "Out of Egypt: Richard Fletcher's Sermon before Elizabeth I after the Execution of Mary Queen of Scots," in *Dissing Elizabeth: Negative Representations of Gloriana,* ed. J. M. Walker, Durham, N.C., and London, pp. 118-49

McCullough, P. E. (1998b). *Sermons at Court: Politics and Religion in Elizabethan and Jacobean Preaching,* Cambridge

Menin, M. (1727). *The Form, Order and Ceremonies of Coronations,* London

Menin, M. (1775). *A Description of the Coronation of the Kings and Queens of France,* London

Merriman, M. (1987). "Mary, Queen of France," *Innes Review* 38, pp. 30-52

Merriman, M. (2000). *The Rough Wooings: Mary Queen of Scots, 1542-1551,* East Linton

Mignet, M. (1852). *Histoire de Marie Stuart,* 2 vols., Paris

Neale, J. E. (1934). *Queen Elizabeth I,* London

Neale, J. E. (1949). *The Elizabethan House of Commons,* London

Neale, J. E. (1953-57). *Elizabeth I and Her Parliaments,* 2 vols., London

Opie, I., and P. Opie (1952). *The Oxford Dictionary of Nursery Rhymes,* Oxford

Parker, G. (2002). "The Place of Tudor England in the Messianic Vision of Philip II of Spain," *Transactions of the Royal Historical Society* 12 (6th series), pp. 167-221

Perry, M. (1990). *The Word of a Prince: A Life of Elizabeth I from Contemporary Documents,* Woodbridge

Phillips, J. E. (1941). "The Background of Spenser's Attitude Toward Women Rulers," *Huntington Library Quarterly* 5, pp. 5-32

Phillips, J. E. (1948-49). "George Buchanan and the Sidney Circle," *Huntington Library Quarterly* 12, pp. 24-25

Phillips, J. E. (1964). *Images of a Queen: Mary Stuart in Sixteenth-Century Literature,* Berkeley and Los Angeles

Plattard, J. (1947). *La Renaissance des Lettres en France de Louis XII à Henri IV,* Paris

Pollitt, R. (1985). "The Defeat of the Northern Rebellion and the Shaping of Anglo-Scottish Relations," *Scottish Historical Review* 64, pp. 1-21

Potter, D. (1983). "The Duc de Guise and the Fall of Calais, 1557-1558," *English Historical Review* 98, pp. 481-512

Potter, D. (1995a). *A History of France, 1460-1560: The Emergence of a Nation State,* London

Potter, D. (1995b). "Foreign Policy," in *The Reign of Henry VIII: Politics, Policy and Piety,* ed. D. MacCulloch, London, pp. 101-34

Read, C. (1913). "Walsingham and Burghley in Queen Elizabeth's Privy Council," *English Historical Review* 28, pp. 34-58

Read, C. (1925a). "The Proposal to Assassinate Mary Queen of Scots at Fotheringhay," *English Historical Review* 40, pp. 234-35

Read, C. (1925b). *Mr. Secretary Walsingham and the Policy of Queen Elizabeth*, 3 vols., Oxford

Read, C. (1955). *Mr. Secretary Cecil and Queen Elizabeth*, London

Read, C. (1960). *Lord Burghley and Queen Elizabeth*, London

Régnier de la Planche, K. (1836). *Histoire de L'Estat de France tant de la République que de la Religion sous le Règne de François II*, Paris

Ritchie, P. E. (2002). *Mary of Guise in Scotland, 1548–1560: A Political Career*, Edinburgh

Rodríguez-Salgado, M. J. (1990). *The Changing Face of Empire: Charles V, Philip II and Habsburg Authority, 1551–1559*, Cambridge

Romier, L. (1913–14). *Les Origines Politiques des Guerres de Religion*, 2 vols., Paris

Rose, D. M. (1905). "Mary Queen of Scots and Her Brother," *Scottish Historical Review* 2, pp. 150–62

Ruble, Alphonse de (1891). *La Première Jeunesse de Marie Stuart*, Paris

Sanderson, M.H.B. (1986). *Cardinal of Scotland: David Beaton, c. 1494–1546*, Edinburgh

Scarisbrick, D. (1995). *Tudor and Jacobean Jewelry*, London

Schiern, F. (1880). *The Life of James Hepburn, Earl of Bothwell*, Edinburgh

Scott, M. (1905). *The Tragedy of Fotheringhay*, Edinburgh and London

Shephard, A. (1994). *Gender and Authority in Sixteenth-Century England: The Knox Debate*, Keele

Skinner, Q. (1978). *The Foundations of Modern Political Thought*, 2 vols., Cambridge

Skinner, Q. (1996). *Reason and Rhetoric in the Philosophy of Hobbes*, Cambridge

Stoddart, J. T. (1908). *The Girlhood of Mary Queen of Scots*, London

Strickland, A. (1888). *Life of Mary Queen of Scots*, 2 vols., London

Strong, R. (1983). *Artists of the Tudor Court: The Portrait Miniature Rediscovered, 1520–1620*, London

Strong, R. and J. T. Oman (1972). *Mary Queen of Scots*, London

Swain, M. (1986). *The Needlework of Mary Queen of Scots*, Bedford

Tannenbaum, S. A., and D. R. Tannenbaum (1944–46). *Marie Stuart, Queen of Scots: A Concise Bibliography*, 3 vols., New York

Taviner, M. (2000). "Robert Beale and the Elizabethan Polity," unpublished St. Andrews Ph.D. dissertation

Thorp, M. R. (1984). "Catholic Conspiracy in Early Elizabethan Foreign Policy," *Sixteenth-Century Journal* 15, pp. 431–48

Thwaites, G., M. Taviner, and V. Gant (1997). "The English Sweating Sickness, 1485–1551," *New England Journal of Medicine* 336, pp. 580–82

Turner, G. (1934). *Mary Stuart: Forgotten Forgeries*, 2nd ed., London

Tytler, P. F. (1828–42). *History of Scotland*, 8 vols., Edinburgh

Villius, H. (1985). "The Casket Letters: A Famous Case Reopened," *Historical Journal* 28, pp. 517–34

Walker, J. M. (1998). "Bones of Contention: Posthumous Images of Elizabeth and Stuart Politics," in *Dissing Elizabeth: Negative Representations of Gloriana*, ed. J. M. Walker, Durham, N.C., and London, pp. 252–76

Way, A., et al. (1859). *Catalogue of Antiquities, Works of Art and Historical Scottish Relics Exhibited in July 1856*, Edinburgh

White, A. (1987). "Queen Mary's Northern Province," *Innes Review* 38, pp. 53–70

Williams, N. (1964). *Thomas Howard, Fourth Duke of Norfolk*, London

Woodward, J. (1997). *The Theater of Death: The Ritual Management of Royal Funerals in Renaissance England, 1570–1625*, Woodbridge

Wormald, J. (1981). *Court, Kirk and Community: Scotland, 1470–1625*, London

Wormald, J. (1985). *Lords and Men in Scotland: Bonds of Manrent, 1442–1603*, Edinburgh

Wormald, J. (1988). *Mary, Queen of Scots: A Study in Failure*, London

Wormald, J. (1993). "Resistance and Regicide in Sixteenth-Century Scotland," *Majestas* 1, pp. 67–87

INDEX

ILLUSTRATION CREDITS

Formerly provost and history professor at the University of St. Andrews in Scotland, Renaissance historian JOHN GUY is now a fellow in history at Clare College, University of Cambridge. He has written several books, including a best-selling textbook, *Tudor England,* and he consults for the BBC. He is currently at work on a biography of Margaret More Roper and the story of her relationship with her father, Sir Thomas More.